DIGITAL TECHNOLOGIES IN OLFACTION

DIGITAL TECHNOLOGIES IN OLFACTION

FUNDAMENTALS TO APPLICATIONS

Edited by

Takamichi Nakamoto
Institute of Integrated Research, Institute of Science Tokyo, Yokohama, Japan

ELSEVIER

Elsevier
Radarweg 29, PO Box 211, 1000 AE Amsterdam, Netherlands
125 London Wall, London EC2Y 5AS, United Kingdom
50 Hampshire Street, 5th Floor, Cambridge, MA 02139, United States

For accessibility purposes, images in electronic versions of this book are accompanied by alt text descriptions provided by Elsevier. For more information, see https://www.elsevier.com/about/accessibility.

Publisher's note: Elsevier takes a neutral position with respect to territorial disputes or jurisdictional claims in its published content, including in maps and institutional affiliations.

Notices
Knowledge and best practice in this field are constantly changing. As new research and experience broaden our understanding, changes in research methods, professional practices, or medical treatment may become necessary.

Practitioners and researchers must always rely on their own experience and knowledge in evaluating and using any information, methods, compounds, or experiments described herein. In using such information or methods they should be mindful of their own safety and the safety of others, including parties for whom they have a professional responsibility.

To the fullest extent of the law, neither the Publisher nor the authors, contributors, or editors, assume any liability for any injury and/or damage to persons or property as a matter of products liability, negligence or otherwise, or from any use or operation of any methods, products, instructions, or ideas contained in the material herein.

ISBN: 978-0-443-15721-9

For Information on all Elsevier publications
visit our website at https://www.elsevier.com/books-and-journals

Publisher: Candice Janco
Acquisitions Editor: Gabriela Capille
Editorial Project Manager: Lindsay Lawrence
Production Project Manager: Bharatwaj Varatharajan
Cover Designer: Christian Bilbow

Typeset by MPS Limited, Chennai, India

Contents

3

Cheminformatics in machine olfaction

7. Fundamentals of chemometrics techniques

TAKAMICHI NAKAMOTO

8. Concept of odor space

DANI PRASETYAWAN AND TAKAMICHI NAKAMOTO

9. Odor impression prediction using multidimensional data analysis

TANOY DEBNATH

10. Odor impression prediction using natural language processing

TANOY DEBNATH

11. Odor reproduction using odor components

DANI PRASETYAWAN

12. Odor reproduction based on several divergences

DANI PRASETYAWAN

13. Analysis for interference removal in odor reproduction

DANI PRASETYAWAN

4

Olfactory displays

14. Fundamental techniques of olfactory displays

TAKAMICHI NAKAMOTO

15. Multicomponent olfactory displays

TAKAMICHI NAKAMOTO

16. Wearable-olfactory displays

TAKAMICHI NAKAMOTO

17. Digital contents with scents

TAKAMICHINAKAMOTO

18. Olfactory VR environment with computational fluid dynamics

TAKAMICHINAKAMOTO

19. Olfactory displays for teleolfaction

TAKAMICHI NAKAMOTO

20. Olfactory Art

NATHAN COHEN

5

Conclusions

21. Conclusions

TAKAMICHI NAKAMOTO

22. Future of machine olfaction

TAKAMICHI NAKAMOTO

List of contributors

Manuel Aleixandre Institute of Integrated Research, Institute of Science Tokyo, Yokohama, Japan

Nathan Cohen Art and Science, University of the Arts London, London, United Kingdom

Tanoy Debnath Yokogawa Electric Corporation, Tokyo, Japan

Hongchao Deng Zhejiang Huanmao Auto-Control Technology Co., Ltd., Hangzhou, Zhejiang, P.R. China; Institute of Integrated Research, Institute of Science Tokyo, Yokohama, Japan

Hsiu-Yi Lu Department of Molecular Genetics and Microbiology, Duke University, Durham, NC, United States

Hiroaki Matsunami Department of Molecular Genetics and Microbiology, Duke University, Durham, NC, United States

Priyanka Meesa Department of Molecular Genetics and Microbiology, Duke University, Durham, NC, United States

Sawit Na Songkhla Department of Electrical Engineering, Faculty of Engineering, Chulalongkorn University, Bangkok, Thailand

Dani Prasetyawan Institute of Integrated Research, Institute of Science Tokyo, Yokohama, Japan

Takamichi Nakamoto Institute of Integrated Research, Institute of Science Tokyo, Yokohama, Japan

About the editor

Takamichi Nakamoto has studied machine olfaction for more than 30 years. He received his BE and ME degrees in 1982 and 1984, respectively, and his PhD degree in electrical and electronic engineering from Tokyo Institute of Technology, Tokyo, Japan. He worked for Hitachi from 1984 to 1987. In 1987 he joined Tokyo Institute of Technology as a Research Associate. In 1993 he became an Associate Professor with the Department of Electrical and Electronics Engineering, Tokyo Institute of Technology. From 1996 to 1997, he was a Visiting Scientist at Pacific Northwest National Laboratory, Richland, WA, United States. He organized digital olfaction society conference in 2014. He was a Vice President of Sensors and Micromachine Society, Institute of Electrical Engineering, Japan from 2015 to 2016. He served as TPC Co-chair of the International Symposium on Olfaction and Electronic Nose in 2019. He is currently a Professor at the Institute of Innovative Research, Tokyo Institute of Technology.

Preface

This book provides a comprehensive overview of digital technologies in olfaction. Digitalization of olfaction is an emerging technology with high impact on broad range of fields such as food, beverage, cosmetics, healthcare, advertisement, ICT, data science, game, web content, and even art. Topics include odor-sensing system, odor impression prediction, odor reproduction, olfactory display, and its contents. This book introduces informatics approach to chemistry where actual experiments are the main parts in most of cases. Although chemometrics has been used in many areas of chemistry, it is typically based on conventional multivariate analysis. This book introduces the extension of the chemometrics in a modernized way such as deep learning and AI from the viewpoint of machine olfaction.

The purpose to publish this book is to let many people know about the emerging digital technologies in olfaction. Although the editor of this book published the edited book of machine olfaction 6 years ago, the new technologies not included in that book have evolved since then. This area has had much progress for recent 6 years.

Readers can gain the up-to-date knowledge of machine olfaction. Products (food, beverage, cosmetics, soap, deodorant, toothpaste, refreshners for breath, room, bathroom, etc.) related to olfaction are mainly handled by chemists who are not always familiar with IT technology. Although olfaction is currently a little far from IT technology, a variety of new products can be generated if digitalization of olfaction is achieved owing to digital olfactory contents such as scented movie, scented animation, scented game, educational tool with scents, scented advertisement, disaster drill with smell, and even teleolfaction. Readers in chemical fields can obtain knowledge related to data science, modernized method of chemometrics, which can be also applicable to other chemical fields since chemometrics is used in many areas of chemistry such as infrared and UV/visible spectroscopy, mass spectrometry, nuclear magnetic resonance, etc.

Readers in IT technology, who are not always familiar with chemistry, have opportunity for finding new areas mainly handled by chemists. Thus readers in both fields can become more interdisciplinary through this book.

The fundamental parts of machine olfaction, that is, mammalian olfaction mechanism, sensor technologies, chemometrics, olfactory display principles, are also included. Thus readers not familiar with machine olfaction can start from those chapters. Then, readers can proceed to chapters oriented to applications.

The editor is grateful to all the contributors of this book, who are the leading experts in this area. He also thanks Ms. Mika Kondo and Ms. Mari Iijima for their assistance. Moreover, the editor appreciates Ms. Lindsay Lawrence of Elsevier for her advice and assistance in the publication process. Finally, a part of the work in several chapters was financially supported in part by JST (Japan Science and Technology Agency) Mirai project (no. 22692685). Their support is highly appreciated.

Takamichi Nakamoto

SECTION 1

Background

CHAPTER

1

Introduction to digital technologies in olfaction

Takamichi Nakamoto

Institute of Integrated Research, Institute of Science Tokyo, Yokohama, Japan

Visual and auditory senses can be handled and processed by digital technology. On the other hand, chemical senses, that is, olfaction and taste senses, have not benefited from digital technologies. In the physical world, all the quantities, such as temperature and light intensity, have analog values. Since analog values themselves cannot be handled and processed by a computer, it is necessary to convert it to digital form.

Digital technologies in this book do not mean just simple binary form of the data. Although it is the first step to convert analog sensor data to digital one via an analog-to-digital converter, digital technologies in this book have more meaning. In digital technologies, the data should be expressed numerically, be reproduced, and be predicted using existing data. In the case of olfaction, it is often difficult to show the data numerically. When we sniff the smell, we can say pleasant/unpleasant smell or other impression of the smell. However, how can we express it numerically? In terms of reproduction, it is nowadays easy to reproduce visual and auditory information. But how about reproduction of olfactory information? In terms of prediction, it is required to show the unexperienced data. Thus it is necessary to construct a model to describe a system behavior.

Although we did not have methods to do them previously, the digital technologies in olfaction are now evolving. We have many types of sensors based on both biological and nonbiological mechanisms to obtain numerical data of the smells. The method to reproduce a smell is becoming feasible using odor components. Moreover, deep learning technique enables us to build a model to predict unseen data.

In the early 1980s Persaud proposed the pattern recognition of sensor-array responses to classify odors (Persaud & Dodd, 1982). Then, many researchers entered the field of odor sensing system, that is, electronic nose. Scientists in machine olfaction field have mainly focused on sensor devices although chemometric techniques, mostly from pattern recognition or multivariate analysis, have been introduced. They are still looking for good sensor devices even if many devices have been so far proposed. Around the middle of the 2000s, a scent presentation device called olfactory display has appeared mainly in virtual reality (VR) field. The papers of olfactory

Digital Technologies in Olfaction
DOI: https://doi.org/10.1016/B978-0-443-15721-9.00001-4

displays have been gradually increasing although its population is still small. Around the middle of the 2010s, the prediction of sensory information, that is, scent impression, has started. Information, more abstract than classification or mixture quantification, can be extracted. Recently, it has been extended even to natural language processing. Since computer science people have also entered this field, many informatic techniques have been so far used.

Digital technologies in olfaction cover three fundamental technologies, as described above, such as odor sensing system, olfactory displays, and sensory information processing. Although only odor sensing system was previously studied, the new technologies are evolving and interacting with each other. Nakamoto, an editor of this book, also served as an editor of the book entitled "Human olfactory displays and interfaces" in 2012 (Nakamoto, 2013) and "Essentials of machine olfaction and taste" in 2016 (Nakamoto, 2016), respectively. Although those books deal with latest technologies at those stages, they cannot cover the latest technologies in 2024 since the progress in technologies is very rapid. Thus a new book covering digital technologies in olfaction is indispensable for people to study artificial olfactory sense from many aspects.

This book consists of background, odor sensing system, cheminformatics in machine olfaction, olfactory displays, and conclusions. Background part has these chapters: Introduction and Physiology of olfactory sense (Chapter 2). Chapter 2 is written by a biologist, Hiroaki Matsunami, who reviews olfaction mechanism so that laymen can understand the biological mechanism of olfaction.

In the part for odor sensing system, many types of sensors are reviewed (Chapter 3). Especially, quartz crystal microbalance sensors (Chapter 4) and odor biosensors (Chapter 5) are described. Moreover, methodology of odor sensing system to quantify mixture composition using active sensing is explained in Chapter 6.

In the part for cheminformatics in machine olfaction, fundamental techniques of multivariate analysis and neural network are explained first in Chapter 7. Then, the concept of odor space (Chapter 8), odor impression prediction from mass spectrum (Chapter 9), and odor impression prediction using natural language processing (Chapter 10) are described. Moreover, odor reproduction techniques using odor components are introduced in Chapter 11. In addition, more sophisticated techniques for odor reproduction are described in Chapters 12 and 13. This part is a kernel of digital technologies in olfaction.

In the part for olfactory displays, fundamental technologies to generate scents and deliver the scents to a user are first explained in Chapter 14. Then, multicomponent olfactory display to blend multiple ingredients is described (Chapter 15). After a wearable olfactory display for VR application, the topic moves to olfactory contents. Although most of contents are described in Chapter 17, the virtual environment based on computational fluid dynamics simulation is especially described in Chapter 18. Moreover, the attempt to realize teleolfaction, which can present the smell at the remote site, is shown in Chapter 19. Furthermore, olfactory display technologies are extended even to olfactory art (Chapter 20).

In the part for Conclusions, what we can do at the current technology level is concluded (Chapter 21). Moreover, future perspective of digital technologies in olfaction is discussed in Chapter 22.

We can establish virtual olfaction in cyber space through the digital technologies described above. In other words, we will have digital olfactory twin. Although the previous technologies just handled odor classification or mixture quantification, the digital technologies in olfaction are moving toward the new stage where we can freely manipulate scents.

References

Nakamoto, T. (2016). *Essentials of machine olfaction and taste*. Wiley. Available from 10.1002/9781118768495.

Nakamoto, T. (2013). *Human olfactory displays and interfaces: Odor sensing and presentation*. IGI Global.

Persaud, K., & Dodd, G. (1982). Analysis of discrimination mechanisms in the mammalian olfactory system using a model nose. *Nature*, *299*(5881), 352–355. Available from https://doi.org/10.1038/299352a0.

CHAPTER

2

Physiology of olfactory sense

Hsiu-Yi Lu, Priyanka Meesa and Hiroaki Matsunami
Department of Molecular Genetics and Microbiology, Duke University, Durham, NC, United States

2.1 Introduction of mammalian olfactory system

2.1.1 Anatomy of the olfactory system

The olfactory system operates as a complex network of cellular systems that collectively facilitate the communication of odor perception to the brain. The system has both central and peripheral components that orchestrate the process of odor detection, transduction, and integration within the brain (Salazar et al., 2019; Smith & Bhatnagar, 2019).

The key structure of the mammalian peripheral olfactory system is the olfactory epithelium, a specialized organ housing an array of distinct cell types, with olfactory sensory neurons (OSNs) being the most relevant. These OSNs extend cilia laden with olfactory receptors to detect odorants. The olfactory epithelium is the initial site of interaction between odorants and the olfactory system, where the process of transduction commences.

Development and regeneration within the olfactory system unfold in a highly orchestrated manner. OSNs arise from progenitor cells located within the olfactory epithelium, with distinct lineage markers guiding their differentiation and maturation. The orchestrated process of cell differentiation within the olfactory epithelium ensures the generation of a diverse population of OSNs equipped with a broad spectrum of olfactory receptors, contributing to the combinatorial coding scheme characteristic of the olfactory system. As these OSNs project their axons to the olfactory bulb, they establish connections with second-order neurons, mitral and tufted cells, forming a sophisticated network that facilitates the transmission of olfactory information to higher brain regions for further analysis and interpretation.

Beneath the olfactory epithelium and across the blood–brain barrier resides the olfactory bulb. The olfactory bulb consists of glomeruli, which comprises bundled olfactory nerves that are essentially axons projected from the olfactory epithelium (Fig. 2.1). These axon bundles meticulously converge within the olfactory bulb, forming synaptic connections with mitral cells and tufted cells. This anatomy within the olfactory bulb represents a crucial convergence point in the olfactory system, where initial

Digital Technologies in Olfaction
DOI: https://doi.org/10.1016/B978-0-443-15721-9.00002-6

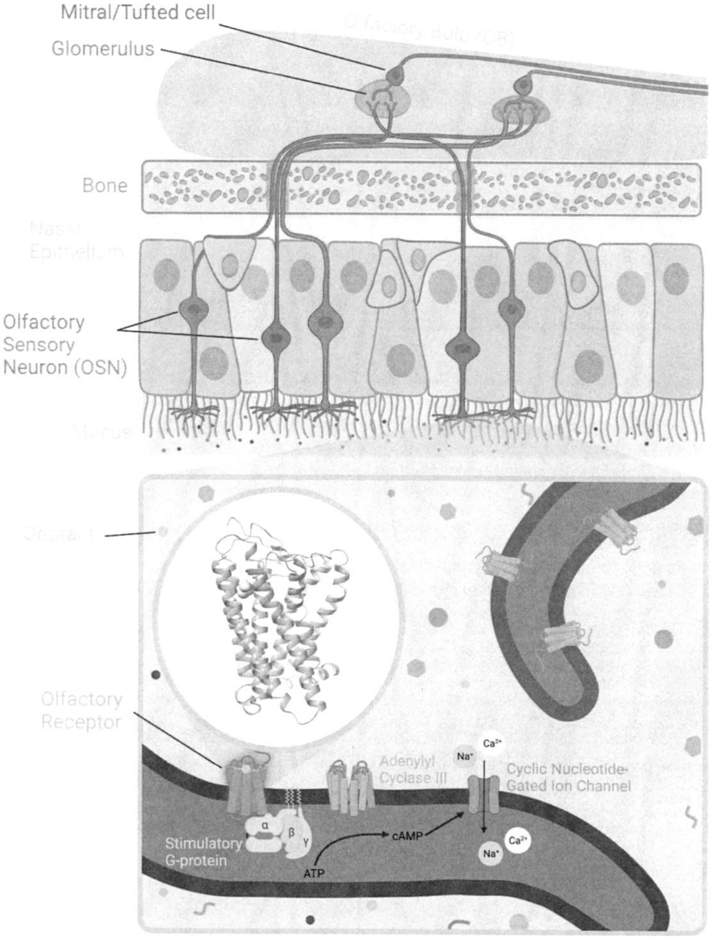

FIGURE 2.1 Olfactory system. The human olfactory system is characterized by a specific anatomical and molecular arrangement. Olfactory sensory neurons, found in the olfactory epithelium, house olfactory receptors, with each neuron expressing a single receptor type. When activated, these neurons transmit signals to well-defined microregions known as glomeruli within the olfactory bulb. Neurons of the same receptor type are dispersed randomly within the olfactory epithelium but converge onto a common glomerulus. Within these glomeruli, olfactory nerve endings excite mitral cells, mediating the transmission of olfactory signals to higher brain regions. Odorants in the nasal mucus directly bind to olfactory receptors located on the cilia's membranes. This binding initiates the activation of stimulatory G-proteins, such as Golf, which subsequently activate adenylyl cyclase, leading to the production of cAMP. cAMP binds and opens a cation-selective channel, allowing the influx of Na^+ and Ca^{2+} ions into the cilia, ultimately resulting in depolarization. *cAMP*, Cyclic AMP.

olfactory information is further processed and transmitted to the central olfactory system.

The central olfactory system is located in the brain and plays a pivotal role in the processing of olfactory information. The system begins with the olfactory bulb. From the bulb, processed signals are transmitted to several areas in the brain, one of which is the pyriform cortex, a region known for its involvement in the analysis and interpretation of olfactory cues. This integration of peripheral and central components within the olfactory system facilitates the perception and interpretation of a diverse range of odors, contributing to the overall olfactory experience and sensory perception.

2.1.2 Olfactory G protein–coupled receptors

The human olfactory system relies on an array of approximately 400 distinct olfactory receptors, including six trace amine associated receptors (TAARs) that respond to amines and ~60 Class I and ~310 Class II odorant receptors (ORs) attuned to carboxylic acids and diverse odorants, respectively (DeMaria & Ngai, 2010; Niimura et al., 2014; Spehr &

Munger, 2009). These receptors serve as the initial gateway for odor perception, where the binding of odorant ligands initiates a cascade of events.

At the molecular level, olfactory receptors, which belong to the G protein–coupled receptor (GPCR) family, share similarities with receptors responsive to many drugs and hormones. Although there exists a diverse array of receptors in OSNs, the downstream molecular targets they influence remain consistent. The activation process begins with the binding of an odorant to the receptor's binding pocket, triggering golf, a stimulatory G protein. Subsequently, adenylyl cyclase type III is activated, leading to the production of cyclic adenosine monophosphate (cAMP), which binds and opens cyclic nucleotide-gated channels, ultimately leading to a depolarization of the cell membrane (Billesbølle et al., 2023; Jones & Reed, 1989; Krautwurst et al., 1998) (Fig. 2.1).

2.1.3 Combinatorial coding in olfaction

Each unique OSN selectively expresses a single unique type of OR when matured. Importantly, each OR is sensitive to a particular set of volatile compounds, and each individual volatile compound can activate a particular set of ORs. This phenomenon is described as combinatorial coding (DeMaria & Ngai, 2010; Malnic et al., 1999) (Fig. 2.2). A greater concentration of compound will activate a greater number of receptors. Therefore, depending on its concentration, a specific environmental compound can activate a varied combination of ORs. However, most of the scents we perceive are derived from complex mixtures of volatile chemicals rather than an individual volatile compound. The interaction of these mixtures with our olfactory receptors is even more complicated because the response is not merely an additive result from each individual component (Chaput et al., 2012; Fukutani et al., 2023; Kurian et al., 2021; Trimmer et al., 2023). The interplay of these diverse receptor activations contributes to the rich and nuanced sensory experience associated with the perception of complex odors. Understanding the underlying principles of combinatorial coding is key to unraveling the intricacies of odor discrimination and the

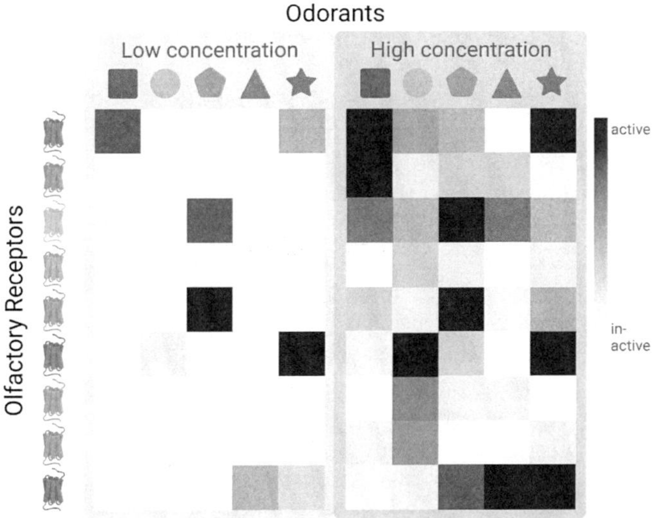

FIGURE 2.2 Combinatorial coding of odors at varied concentrations. In the olfactory system, individual olfactory receptors can respond to a variety of odorants; conversely, a single odorant can also activate a wide range of olfactory receptors. Additionally, as the concentration of a given odorant increases, the number of receptor responded can also change.

remarkable diversity of scents discernible to the human olfactory system.

2.1.4 Understanding olfactory receptor activation: a bioassay perspective

Understanding the intricate world of olfactory receptor activation demands an in-depth exploration of the complex interplay between odorants and receptor proteins. A common avenue for understanding these mechanisms lies in the development of receptor-based bioassays (Bavan et al., 2014; Saito et al., 2009; Zhuang & Matsunami, 2008). However, pioneering insights into the complexities of olfactory receptor activation underscore the challenges associated with the cell surface and functional expression of these receptors in heterologous cell systems. Before the discovery of accessory proteins, receptor-transporting proteins (RTPs), there were formidable obstacles to achieving a successful expression of olfactory receptors in nonnative cellular environments.

A critical breakthrough in this field emerged with the identification of accessory proteins RTP1 and RTP2, which enabled the functional expression of numerous ORs in heterologous cells (Saito et al., 2004; Sharma et al., 2017; Zhuang & Matsunami, 2007). Despite the discovery, achieving functional expression in vitro for all ORs remains limited, with only a fraction of receptors successfully translatable to a nonnative cellular environment. The complexities of protein folding, particularly in the endoplasmic reticulum, contribute significantly to this bottleneck, underscored by the divergence in amino acid sequences.

Moreover, the impact of nasal mucus on receptor function is important to consider. The mucus not only serves as a potential vehicle for transporting volatiles to the receptors but also hosts enzymatic activity that could modify odorants before they reach their respective receptors (Kurian et al., 2021). As a result, the environmental molecules interacting with the receptors might not entirely mirror the chemical composition of the odorant stimulus once passing through the nasal mucosa.

Furthermore, in vivo bioassays using model organisms, such as mice provide valuable insights. Techniques that genetically tag OSNs with fluorescent reporter proteins enable researchers to monitor the activation of specific types of olfactory receptors in vivo (Bozza et al., 2009; Zhu et al., 2022). Furthermore, emerging technologies that identify olfactory receptors that are activated by odor stimulation enable to comprehensively identify a set of olfactory receptors activated by tested odors, offering promising prospects. Yet the applicability of findings from model organisms to humans necessitates careful consideration.

2.1.5 Olfactory receptor structure: insights from structural analysis

The elucidation of the structural details of the diverse family of ORs has long remained a challenging pursuit. However, breakthroughs in the field have provided a glimpse into the intricate process of chemical recognition between an odorant and a specific human OR. The cryo-electron microscopy (cryo-EM) analysis of the active human odorant receptor OR51E2, in complex with the fatty acid propionate and with structural elucidation of murine TAARs (Guo et al., 2023), has unveiled critical insights into the molecular interactions governing the binding of odorants to the receptor (Billesbølle et al., 2023).

Billesbølle et al. (2023) have provided a clear understanding of the location and specific amino acid interactions within the receptor's binding pocket and demonstrated the remarkable selectivity of the receptor in response to certain carbon chain lengths of fatty acids. Utilizing site-directed mutagenesis, the research team engineered a modified OR51E2 with a tailored binding pocket, showcasing the dynamic adaptability of the receptor to accommodate varying ligand sizes. Notably, the conformational changes observed in the receptor following activation by the odorant underscore the complex interplay between the receptor and its ligand.

While this structural analysis represents a significant step forward, it also highlights the complexity and diversity inherent within the broader family of ORs. The findings pave the way for a deeper understanding of other olfactory receptors and lay the foundation for unraveling the dynamics governing the binding relationship between odorants and receptors. Further studies building on these structural insights hold the potential to unlock the underlying mechanisms driving olfactory sensory perception.

2.2 Leveraging computational and machine learning techniques for olfactory research

Understanding and predicting the perceptual properties of odorants based on their chemical structures, like how wavelengths encode color properties, has long been a challenging pursuit. However, advancements in computational and machine learning techniques hold promising prospects for unraveling the complexities of olfactory perception. There have been numerous efforts from a variety of research groups to better understand the complexities by applying machine learning to encode volatile molecule to scent (Khan et al., 2007; Lötsch et al., 2019; Mayhew et al., 2022; Nozaki & Nakamoto, 2018). In a notable development, Lee et al. (2023) utilized the power of graph neural networks (GNN) to create an improved principal odor map.

This principal odor map utilizes a high-dimensional framework to represent an extensive catalog of odorants and their corresponding perception. This map effectively captures known perceptual distances and demonstrates superior predictive capabilities compared to previous models. Lee et al.'s approach represents a significant stride in digitalizing olfaction, enabling a comprehensive understanding of the intricate relationships within the molecular and odor spaces.

As the field of computational olfaction continues to progress, it is crucial to acknowledge the existing limitations that shape the path for future advancements. While methodologies like GNNs have shown promise, the absence of experimentally verified structures for the vast majority of human olfactory receptors remains a significant obstacle. Alphafold (Jumper et al., 2021), a deep learning system that has demonstrated remarkable success in protein structure prediction, presents an exciting prospect for overcoming this challenge. However, there are some notable differences between the Cryo-EM and Alphafold structures of OR51E2 in domains critical for odorant binding and OR activation. While the Cryo-EM structure is an active conformation complexed with the active ligand, the corresponding Alphafold structural model is likely to mimic more inactive conformations (Choi et al., 2023). Future integration of experimental structures and computational frameworks could provide more accurate and reliable predictions of olfactory receptor structures complexed with their ligands. This would enhance our understanding of the intricate mechanisms underlying olfactory perception.

Additionally, it is important to note that existing prediction framework often overlooks the critical factor of odor concentration. The perceived intensity of an odor is closely linked to its concentration, a parameter that significantly influences receptor activation (Yang et al., 2023). Neglecting this crucial aspect can limit the ability of existing prediction framework to precisely predict the complete olfactory experience. Future research endeavors should incorporate concentration-dependent factors into computational models, enabling a more comprehensive understanding of how odorant molecules interact with olfactory receptors at varying concentrations. By addressing these challenges, the future of computational olfaction holds the potential to unravel the complex interplay between odorant properties, receptor activation, and perceptual responses. This would lead to a more holistic understanding of the intricacies of the sense of smell. Based on recent progress in computational approaches using nonolfactory GPCR family for drug screening and design (Hauser et al., 2017), similar computational techniques and machine learning methodologies in olfactory research could virtually screen a vast number of volatiles to identify novel ligands for olfactory receptors. This would yield insights and transformative outcomes in the quest to unravel the mysteries of the sense of smell.

In conclusion, the integration of computational strategies into olfactory research is an evolving landscape that is poised to illuminate the complexities of scent perception and provide a solid foundation for further advancements in the field. Improving and perfecting the principle odor map is a crucial step forward, opening up new avenues for understanding and manipulating olfactory experiences.

2.2.1 Deciphering olfaction through olfactory receptors and implications for the future

Decoding the olfactory system and better understanding its implications for the future of sensory experiences offer a fascinating avenue for exploration. The multidimensional interplay between olfactory receptors and the perception of flavors suggests the significance of this sensory pathway in our daily lives. An excellent example of this phenomenon is the collective experience of flavor that occurs when savoring a bowl of ramen, where the gustatory system discerns salty, umami, sweet, sour, and bitterness, while the olfactory system discerns the distinctive notes of various ingredients, contributing to the overall flavor experience. Such sensory comprehension has paved the way for groundbreaking developments like meat-imitation vegetarian burgers that utilize olfaction to mimic the scent and taste of real meat. Illustrating how our olfactory experiences shape our perception of taste and flavor. This understanding holds profound implications for the future, particularly in the realms of food technology and flavor design.

The prospect of mapping a comprehensive array of receptors in relation to significant odorants opens a realm of possibilities where the creation of flavors can be approached with scientific precision. Imagine a future where the formulation of ingredients and flavors is guided not merely by subjective assessments but rather by an understanding of receptor activities. Envision a high-throughput virtual and experimental screening process that integrates receptor-based interactions, allowing for the rational design of flavors and additives tailored to specific preferences and palates. Such a paradigm shift could revolutionize the food, flavor, and fragrance industry, facilitating the seamless replication of sensory experiences.

Despite promising advancements, challenges persist. Identifying all key odorants remains a daunting task, particularly given the prevalence of trace molecules that evade conventional detection methods. Moreover, the complexity of odor perception, often stemming from a fusion of multiple odor components, presents an additional hurdle. In the pursuit of a more comprehensive understanding, it is imperative to delineate interaction maps that encapsulate the amalgamation of natural odor mixtures, thereby enabling the accurate reconstruction of complex flavor profiles from their elemental constituents.

The convergence of atomic-level insights into odor-receptor dynamics, bolstered by the integration of cutting-edge computational and machine learning techniques, offers an optimistic trajectory for the development of comprehensive odor maps. Such endeavors hold the promise of unraveling the complex nuances of sensory perception, leading to an enhanced and tailored odor experience that resonates with diverse palates and preferences.

2.3 AI disclosure

During the preparation of this work, the author(s) used ChatGPT to make writing more fluid. After using this tool/service, the author (s) reviewed and edited the content as needed and take(s) full responsibility for the content of the publication.

References

Bavan, S., Sherman, B., Luetje, C. W., & Abaffy, T. (2014). Discovery of novel ligands for mouse olfactory receptor MOR42-3 using an in silico screening approach and in vitro validation. *PLoS One*, *9*(3). Available from https://doi.org/10.1371/journal.pone.0092064, http://www.plosone.org/article/fetchObject.action?uri = info%3Adoi%2F10.1371%2Fjournal.pone.0092064&representation = PDF.

Billesbølle, C. B., de March, C. A., van der Velden, W. J. C., Ma, N., Tewari, J., del Torrent, C. L., Li, L., Faust, B., Vaidehi, N., Matsunami, H., & Manglik, A. (2023). Structural basis of odorant recognition by a human odorant receptor. *Nature*, *615*(7953), 742–749. Available from https://doi.org/10.1038/s41586-023-05798-y, https://www.nature.com/nature/.

Bozza, T., Vassalli, A., Fuss, S., Zhang, J. J., Weiland, B., Pacifico, R., Feinstein, P., & Mombaerts, P. (2009). Mapping of Class I and Class II odorant receptors to glomerular domains by two distinct types of olfactory sensory neurons in the mouse. *Neuron*, *61*(2), 220–233. Available from https://doi.org/10.1016/j.neuron.2008.11.010.

Chaput, M. A., El Mountassir, F., Atanasova, B., Thomas-Danguin, T., Le Bon, A. M., Perrut, A., Ferry, B., & Duchamp-Viret, P. (2012). Interactions of odorants with olfactory receptors and receptor neurons match the perceptual dynamics observed for woody and fruity odorant mixtures. *European Journal of Neuroscience*, *35*(4), 584–597. Available from https://doi.org/10.1111/j.1460-9568.2011.07976.x.

Choi, C., Bae, J., Kim, S., Lee, S., Kang, H., Kim, J., Bang, I., Kim, K., Huh, W. K., Seok, C., Park, H., Im, W., & Choi, H. J. (2023). Understanding the molecular mechanisms of odorant binding and activation of the human OR52 family. *Nature Communications*, *14*(1). Available from https://doi.org/10.1038/s41467-023-43983-9, https://www.nature.com/ncomms/.

DeMaria, S., & Ngai, J. (2010). The cell biology of smell. *Journal of Cell Biology*, *191*(3), 443–452. Available from https://doi.org/10.1083/jcb.201008163.

Fukutani, Y., Abe, M., Saito, H., Eguchi, R., Tazawa, T., de March, C. A., Yohda, M., & Matsunami, H. (2023). Antagonistic interactions between odorants alter human odor perception. *Current Biology*, *33*(11), 2235–2245.e4. Available from https://doi.org/10.1016/j.cub.2023.04.072, http://www.elsevier.com/journals/current-biology/0960-9822.

Guo, L., Cheng, J., Lian, S., Liu, Q., Lu, Y., Zheng, Y., Zhu, K., Zhang, M., Kong, Y., Zhang, C., Rong, N., Zhuang, Y., Fang, G., Jiang, J., Zhang, T., Han, X., Liu, Z., Xia, M., Liu, S., & Sun, J. P. (2023). Structural basis of amine odorant perception by a mammal olfactory receptor. *Nature*, *618*(7963), 193–200. Available from https://doi.org/10.1038/s41586-023-06106-4, https://www.nature.com/nature/.

Hauser, A. S., Attwood, M. M., Rask-Andersen, M., Schiöth, H. B., & Gloriam, D. E. (2017). Trends in GPCR drug discovery: New agents, targets and indications. *Nature Reviews Drug Discovery*, *16*(12), 829–842. Available from https://doi.org/10.1038/nrd.2017.178, http://www.nature.com/nrd/index.html.

Jones, D. T., & Reed, R. R. (1989). Golf: An olfactory neuron specific-G protein involved in odorant signal transduction. *Science (New York, N.Y.)*, *244*(4906), 790–795. Available from https://doi.org/10.1126/science.2499043.

Jumper, J., Evans, R., Pritzel, A., Green, T., Figurnov, M., Ronneberger, O., Tunyasuvunakool, K., Bates, R., Žídek, A., Potapenko, A., Bridgland, A., Meyer, C., Kohl, S. A. A., Ballard, A. J., Cowie, A., Romera-Paredes, B., Nikolov, S., Jain, R., Adler, J., & Hassabis, D. (2021). Highly accurate protein structure prediction with AlphaFold. *Nature*, *596*(7873), 583–589. Available from https://doi.org/10.1038/s41586-021-03819-2, http://www.nature.com/nature/index.html.

Khan, R. M., Luk, C. H., Flinker, A., Aggarwal, A., Lapid, H., Haddad, R., & Sobel, N. (2007). Predicting odor pleasantness from odorant structure: Pleasantness as a reflection of the physical world. *Journal of Neuroscience*, *27*(37), 10015–10023. Available from https://doi.org/10.1523/JNEUROSCI.1158-07.2007, http://www.jneurosci.org/cgi/reprint/27/37/10015.

Krautwurst, D., Yau, K. W., & Reed, R. R. (1998). Identification of ligands for olfactory receptors by functional expression of a receptor library. *Cell*, *95*(7), 917–926. Available from https://doi.org/10.1016/

S0092-8674(00)81716-X, https://www.sciencedirect.com/journal/cell.

Kurian, S. M., Naressi, R. G., Manoel, D., Barwich, A. S., Malnic, B., & Saraiva, L. R. (2021). Odor coding in the mammalian olfactory epithelium. *Cell and Tissue Research*, *383*(1), 445–456. Available from https://doi.org/10.1007/s00441-020-03327-1, link.springer.de/link/service/journals/00441/index.htm.

Lee, B. K., Mayhew, E. J., Sanchez-Lengeling, B., Wei, J. N., Qian, W. W., Little, K. A., Andres, M., Nguyen, B. B., Moloy, T., Yasonik, J., Parker, J. K., Gerkin, R. C., Mainland, J. D., & Wiltschko, A. B. (2023). A principal odor map unifies diverse tasks in olfactory perception. *Science (New York, N.Y.)*, *381*(6661), 999–1006. Available from https://doi.org/10.1126/science.ade4401.

Lötsch, J., Kringel, D., & Hummel, T. (2019). Machine learning in human olfactory research. *Chemical Senses.*, *44*(1), 11–22. Available from https://doi.org/10.1093/chemse/bjy067.

Malnic, B., Hirono, J., Sato, T., & Buck, L. B. (1999). Combinatorial receptor codes for odors. *Cell*, *96*(5), 713–723. Available from https://doi.org/10.1016/S0092-8674(00)80581-4, http://www.cell.com.

Mayhew, E. J., Arayata, C. J., Gerkin, R. C., Lee, B. K., Magill, J. M., Snyder, L. L., Little, K. A., Yu, C. W., & Mainland, J. D. (2022). Transport features predict if a molecule is odorous. *Proceedings of the National Academy of Sciences of the United States of America*, *119*(15). Available from https://doi.org/10.1073/pnas.2116576119, https://www.pnas.org/doi/10.1073/pnas.2116576119.

Niimura, Y., Matsui, A., & Touhara, K. (2014). Extreme expansion of the olfactory receptor gene repertoire in African elephants and evolutionary dynamics of orthologous gene groups in 13 placental mammals. *Genome Research*, *24*(9), 1485–1496. Available from https://doi.org/10.1101/gr.169532.113.

Nozaki, Y., & Nakamoto, T. (2018). Predictive modeling for odor character of a chemical using machine learning combined with natural language processing. *PloS one*, *13*. Available from https://doi.org/10.1371/journal.pone.0198475.

Saito, H., Chi, Q., Zhuang, H., Matsunami, H., & Mainland, J. D. (2009). Odor coding by a mammalian receptor repertoire. *Science Signaling*, *2*(60). Available from https://doi.org/10.1126/scisignal.2000016, http://stke.sciencemag.org/about/.

Saito, H., Kubota, M., Roberts, R. W., Chi, Q., & Matsunami, H. (2004). RTP family members induce functional expression of mammalian odorant receptors. *Cell*, *119*(5), 679–691. Available from https://doi.org/10.1016/j.cell.2004.11.021, https://www.sciencedirect.com/journal/cell.

Salazar, I., Sanchez-Quinteiro, P., Barrios, A. W., López Amado, M., & Vega, J. A. (2019). Anatomy of the olfactory mucosa. *Handbook of Clinical Neurology*, *164*. Available from https://doi.org/10.1016/B978-0-444-63855-7.00004-6, http://www.sciencedirect.com/science/journal/00729752.

Sharma, R., Ishimaru, Y., Davison, I., Ikegami, K., Chien, M. S., You, H., Chi, Q., Kubota, M., Yohda, M., Ehlers, M., & Matsunami, H. (2017). Olfactory receptor accessory proteins play crucial roles in receptor function and gene choice. *eLife*, *6*. Available from https://doi.org/10.7554/eLife.21895, https://elifesciences.org/content/6/e21895-download.pdf.

Smith, T. D., & Bhatnagar, K. P. (2019). Anatomy of the olfactory system. *Handbook of Clinical Neurology*, *164*. Available from https://doi.org/10.1016/B978-0-444-63855-7.00002-2, http://www.sciencedirect.com/science/journal/00729752.

Spehr, M., & Munger, S. D. (2009). Olfactory receptors: G protein-coupled receptors and beyond. *Journal of Neurochemistry*, *109*(6), 1570–1583. Available from https://doi.org/10.1111/j.1471-4159.2009.06085.x.

Trimmer, C., Arroyave, R., Vuilleumier, C., Wu, L., Dumer, A., DeLaura, C., Kim, J., Pierce, G. M., Borisovska, M., De Nanteuil, F., Emberger, M., Varganov, Y., Margot, C., Rogers, M. E., & Pfister, P. (2023). Allosteric modulation of a human odorant receptor. *Current Biology*, *33*(8), 1523–1534.e4. Available from https://doi.org/10.1016/j.cub.2023.03.016, http://www.elsevier.com/journals/current-biology/0960-9822.

Yang, J. Y., O'Connell, T. F., Hsu, W. M. M., Bauer, M. S., Dylla, K. V., Sharpee, T. O., & Hong, E. J. (2023). Restructuring of olfactory representations in the fly brain around odor relationships in natural sources. *bioRxiv*. Available from https://www.biorxiv.org, http://doi.org/10.1101/2023.02.15.528627.

Zhu, K. W., Burton, S. D., Nagai, M. H., Silverman, J. D., de March, C. A., Wachowiak, M., & Matsunami, H. (2022). Decoding the olfactory map through targeted transcriptomics links murine olfactory receptors to glomeruli. *Nature Communications*, *13*(1). Available from https://doi.org/10.1038/s41467-022-32267-3, https://www.nature.com/ncomms/.

Zhuang, H., & Matsunami, H. (2007). Synergism of accessory factors in functional expression of mammalian odorant receptors. *Journal of Biological Chemistry*, *282*(20), 15284–15293. Available from https://doi.org/10.1074/jbc.m700386200.

Zhuang, H., & Matsunami, H. (2008). Evaluating cell-surface expression and measuring activation of mammalian odorant receptors in heterologous cells. *Nature Protocols*, *3*(9), 1402–1413. Available from https://doi.org/10.1038/nprot.2008.120.

SECTION 2

Odor sensing system

CHAPTER

3

Overview of olfactory sensors

Takamichi Nakamoto

Institute of Integrated Research, Institute of Science Tokyo, Yokohama, Japan

3.1 Odor sensing system

The odor sensing system called electronic nose has been studied for more than three decades. Persaud proposed the system made up of array of metal oxide sensors followed by pattern recognition in 1982 (Persaud & Dodd, 1982). Nakamoto and Moriizumi (1988) proposed the odor sensing system using a neural network. Since then, a variety of odor sensing system has been so far proposed (Pearce et al., 2003).

The principle of odor sensing system is illustrated in Fig. 3.1. Although every sensor responds to both odor A and odor B, the response pattern of odor A is different from that of odor B. An array of multiple sensors with partially overlapping specificities is used. Its output pattern is recognized by pattern recognition engine such as multivariate analysis or neural network. The principle based on pattern recognition is the same as that of the biological system.

Various sensors instead of olfactory receptors (ORs) are available. Those sensors are introduced in the next section.

3.2 Classification of olfactory sensor devices

Sensors used in odor sensing system are classified into acoustic wave devices, conductive devices, electrochemical devices, and optical ones. Those are shown in Table 3.1.

Acoustic wave devices are QCMs (quartz crystal microbalances; Ema et al., 1989; King, 1964; Sauerbrey, 1959), SAW (surface acoustic wave) devices (Grate et al., 1988; Rapp et al., 2000; Wohltjen, 1984), FBAR (film bulk acoustic resonator; Gabl et al., 2003; Johnston et al., 2010), cantilever sensors (Lang et al., 1999; Maute et al., 1999), and MSS sensors (Yoshikawa et al., 2012). Moreover, FPW device was applied to gas sensing (Cai et al., 2000). Each device is coated with sensing film and mass loading at the sensing film causes the frequency or phase shift from the air level.

Conductive devices are metal oxide sensors (Kaneyasu et al., 1987; Shurmer & Gardner, 1992; Yamazoe et al., 1983), micro hotplate sensors (Afridi et al., 2002; Graf et al., 2004), conducting polymers (Hatfield et al., 1994; Wilson et al., 2013), and carbon black

Digital Technologies in Olfaction
DOI: https://doi.org/10.1016/B978-0-443-15721-9.00003-8

composites (Kim et al., 2005; Lonergan et al., 1996). The sensor outputs are conductance change from the air level.

Electrochemical sensor utilizes the electrochemical reaction in the electrolyte (Stetter & Li, 2008). Its output is the current change from the air level. Moreover, the various types of optical sensors are used. Color change, fluorescence, chemiluminescence, and photoacoustic effect are used to detect gas/odor. Moreover, biological material, that is, OR, can be used as a sensor.

Those sensors are explained except QCMs sensor and OR in the subsequent sections. QCM and OR are explained in Chapters 4 and 5, respectively.

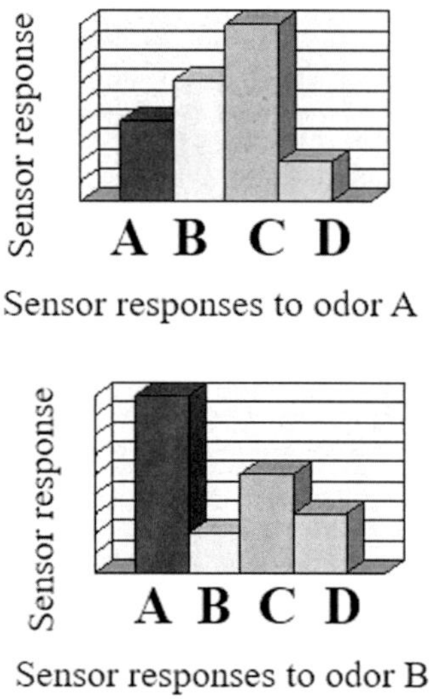

FIGURE 3.1 Principle of odor sensing system output pattern from an array made up of multiple sensors with partially overlapping specificities is recognized.

3.3 Gravimetric sensors

3.3.1 Surface acoustic wave sensors

A SAW device is one of acoustic wave devices. It uses Rayleigh wave, that is, SAW. In SAW device, the wave propagates along the substrate surface, whereas a QCM is classified as bulk wave device. Its behavior can be theoretically analyzed and has been studied well (Auld & Alexander, 1973). Although its main application is RF (radio frequency) filter, it is also used as gas/odor sensor.

TABLE 3.1 Classification of olfactory sensors.

Sensor type	Output format
Quartz crystal microbalance	Frequency shift
Surface acoustic wave device	Frequency shift
Metal oxide gas sensor	Resistance change
Conducting polymer	Resistance change
MOS FET	Change of threshold voltage
Optical method	Fluorescence luminescence light absorption, SPR
Electrochemical sensor	Current change
Odor biosensor	Voltage, current, light, etc.
Gas chromatography	Retention time
Mass spectrum	Mass

SPR, Surface plasmon resonance.

Fig. 3.2 shows the schematic of SAW sensor. SAW devices have a pair of IDTs (interdigital transducers). One works as transmitter and the other works as receiver. Since the electric field generated by IDT is applied to the piezoelectric substrate, the substrate surface deforms and then SAW is generated. It propagates along the substrate surface and then causes the deformation at the receiver IDT, which generates the electrical signal again with some delay.

The surface of SAW device along the propagation path is coated with thin organic film. When gas/odorant molecules come, it adsorbs them and SAW velocity changes. Although SAW velocity can be numerically calculated (Campbell & Jones, 1970), perturbation theory is often used to obtain the relationship between amount of adsorbed quantity and SAW velocity change (Auld & Alexander, 1973; Wohltjen, 1984).

Using the perturbation theory, SAW velocity change ΔV_R is obtained as follows:

$$\frac{\Delta V_R}{V_R} = (k_1 + k_2) f h \rho' - k_2 f h \frac{4\mu'}{{V_R}^2}\left(\frac{\lambda' + \mu'}{\lambda' + 2\mu'}\right) \quad (3.1)$$

where k_1 and k_2 are constants related to material, cut, and propagation direction, h the coating thickness, ρ' coating density, λ' and μ' are coating Lamé constants, and V_R the Rayleigh wave velocity. First term is related to mass change, whereas the second term is related to mechanical property of the thin film.

For SAW devices, several piezoelectric materials are available. ST-Cut quartz is very stable since its temperature coefficient is small. However, its electromechanical coupling coefficient is small. On the other hand, lithium niobate has large electromechanical coupling coefficient, whereas its temperature coefficient is large. Lithium tantalite is located between quartz and lithium niobate. Moreover, lead zirconate titanate is sometimes used. For the application to chemical sensor, ST-Cut quartz or Lithium tantalite is preferred because of low noise owing to low temperature coefficient.

For the actual measurement, there are two methods. One is to use vector network analyzer and the other is to use oscillator. Conventionally, the vector network analyzer is more expensive than the oscillator, whereas its measurement is more accurate. However, recently a small and cheap network analyzer is commercially available and it can be applied to SAW gas sensor.

As mentioned above, the piezoelectric material has temperature coefficient. Thus, SAW sensor sometimes suffers from large drift and aging problems. Dual type SAW device is used to compensate for drift and aging problems. Its structure is shown in Fig. 3.3. Two pairs of IDTs are fabricated on the same piezoelectric substrate. One is

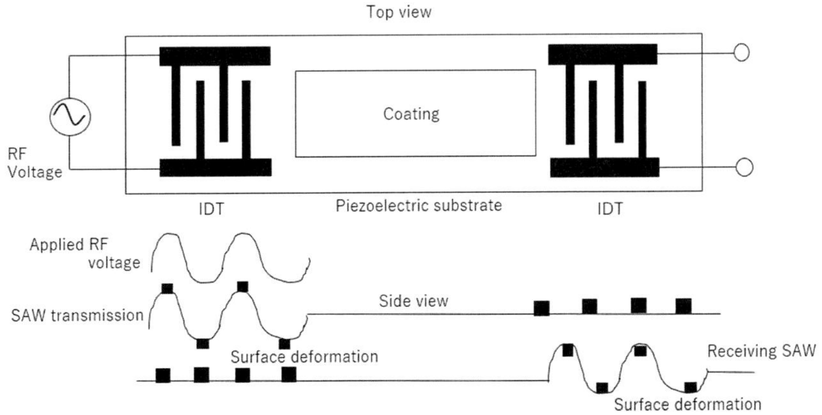

FIGURE 3.2 Schematic of SAW sensor SAW velocity changes owing to mass loading and viscoelastic effect. *SAW*, Surface acoustic wave.

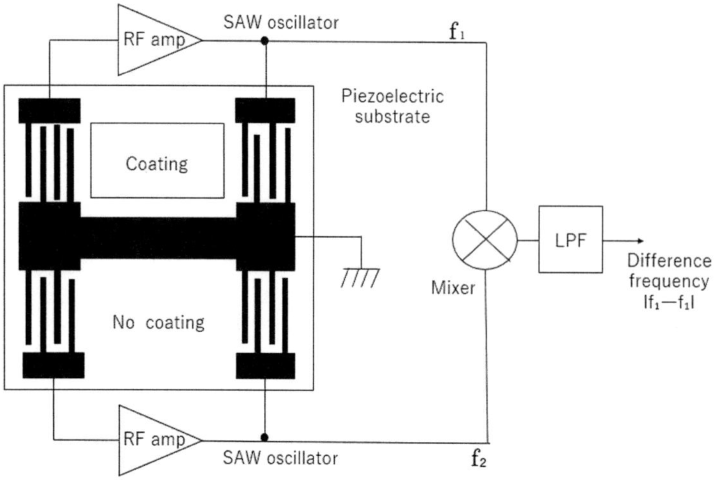

FIGURE 3.3 Dual type SAW sensor one with coating and the other without coating. *SAW*, Surface acoustic wave.

coated with a sensing film, whereas the other is bare substrate. The two oscillators connected to two pairs of IDTs are used. Two outputs of SAW oscillators are input to a mixer. Since the mixer works as a multiplier, its output V_{mixer} includes two components with the frequency $|f_1 - f_2|$ and $|f_1 + f_2|$ as is shown in Eq. (3.2). A low pass filter after the mixer eliminates the component with the frequency $|f_1 + f_2|$ and only the component with the frequency $|f_1 - f_2|$ is obtained. Thus, drift originating from the piezoelectric substrate can be compensated using the dual SAW device if two SAW devices are identical.

$$V_{mixer} \propto \cos 2\pi f_1 t \cos 2\pi f_2 t = \frac{1}{2}\cos 2\pi (f_1 - f_2)t - \frac{1}{2}\cos 2\pi (f_1 + f_2)t. \quad (3.2)$$

M. Rapp studied an electronic nose based on SAW sensor array (Barié et al., 2006). He developed the multiplexing technique to avoid interference between channels. He made a miniaturized sensor array (60×30 mm^2) composed of eight sensors with the frequencies of 433 MHz (Fig. 3.4).

Other SAW gas sensor example is zNose (Staples & Viswanathan, 2005). An uncoated SAW device is used as a detector of gas chromatography (GC). It is cooled as low as 0°C to improve its sensitivity and is heated as high as 150°C to desorb the gas molecules. Since there is no coating on SAW device, its Q factor can be kept high. Thus, good SNR (signal-to-noise ratio) is obtained.

3.3.2 Film bulk acoustic resonator sensors

FBAR (Film bulk acoustic resonator) is one type of bulk acoustic wave as well as QCM but the thickness of piezoelectric material is very thin. Thus, its resonance frequency is very high and it works in the range of GHz.

Fig. 3.5 shows the structure of FBAR. A thin piezoelectric material is between two electrodes. Since its thickness is very thin, the wavelength of standing wave is very short. The active area is isolated from the substrate and the generated acoustic wave does not radiate into it. Thus, its resonance is maintained.

Piezoelectric materials often used for FBAR are zinc oxide (ZnO) (Yan et al., 2007) and AlN (Wingqvist, 2010). The crystallographic orientation and surface morphology are important. Since AlN offers high acoustic wave velocity, it yields high frequency FBAR device. Temperature coefficient frequencies of AlN and ZnO are 25 and 40 ppm/°C, respectively, and they are sensitive to temperature drift (Mujahid et al., 2019).

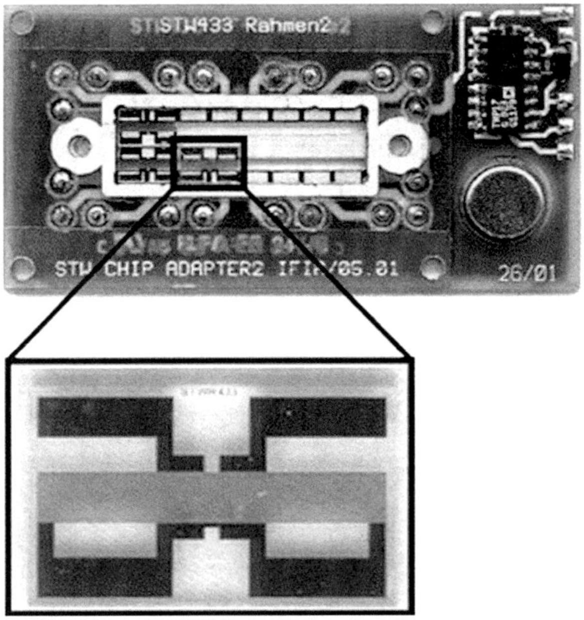

FIGURE 3.4 Photograph of the microarray with three sensor devices inserted (top view, cover removed) Eight SAW sensors can be put into a chamber. *SAW*, Surface acoustic wave. Source: *From Barié, N., Bücking, M., & Rapp, M. (2006). A novel electronic nose based on miniaturized SAW sensor arrays coupled with SPME enhanced headspace-analysis and its use for rapid determination of volatile organic compounds in food quality monitoring.* Sensors and Actuators B: Chemical, *114(1), 482–488. https://doi.org/10.1016/j.snb.2005.06.051.*

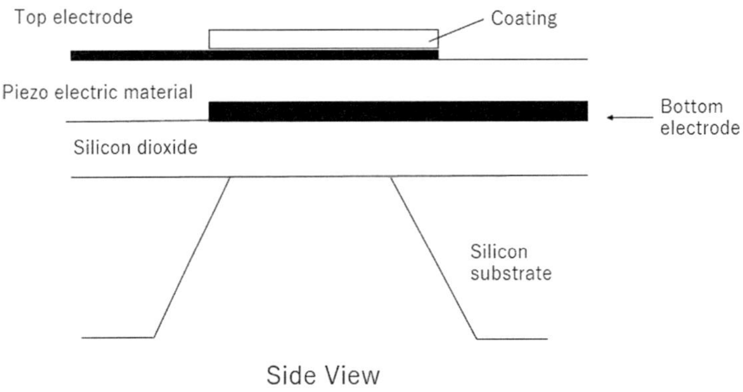

FIGURE 3.5 Structure of FBAR Thin piezoelectric material is located between two electrodes. It is supported by silicon substrate. *FBAR*, Film bulk acoustic resonator.

The application of FBAR to gas sensor was reported by Benetti et al. (2005). Thin Pd film for hydrogen detection and thin co-tetra-phenyl-porphyrin for ethanol and carbon monoxide were used as sensing films. They could detect 2 ppm H_2, 40 ppm CO, and 500 ppm ethanol.

SAM (self-assembled monolayer) is suitable for sensing film of FBAR because it is very thin. Chang et al. (2016) developed an array of FBAR coated with different SAMs and detected methanol, ethanol, acetone, and so on and reported that they obtained concentration-independent response patterns. Although FBAR can a have high resonance frequency, its issue is maintaining a high *Q* factor and a high SNR.

3.3.3 Cantilever sensors

A cantilever with rectangular bar of Si less than 1 μ m thick can be used as gas sensor when it is coated with organic film (Lang et al., 2005).

Adsorption of odorant molecules on the cantilever surface results in bending the cantilever due to the surface stress. A small-size sensor can be realized using the cantilever. Its deflection is detected by laser beam. When cantilever sensor array is used, each cantilever is coated with different sensing film as is shown in Fig. 3.6.

Cantilever sensor can be operated in several modes such as static and dynamic modes. In dynamic mode, the cantilever is oscillated at its resonant frequency as is shown in Fig. 3.7A (Battiston et al., 2001). Its resonant frequency is shifted to a lower frequency due to mass change caused by odorant adsorption (Battiston et al., 2001). In static mode, the surface stress due to adsorption of odorant molecules results in a static bending of the cantilever as is shown in Fig. 3.7B (Berger et al., 1997). Moreover, the photo of cantilever array is shown in Fig. 3.7C. Lang et al. (2007) performed the experiment on discriminating perfumes using the array of cantilever sensors coated with polymer films.

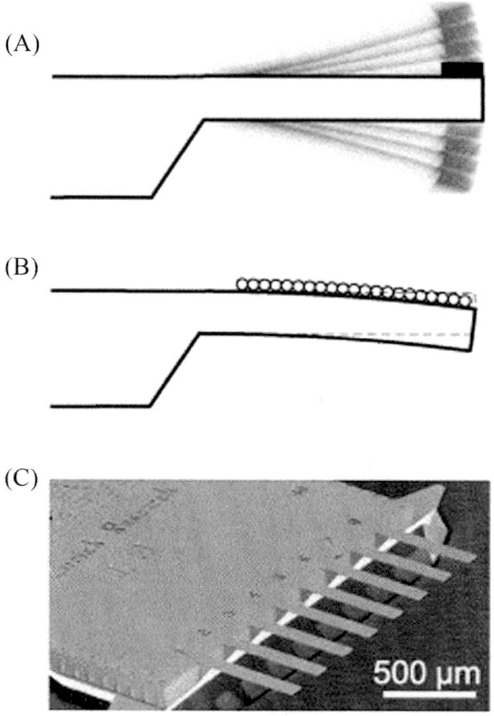

FIGURE 3.7 Modes of cantilever sensor (A) dynamic mode, (B) static mode, and (C) photo of cantilever array. Source: *From Battiston, F. M., Ramseyer, J.-P., Lang, H. P., Baller, M. K., Gerber, C. H., Gimzewski, J. K., Meyer, E., & Güntherodt, H.-J. (2001). A chemical sensor based on a microfabricated cantilever array with simultaneous resonance-frequency and bending readout.* Sensors and Actuators B: Chemical, *77(1–2), 122–131. https://doi.org/10.1016/s0925-4005(01)00683-9.*

3.3.4 Membrane-type surface stress sensors

A membrane-type surface stress (MSS) sensor is an extension of cantilever sensor (Minami et al., 2022; Yoshikawa et al., 2012). MSS sensor does not need bulky laser system to detect cantilever deflection. Instead of it, the resistance change due to piezoresistive effect caused by the stress is read out. However, the sensitivity of piezoresistive cantilever is not so high.

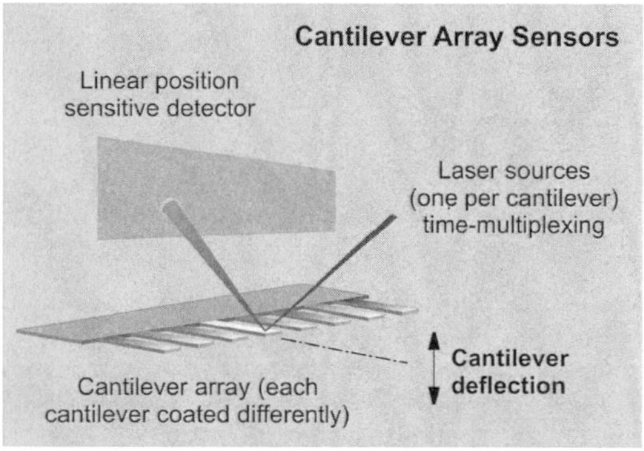

FIGURE 3.6 Principle of cantilever gas sensor cantilever coated with organic thin film works as a gas sensor. Source: *From Lang, H. P., Hegner, M., & Gerber, C. (2005). Cantilever array sensors.* Materials Today, *8(4), 30–36. https://doi.org/10.1016/s1369-7021(05)00792-3.*

Yoshikawa et al. developed the MSS sensor as is shown in Fig. 3.8. The surface stress on the membrane coated with sensing film is efficiently transduced onto the four beams since the stress is concentrated at the narrow beams. Moreover, this configuration enhances the thermal stability by self-compensation. The resistance change can be read out using a full Wheatstone bridge.

3.4 Conductance based sensors

3.4.1 Metal oxide sensors

Chemoresistor sensors are by far the simplest and most widespread technology used in olfactory sensors. Among them, the metal oxide semiconductor (MOX) sensors have been known and used long before the invention of electronic nose (Dey, 2018). MOX sensors are mainly classified into bulk and film types. In the film type MOX sensors, a layer of the metal oxides (usually SnO_2) is formed on a substrate. Metal oxides chemisorb oxygen atoms from the atmosphere, which results in the increase in electrical resistance. The odorant molecules are adsorbed in the metal oxide and react with chemisorbed oxygen species. This modulates the sensor resistance. The metal oxides that are intrinsically n-type semiconductors respond to reducible gases (H_2, CH_4, CO, and H_2S). The MOX sensors are operated at high temperature. The sensor temperature often alters the sensor characteristic, which allows for modulation of sensitivity and acquisition of more information.

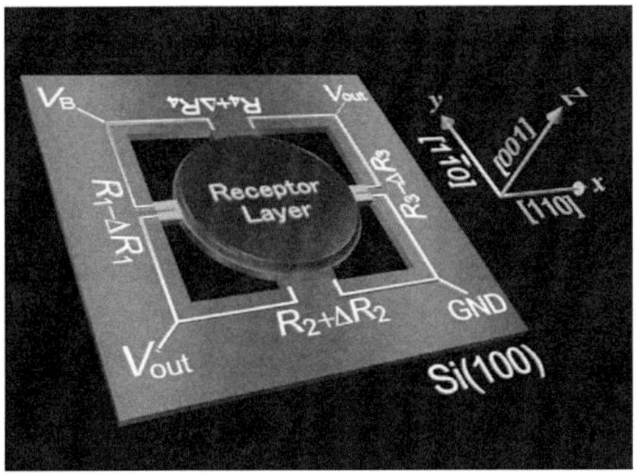

FIGURE 3.8 Structure of membrane-type surface stress sensor the whole surface stress induced on the round center membrane is efficiently detected by piezo resistors (red and blue colored parts) embedded in the constricted beams. Source: *From Yoshikawa, G., Akiyama, T., Loizeau, F., Shiba, K., Gautsch, S., Nakayama, T., Vettiger, P., Rooij, N., & Aono, M. (2012). Two dimensional array of piezoresistive nanomechanical membrane-type surface stress sensor (MSS) with improved sensitivity.* Sensors, 12*(11), 15873–15887. https://doi.org/10.3390/s121115873.*

The MOX materials are often doped with catalytic metals to increase sensitivity and selectivity of the sensors (Yamazoe et al., 1983). The MOX sensors are readily available and are often used in odor sensing systems. For that reason, this type of sensors dominates in the applications to evaluation of food and beverage.

Schematic depiction of the SnO_2 chemo resistive sensors' principle of operation is shown in Fig. 3.9. MOX does not consist of a single crystal but many grains. When the sensor is exposed to reducible gas, the potential barrier at the SnO_2 grain boundary decreases. Thus, electron can easily move across the grain boundary, resulting in conductivity increase. The sensors can be very sensitive to lower homologes of alcohols. Thus, it is often used as alcohol detector to prevent drunk driving. Moreover, an array of MOX sensors with different characteristics is often used as odor sensing system (Gardner et al., 2000). There is typically a significant cost in energy used to operate the sensors at elevated

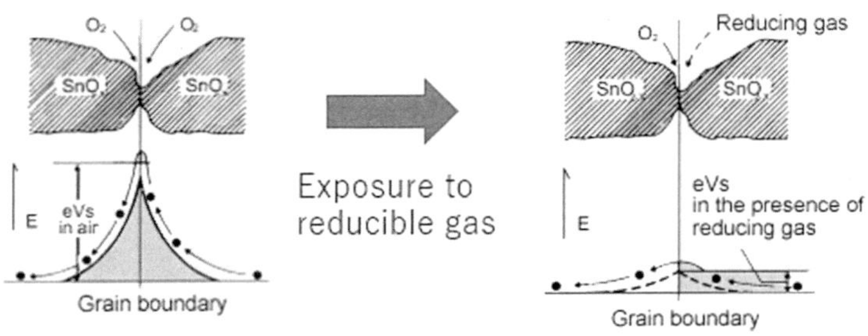

FIGURE 3.9 Principle of metal oxide gas sensor potential barrier decreases owing to reducible gas.

temperatures. However, MEMS-based MOX sensor has advantages in terms of consumption power and size. It will be described in the next section.

3.4.2 Micro hot plate sensors

Micro hot plate gas sensor is MEMS-based metal oxide gas sensor as is shown in Fig. 3.10 (Simon et al., 2001). It is a suspended structure including a heating element made of polysilicon layer (Afridi, Suehle, Zaghloul, et al., 2002; Meier et al., 2007). The micro hot plate can be fabricated using a standard commercial CMOS (complementary metal oxide semiconductor) process. The isolated micro hot plate structure enables high thermal efficiency and low thermal constant. After making the micro hot plate structure, metal oxide is deposited over the sensing electrode on the micro hot plate. Since the size of micro hot plate is small and can be fabricated using standard CMOS technology, a sensor array can be realized on a single chip. Peripheral circuits, including heater drive circuit, readout circuit, and analog-to-digital converter, are put into the same chip (Graf et al., 2004). Thus, it is suitable for portable odor sensing system. Moreover, its temperature can be rapidly changed due to short thermal time constant around millisecond. Since the sensor characteristic depends on its operation temperature, a single sensor equivalent to multiple sensors can be made when its temperature is temporarily changed. Thus, it is possible to optimize temperature-pulse sequence to enhance the discrimination capability of micro hot plate sensor array (Cavicchi et al., 1996).

3.4.3 Conducting polymer sensors

Although most of polymers do not have conductivity, there are several ones with conductivity, which are called conducting polymers. It can be also used as chemical sensor since its resistance changes upon gas exposure (Hatfield et al., 1994; Persaud, 2005). Materials are of low cost to manufacture and can be prepared as films. In comparison with metal oxide gas sensor, its response speed is fast even at room temperature. Its power consumption is very low (order of micro watts) because no heater is required. Moreover, the sensor is resilient to poisoning by compounds, which normally cause poisoning of metal oxide gas

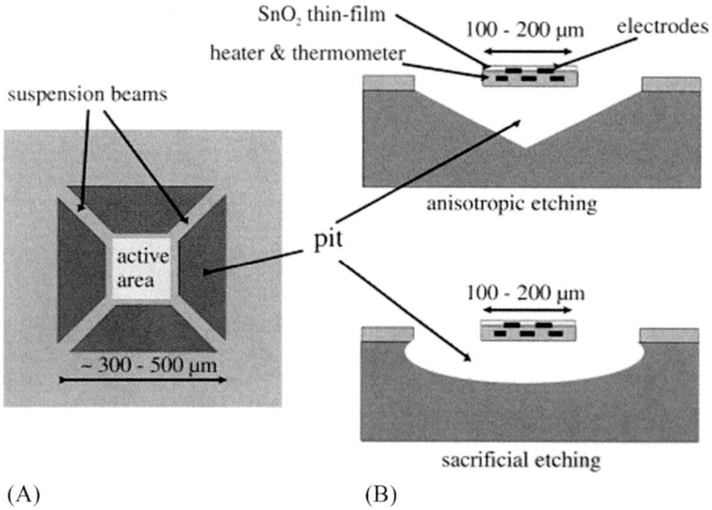

FIGURE 3.10 Structure of micro hot plate (A) top view and (B) side view. Source: *From Simon, I., Bârsan, N., Bauer, M., & Weimar, U. (2001). Micromachined metal oxide gas sensors: Opportunities to improve sensor performance.* Sensors and Actuators B: Chemical, *73(1), 1–26. https://doi.org/10.1016/s0925-4005(00)00639-0.*

sensor. The dependency of vapor concentration upon its resistance is linear, while it is nonlinear in the case of metal oxide gas sensor.

Resistance of conducting polymer is high even if it has conductivity. Thus, interdigitated electrode illustrated in Fig. 3.11 is often used. Conducting polymer is deposited over it. Distance between fingers should be small and number of finger pairs should be large to increase the conductivity. Since the interdigitated electrodes are fabricated using photolithography, it is possible to make small devices.

3.4.4 Carbon black composites

Although normal polymer does not have conductivity, carbon black is dispersed over the polymer to have the conductivity. The polymer swells when it adsorbs vapor. Then, its resistance increases during the vapor exposure (Llobet, 2013). Its principle is illustrated in Fig. 3.12. Since conducting path becomes longer or disrupted due to polymer swelling, its resistance increases during vapor exposure. Any polymer that adsorbs vapor can be used even if polymer itself does not have conductivity. Although polymer is used, its principle is different from that of conducting polymer. Rather than conducting polymer, its principle is close to QCM since amount of adsorbed molecules is measured in both cases. It can be used for portable electronic nose system (Kim et al., 2005). The array of carbon-black composite sensors was commercialized and it is known as cyranose (Muñoz et al., 1999).

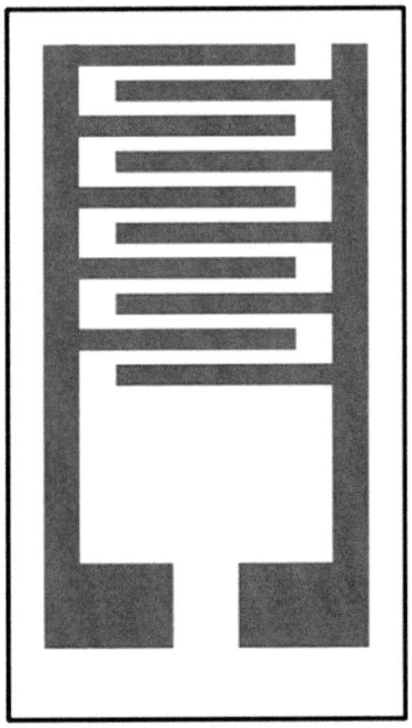

FIGURE 3.11 Interdigitated electrodes for conducting polymer sensor conducting polymer is deposited over interdigitated electrodes (Bai & Shi, 2007). Source: *From Bai, H., & Shi, G. (2007). Gas sensors based on conducting polymers.* Sensors, *7(3), 267–307. https://doi.org/10.3390/s7030267.*

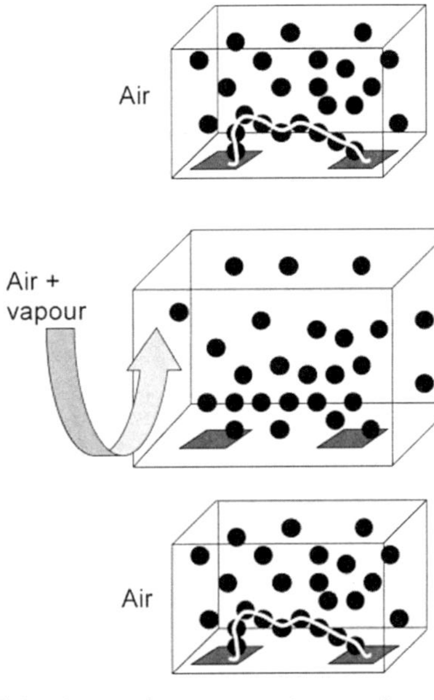

FIGURE 3.12 Gas sensing mechanism for carbon black composite gas sensor the subplot on top shows that conductive paths (white line) exist between the electrodes, which determine the baseline resistance of the sensor. In the presence of a vapor the polymer matrix swells, conduction paths are disrupted and sensor resistance increases (middle subplot). If the vapor is removed from the environment, the polymer regains its original shape, the conductive paths are restored and so is the original value of the sensor baseline resistance (bottom). Source: *From Llobet, E. (2013). Gas sensors using carbon nanomaterials: A review.* Sensors and Actuators B: Chemical, *179, 32–45. https://doi.org/10.1016/j.snb.2012.11.014.*

3.5 Electrochemical sensors

3.5.1 Typical electrochemical sensor

An electrochemical sensor utilizes electrochemical reaction to detect gases (Stetter & Li, 2008). Its structure is shown in Fig. 3.13. Three electrodes, such as WE (working electrode), CE (counter electrode), and RE (reference electrode), are immersed in electrolyte. The vapor passes through a permeable membrane and enters the electrolyte. When the vapor is dissolved into electrolyte, the current between WE and CE changes due to electrochemical reaction under the voltage between WE and RE. The current should not flow at RE since the voltage applied to bulk electrolyte changes according to the current. A potentiostat using an operational amplifier is used to set the voltage of the WE against the RE. Recently AFE (analog front end) IC chip including potentiostat is commercially available (Corva et al., 2022).

As shown in Fig. 3.13, the WE is deposited beneath the permeable membrane. The permeable membrane confines the electrolyte inside the chamber, whereas it is porous enough for vapor to pass through it. Stetter et al. (1986) studied an array of electrochemical sensors to discriminate among VOCs (volatile organic compounds).

3.5.2 Electrochemical gas sensor with atomic gold decorated polyaniline

Catalysts such as noble metals at WE are often used to enhance sensitivity and selectivity of the gas sensor. As the size of metal cluster decreases, its surface area increases, resulting in its activity enhancement. Although bulk metal is inserted, catalytic activity appears in nanostructure. Ultimate size is atomic size. Superior catalytic activity is expected in atomic metal as is illustrated in Fig. 3.14. Much improvement is expected when we can use it (Janata & Nakamoto, 2016).

Janata et al. have studied polyaniline doped with atomic gold (Jonke et al., 2011). Atomic metal is isolated and supported by polyaniline as is shown in Fig. 3.15. Platinum WE with atomic gold decorated polyaniline was successfully applied to gas sensing although gas was bubbled in KOH electrolyte (Chakraborty et al., 2020).

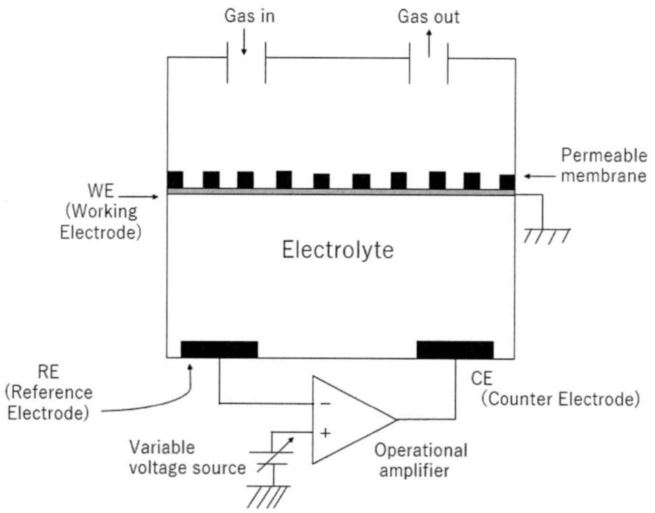

FIGURE 3.13 Principle of electrochemical sensor three electrodes, that is, WE, CE, and RE, are used. *CE*, Counter electrode; *RE*, reference electrode; *WE*, working electrode.

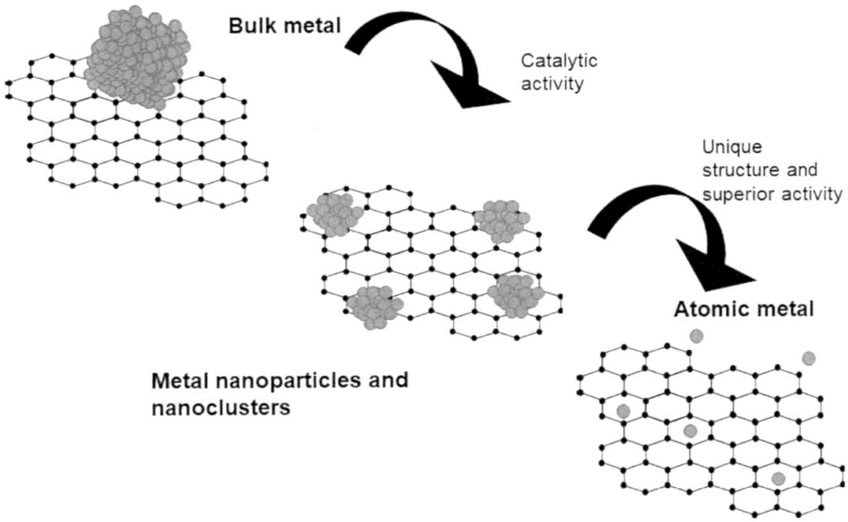

FIGURE 3.14 Change of metal cluster size and its catalyst activity from bulk metal to atomic metal.

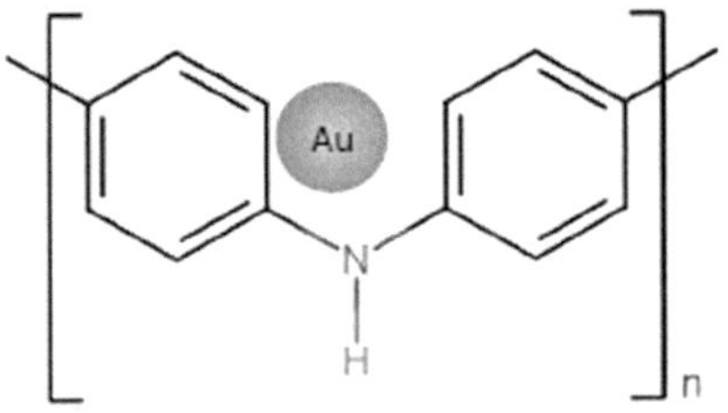

FIGURE 3.15 Atomic gold decorated polyaniline Atomic gold (Au1).

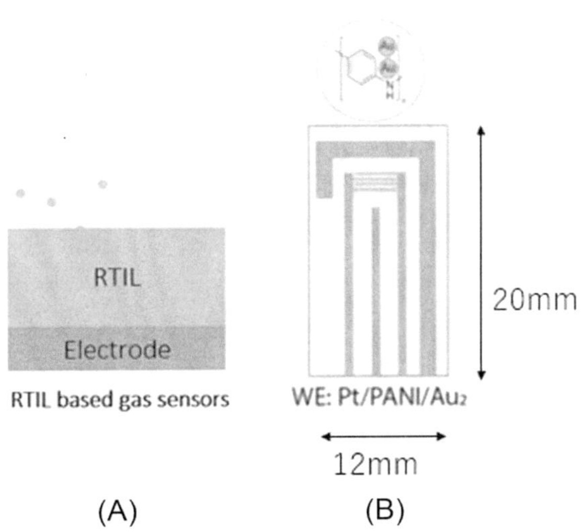

FIGURE 3.16 Electrochemical sensor coated with RTIL (A) amperometric gas sensor with RTIL and (B) amperometric gas sensor with atomic metal in RTIL. *RTIL*, Room temperature ionic liquid.

3.5.3 Size reduction of electrochemical gas sensor

The electrochemical gas sensor is typically bulky since a chamber containing electrolyte is indispensable. Thus, RTIL (room temperature ionic liquid) was used instead of liquid electrolyte. Since RTIL is viscous electrolyte liquid, the planar substrate can be coated with RTIL as is shown in Fig. 3.16A. Electrolyte can be thin film and gas permeable membrane is not necessary. Higher sensitivity and faster sensor responses are expected since film thickness is much smaller than bulky electrolyte of the conventional electrochemical sensor. Moreover, the size of electrode patterns on the planar substrate can be miniaturized (Paul et al., 2019).

Faricha et al. studied both electrochemical sensor and QCM sensor using same RTIL films. They claimed that two different sensing mechanisms of amperometric and gravimetric sensors enhanced the discrimination capability even if the same RTIL films were used (Faricha et al., 2022).

3.5.4 Electrochemical gas sensor with atomic gold decorated polyaniline covered with room temperature ionic liquid

Further improvement of the sensing capability is expected if atomic gold is included in the electrochemical sensor coated with RTIL. Thus, atomic gold decorated polyaniline was deposited on the planar substrate as is shown in Fig. 3.16B (Chakraborty et al., 2023). In this study, bi-atomic gold electrocatalyst has been successfully decorated on a polyaniline matrix. Then, RTIL was dropped on the atomic gold decorated polyaniline on the substrate. The sensor with bi-atomic gold showed the responses to two propanol isomers with good selectivity. It can be extended to the array of electrochemical sensors with small size in the future.

3.6 Optical sensors

3.6.1 Colorimetric sensor

Some dyes change their colors when they are exposed to vapors. Thus, colorimetric method is used to detect odors. Di Natale et al. (2000) used the porphyrin-based sensors to detect VOCs. They used LED to illuminate transparent wall with different dyes and photodiode array to measure the light intensity changes. Nakamoto et al. (2006) used multiLED and single dye-coated plate to discriminate among vapors. Suslick et al. studied an array of metal porphyrin dyes (Rakow & Suslick, 2000). Since the dye spots are formed on the plate, the sensor array can be miniaturized. Moreover, its image can be easily captured by a digital camera. Fig. 3.17 shows the typical example of their sensor responses. Left part is the original photos and right four figures show the differences of the images between the images after vapor exposure and original ones. They call that system "Smell seeing." The color change is often irreversible. The dye plate is typically disposable since its cost is cheap.

Other method is to use gas detector tube. When the gas passes through the gas detector tube (Bäther, 1998), its color changes from the inlet. If the concentration is higher, length of color zone becomes longer. Although color zone was conventionally read out by manual operation, the system to obtain the vapor concentration using optical scanner or mobile phone camera was developed (Tanaka et al., 2006). Its selectivity is based on chemical

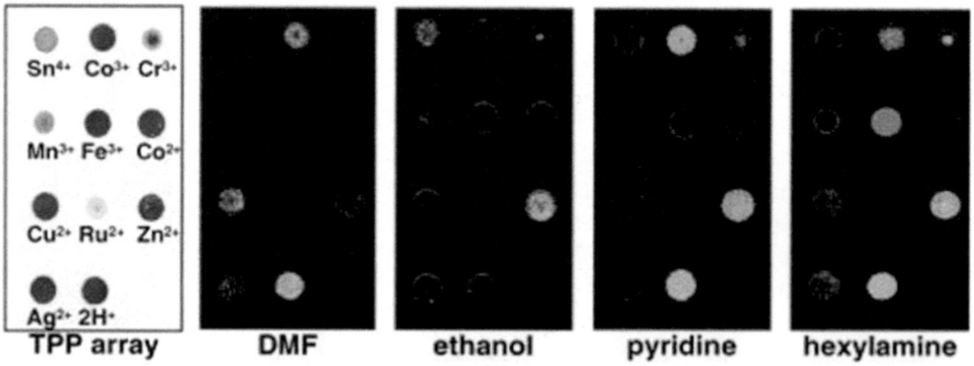

FIGURE 3.17 Profile of dye color change due to vapor exposure when metal porphyrin dyes were used TPP: 5, 10, 15, 20-tetraphenylporphyrinate(−2). Source: *From Rakow, N. A., & Suslick, K. S. (2000). A colorimetric sensor array for odour visualization.* Nature, *406(6797), 710–713. https://doi.org/10.1038/35021028.*

reaction and is high to some degree. Its cost is cheap and a variety of gas detector tubes for many analytes are commercially available. Since its cost is cheap, the sensor network to obtain bad smell distribution was developed (Ninh et al., 2007). However, the gas detector should be replaced with new one when the color change reaches the outlet because of irreversible chemical reaction.

3.6.2 Surface plasmon resonance

SPR (surface plasmon resonance) is an optical phenomenon where an incident light excites a charge-density wave at the interface between conductive metal and dielectric material. SPR is widely used as an analytical tool to measure small change of refractive index of thin region close to the metal surface.

Fig. 3.18A shows schematic of the SPR sensor. It consists of a light source, SPR coupler (prism), and photo detector (Piliarik & Homola, 2009). Light beam illuminated through SPR coupler excites a surface plasmon. Then, the reflected light is detected by the photo detector. As a change in the refractive index at the metal/dielectric interface result in the effective index of the surface plasmon, the refractive index change causes the intensity change or angular spectrum as is shown in Fig. 3.18B. The shift of the angular spectrum is caused by the refractive index change. Thus, the change of the angle at the minimum reflected intensity or the intensity change at the fixed angle is typically measured. SPR sensor can be used as a gas sensor when thin organic film is deposited on the metal surface (Nanto & Stetter, 2002). The refractive index change occurs due to the dielectric constant change during the gas exposure.

In addition to SPR sensor, LSPR (localized surface plasmon resonance) has been used for gas sensing. LSPR occurs when the illuminated light frequency matches the electromagnetic frequency of the noble-metal nanoparticle. LSPR causes the strong light absorption or scattering. Since light absorption caused by LSPR is a physical phenomenon, the response speed of LSPR sensor is fast. Thus, LSPR sensor can be used for gas source localization using a mobile robot (Yang et al., 2018).

3.6.3 Optical absorption

Many important organic and inorganic molecules absorb light in the mid-infrared

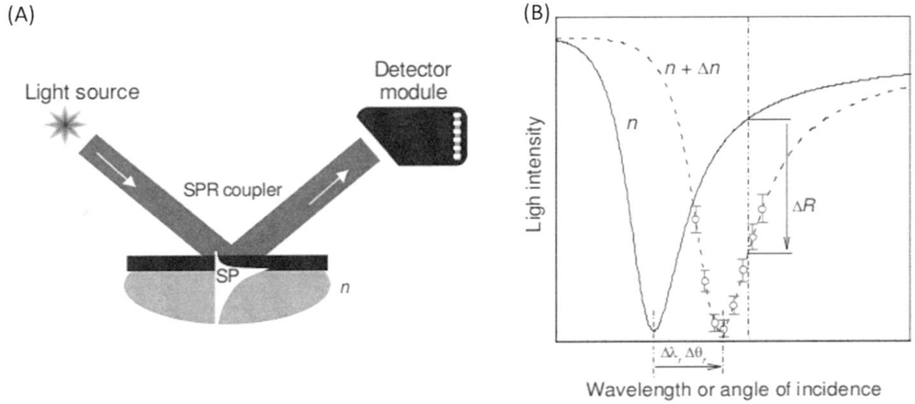

FIGURE 3.18 Principle of SPR measurement (A) schematic of measurement setup and (B) angular spectrum. *SPR*, Surface plasmon resonance. Source: *From Piliarik, M., & Homola, J. (2009). Surface plasmon resonance (SPR) sensors: Approaching their limits?* Optics Express, 17(19). *https://doi.org/10.1364/oe.17.016505.*

spectral region (2–20 μ m) (Bogue, 2015). Thus, we can detect gas/odorant molecules when we measure light intensity change as is shown in Fig. 3.19. That method has the advantage of long-term stability and short response time. Gas/odor sensor based on light absorption works according to Beer–Lambert law, that is, $I = I_0 \exp(-\alpha L)$, where I is the light intensity after passing sensor cell, I_0 in the incident-light intensity, α is the absorption coefficient of the gas, and L the length of optical path. The larger the optical path length is, the larger the light attenuation is. Therefore, the sensitivity to gas depends on the optical path length according to Beer–Lambert law.

Optical absorption caused by gas molecule occurs at wavelength specific to the functional group of gas molecule. In nondispersive infrared sensor, a broad-band light source and dual detectors are used. A filter covering absorption band of the target gas is placed before one detector, whereas another filter covering non-absorption band of the target gas is placed before another detector as is shown in Fig. 3.20. The latter detector can be used as a reference. Its typical example is carbon dioxide sensor, which is commercially available.

Although it is necessary to increase the optical path length to increase the sensitivity, the device size should be kept small. Thus, that path length can be increase without increasing the device size if the multiple reflections are adopted. Thus, the interferometer was used for that purpose (Shamy et al., 2020). Moreover, several researchers have made MEMS/integrated circuit system for gas sensors based on optical absorption (Popa & Udrea, 2019). The sensors used in this subsection are often used for gas sensing rather than odor sensing.

3.6.4 Photoacoustic sensor

Photoacoustic method is an elegant measurement method used in many fields (Tam, 1986). Optical absorption by gas is captured finally by a microphone as is shown in Fig. 3.21 (Elia et al., 2009). When light is absorbed by gas molecules, heat is generated, followed by thermal expansion of the gas region. When the light beam is modulated, thermal expansion and contraction periodically occur, and then sound is generated synchronously with the light modulation. This

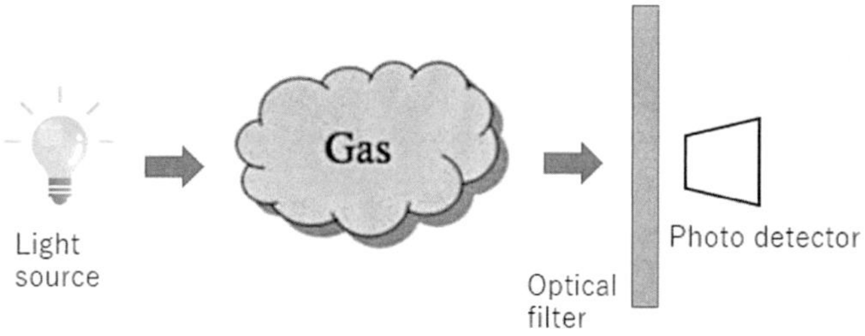

FIGURE 3.19 Principle of gas sensing based on optical absorption the attenuation due to gas absorption is a sensor output.

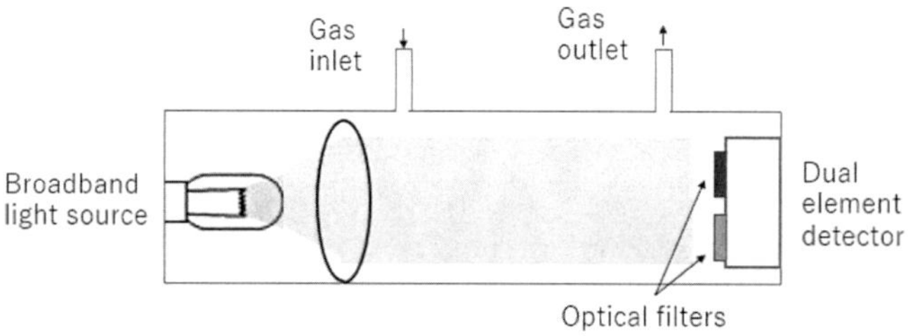

FIGURE 3.20 Principle of NDIR gas sensor. *NDIR*, Nondispersive infra-red.

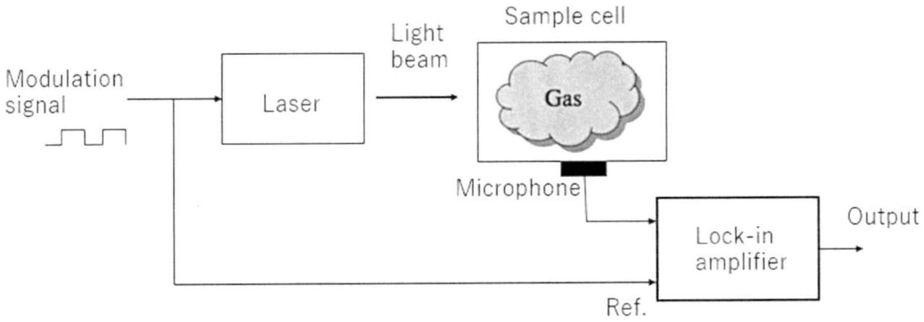

FIGURE 3.21 Principle of gas sensing based upon photoacoustic effect modulated light beam generates sound caused by periodical thermal expansion and contraction due to light absorption by gas molecules.

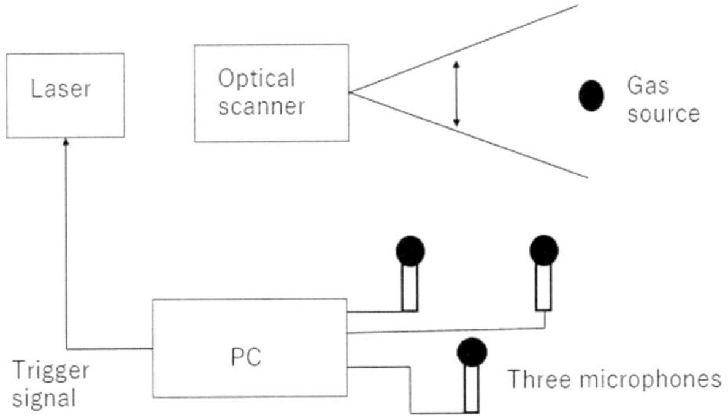

FIGURE 3.22 Photoacoustic measurement system to obtain gas distribution gas source becomes a sound source, which is localized by multiple microphones.

sound is captured by the microphone. The measurement of the photoacoustic signal from the microphone can be done using lock-in method, which can extract only the photoacoustic signal even buried under various sounds. Thus, the soundless environment is not necessary. The photoacoustic signal can be enhanced by tuning the modulation frequency to one of acoustic resonant frequencies of the sample cell.

If multiple microphones are placed in the open space, the gas distribution can be obtained by solving the sound-source localization problem (Ochiai & Kuroki, 1998). The setup for gas distribution measurement is shown in Fig. 3.22. The laser beam is two-dimensionally scanned and the photoacoustic signal is detected by three microphones. The location of the sound source, that is, gas source, can be localized using the time differences among three microphones' signals. This method is elegant since it is not necessary for a user to move the actual gas sensor to search for the gas source.

3.6.5 Fluorescence

Fluorescence is the emission of light by a substance that has absorbed excitation light. The emitted light typically has a longer wavelength than the absorbed radiation.

Fluorescence technique is often used by biologists and Chapter 5 (odor biosensor) includes this topic. Therefore, odor sensing using fluorescence except odor biosensor is explained here. Walt et al. studied fiber-optic sensors with their tips coated with different fluorescent dyes (Aernecke & Walt, 2009). The schematic of the measurement setup is illustrated in Fig. 3.23. Excitation light directed to the optical fiber through dichroic mirror and

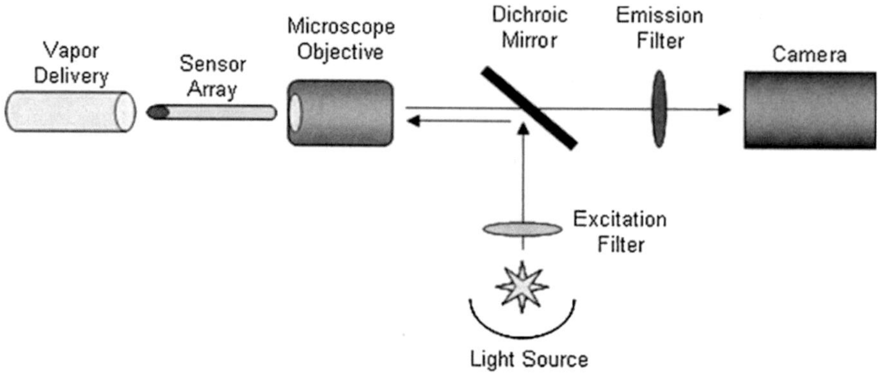

FIGURE 3.23 Setup of odor measurement system using fluorescence at dye located at the tip of optical fiber is captured by a camera. Source: *From Aernecke, M. J., & Walt, D. R. (2009). Optical-fiber arrays for vapor sensing.* Sensors and Actuators B: Chemical, 142*(2), 464–469. https://doi.org/10.1016/j.snb.2009.06.054.*

microscope objective lens. The fluorescence from the dyes at the tip of the optical fiber is collected and transmitted back through optical fiber, dichroic mirror, and emission filter to the camera. The bundle of optical fibers is used and the tip of each fiber is coated with different fluorescent dye.

Hayashi et al. developed fluorescent imaging sensor to detect human body sweat odor (Liu et al., 2013). UV light was illuminated to fluorescent film on a glass sheet, followed by fluorescent light emission, which was captured by CCD camera. They claimed that imaging of the human body odor was achieved.

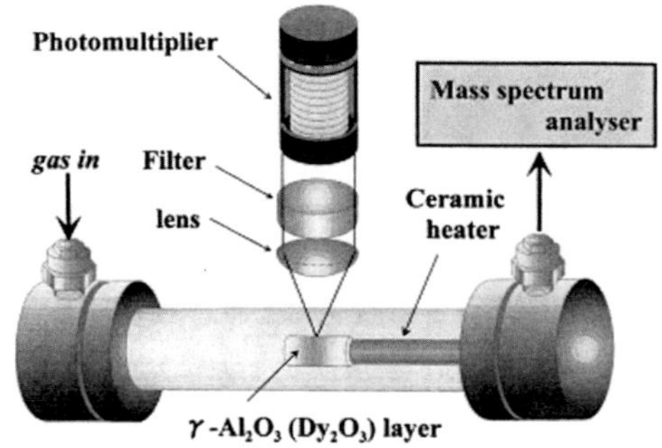

FIGURE 3.24 Experimental setup for measuring cataluminescence is chemiluminescence during catalytic oxidation of combustible gases. Source: *From Okabayashi, T., Toda, T., Yamamoto, I., Utsunomiya, K., Yamashita, N., & Nakagawa, M. (2001). Temperature-programmed chemiluminescence measurements for discrimination and determination of fragrance.* Sensors and Actuators B: Chemical, *74(1–3), 152–156. https://doi.org/10.1016/s0925-4005(00)00725-5.*

3.6.6 Chemiluminescence

Chemiluminescence is the emission of light as a result of chemical reaction. Chemiluminescence sensor is highly selective and sensitive and its application to explosive detection was studied (Jimenez, 2004). Nakagawa et al. studied discrimination among fragrances using the temperature-programed chemiluminescence method (Okabayashi et al., 2001). Fig. 3.24 shows the measurement setup. γ-Al_2O_3 with Dy_2O_3 catalyst was exposed to the gas flow. A ceramic heater was used to raise the temperature up to 700°C. Emitted light was detected by a photomultiplier. Mass spectrometer was used to detect the reaction product. They measured cataluminescence, which is chemiluminescence emitted during catalytic oxidation of combustible gases. They measured cataluminescence under elevated temperature. They claimed the possibility of discrimination among the fragrances using PCA diagram.

3.6.7 Photo ionization detector

Photo ionization detector is often used to measure vapor concentration (Spinelle et al., 2017). Since it does not have selectivity, it can be used only when the analyte compound is known. It is commercially available portable device and the conversion factor from the readout value to actual concentration is prepared by a manufacturer. It has high sensitivity covering the concentration range from ppb level to percent level.

When gas molecules enter the detector, they are bombarded by high-energy UV photons and then ionized, resulting in ejection of electrons and formation of positively charged ions. The electric current, which is the signal output of the detector, flows due to the ions. Photo ionization detector only responds to compounds with ionization energy lower than the energy of photons from the UV lamp.

3.7 Nose chip

Several researchers have developed IC chips to realize a portable odor sensing system. Baltes's group studied the implementation of several different types of gas sensors into a single chip (Hagleitner et al., 2001). They developed CMOS IC chip including cantilever sensor, calorimetric sensor, and capacitive sensor into one chip. The peripheral circuits, including AFE, analog-to-digital conversion, and digital bus interface, were also implemented into the chip. Its photo is shown in Fig. 3.25.

The calorimetric sensor is based on thermoelectric effect, that is, Seebeck effect. It detects temperature change due to adsorption and desorption of the coated polymer. Capacitive sensor detects the dielectric constant change of the coated polymer. The analog-to-digital converter adopted one was sigma-delta one (Norsworthy et al., 1996).

Nakamoto et al. developed the odor recognition chip where sensor measurement circuits and neural network circuit for the pattern recognition of sensor-array signals were implemented (Yoshino & Nakamoto, 2014). The chip included only digital circuits and analog circuit was not necessary since off-chip sensors were QCM sensors. On-chip training was achieved using this system. Tang et al. (2011) developed low power e-nose signal processing chip for conducting

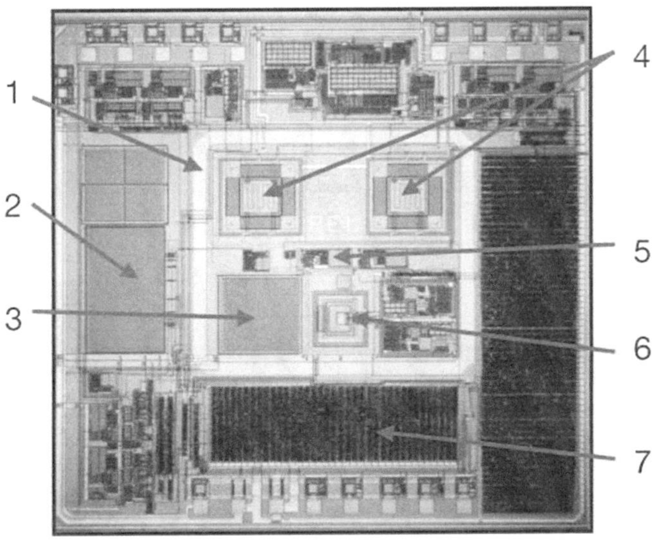

FIGURE 3.25 Photo of smart sensor chip 1, flip-chip frame; 2, reference capacitor; 3, sensing capacitor; 4, calorimetric sensor and reference; 5, temperature sensor; 6, mass-sensitive resonant cantilever; and 7, digital interface. Source: *From Hagleitner, C., Hierlemann, A., Lange, D., Kummer, A., Kerness, N., Brand, O., & Baltes, H. (2001). Smart single-chip gas sensor microsystem.* Nature, *414(6861), 293–296. https://doi.org/10.1038/35104535.*

polymer sensor array. The signal processing chip included interface circuit, memory module, and microprocessor embedded with pattern recognition algorithm. Moreover, the papers related to portable odor sensing systems were summarized in the literature (Cheng et al., 2021).

3.8 Analytical apparatus approach

GC is an apparatus for gas analysis. Since compounds are temporarily separated by the column, it typically takes much time to analyze a sample. Thus, typical GC is not classified as sensor. However, fast GC has been developed since it was intended to be used in the way similar to sensors. In fast GC, the analysis time range is the order of minutes. The trend of fast GC was summarized in the literature (Matisová & Dömötörová, 2003).

Mass spectrometer (Gross, 2006) is often used together with GC. Its measurement time is much shorter than GC, it is sometimes used standalone. Since electronic ionization is adopted in most of mass spectrometers, many fragment ions are generated from an original compound. Then, we can have fingerprint pattern in the mass spectrum. Example of mass spectrum is shown in Fig. 3.26. Since each *m/z* (mass-to-charge ratio) can be regarded as a sensor element in a sensor array, the mass spectrum is identical to the sensor-array output pattern. This system is called MS-based electronic nose (Ampuero et al., 2002; Saevels et al., 2004).

The dimension of mass spectrum is typically more than 200 and it has much richer information than typical sensor array. Moreover, linear superposition is valid in mass spectrum. Thus, it is very powerful tool to quantify mixture compared with typical sensors (Mamun & Nakamoto, 2008). The application of mass spectrometry to odor reproduction will be described in Chapters 11–13.

Ion mobility spectrometry is located between sensor and mass spectrometry. Odorant molecules are ionized in the air. When an electric field is applied, those molecules are separated due to their mobility difference (Shvartsburg, 2008).

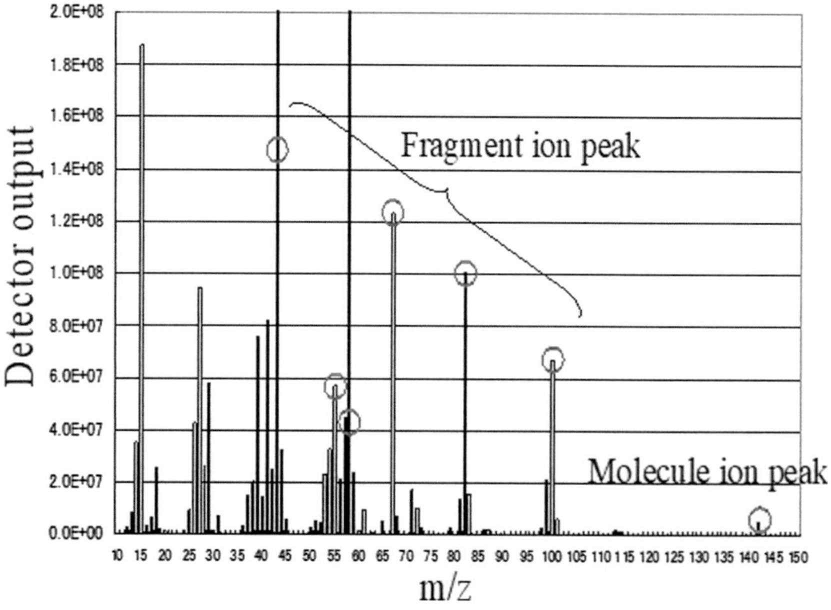

FIGURE 3.26 Example of mass spectrum sample: trans-2-hexenyl acetate.

Especially, FAIMS (Field Asymmetric Ion Mobility Spectrometry) was used by several researchers (Li et al., 2016; Yokoshiki & Nakamoto, 2019). Although FAIMS is highly sensitive, its behavior is sometimes nonlinear. Sophisticated analysis method is required.

3.9 Summary

A variety of sensor devices have been introduced in this chapter. In gravimetric sensor, QCM sensor was not described here. It will be described in Chapter 4. Moreover, odor biosensor will be described in Chapter 5.

Since sensor devices are small, cheap, and able to be used in real time, those can be frequently used if sufficient performance is obtained. In some demonstration such as simple odor classification, those sensors work. However, the practical situations are more severe in terms of sensitivity, selectivity, response/recovery time, robustness against drift, and aging. Although mass spectrometry has good sensitivity, selectivity, and stability, it is expensive and bulky. R&D for sensor devices is still required to apply them to digital olfaction.

References

Aernecke, M. J., & Walt, D. R. (2009). Optical-fiber arrays for vapor sensing. *Sensors and Actuators B: Chemical, 142*(2), 464–469. Available from https://doi.org/10.1016/j.snb.2009.06.054.

Afridi, M. Y., Suehle, J. S., Zaghloul, M. E., Berning, D. W., Hefner, A. R., Cavicchi, R. E., Semancik, S., Montgomery, C. B., & Taylor, C. J. (2002). A monolithic CMOS microhotplate-based gas sensor system. *IEEE Sensors Journal, 2*(6), 644–655. Available from https://doi.org/10.1109/JSEN.2002.807780.

Ampuero, S., Zesiger, T., Gustafsson, V., Lundén, A., & Bosset, J. (2002). Determination of trimethylamine in milk using an MS based electronic nose. *European Food Research and Technology, 214*(2), 163–167. Available from https://doi.org/10.1007/s00217-001-0463-0.

Auld, B., & Alexander. (1973). *Acoustic fields and waves in solids*. Рипол Классик.

Bai, H., & Shi, G. (2007). Gas sensors based on conducting polymers. *Sensors, 7*(3), 267–307. Available from https://doi.org/10.3390/s7030267.

Barié, N., Bücking, M., & Rapp, M. (2006). A novel electronic nose based on miniaturized SAW sensor arrays coupled with SPME enhanced headspace-analysis and its use for rapid determination of volatile organic compounds in food quality monitoring. *Sensors and Actuators B: Chemical, 114*(1), 482–488. Available from https://doi.org/10.1016/j.snb.2005.06.051.

Battiston, F. M., Ramseyer, J.-P., Lang, H. P., Baller, M. K., Gerber, C. H., Gimzewski, J. K., Meyer, E., & Güntherodt, H.-J. (2001). A chemical sensor based on a microfabricated cantilever array with simultaneous resonance-frequency and bending readout. *Sensors and Actuators B: Chemical, 77*(1–2), 122–131. Available from https://doi.org/10.1016/s0925-4005(01)00683-9.

Benetti, M., Cannatà, D., Di Pietrantonio, F., Foglietti, V., & Verona, E. (2005). Microbalance chemical sensor based on thin-film bulk acoustic wave resonators. *Applied Physics Letters, 87*(17). Available from https://doi.org/10.1063/1.2112187.

Berger, R., Delamarche, E., Lang, H. P., Gerber, C., Gimzewski, J. K., Meyer, E., & Güntherodt, H.-J. (1997). Surface Stress in the Self-Assembly of Alkanethiols on Gold. *Science (New York, N.Y.), 276*(5321), 2021–2024. Available from https://doi.org/10.1126/science.276.5321.2021.

Bogue, R. (2015). Detecting gases with light: A review of optical gas sensor technologies. *Sensor Review, 35*(2), 133–140. Available from https://doi.org/10.1108/sr-09-2014-696.

Bäther, W. (1998). Detector tube technology. *Sensors Update, 4*(1), 81–108. Available from https://doi.org/10.1002/1616-8984(199808)4:1 < 81::aid-seup81 > 3.0.co;2-w.

Cai, Q. Y., Park, J., Heldsinger, D., Hsieh, M. D., & Zellers, E. T. (2000). Vapor recognition with an integrated array of polymer-coated flexural plate wave sensors. *Sensors and Actuators, B: Chemical, 62*(2), 121–130. Available from https://doi.org/10.1016/S0925-4005(99)00381-0.

Campbell, J. J., & Jones, W. R. (1970). Propagation of surface waves at the boundary between a piezoelectric crystal and a fluid medium. *IEEE Transactions on Sonics and Ultrasonics, 17*(2), 71–76. Available from https://doi.org/10.1109/TSU.1970.7404095.

Cavicchi, R. E., Suehle, J. S., Kreider, K. G., Gaitan, M., & Chaparala, P. (1996). Optimized temperature-pulse sequences for the enhancement of chemically specific response patterns from micro-hotplate gas sensors. *Sensors and Actuators B: Chemical, 33*(1–3), 142–146. Available from https://doi.org/10.1016/0925-4005(96)01821-7.

Chakraborty, P., Chien, Y.-A., Chiu, W.-T., Mark Chang, T.-F., Sone, M., Nakamoto, T., Josowicz, M., & Janata, J. (2020). Design and development of amperometric gas sensor with atomic Au–polyaniline/Pt composite. *IEEE*

Sensors Journal, 20(21), 12479–12487. Available from https://doi.org/10.1109/jsen.2020.3002822.

Chakraborty, P., Faricha, A., Okamoto, K., Kawakami, H., Mark Chang, T.-F., Sone, M., & Nakamoto, T. (2023). Towards planar atomic-gold decorated polyaniline gas sensors for enhanced electrochemical sensing. *IEEE Sensors Journal, 23*, 6481–6488. Available from https://ieeexplore.ieee.org/stamp/stamp.jsp?tp = &arnumber = 10041840.

Chang, Y., Tang, N., Qu, H., Liu, J., Zhang, D., Zhang, H., Pang, W., & Duan, X. (2016). Detection of volatile organic compounds by self-assembled monolayer coated sensor array with concentration-independent fingerprints. *Scientific Reports, 6*(1). Available from https://doi.org/10.1038/srep23970.

Cheng, L., Meng, Q.-H., Lilienthal, A. J., & Qi, P.-F. (2021). Development of compact electronic noses: A review. *Measurement Science and Technology, 32*(6). Available from https://doi.org/10.1088/1361-6501/abef3b.

Corva, D. M., Scott, D., Adams., Egan, H., Doeven, P., Hashemi, A. Z., & Kouzani. (2022). *TinyStat: A miniaturised potentiostat for portable electrochemical measurements. 2022 29th IEEE International Conference on Electronics, Circuits and Systems (ICECS)* (pp. 1–4). IEEE. Available from https://ieeexplore.ieee.org/document/9970877.

Dey, A. (2018). Semiconductor metal oxide gas sensors: A review. *Materials Science and Engineering: B, 229*, 206–217. Available from https://doi.org/10.1016/j.mseb.2017.12.036.

Di Natale, C., Salimbeni, D., Paolesse, R., Macagnano, A., & D'Amico, A. (2000). Porphyrins-based opto-electronic nose for volatile compounds detection. *Sensors and Actuators B: Chemical, 65*(1–3), 220–226. Available from https://doi.org/10.1016/s0925-4005(99)00316-0.

Elia, A., Mario Lugarà, P., Di Franco, C., & Spagnolo, V. (2009). Photoacoustic techniques for trace gas sensing based on semiconductor laser sources. *Sensors, 9*(12), 9616–9628. Available from https://doi.org/10.3390/s91209616.

Ema, K., Yokoyama, M., Nakamoto, T., & Moriizumi, T. (1989). Odour-sensing system using a quartz-resonator sensor array and neural-network pattern recognition. *Sensors and Actuators, 18*(3–4), 291–296. Available from https://doi.org/10.1016/0250-6874(89)87036-2.

Faricha, A., Chakraborty, P., Okamoto, K., Mark Chang, T.-F., Sone, M., & Nakamoto, T. (2022). Microgravimetric and amperometric sensor coated with room temperature ionic liquid to enhance butanol isomers gas separation. *IEEE Sensors Journal, 22*(24), 24471–24478. Available from https://doi.org/10.1109/jsen.2022.3221114.

Gabl, R., Green, E., Schreiter, M., Feucht, H. D., Zeininger, H., Primig, R., Pitzer, D., Eckstein, G., Wersing, W., Reichl, W., & Runck, J. (2003). Novel integrated FBAR sensors: A universal technology platform for bio- and gas-detection. *Proceedings of IEEE Sensors, 2*, 1184–1188. Available from https://ieeexplore.ieee.org/document/1279132.

Gardner, J. W., Woo Shin, H., & Hines, E. L. (2000). An electronic nose system to diagnose illness. *Sensors and Actuators B: Chemical, 70*(1–3), 19–24. Available from https://doi.org/10.1016/s0925-4005(00)00548-7.

Graf, M., Barrettino, D., Zimmermann, M., Hierlemann, A., Baltes, H., Hahn, S., Bârsan, N., & Weimar, U. (2004). CMOS monolithic metal-oxide sensor system comprising a microhotplate and associated circuitry. *IEEE Sensors Journal, 4*(1), 9–16. Available from https://doi.org/10.1109/JSEN.2003.820367.

Grate, J. W., Snow, A., Ballantine, D. S., Wohltjen, H., Abraham, M. H., Mcgill, R. A., & Sasson, P. (1988). Determination of partition coefficients from surface acoustic wave vapor sensor responses and correlation with gas—Liquid chromatographic partition coefficients. *Analytical Chemistry, 60*(9), 869–875. Available from https://doi.org/10.1021/ac00160a010.

Gross, J. H. (2006). *Mass spectrometry: A textbook*. Springer Science & Business Media.

Hagleitner, C., Hierlemann, A., Lange, D., Kummer, A., Kerness, N., Brand, O., & Baltes, H. (2001). Smart single-chip gas sensor microsystem. *Nature, 414*, 293–296. Available from https://doi.org/10.1038/35104535.

Hatfield, J. V., Neaves, P., Hicks, P. J., Persaud, K., & Travers, P. (1994). Towards an integrated electronic nose using conducting polymer sensors. *Sensors and Actuators B: Chemical, 18*(1–3), 221–228. Available from https://doi.org/10.1016/0925-4005(94)87086-1.

Janata, J., & Nakamoto, T. (2016). Vision of new olfactory sensing array. *IEEJ Transactions on Electrical and Electronic Engineering, 11*(3), 261–267. Available from https://doi.org/10.1002/tee.22214.

Jimenez, A. (2004). Chemiluminescence detection systems for the analysis of explosives. *Journal of Hazardous Materials, 106*(1), 1–8. Available from https://doi.org/10.1016/j.jhazmat.2003.07.005.

Johnston, M. L., Kymissis, I., & Shepard, K. L. (2010). FBAR-CMOS oscillator array for mass-sensing applications. *IEEE Sensors Journal, 10*(6), 1042–1047. Available from https://doi.org/10.1109/JSEN.2010.2042711.

Jonke, A. P., Josowicz, M., & Janata, J. (2011). Polyaniline doped with atomic gold. *Journal of the Electrochemical Society, 158*(12), E147–E151. Available from https://doi.org/10.1149/2.052112jes.

Kaneyasu, M., Ikegami, A., Arima, H., & Iwanaga, S. (1987). Smell identification using a thick-film hybrid gas sensor. *IEEE Transactions on Components, Hybrids, and*

Manufacturing Technology, 10(2), 267–273. Available from https://doi.org/10.1109/TCHMT.1987.1134730.

Kim, Y. S., Ha, S. C., Yang, Y., Kim, Y. J., Cho, S. M., Yang, H., & Kim, Y. T. (2005). Portable electronic nose system based on the carbon black-polymer composite sensor array. *Sensors and Actuators, B: Chemical, 108*(1–2)), 285–291. Available from https://doi.org/10.1016/j.snb.2004.11.067.

King, W. H. (1964). Piezoelectric sorption detector. *Analytical Chemistry, 36*(9), 1735–1739. Available from https://doi.org/10.1021/ac60215a012.

Lang, H. P., Hegner, M., & Christoph, Gerber (2005). Cantilever array sensors. *Materials Today, 8*(4), 30–36. Available from https://doi.org/10.1016/s1369-7021(05)00792-3.

Lang, H. P., Ramseyer, J. P., Grange, W., Braun, T., Schmid, D., Hunziker, P., Jung, C., Hegner, M., & Gerber, C. (2007). An artificial nose based on microcantilever array sensors. *Journal of Physics: Conference Series, 61*(1), 663–667. Available from https://doi.org/10.1088/1742-6596/61/1/133.

Lang, H. P., Baller, M. K., Berger, R., Gerber, C., Gimzewski, J. K., Battiston, F. M., Fornaro, P., Ramseyer, J. P., Meyer, E., & Güntherodt, H. J. (1999). An artificial nose based on a micromechanical cantilever array. *Analytica Chimica Acta, 393*(1–3), 59–65. Available from https://doi.org/10.1016/S0003-2670(99)00283-4.

Li, J., Gutierrez-Osuna, R., Hodges, R. D., Luckey, G., Crowell, J., Schiffman, S. S., & Nagle, H. T. (2016). Using field asymmetric ion mobility spectrometry for odor assessment of automobile interior components. *IEEE Sensors Journal, 16*(14), 5747–5756. Available from https://doi.org/10.1109/JSEN.2016.2568209.

Liu, C., Furusawa, Y., & Hayashi, K. (2013). Development of a fluorescent imaging sensor for the detection of human body sweat odor. *Sensors and Actuators B: Chemical, 183*, 117–123. Available from https://doi.org/10.1016/j.snb.2013.03.111.

Llobet, E. (2013). Gas sensors using carbon nanomaterials: A review. *Sensors and Actuators B: Chemical, 179*, 32–45. Available from https://doi.org/10.1016/j.snb.2012.11.014.

Lonergan, M. C., Severin, E. J., Doleman, B. J., Beaber, S. A., Grubbs, R. H., & Lewis, N. S. (1996). Array-based vapor sensing using chemically sensitive, carbon black-Polymer resistors. *Chemistry of Materials, 8*(9), 2298–2312. Available from https://doi.org/10.1021/cm960036j, http://pubs.acs.org/journal/cmatex.

Mamun, A. L., & Nakamoto, T. (2008). Recipe estimation using mass spectrometer and large-scale data -odor recipe. *IEEJ Transactions on Sensors and Micromachines, 128*(12), 467–471. Available from https://doi.org/10.1541/ieejsmas.128.467.

Matisová, E., & Dömötörová, M. (2003). Fast gas chromatography and its use in trace analysis. *Journal of Chromatography. A, 1000*(1–2), 199–221. Available from https://doi.org/10.1016/s0021-9673(03)00310-8.

Maute, M., Raible, S., Prins, F. E., Kern, D. P., Weimar, U., & Göpel, W. (1999). Fabrication and application of polymer coated cantilevers as gas sensors. *Microelectronic Engineering, 46*(1), 439–442. Available from https://doi.org/10.1016/S0167-9317(99)00128-8.

Meier, D. C., Semancik, S., Button, B., Strelcov, E., & Kolmakov, A. (2007). Coupling nanowire chemiresistors with MEMS microhotplate gas sensing platforms. *Applied Physics Letters, 91*(6). Available from https://doi.org/10.1063/1.2768861.

Minami, K., Imamura, G., Tamura, R., Shiba, K., & Yoshikawa, G. (2022). Recent advances in nanomechanical membrane-type surface stress sensors towards artificial olfaction. *Biosensors, 12*(9). Available from https://doi.org/10.3390/bios12090762.

Mujahid, A., Afzal, A., & Dickert, F. L. (2019). An overview of high frequency acoustic sensors—QCMs, SAWs and FBARs—Chemical and biochemical applications. *Sensors, 19*(20). Available from https://doi.org/10.3390/s19204395.

Muñoz, B. C., Steinthal, G., & Sunshine, S. (1999). Conductive polymer-carbon black composites-based sensor arrays for use in an electronic nose. *Sensor Review, 19*(4), 300–305. Available from https://doi.org/10.1108/02602289910294745.

Nakamoto, T., & Moriizumi, T. (1988). Odor sensor using quartz-resonator array and neural-network pattern recognition. Ultrasonics symposium proceedings (Vol. 1, pp. 613–616). New York City, USA: IEEE. Available from https://ieeexplore.ieee.org/document/49450.

Nakamoto, T., Yosihioka, M., Tanaka, Y., Kobayashi, K., Moriizumi, T., Ueyama, S., & Yerazunis, W. S. (2006). Colorimetric method for odor discrimination using dye-coated plate and multiLED sensor. *Sensors and Actuators B: Chemical, 116*(1–2), 202–206. Available from https://doi.org/10.1016/j.snb.2005.11.076.

Nanto, H., & Stetter, J. R. (2002). Introduction to Chemosensors. *Handbook of machine olfaction:Electronic nose technology* (pp. 79–104). John Wiley & Sons.

Ninh, H. P., Tanaka, Y., Nakamoto, T., & Hamada, K. (2007). A bad-smell sensing network using gas detector tubes and mobile phone cameras. *Sensors and Actuators B: Chemical, 125*(1), 138–143. Available from https://doi.org/10.1016/j.snb.2007.01.049.

Norsworthy, R. S., Schreier, R., & Temes, C. G. (1996). *Delta-Sigma Data Converters: Theory, Design, and Simulation*. Wiley-IEEE Press.

Ochiai, M., & Kuroki, M. (1998). *Presented at technical meeting on chemical sensor*. IEE of Japan.

Okabayashi, T., Toda, T., Yamamoto, I., Utsunomiya, K., Yamashita, N., & Nakagawa, M. (2001). Temperature-programmed chemiluminescence measurements for discrimination and determination of fragrance. *Sensors and Actuators B: Chemical*, *74*(1–3), 152–156. Available from https://doi.org/10.1016/s0925-4005(00)00725-5.

Paul, A., Muthukumar, S., & Prasad, S. (2019). Room-temperature ionic liquids for electrochemical application with special focus on gas sensors. *Journal of the Electrochemical Society*, *167*(3). Available from https://doi.org/10.1149/2.0112003JES.

Pearce, T. C., Schiffman, S. S., & Nagle, H. T. (2003). *Gardner, handbook of machine olfaction*. Wiley-VCH.

Persaud, K., & Dodd, G. (1982). Analysis of discrimination mechanisms in the mammalian olfactory system using a model nose. *Nature*, *299*(5881), 352–355. Available from https://doi.org/10.1038/299352a0.

Persaud, K. C. (2005). Polymers for chemical sensing. *Materials Today*, *8*(4), 38–44. Available from https://doi.org/10.1016/s1369-7021(05)00793-5.

Piliarik, M., & Homola, J. (2009). Surface plasmon resonance (SPR) sensors: Approaching their limits? *Optics Express*, *17*(19). Available from https://doi.org/10.1364/oe.17.016505.

Popa, D., & Udrea, F. (2019). Towards integrated mid-infrared gas sensors. *Sensors*, *19*(9). Available from https://doi.org/10.3390/s19092076.

Rakow, N. A., & Suslick, K. S. (2000). A colorimetric sensor array for odour visualization. *Nature*, *406*, 710–713. Available from https://doi.org/10.1038/35021028.

Rapp, M., Reibel, J., Voigt, A., Balzer, M., & Bülow, O. (2000). New miniaturized SAW-sensor array for organic gas detection driven by multiplexed oscillators. *Sensors and Actuators, B: Chemical*, *65*(1), 169–172. Available from https://doi.org/10.1016/S0925-4005(99)00321-4.

Saevels, S., Lammertyn, J., Berna, A. Z., Veraverbeke, E. A., Di Natale, C., & Nicolaï, B. M. (2004). An electronic nose and a mass spectrometry-based electronic nose for assessing apple quality during shelf life. *Postharvest Biology and Technology*, *31*(1), 9–19. Available from https://doi.org/10.1016/s0925-5214(03)00129-7.

Sauerbrey, G. (1959). Verwendung von Schwingquarzen zur Wägung dünner Schichten und zur Mikrowägung. *Zeitschrift für Physik*, *155*(2), 206–222. Available from https://doi.org/10.1007/bf01337937.

Shamy, E., Raghi, S., Khalil, D., & Swillam, M. A. (2020). Mid infrared optical gas sensor using plasmonic Mach-Zehnder interferometer. *Scientific Reports.*, *10*. Available from https://doi.org/10.1038/s41598-020-57538-1.

Shurmer, H. V., & Gardner, J. W. (1992). Odour discrimination with an electronic nose. *Sensors and Actuators: B. Chemical*, *8*(1), 1–11. Available from https://doi.org/10.1016/0925-4005(92)85001-D.

Shvartsburg, A. A. (2008). *Differential ion mobility spectrometry: Nonlinear ion transport and fundamentals of FAIMS* (pp. 1–301). CRC Press. Available from http://www.tandfebooks.com/doi/book/10.1201/9781420051070, 10.1201/9781420051070.

Simon, I., Bârsan, N., Bauer, M., & Weimar, U. (2001). Micromachined metal oxide gas sensors: Opportunities to improve sensor performance. *Sensors and Actuators B: Chemical*, *73*(1), 1–26. Available from https://doi.org/10.1016/s0925-4005(00)00639-0.

Spinelle, L., Gerboles, M., Kok, G., Persijn, S., & Sauerwald, T. (2017). Review of portable and low-cost sensors for the ambient air monitoring of benzene and other volatile organic compounds. *Sensors*, *17*(7). Available from https://doi.org/10.3390/s17071520.

Staples, E. J., & Viswanathan, S. (2005). Ultrahigh-speed chromatography and virtual chemical sensors for detecting explosives and chemical warfare agents. *IEEE Sensors Journal*, *5*(4), 622–631. Available from https://doi.org/10.1109/JSEN.2005.850990.

Stetter, J. R., Jurs, P. C., & Rose, S. L. (1986). Detection of hazardous gases and vapors: Pattern recognition analysis of data from an electrochemical sensor array. *Analytical Chemistry*, *58*(4), 860–866. Available from https://doi.org/10.1021/ac00295a047.

Stetter, J. R., & Li, J. (2008). Amperometric gas sensors—A review. *Chemical Reviews*, *108*(2), 352–366. Available from https://doi.org/10.1021/cr0681039.

Tam, A. C. (1986). Applications of photoacoustic sensing techniques. *Reviews of Modern Physics*, *58*(2), 381–431. Available from https://doi.org/10.1103/revmodphys.58.381.

Tanaka, Y., Nakamoto, T., & Moriizumi, T. (2006). Study of highly sensitive smell sensing system using gas detector tube combined with optical sensor. *Sensors and Actuators B: Chemical*, *119*(1), 84–88. Available from https://doi.org/10.1016/j.snb.2005.11.068.

Tang, K.-T., Chiu, S.-W., Chang, M.-F., Hsieh, C.-C., & Shyu, J.-M. (2011). A low-power electronic nose signal-processing chip for a portable artificial olfaction system. *IEEE Transactions on Biomedical Circuits and Systems*, *5*(4), 380–390. Available from https://doi.org/10.1109/tbcas.2011.2116786.

Wilson, A. D., Oberle, C. S., & Oberle, D. F. (2013). Detection of off-flavor in catfish using a conducting polymer electronic-nose technology. *Sensors (Switzerland)*, *13*(12), 15968–15984. Available from https://doi.org/10.3390/s131215968, http://www.mdpi.com/1424-8220/13/12/15968/pdf.

Wingqvist, G. (2010). AlN-based sputter-deposited shear mode thin film bulk acoustic resonator (FBAR) for biosensor applications—A review. *Surface and Coatings Technology*, *205*(5), 1279–1286.

Wohltjen, H. (1984). Mechanism of operation and design considerations for surface acoustic wave device vapour sensors. *Sensors and Actuators*, *5*(4), 307–325. Available from https://doi.org/10.1016/0250-6874(84)85014-3.

Yamazoe, N., Kurokawa, Y., & Seiyama, T. (1983). Effects of additives on semiconductor gas sensors. *Sensors and Actuators*, *4*(C), 283–289. Available from https://doi.org/10.1016/0250-6874(83)85034-3.

Yan, Z., Zhou, X. Y., Pang, G. K. H., Zhang, T., Liu, W. L., Cheng, J. G., Song, Z. T., Feng, S. L., Lai, L. H., Chen, J. Z., & Wang, Y. (2007). ZnO-based film bulk acoustic resonator for high sensitivity biosensor applications. *Applied Physics Letters*, *90*(14). Available from https://doi.org/10.1063/1.2719149.

Yang, Z., Sassa, F., & Hayashi, K. (2018). A robot equipped with a high-speed LSPR gas sensor module for collecting spatial odor information from on-ground invisible odor sources. *ACS Sensors*, *3*(6), 1174–1181. Available from https://doi.org/10.1021/acssensors.8b00214.

Yokoshiki, Y., & Nakamoto, T. (2019). Ternary gas mixture quantification using field asymmetric ion mobility spectrometry (FAIMS). *Sensors*, *19*(13). Available from https://doi.org/10.3390/s19133007.

Yoshikawa, G., Akiyama, T., Loizeau, F., Shiba, K., Gautsch, S., Nakayama, T., Vettiger, P., Rooij, N., & Aono, M. (2012). Two dimensional array of piezoresistive nanomechanical membrane-type surface stress sensor (MSS) with improved sensitivity. *Sensors (Switzerland)*, *12*(11), 15873–15887. Available from https://doi.org/10.3390/s121115873, http://www.mdpi.com/1424-8220/12/11/15873/pdf.

Yoshino, F., & Nakamoto, T. (2014). Odor recognition system using embedded leaning vector quantization circuit. *Sensors and Materials*, *26*(3), 137–147.

CHAPTER

4

Quartz crystal microbalance sensors

Sawit Na Songkhla

Department of Electrical Engineering, Faculty of Engineering, Chulalongkorn University, Bangkok, Thailand

4.1 Introduction

The quartz crystal microbalance (QCM) has emerged as a popular choice for chemical and biosensing applications, including odorant sensing. This sensor operates by detecting changes in the resonant frequency of a crystal caused by the adsorption or desorption of mass on its surface. The key advantages of QCM include its ability to function in both liquid and gas mediums, easy availability, and cost-effectiveness. However, it is important to note that QCM exhibits relatively lower sensitivity compared to its acoustic device counterparts such as surface acoustic wave devices.

In the context of electronic noses (E-nose), which typically require an array of sensors with different selective coating molecules to mimic the odorant receptors found in animals, QCM proves advantageous due to its sensing mechanism being based on weight attached to its surface, independent of other physical properties.

Nevertheless, it is crucial to thoroughly study the advantages, disadvantages, and limitations of QCM, like any other type of sensor, before applying it in real-world applications. This approach helps to mitigate unexpected issues that may arise during the implementation of the sensor. As for the advantages and disadvantages of QCM, a comparison table in Fig. 4.1 from a previous review paper can provide a comprehensive comparison (Mujahid & Adeel Dickert, 2019).

4.2 Principle and method

4.2.1 Piezoelectric effect

The piezoelectric phenomenon is observed in certain materials that exhibit electric polarization when subjected to mechanical strain. These materials possess a crystal structure where the atoms can shift position, creating electrical dipoles and a net macroscopic moment. This effect occurs only in crystalline materials that lack inversion symmetry and are reversible, meaning that an applied electric field can induce strain in the material, known as the converse piezoelectric effect. It is this fundamental principle of piezoelectricity that underlies the functioning of QCM.

Digital Technologies in Olfaction
DOI: https://doi.org/10.1016/B978-0-443-15721-9.00004-X

	Quartz Crystal Microbalance (QCM)	Surface Acoustic Wave (SAW)	Film Bulk Acoustic Resonator (FBAR)
Acoustic Wave Type	Bulk	Surface	Bulk
Operating Frequency	5–200 MHz	100–1500 MHz	1.0–4.5 GHz
Sensitivity	Low	Intermediate	High
Common Piezoelectric Materials	Quartz	Quartz, $LiNbO_3$, $LiTaO_3$	AlN, ZnO
Dielectric Constant	Quartz = 3.8	$LiNbO_3$ = 29 $LiTaO_3$ = 43	AlN = 8.5 ZnO = 8.8
Acoustic Wave Velocity (m/s)	Quartz = 3158	$LiNbO_3$ = 3488–3992 $LiTaO_3$ = 3230–3394	AlN = 5600 ZnO = 2558
Temperature Coefficient of Frequency (TCF) (ppm/°C)	Quartz ≅ 0	$LiNbO_3$ = 75–94 $LiTaO_3$ = 35–64	AlN = 25 ZnO = 40
Working Medium	Liquids and Gases	Gases	Liquids and Gases
Availability	Common and Easily Available	Common and Easily Available	Not Easily available
Manufacturing Costs	Low	Moderate	High
Commercial Adaptability	Medium	High	-
General Comments	Most common and widely used acoustic transducer for sensor applications. QCMs have low fabrication costs. But has comparatively low sensitivity due to low resonance frequency.	Well known and widely used for gas phase sensor applications and found a number of commercial applications. However, not commonly used for liquid phase sensing.	Highly sensitive acoustic transducer for gaseous and liquid phase sensing. Not commonly available as advance fabrication techniques are need for their manufacturing thus, has high costs.

Acoustic wave velocity and TCF values for quartz, $LiNbO_3$ and $LiTaO_3$ are taken from [23]. The dielectric constant value of quartz is from [102] while for $LiNbO_3$, $LiTaO_3$, AlN and ZnO are from [147]. Acoustic wave velocity of AlN and ZnO are taken from [148] while TCF values are from [132].

FIGURE 4.1 An overview of high-frequency acoustic devices for chemical/biochemical sensor applications. Comparative analysis of acoustic wave sensors, highlighting their operating frequencies, material properties, and application adaptability. Source: *From Mujahid, A. A., & Adeel Dickert, F. L. (2019). An overview of high frequency acoustic sensors—QCMs, SAWs and FBARs—chemical and biochemical applications,* Sensors (Switzerland), *19, 20. https://doi.org/10.3390/s19204395.*

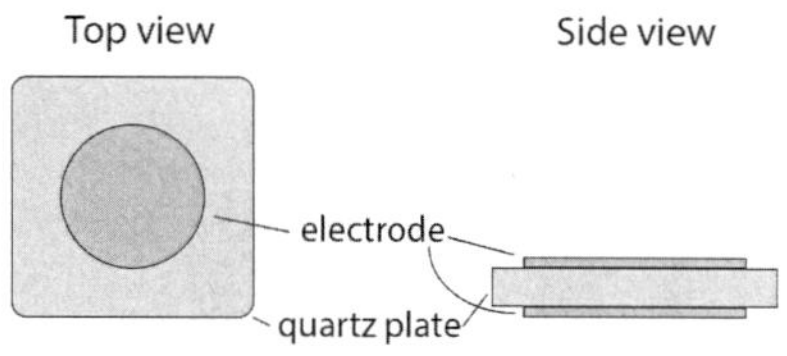

FIGURE 4.2 A structure of QCM. QCM structure consists of quartz plate coupled with metal electrode. *QCM,* Quartz crystal microbalance.

4.2.2 Physical structure

The typical structure of a QCM involves sandwiching a crystal between two metal electrodes, as shown in Fig. 4.2. The crystal can exhibit various modes of resonance, including thickness-shear mode, flexural mode, and extensional mode, depending on the angle at which it is operated. The resonant frequency of a QCM typically falls within the range of 500 Hz–300 MHz (Janshoff et al., 2000).

The AT-cut crystal is the most widely used cut in QCM applications. It operates in the thickness shear mode and exhibits a temperature coefficient that is relatively close to zero within the temperature range of 0°C–50°C. In the thickness shear mode, the acoustic wave behavior of the QCM can be described by expressing the crystal thickness (d) as half of the shear wave wavelength:

$$d = \frac{\lambda}{2}. \tag{4.1}$$

Let soundwave velocity be v and its frequency f:

$$v = f\lambda. \tag{4.2}$$

Therefore

$$f = \frac{v}{2d}. \quad (4.3)$$

The additional mass contributes to an increase in quartz thickness per unit area (Δd), which will result in a corresponding change in resonant frequency (Δf) according to Eq. (4.4). Consequently, the inclusion of this mass leads to a decrease in the resonant frequency:

$$(f + \Delta f) = \frac{v}{2(d + \Delta d)}. \quad (4.4)$$

One limitation of this approximation arises when the film thickness is relatively thin, typically limited to 2% of the resonant frequency of the unloaded crystal (Lu & Lewis, 1972). More accurate calculations based on acoustic properties, using stress equations, offer significantly improved accuracy. These precise calculations are based on the Mason equivalent circuit, incorporating boundary conditions for film deposition, and they reveal deviations from the linear approximations in the equations (Nakamoto & Moriizumi, 1990).

The research findings indicate that the QCM exhibits the highest sensitivity to mass at its center, and this sensitivity gradually decreases as you move away from the center. The sensitivity pattern can be approximated using a Gaussian curve (Huang et al., 2017; Ward & Delawski, 1991).

When a nonrigid layer is deposited on the QCM, it is expected to introduce additional energy loss within the deposited material. Consequently, this leads to a decrease in the Q factor or an increase in the dissipation factor. Existing equations commonly used cover scenarios involving lossless film deposition and bulk liquid deposition but do not account for the deposition of thin, viscous films, which is common in many applications.

4.2.3 Behavior of quartz crystal microbalance sensor

The QCM sensor is versatile and can be used in both gas and liquid phases. Typically, as an odorant and gas sensor, bare QCM (uncoated) is not used in these applications due to its nonspecificity to any molecular structure or chemical property, only mass and physical properties such as viscosity are being sensed. The simplified responses are mass loading, which is the attachment of rigid mass to QCM, and viscous loading, which represents unbound Newtonian liquid in contact with QCM. Also, the combination of mass loading and viscous loading, or the bounded viscous loading, can also be derived mathematically and can be useful for explaining the QCM responses.

4.2.3.1 Quartz crystal microbalance behavior in gas phase

The most simple behavior of QCM is the frequency change due to mass deposited or attached uniformly on the QCM surface. The frequency change behavior was experimentally described by Sauerbrey (1959). The assumption was made in the formulation of the equation, linking the deposited mass to the extension of QCM thickness. This simplification allowed the relationship between frequency and mass to be unaffected by the geometry of the electrode. The Sauerbrey equation can be defined as follows:

$$\Delta f = -\frac{2{f_0}^2}{A\sqrt{\rho_q \eta_q}}\Delta m \quad (4.5)$$

where f_0 is the resonant frequency of the fundamental mode (Hz), Δf is the frequency change (Hz), Δm is the mass change (g), A is the electrode area (cm^2), ρ_q is the density of quartz (g/cm^3), and η_q is the shear modulus if quartz AT-cut crystal ($g/cm/s^2$).

However, in order for the Sauerbrey equation to be applicable, three conditions must be satisfied. First, the deposited mass must be rigid. Second, the mass must be distributed evenly. Lastly, the frequency change resulting from the mass deposition should be less than

2%−5% of $\Delta f/f$, or else the predicted frequency shift (Δf) becomes less accurate. This frequency change caused by mass deposition is known as the "mass loading" effect. Generally, the mass loading effect does not impact the Q factor of the QCM. This has the advantage of maintaining a sharp frequency characteristic, making it easy to determine the resonant frequency.

Sauerbrey's equation is particularly useful for measuring small amounts of deposition and can be employed to monitor the thickness of coating films. This is because, at small film thicknesses, the impact of viscosity-induced losses is less significant compared to the deposited mass itself, which represents the film thickness.

4.2.3.2 *Quartz crystal microbalance behavior in liquid phase*

Another commonly observed phenomenon is the behavior of QCM when immersed in a liquid medium. The study of QCM behavior in liquids was carried out by Kanazawa and Gordon (1985). His equation describes the relationship between the liquid shear velocity and the interaction with the quartz plate. His equation is defined as

$$\Delta f = -f_0^{3/2}\left(\frac{\eta_l \rho_l}{\pi \rho_q \mu_q}\right)^{1/2} \quad (4.6)$$

where f_0 is the resonant frequency of the fundamental mode (Hz), η_l is the liquid viscosity (cP), ρ_l is the liquid density (g/cm^3), ρ_q is the density of quartz (g/cm^3), and μ_q is the shear modulus if quartz AT-cut crystal (g/cm/s^2).

The equation is derived by considering the disappearance of the shear wave within the liquid. In Fig. 4.3, when an acoustic wave is generated by the QCM, it dissipates into the liquid medium, and its amplitude decreases as it propagates. The depth of penetration, denoted as lambda (λ), indicates the distance from the quartz plate, where the amplitude attenuates to 1/e of its original value:

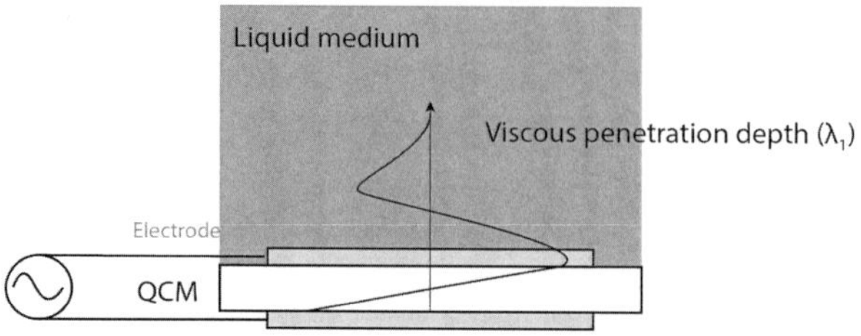

FIGURE 4.3 Penetration of Waves in a QCM submerged in a liquid medium. Schematic of QCM sensor operation in liquid medium: demonstrating the viscous penetration depth from the oscillating electrode surface into the liquid. *QCM*, Quartz crystal microbalance.

$$\lambda = \frac{1}{2}\left(\sqrt{\frac{\rho_L \omega}{2\eta}}\right)^{-1}.$$

The concept of penetration depth assists in estimating the maximum thickness of the liquid layer on the QCM, which still influences the frequency shift. Numerous researchers have found that the responses obtained from the QCM are considerably more complex. In some cases, the QCM frequency shifts even when the disturbance is beyond the penetration depth range. An interesting observation was the occurrence of a positive frequency shift when exposed to a specific measurement parameter. Kanazawa's equation, although simplistic, failed to consider the importance of the coating layer thickness, which plays a critical role in the sensor's response. Furthermore, similar to Sauerbrey's equation, Kanazawa's equation cannot account for the positive frequency change.

4.2.4 Quartz crystal microbalance equivalent circuit

To harness the full potential of QCM sensors, it is crucial to investigate and comprehend the conversion between mechanical and electrical parameters. Mason proposed a

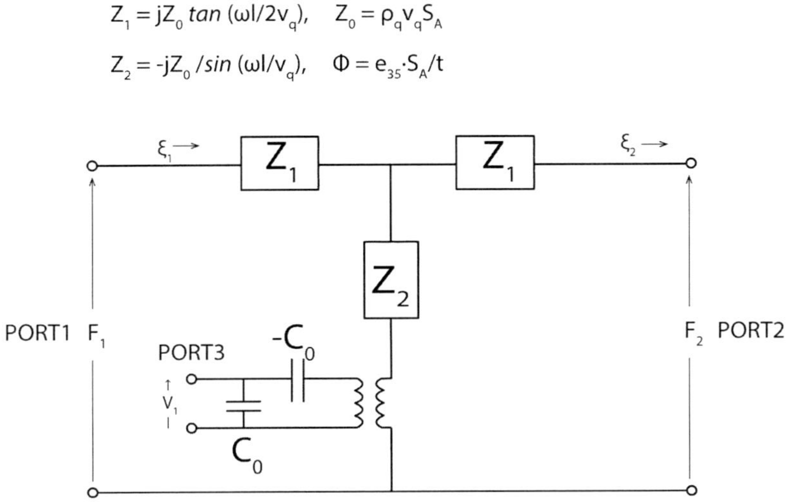

FIGURE 4.4 Mason equivalent circuit. Mason's equivalent circuit model for piezoelectric resonators: diagram depicting mechanical ports with impedances and a central electrical port with capacitance and inductance, including mathematical expressions for impedance calculation.

one-dimensional model, known as the acoustic wideband model, to facilitate this understanding. The Mason model comprises two acoustic ports (port1 and port2) and one electrical port (port3), as shown in Fig. 4.4 (Nakamoto & Moriizumi, 1990). Then, after several mathematical simplifications, the model can be simplified into an equivalent circuit of lumped elements by shorting one of its mechanical ports and treating the other port as a stress-free surface. This simplified circuit is particularly useful near the resonant frequency called the Butterworth–van Dyke (BVD) circuit (Fig. 4.5). This BVD model was vastly ubiquitous among many pieces of research on QCM and very useful for response interpretation since the information mostly comes from near resonant frequency and the BVD model accounts for that.

The BVD circuit consists of resonant elements R_1, L_1, and C_1 connected in parallel with C_0. Typically, when the circuit undergoes acoustic loading, the resonant components are influenced by additional resistance and reactance. On the other hand, C_0 represents an inherent capacitance resulting from the connection between an electrode and the

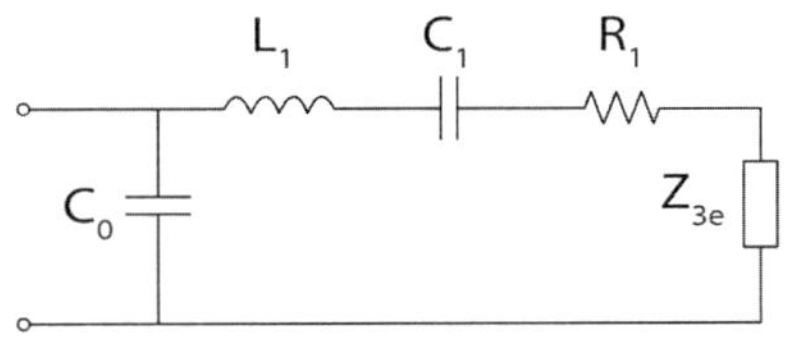

FIGURE 4.5 The equivalent circuit of the QCM, the Butterworth–van Dyke model with viscous loading. Butterworth–van Dyke model of a piezoelectric resonator: an electrical circuit with a parallel capacitor (C_0), series inductor (L_1) and capacitor (C_1), and an end complex impedance (Z_{3e}). *QCM*, Quartz crystal microbalance.

piezoelectric material, possessing a defined dielectric constant.

The admittance of the BVD circuit is expressed as

$$Y = \frac{1}{R_1 + j\omega L_1 - \frac{1}{j\omega C_1} + Z_{eL}} + j\omega C_0. \qquad (4.7)$$

The admittance plot near the resonant frequency is depicted in Fig. 4.6. The plot forms a nearly circular shape with the Y-offset approximately equal to omega C_0, which remains relatively constant near the resonant frequency. The diameter of this circle corresponds to the reciprocal of R_1. The rightmost point on the

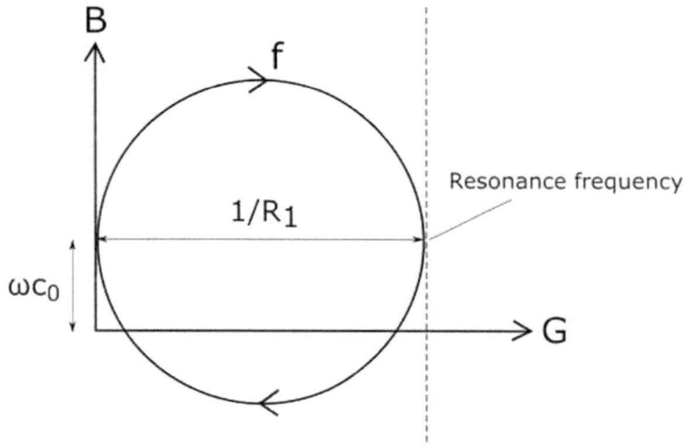

FIGURE 4.6 **Admittance plot of QCM**. Admittance circle for a QCM resonator: illustrating the relationship between conductance (G) and susceptance (B) with a circle that defines the resonance frequency on the complex admittance plane. *QCM*, Quartz crystal microbalance.

admittance plot is referred to as the maximum series resonant frequency, representing the resonance point in the motional admittance branch (the branch consisting of R_1, L_1, and C_1). When mass loading occurs, it adds additional inductance to L_1 without increasing the value of R_1 in the BVD model. On the other hand, viscous loading contributes to an increase in both R_1 and L_1. The BVD model with loading, illustrated in Fig. 4.5, also has the extra impedance (Z_{eL}) added to the motional admittance branch. With this notation, we can explore many boundary conditions, such as mass loading (rigid film), viscous loading (bulk liquid), and viscous film. They are useful since many sensitive coating materials are viscous.

4.2.5 Interpretation of viscous sensing film using equivalent circuit

We still haven't mentioned how we derived the QCM loading impedance (Z_{eL}) from the deposited viscous load. This condition is the main interest in applying the QCM sensor in gas sensing and odor sensing applications. As mentioned previously, QCM itself is a mass

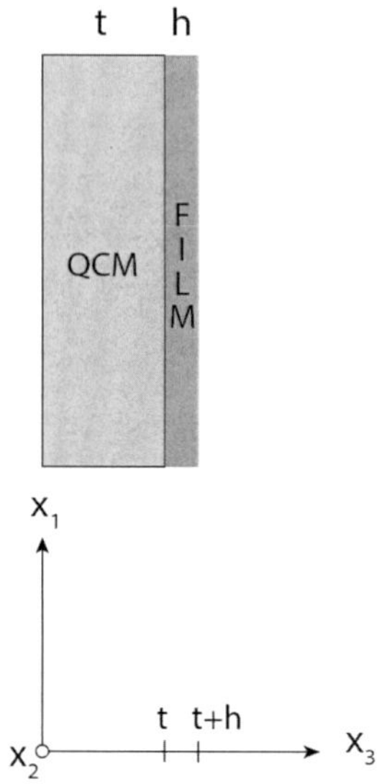

FIGURE 4.7 **Model of viscous film**. One-dimensional model of a QCM with an adjacent film layer, indicating the thicknesses of the QCM (t) and the film (h) on the coordinate axes. *QCM*, Quartz crystal microbalance. Source: *From Songkhla, S. N., & Nakamoto, T. (2021). Interpretation of quartz crystal microbalance behavior with viscous film using a mason equivalent circuit.* Chemosensors, *9(1). https://doi.org/10.3390/chemosensors9010009.*

sensor, which means it has no specific binding to any molecule; we need to add a sensitive coating to the QCM to act as an adsorption site for the target molecules that need detections. We can use a one-dimensional model for the calculation (Nakamoto & Moriizumi, 1990; Songkhla & Nakamoto, 2021) (see Fig. 4.7). The model defines QCM with a thickness of h and film with the thickness of t. One condition that can be explored with the use of an equivalent circuit is to apply the boundary conditions of various material conditions to the model.

Let's examine a scenario involving a thin film with viscosity, in which we have variables: angular frequency denoted as ω, shear modulus as μ, and shear wave velocity as v_{A1}. The particle displacement u_1 and shear stress T_5 can be expressed as follows:

$$u_1 = Ae^{-j(\omega/v_{A1})x_3} + Be^{j(\omega/v_{A1})x_3} \quad (4.8)$$

and

$$T_5 = \mu S_5 = \mu \frac{\partial u_1}{\partial x_3} = j\frac{\omega}{v_{A1}}\mu\left(-Ae^{-j(\omega/v_{A1})x_3} + Be^{j(\omega/v_{A1})x_3}\right). \quad (4.9)$$

A and B are constant. At the film surface ($x_3 = t + h$), the stress is assumed to be unrestricted ($T_5 = 0$). The proportion of A to B can be determined as

$$\frac{A}{B} = e^{j(2\omega/v_{A1})(t+h)}. \quad (4.10)$$

Using Eqs. (4.8) and (4.9), the solution to the piezoelectric stress equation yields the acoustic impedance Z_L:

$$Z_L = -\frac{T_5}{j\omega u_1}\bigg|_{x_3=t} = \frac{\mu}{v_{A1}}\frac{1-e^{j(2\omega/v_{A1})}}{1+e^{j(2\omega/v_{A1})}} = j\rho_L v_{A1}\tan\left(\frac{\omega h}{v_{A1}}\right) \quad (4.11)$$

where ρ_L is film density and h is the film thickness. In the context of liquids or the viscous medium, the notion of strain does not apply. Instead, strain velocity is employed in the wave equations. The equation of motion can be derived by substituting strain with strain velocity in the strain–stress relationship used for solids. The fundamental equations governing shear waves in liquids are as follows:

$$\rho_L \frac{\partial^2 u_1}{\partial t^2} = \frac{\partial T_5}{\partial x_3} \quad (4.12)$$

and

$$T_5 = j\omega\eta_L S_5 \quad (4.13)$$

where η_L is the film viscosity. Substituting Eq. (4.13) into Eq. (4.12) yields

$$v_{A1} = \sqrt{\frac{j\omega\eta_L}{\rho_L}}. \quad (4.14)$$

Substituting Eq. (4.14) into Eq. (4.11), we obtain

$$Z_L = j\sqrt{j\rho_L\eta_L\omega}\tan\left(\omega h\sqrt{\frac{\rho_L}{j\omega\eta_L}}\right). \quad (4.15)$$

Electrical impedance of mason equivalent is expressed as

$$Z_{eL} = \frac{t^2 Z_L}{4e^2{}_{35}S_A} \quad (4.16)$$

where t is the quartz thickness, e_{35} is the piezoelectric constant, and S_A is the QCM electrode area. Substituting Eq. (4.15) into Eq. (4.16), we obtain

$$Z_{3e} = \frac{t^2}{4e^2{}_{35}S_A} j\sqrt{j\rho_L\eta_L\omega}\tan\left(\omega h\sqrt{\frac{\rho_L}{j\omega\eta_L}}\right) \quad (4.17)$$

which is the additional impedance from the viscous film in Fig. 4.5. As illustrated in Fig. 4.8A, the frequency variation was obtained by substituting Eq. (4.11) into Eq. (4.1), considering the impedance change caused by the effect of viscous loading on the overall impedance. From the model frequency, change can be calculated from the imaginary part of the impedance, and the resistance is calculated from the real part of the impedance. Three factors were considered concerning the coating film's properties: film thickness (h), film viscosity (η_L), and film density (ρ_L). Assuming the film density remains relatively constant at 1 g/cm^3, Fig. 4.8A illustrates the frequency shift as a function of the deposited film thickness and film viscosity. Similar frequency changes are represented by the same color.

The advantage of this model is its capability to explain various responses that cannot be explained by conventional equations such as Sauerbrey's equation and Kanazawa's equation. For example, Sauerbrey's equation explains that the frequency change of QCM is always decreasing when extra depositions occur on the QCM surface. However, many investigations have found that it is not always the case. The investigation on molecular imprinting, aimed at creating artificial antibodies, also demonstrated a positive frequency shift during the detection of tobacco mosaic virus (TMV) (Hayden et al., 2006). At a lower

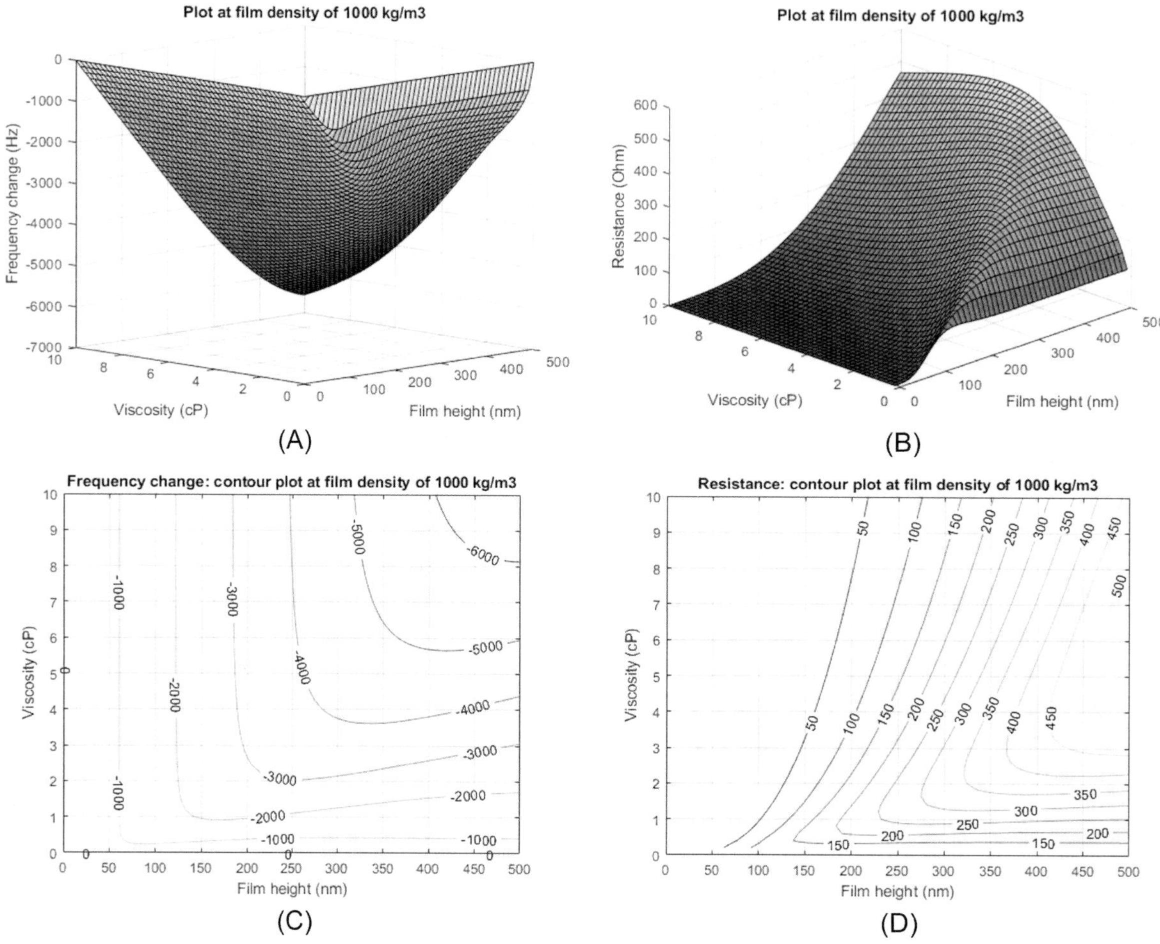

FIGURE 4.8 3D plot and contour plot of the frequency shift and resistance at a film density of 1000 kg/m^3. (A) 3D plot of the calculated frequency shift. (B) 3D plot of the calculated resistance shift. (C) Contour plot of the frequency change. (D) Contour plot of the resistance. The simulation involves specific parameters for the QCM, which are defined as follows: $t = 181$ μm (thickness), $e_{35} = -0.095$ (electric charge constant), $S_A = 1.96 \times 10^{-5}$ m^2 (electrode surface area), $L_1 = 0.0126$ H (inductance), and $C_1 = 2.47 \times 10^{-14}$ F (capacitance). These parameters correspond to the 9 MHz AT-Cut QCM. *QCM*, Quartz crystal microbalance. Source: *From Songkhla, S. N., & Nakamoto, T. (2021). Interpretation of quartz crystal microbalance behavior with viscous film using a mason equivalent circuit.* Chemosensors, *9(1). https://doi.org/10.3390/chemosensors9010009.*

concentration of TMV (10 μg/mL), the positive frequency alteration occurred when the polyurethane film was not imprinted. The author concluded that the virus did not attach firmly and tended to slip from the surface, causing the observed positive frequency change. Additionally, the author explained that, at a higher concentration of 100 μg/mL, the viruses formed larger aggregates that adhered strongly to the surface, resulting in the typical negative frequency change, μ. At a lower concentration of TMV (10 μg/mL), the positive frequency alteration occurred when the polyurethane film was not imprinted. The author concluded that the virus did not attach firmly and tended to slip from the surface, causing the observed positive frequency change. Additionally, the author explained that, at a higher concentration of 100 μg/mL, the viruses formed larger aggregates that adhered strongly to the surface, resulting in the typical negative frequency change, μ. At a lower concentration of TMV (10 μg/mL), the positive frequency alteration occurred when the polyurethane film was not imprinted. The author concluded that the virus did not attach firmly and tended to slip from the surface, causing the observed positive frequency change. Additionally, the author explained that, at a higher concentration of 100 μg/mL, the viruses formed larger aggregates that adhered strongly to the surface, resulting in the typical negative frequency change.

4.3 Quartz crystal microbalance measurement methods

Regarding the measurement methods, as mentioned earlier, the equivalent circuit of the QCM consists of the motional admittance branch representing the resonant frequency and the parallel static capacitance. Due to the similarity of the QCM structure to crystal oscillators used in microcomputer systems, the QCM can function as an oscillator. The simplest circuit used in conjunction with QCM is known as a Pierce oscillator (Fig. 4.9). The effect of mass loading can be estimated from the oscillating frequency rather than the resonant frequency. However, if the damping of the QCM increases due to viscous coating or liquid loading, it negatively affects the Q factor, rendering the QCM unsuitable for oscillation as it cannot maintain a positive feedback loop. To overcome this limitation, several methods can be employed, such as dedicated

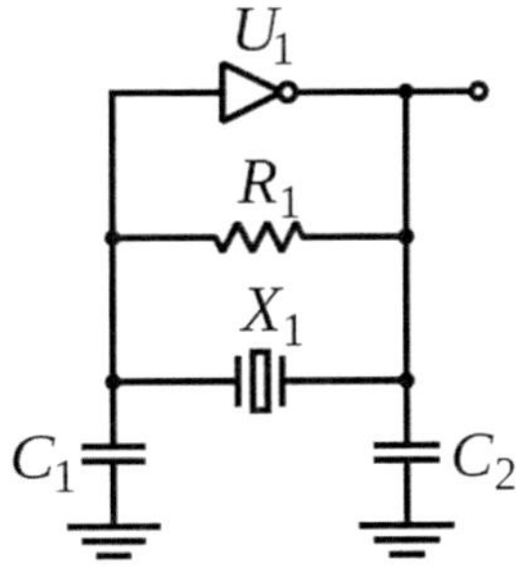

FIGURE 4.9 **Pierce Oscillator Circuit.** Circuit diagram of a pierce oscillator: Featuring an inverter connected to a parallel resonant circuit with a resistor, an inductor, and two capacitors. Source: *From Creative Commons Attribution-Share Alike 3.0 Unported License.*

circuit designs or the careful design and characterization of an oscillator circuit. Alternatively, one can consider the use of a vector network analyzer (VNWA), as portable and relatively inexpensive options are available nowadays.

4.3.1 Dedicated circuits

With the increasing use of QCM in diverse sensing applications, the need for coatings to enhance QCM sensitivity has become inevitable. Regrettably, these coatings tend to be viscous, causing instability in traditional oscillator circuits. To address this issue, numerous specialized circuits have been developed. Additionally, the dissipation or loss in the QCM is a crucial parameter that cannot be obtained solely from the oscillator circuit and requires the use of an additional circuit.

By combining an automatic gain control circuit with frequency sweeping, it becomes feasible to extract both the resonant frequency and Q factor, as depicted in Fig. 4.10 (Nakamoto & Kobayashi, 1994). Nonetheless, this circuit is composed of numerous intricate electrical components, and accurately determining the resonant frequency remains challenging due to difficulties in tracking the peak conductance.

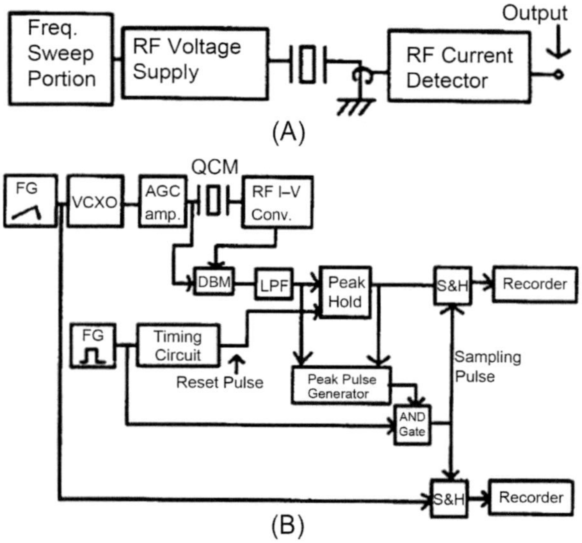

FIGURE 4.10 **Circuit diagram of the system, which measures both the *Q* factor and resonant frequency.** (A) Basic RF characterization circuit, (B) advanced QCM analysis circuit with signal processing and recording modules. *QCM*, Quartz crystal microbalance; *RF*, random forest. Source: *From Nakamoto, T., & Kobayashi, T. (1994). Development of circuit for measuring both Q variation and resonant frequency shift of quartz crystal microbalance.* IEEE Transactions on Ultrasonics, Ferroelectrics, and Frequency Control, *41(6). https://doi.org/10.1109/58.330261.*

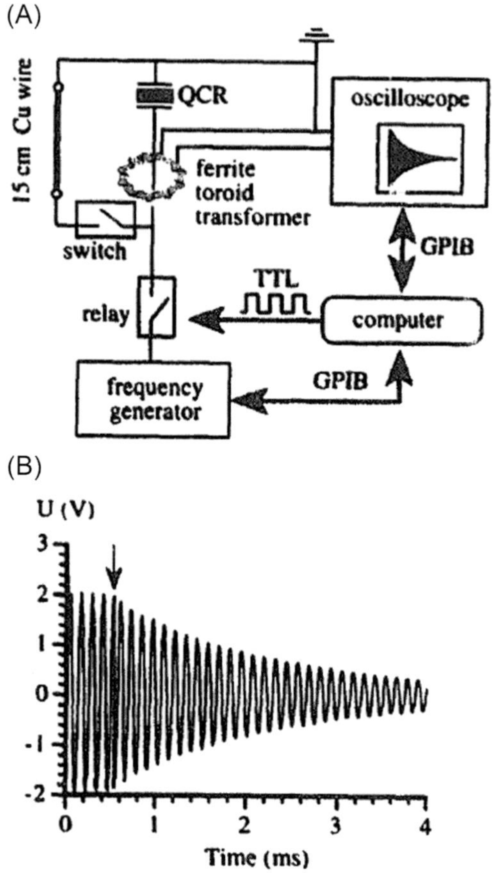

FIGURE 4.11 **Block diagram of the circuit that can measure both dissipation factor and resonant frequency.** (A) Schematic diagram; (B) circuit diagram in detail. Source: *From Rodahl, M., & Kasemo, B. (1996). Frequency and dissipation-factor responses to localized liquid deposits on a QCM electrode.* Sensors and Actuators, B: Chemical, *37 (1–2). https://doi.org/10.1016/S0925-4005(97)80077-9.*

An alternative method was suggested to detect both the resonant frequency and loss in the QCM (Rodahl & Kasemo, 1996). This approach involves employing an oscillator circuit in conjunction with a relay switch to isolate the QCM from the oscillator circuit, as illustrated in Fig. 4.11. Consequently, the oscillation signal gradually diminishes over time, and this decay is inversely proportional to the dissipation factor. The system's performance was assessed by testing it with a droplet of an aqueous glycerol solution.

To simplify the QCM measurement system and reduce costs, there was an approach involving utilizing a direct digital synthesizer to generate multiple reference frequencies, as shown in Fig. 4.12 (Addabbo et al., 2014). The frequency change is then measured using heterodyne demodulation from the oscillating signal, simplifying the generation of reference frequencies. However, the dissipation is not measured in this particular research.

Dedicated circuits designed for QCM measurements offer tailored solutions for accurately analyzing QCM sensors. They enable operation at multiple frequencies, handle viscous loading effects, and provide enhanced

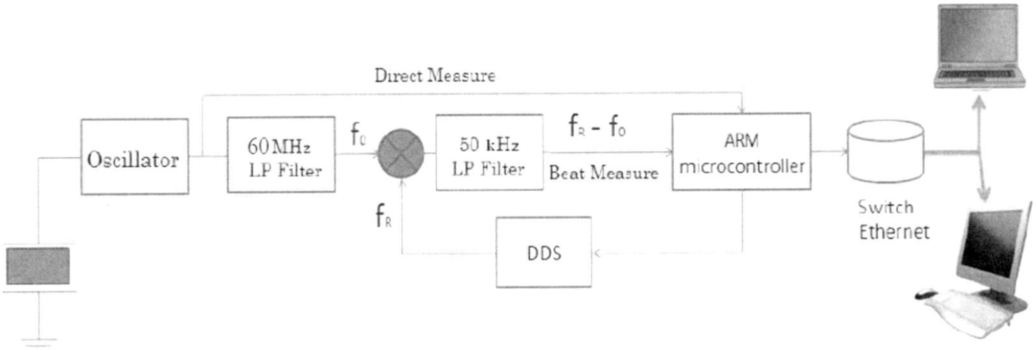

FIGURE 4.12 **QCM frequency measurement using a DDS**. Block diagram of a frequency measurement system: showing signal processing from an oscillator through low-pass filters and a DDS. *DDS*, Direct digital synthesizer; *QCM*, quartz crystal microbalance. Source: *From Addabbo, T., Bertocci, F., Fort, A., Mugnaini, M., Shahin, L., Vignoli, V., & Rocchia, S. (2014). A DDS-based multi-harmonic frequency meter for QCM sensor applications.* Procedia Engineering, *87. https://doi.org/10.1016/j.proeng.2014.11.664.*

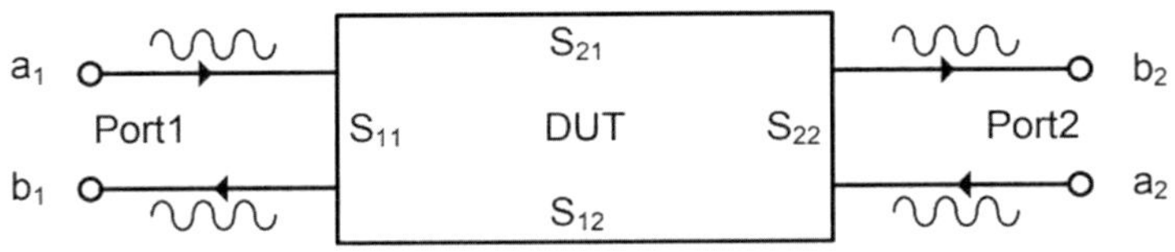

FIGURE 4.13 **Basic concept of a vector network analyzer**. Schematic of a two-port network testing a DUT with scattering parameters (*S*-parameters). Indicated for incident and reflected waves. *DUT*, Device under test.

measurement capabilities. These circuits contribute to advancing the understanding and application of QCM technology in diverse fields.

4.3.2 Vector network analyzer

The VNWA is utilized to analyze and evaluate two-port networks, including amplifiers and filters. Its operation is based on the measurement of transmitted and reflected waves as a signal passes through a device under test (DUT). The VNWA measures both the reflected signal originating from the input side and the transmitted signal through the DUT to the output side. The measurements obtained through the VNWA are commonly expressed in terms of S parameters, which indicate the ratio of transmitted or reflected voltage to the incident voltage. In the context of QCM measurement, the S_{11} parameter is frequently employed. It represents the ratio of the reflected signal from port1 (b_1/a_1), where a_1 denotes the incident voltage at port 1 and b_1 represents the reflected voltage at port 1. The basic illustration of the VNWA can be seen in Fig. 4.13.

A notable benefit of using VNWA is its ability to directly determine the resonant frequency of the QCM through measurement. The frequency of oscillation may not always align with the actual change in resonant frequency, especially when there is an increase in loss. In such cases, the oscillation frequency tends to be lower (Nakamoto et al., 1996). This phenomenon is depicted in Fig. 4.14. While it is challenging to avoid loss caused by viscous loading, the use of VNWA offers a solution to overcome this issue.

QCM measurement techniques generally involve assessing frequency and Q factor. Frequency measurement often relies on oscillator circuits and frequency sweep methods, which are prominent. On the other hand,

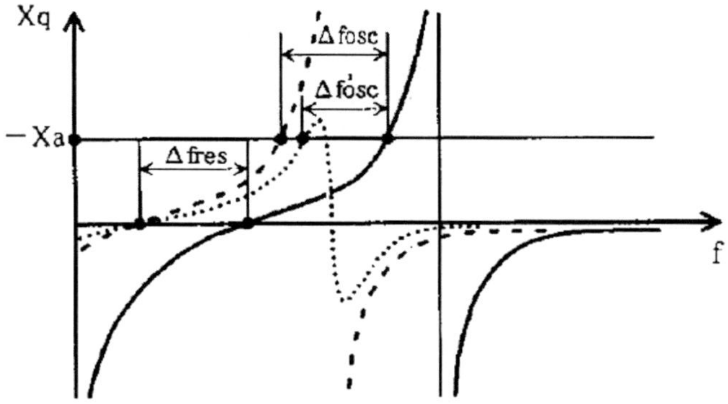

FIGURE 4.14 Dependency of reluctance upon the frequency. Solid line: no coating and no sorption, thick dashed line: only mass loading without loss increase, thin dash ling: mass loading with additional loss. Oscillation frequency deviates from actual resonant frequency change when loss increases. Source: *From Nakamoto, T., Nakamura, K., & Moriizumi, T. (1996). Proceedings of the IEEE Ultrasonics Symposium, Study of oscillator-circuit behavior for QCM gas sensor 1. https://doi.org/10.1109/ultsym.1996.583990.*

measuring the *Q* factor offers a range of approaches. Additionally, the use of small VNWAs has gained popularity in recent years due to multiple manufacturers offering these devices. The summary of various QCM measurement methods is shown in Table 4.1.

4.4 Application area

The applications of E-nose technology are extensive, although they are not yet widely prevalent. This could be attributed to its limited sensitivity, as the substances being detected are often present in low concentrations and are further diluted in the surrounding atmosphere. In the field of food quality control, E-noses can be used to assess the freshness of food products, detect spoilage or contamination, and monitor the ripeness of fruits. Industrial quality control and environmental monitoring applications include detecting and identifying pollutants in the air, monitoring indoor air quality, and assessing the emissions from industrial processes. In the medical field, E-noses with QCM sensors have shown promise in diagnosing diseases through breath analysis. They can detect specific volatile biomarkers associated with various medical conditions, such as lung cancer, diabetes, and bacterial infections. E-noses can also be employed in the pharmaceutical industry for quality control of drugs and detecting counterfeit products. Furthermore, E-noses using QCM sensors find applications in explosive detection as the use of nonmanual inspection can be dangerous.

4.4.1 Food and beverage

Food safety is a multifaceted matter encompassing various elements, such as food production, processing, storage, and handling. Initially, brand identification or foreign substance detection was studied (Ema et al., 1989; Nakamoto et al., 1993). Then, foodborne illnesses pose a significant public health concern, leading to numerous cases of sickness and fatalities annually. Inadequate or prolonged storage can result in food spoilage. Furthermore, the aroma and taste of food play a crucial role in consumer enjoyment and serve as indicators of freshness.

4.4.1.1 *Fruit ripeness*

The freshness and ripeness of fruits are closely tied to the unique blend and quantity of aroma they emit. Therefore alterations in the fruit's aroma provide a trustworthy cue for consumers to assess its freshness and ideal ripeness. It is essential to proactively identify the specific vapors linked to the fruit's attributes, particularly during the stages of ripening and storage.

TABLE 4.1 Overview measurement methods for **quartz crystal microbalance** (QCM) sensor.

Summary	Principles	Advantages	Disadvantages	Ref.
Development of circuit for measuring both Q variation and resonant frequency shift of quartz crystal microbalance	Frequency sweep and automatic gain control amplifier	Complete design, test results agreeable with impedance analyzer	Complicate circuit design. Difficulty in peak detection at low Q factor	Nakamoto and Kobayashi (1994)
Responses in frequency and dissipation factor to localized liquid deposits on a QCM electrode	Oscillator circuit, measurement of decay signal to calculate dissipation factor	Relatively simple circuit, dissipation factor can be calculated easily from time constant	At high dissipation factor, the time constant becomes shorter and harder to measure	Rodahl and Kasemo (1996)
An oscillator circuit for dual-harmonic tracking of frequency and resistance in QCM sensors	Dual-harmonic oscillator and automatic gain control amplifier	Fundamental and third harmonic operate at the simultaneously	Need compensating capacitor to cancel out C_0, circuit design a little bit complicated	Ferrari and Ferrari (2009)
A DDS-based multiharmonic frequency meter for QCM sensor applications	Oscillator using DDS to synthesize reference frequency and measured frequency change using heterodyne	Fully embedded measurement system. Operate at three harmonic frequency	Measure only frequency change, not measure Q factor	Addabbo et al. (2014)
Stability enhanced, repeatability improved parylene-C passivated on QCM sensor for aPTT measurement	Bench-top VNWA: peak-to-peak value of s parameter was used to evaluate hydrophobicity	Interesting choice of parameter to evaluate the result	Did not utilize other parameters such as resonant frequency and dissipation factor	Yang et al. (2017)
Verification of quartz crystal microbalance array using vector network analyzer and OpenQCM	Portable VNWA	Comparison between portable VNWA and bench-top VNWA	The contribution is quite minimal	Zainuddin et al. (2018)
Simple and ultrafast resonance frequency and dissipation shift measurements using a fixed frequency drive	Fixed frequency drive and network analyzer	Operate at a single frequency ensures very fast measurement time of 112 US	The fast measurement time may not be as important as the more data gained from frequency sweep using VNWA	Guha et al. (2019)

DDS, Direct digital synthesizer; *VNWA*, vector network analyzer.

Ghatak et al. synthesized a molecularly imprinted polymer derived from a copolymer of methacrylic acid and divinylbenzene. The template was imprinted by adding 3-carene to the polymers. 3-Carene is volatile organic compound (VOC) that is among the most prominent and dominant terpene hydrocarbons identified specifically for their flavor attribute. The developed sensor resulted in an improved sensitivity of 0.11 Hz/ppm (Ghatak et al., 2018).

β-Caryophyllene is a compound prevalently present in mango aroma. By utilizing gold nanoparticles embedded in 2-mercaptobenzothiazole on a QCM, the sensing mechanism is shown in Fig. 4.15A. The researchers achieved a sensitivity

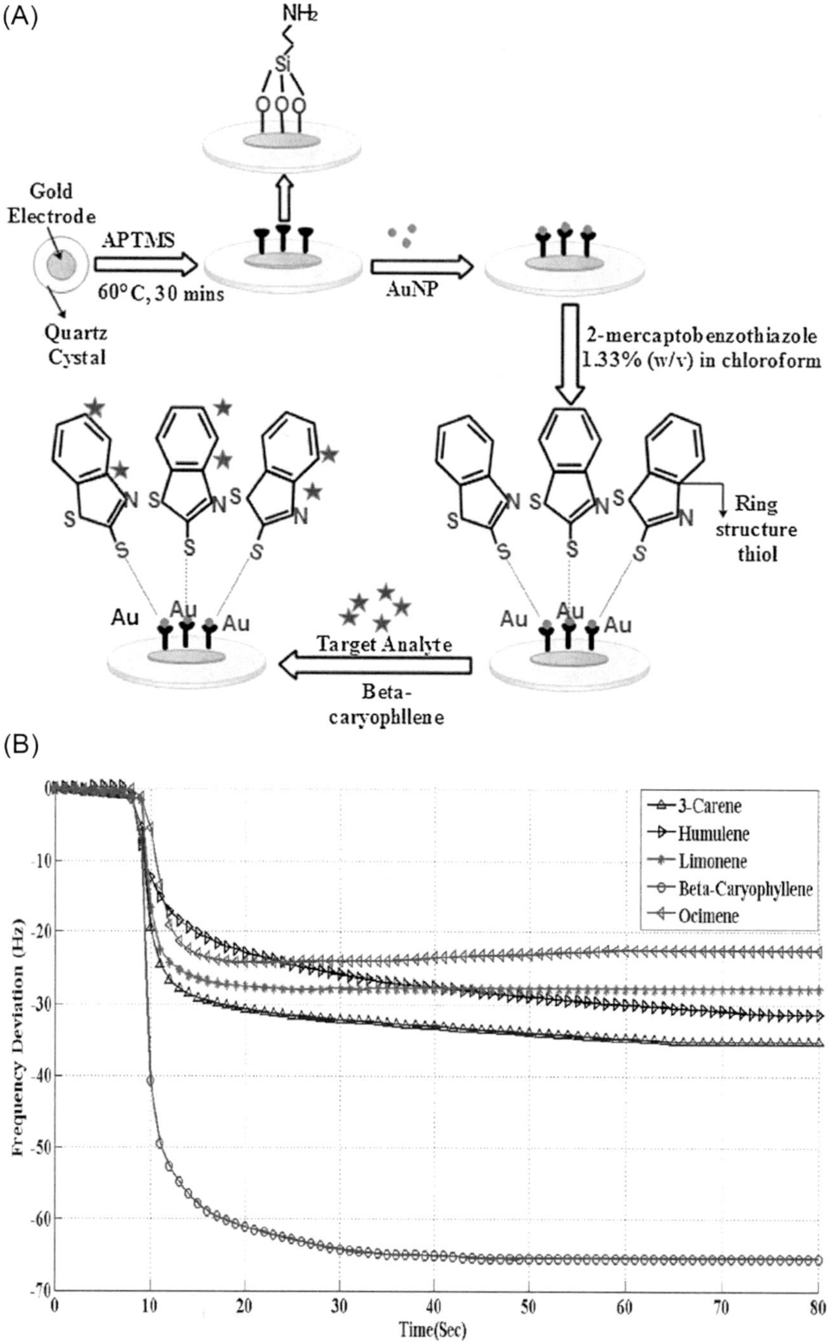

FIGURE 4.15 QCM sensor for detecting beta-caryophyllene. (A) Schematic representation of sensing mechanism. (B) Response profile of sensor for all volatiles at 300 ppm. *QCM*, Quartz crystal microbalance.

of 0.17 Hz/ppm for β-caryophyllene detection (Fig. 4.15B). The correlation between the QCM response and the results obtained from gas chromatography–mass spectrometry (GC–MS) was determined to be 0.453 (Ali et al., 2018).

Ethylene gas plays a crucial role in regulating fruit ripeness, particularly in situations where ventilation is limited, such as during storage and transportation. Monitoring fruit ripeness using an E-nose capable of detecting ethylene has been reported. A QCM ethylene gas sensor was developed using silver(I)/polymer composites, exhibiting a relatively high sensitivity of 54 Hz/ppm and a low detection limit down to 420 ppb (Tolentino et al., 2018).

Bananas and pears exhibited a strong response due to their high ethylene emissions. Another study by Juncai et al. (2014) focused on developing an E-nose system using an array of QCM sensors coated with four different materials (ethyl cellulose, cellulose acetate, 1,2-dioleoyl-sn-glycero-3-[phosphor-L-serine], and galactosylceramide) through the drop coating technique. The purpose was to detect various VOCs, including ethylene, isobutyl alcohol, and ethyl acetate. The results indicated that the cellulose acetate-coated sensor exhibited the highest response to all of the VOCs.

4.4.1.2 Quality grading and classification

Each type of food is likely to have its own unique aroma, with volatile components playing a significant role in defining the aroma's quality. Researchers have made efforts to utilize E-noses in the assessment and classification of food products, particularly alcoholic beverages.

A QCM-based E-nose system, employing eight different coatings, was utilized to analyze the flavor profiles of Chinese liquor. Symmetrical coatings were applied on both sides of each QCM transducer, incorporating a variety of materials, such as polyvinyl chloride (PVC), polyamide, polyethylene (PE)/silver chloride (AgCl), polytef, AgCl, azithromycin, copper chloride ($CuCl_2$)/PE, and $CuCl_2$/AgCl/PE. The sensor array was mounted in the sensor box with the inlet and outlet pipes; the sensor box is shown in Fig. 4.16. The obtained data were

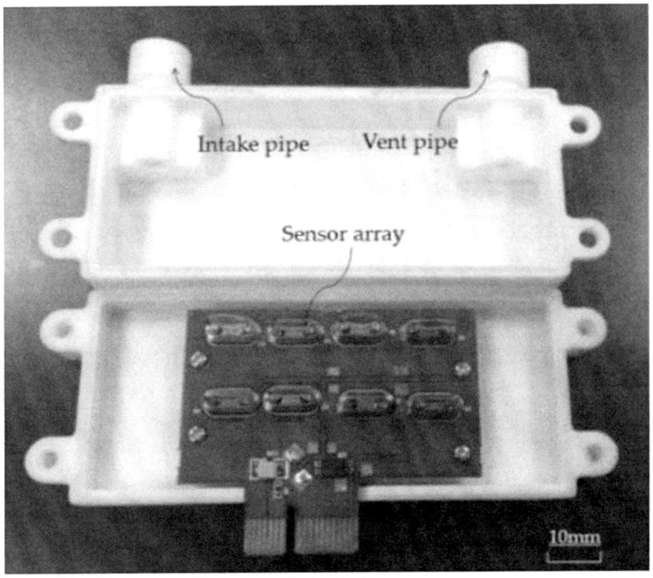

FIGURE 4.16 Photo of the sensor array for assessing Chinese liquor. Sensor box with array and flow pipes: an array of sensors mounted on a circuit board inside a box, equipped with an intake pipe and a vent pipe for liquor analysis. Source: *From Li, Q., Gu, Y., & Wang, N. F. (2017). Application of random forest classifier by means of a QCM-based E-nose in the identification of Chinese liquor flavors.* IEEE Sensors Journal, *17(6). https://doi.org/10.1109/JSEN.2017.2657653.*

optimized using a random forest (RF) classifier. When compared to other classifiers, such as support vector machine (SVM), linear discriminant analysis (LDA), and back-propagation artificial neural network (BP-ANN), the RF classifier demonstrated a remarkable prediction accuracy of 100% for distinguishing among 12 different types of Chinese liquors (Li et al., 2017).

Interestingly, it is possible to indicate the sugar content of fruit by using the E-nose. Saevels et al. (2003) attempted to evaluate the sugar content index in apple samples. The commercial E-nose system called "Libra nose," consisting of five QCMs, was utilized to predict the optimal harvest date of apples (Fig. 4.17). The prediction was based on the partial least squares (PLS) model, which considered parameters, such as firmness, soluble solid content, and the stage of starch transition. The Strife index, which decreases during the preclimacteric stage of fruit development, was used. "Jonagold" and "Braeburn" apples were assessed over 2 consecutive harvest years using the E-nose. In Fig. 4.18, the PLS model demonstrated cross-validation correlations of 0.89 for "Jonagold" and 0.92 for "Braeburn." The calibration models for maturity prediction, expressed as the number of days before harvest, also exhibited reasonable accuracy, with validation correlations of 0.91 for "Jonagold" and 0.84 for "Braeburn."

The assessment of olive oil quality using an E-nose is widely practiced. Escuderos et al. (2010) conducted a study where they utilized five GC stationary phases (OV-17, OV-275, polyethylene glycol (PEG), Span 80, and Vaseline) as sensitive coatings on QCM sensors. The objective was to differentiate olive oil into three categories: extra virgin olive oil, virgin olive oil, and lampante olive oil. Fig. 4.19 shows the results obtained through principal component analysis (PCA) indicated that using four sensors (OV-17, OV-275, PEG, and Span 8) was sufficient to accurately classify the samples into the three categories (extra virgin, virgin, and lampante) with 91.7% accuracy, aligning with the official method of panel testing.

4.4.1.3 Spoilage and microorganism detection

The existence of detrimental bacteria in food products presents a notable obstacle to the preservation and consumption of food. The E-nose can detect volatile metabolites produced by microorganisms, which are often responsible for the characteristic odors associated with microbial growth. By analyzing the volatile profiles, the E-nose can identify the presence of specific microorganisms or microbial contamination in samples. This enables early detection of microbial growth, allowing for timely interventions such as sanitation and

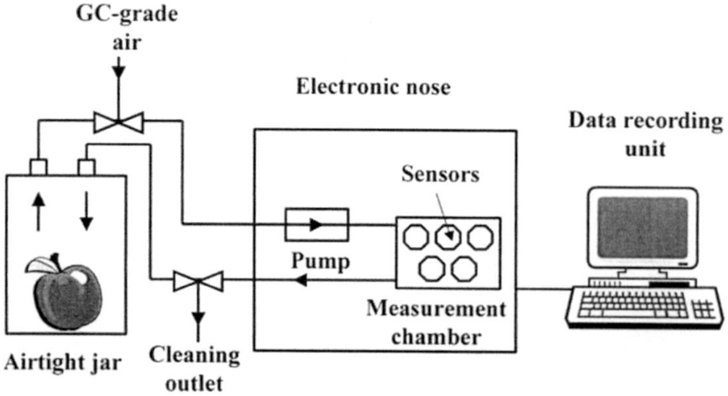

FIGURE 4.17 **Schematic diagram of the E-nose measurements during the experiment**. Diagram of an electronic nose system: the airflow from a sample jar through sensor arrays to a data recording unit for scent analysis. Source: *From Saevels, S., Lammertyn, J., Berna, A. Z., Veraverbeke, E. A., Natale, C. D., & Nicolai, B. M. (2003). Electronic nose as a non-destructive tool to evaluate the optimal harvest date of apples.* Postharvest Biology and Technology, *30(1). https://doi.org/10.1016/S0925-5214(03)00059-0.*

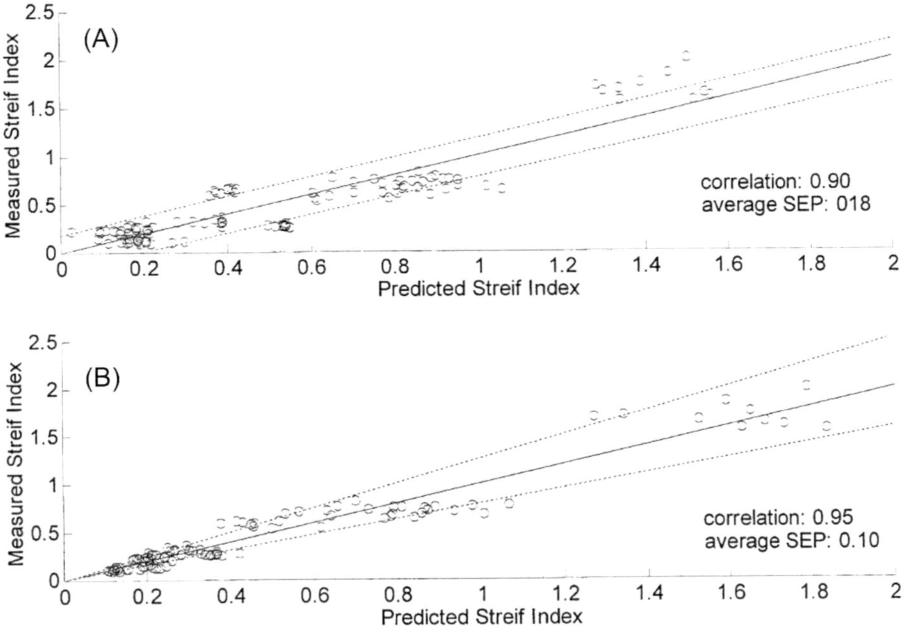

FIGURE 4.18 **Plots of the predicted (PLS models based on the E-nose measurements) and measured Streif Index for the "Braeburn" data of the year 2001**. The PLS model is based on the Streif indices has a constant standard error of prediction (SEP), indicated by the dotted lines (A). The SEP of the PLS model based on the logarithmically transformed Streif indices (B). *E-nose*, Electronic noses. Source*: From Saevels, S., Lammertyn, J., Berna, A. Z., Veraverbeke, E. A., Natale, C. D., & Nicolai, B. M. (2003). Electronic nose as a non-destructive tool to evaluate the optimal harvest date of apples.* Postharvest Biology and Technology, *30(1). https://doi.org/10.1016/S0925-5214(03)00059-0.*

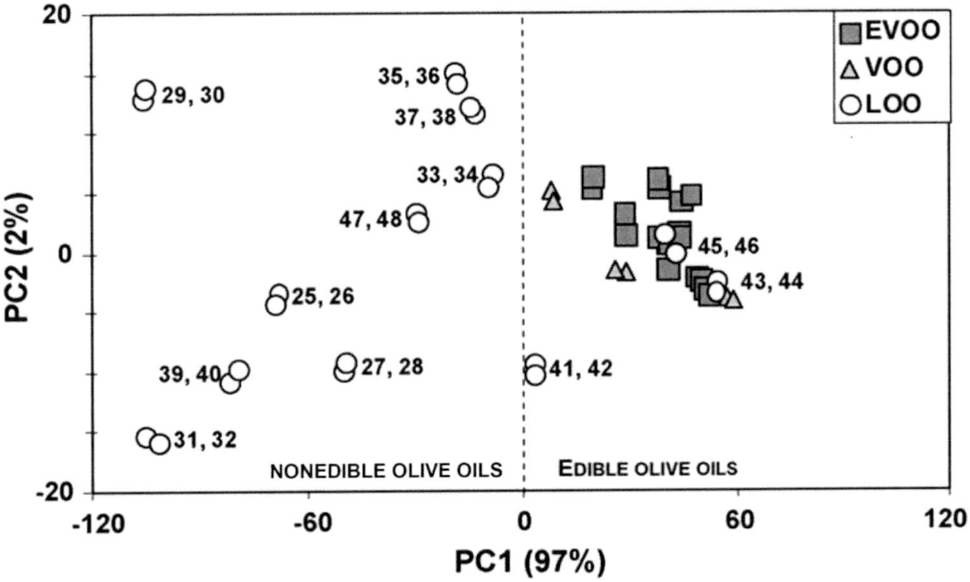

FIGURE 4.19 **PCA score plot of olive oil samples**. Sample codes: EVOO, extra-virgin olive oil; VOO, virgin olive oil; LOO, lampante olive oil. LOO samples have been numbered to differentiate between olive oil defects. *PCA*, Principal component analysis. Source*: From Escuderos, M. E., Sebastián, S., & Jiménez, A. (2010). Virgin olive oil sensory evaluation by an artificial olfactory system, based on quartz crystal microbalance (QCM) sensors.* Sensors and Actuators, B: Chemical, *147(1). https://doi.org/10.1016/j.snb.2010.01.071.*

disinfection measures to prevent the spread of contaminants.

Papadopoulou et al. (2013) employed a commercial E-nose called "Libra nose" to monitor the headspace air of meat samples. The collected data were then compared with the total viable count (TVC) of various microbes, such as *Pseudomonas* spp., *Brochothrix thermosphacta*, lactic acid bacteria, and Enterobacteriaceae. The qualitative interpretation of the E-nose data involved classifying the samples into three quality categories: fresh, semifresh, and spoiled. The collected data were then subjected to SVM classification using a radial basis kernel function. The results obtained demonstrated effective discrimination among all three quality classes, with accuracy exceeding 89% for each class. Fig. 4.20 shows the comparison between observed and predicted counts of TVC.

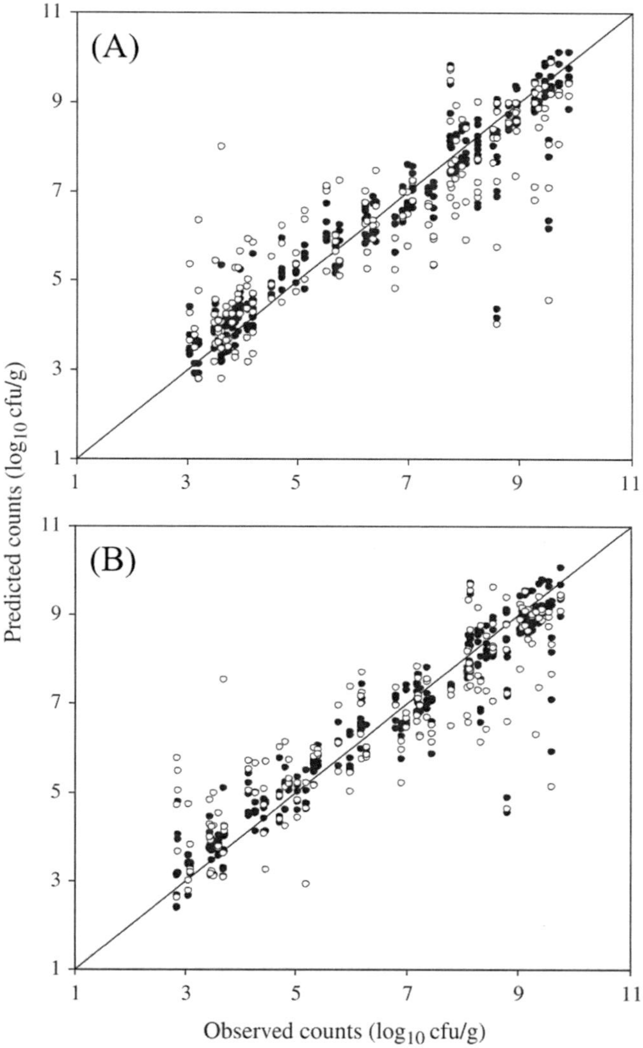

FIGURE 4.20 Comparison between observed and predicted counts of total viable counts. (A) and *Pseudomonas* spp. (B) by the SVM regression model based on E-nose sensor array data (solid symbols: training data; open symbols: test data). *E-nose*, Electronic noses; *SVM*, support vector machine. Source: *From Papadopoulou, O. S., Panagou, E. Z., Mohareb, F. R., John G., & Nychas, E. (2013). Sensory and microbiological quality assessment of beef fillets using a portable electronic nose in tandem with support vector machine analysis.* Food Research International, *50(1). https://doi.org/10.1016/j.foodres.2012.10.020.*

In addition, odorant-binding protein (OBP)-based sensors were developed for the detection of alcohols by Sankaran et al. (2011). These alcohols are part of a group of compounds that may be released when packaged beef is contaminated with *Salmonella*. The OBP named LUSH from the fruit fly *Drosophila melanogaster*, known for its high sensitivity and selectivity to alcohols, was deposited onto QCM sensors using the self-assembly of the thiol group. These sensors were employed to detect alcohols such as 3-methyl-1-butanol and 1-hexanol, achieving lower detection limits of less than 5 ppm (Fig. 4.21).

We explore the applications of QCM in the field of food and beverage. Fruit ripeness has been one of the major interests in E-nose technology, as nondestructive detection of fruit ripeness can be a major revolution in the agriculture industry, especially for highly valued fruits. Additionally, the quality grading and classification of fruit and beverages can be crucial for producers, as product quality can impact pricing and the reputation of the producer. Lastly, the use of spoilage and microorganism detection opens the door for industries to test products without the slow process of traditional lab testing. In Table 4.2, we show the extensive applications of using QCM in food and beverage products.

4.4.2 Industrial quality control and environmental monitoring

This section extensively explores hazardous gases, emphasizing the need for ongoing vigilance and management. Beyond understanding the gases' origins, properties, and impacts, the chapter highlights the QCM sensor as an indispensable ally in gas detection. Also, humidity is a crucial yet often overlooked factor in gas monitoring. The QCM sensor's sensitivity to humidity fluctuations aids environmental monitoring for corrosion prevention, indoor air quality assessment, and artifact preservation. Since the application of using QCM as a gas sensor was so abundant, there are summary tables for all of the subsections.

4.4.2.1 Common hazardous gases

Various gases, including CO_2, CO, SO_2, NO_2, O_3, NH_3, CH_4, and others, require monitoring due to their environmental and health impacts. Certain gases can pose serious health risks, while others contribute to environmental damage and play a significant role in climate change. Air quality standards have been implemented for numerous industries. To effectively monitor these gases, highly specific and accurate gas sensors are necessary. The QCM sensor offers the advantage of being able to choose its coating materials freely, as its principle is based on mass precipitation, making it completely customizable. In this section, we will delve into the utilization of QCM technology for sensing these commonly hazardous gases.

Ammonia (NH_3) is released from agricultural practices, industrial operations, and natural sources. With its potential to contribute to soil acidification and air pollution, NH_3 monitoring assumes significance. Minh et al. (2013) were fabricating an NH_3 sensor using ZnO nanorods. They varied the nanorod length in four lengths: 1200, 2000, 2500, and 3000 nm (Fig. 4.22). It was found that a longer length of ZnO nanorod also has a greater frequency shift. The frequency shift of each sensor linearly increases with an increase in NH_3 gas concentration in the range of 50–800 ppm, as shown in Fig. 4.23.

Another example of a QCM gas sensor is to detect methane gas (CH_4), which is a combustible gas that is abundant in nature. The mixture of 5%–14% of methane gas with air is explosive. The materials called cryptophane-A and cryptophane are supramolecular compounds that look like two

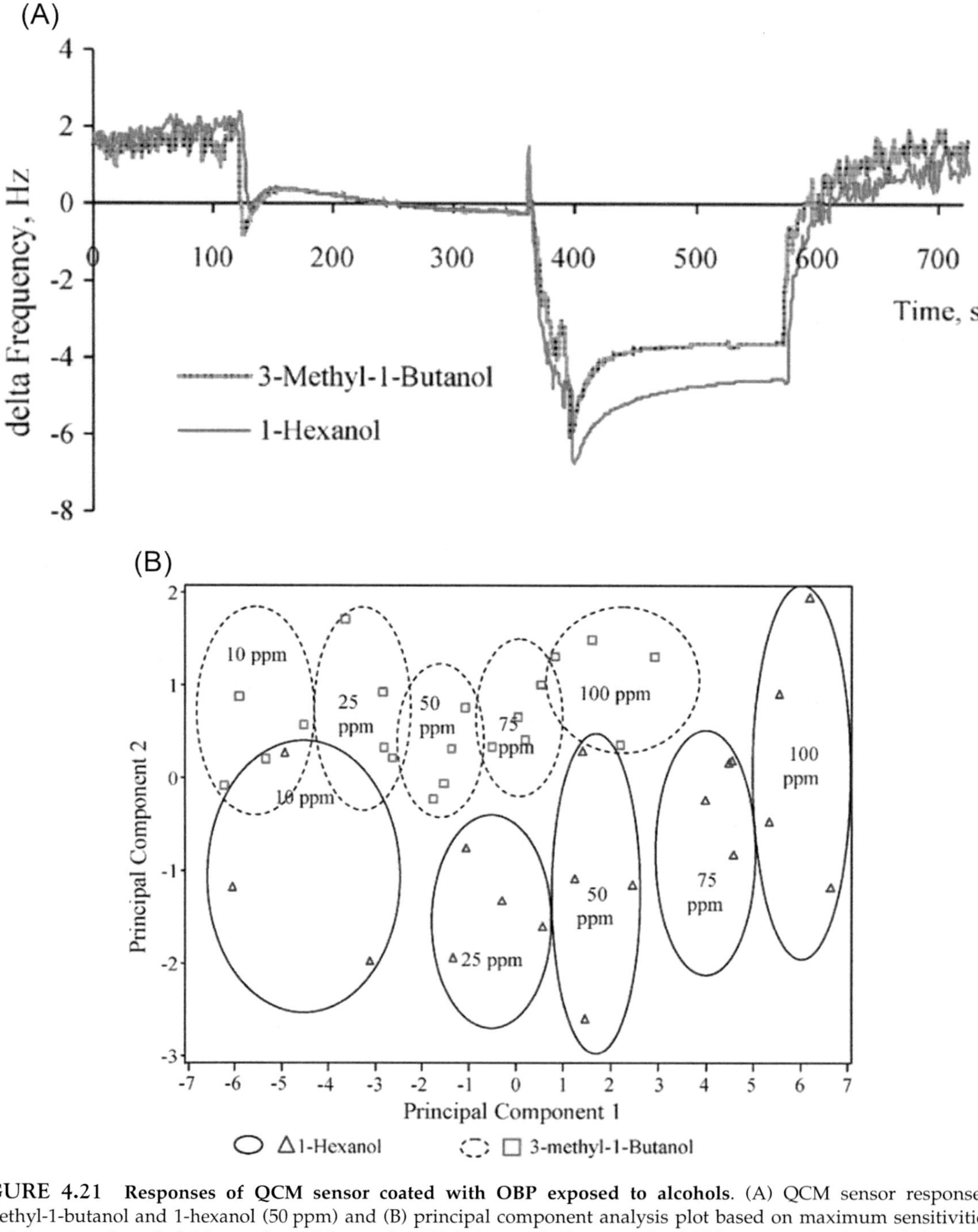

FIGURE 4.21 Responses of QCM sensor coated with OBP exposed to alcohols. (A) QCM sensor responses to 3-methyl-1-butanol and 1-hexanol (50 ppm) and (B) principal component analysis plot based on maximum sensitivities of QCM sensors to 3-methyl-1-butanol and 1-hexanol. *OBP*, Odorant binding protein; *QCM*, quartz crystal microbalance.

TABLE 4.2 Overview application of electronic noses (E-nose) based on quartz crystal microbalance (QCM) sensor in food and beverage.

Application	Category	Coating material	Target substance	Ref.
Detection of *Salmonella* in beef by alcohol vapor generated from bacteria	Spoilage	OBP LUSH	Alcohols: 3-methyl-1-butanol and 1-hexanol	Sankaran et al. (2011)
Evaluation of apple maturity using E-nose data for "Jonagold" and "Braeburn" apples	Quality grading	Modified metallo-porphyrins	Alcohols and amines	Saevels et al. (2003)
Classification between edible and nonedible olive oils	Quality grading	OV-17, OV-275, PEG, Span 80 and Vaseline	Headspace odor	Escuderos et al. (2010)
Categorize three herbs (rosemary, basil, sage) between fresh and dried	Quality grading	MIP—six molecules	Terpenes, α-pinene, β-pinene, α-terpenes, α-pinene, β-pinene βterpenes, α-pinene, β-pinene	Iqbal et al. (2010)
Differentiate between headspace odors of different cinnamon varieties	Quality grading	RTIL	Headspace odor	Toniolo et al. (2013)
QCM sensor system is developed to detect headspace linalool gas in black tea	Quality grading	PEG200	Linalool	Sharma et al. (2014)
The sensor employs a coating made from β-CD to detect black tea constituents	Quality grading	β-CD	Geraniol	Sharma et al. (2015)
Identification between six different citrus scents using sensors array coated with RTIL	Quality grading	RTIL	Lemon, lemon *Eucalyptus*, Lemongrass, Lime, Orange	Speller et al. (2016)
Detection of methyl salicylate in black tea	Quality grading	Maltodextrin	Methyl salicylate	Sharma et al. (2016)
Identifying eight different Chinese liquors using E-nose	Quality grading	PVC, polyamide, PE, Polytef, AgCl, Azithromycin, $CuCl_2$	Headspace odor	Li et al. (2017)
Assess three varieties of white ginger	Quality grading	MIP - PAA	Borneol, citral, and geraniol	Hardoyono et al. (2019)
Detection of active compounds in turmeric (*Curcuma longa* L.)	Quality grading	MIP—PAA	ar-Turmerone, curlone, ethyl-*p*-methoxycinnamate, and tumerone	Hardoyono and Windhani (2021)
Evaluate different mint species	Quality grading	ZnO, AZO, SnO_2, ITO, PVA, PEDOT:PSS, PB, SW, PFO, SB, SG, Cu(BPDC)	Headspace odor	Okur et al. (2021)
Detection of β-pinene in Indian cardamom using various vegetable oils	Quality grading	Olive oil	β-pinene	Debabhuti et al. (2022)

(Continued)

TABLE 4.2 (Continued)

Application	Category	Coating material	Target substance	Ref.
Detection of isobutyl alcohol, ethyl acetate, and ethylene for fruit ripeness application	Ripeness	Ethyl cellulose, cellulose acetate, 1,2-dioleoyl-sn-glycero-3-[phosphor-L-serine], and galactosyl ceramide	Isobutyl alcohol, ethyl acetate, and ethylene	Juncai et al. (2014)
Fabrication of a sensor to detect 3-carene, which can indicate mango ripeness	Ripeness	MIP—methacrylic acid and divinylbenzene	3-Carene	Ghatak et al. (2018)
Detect *trans* 2-hexenal, which can indicate tomato ripeness	Ripeness	PEG6000, PFBBr, tri-ethyl amine	*Trans*-2-hexenal	Debabhuti et al. (2017)
Development of a ripeness sensor to test with four different varieties of mango	Ripeness	Gold nanoparticle embedded 2-mercaptobenzothiazole	β-caryophyllene	Ali et al. (2018)
Development of an ethylene gas monitor for fruit ripeness monitoring application	Ripeness	AgBF4/Polyvinylpyrrolidone composite	Ethylene, headspace odor	Tolentino et al. (2018)
Detection of ethyl butyrate using cyclodextrin films for fruit fragrance detection	Ripeness	CDs	Ethyl butyrate, limonene	Sasaki et al. (2018)
Categorize beef into three quality classes (fresh, semifresh, and spoiled)	Spoilage	Metallo-porphyrins	Organic acids, alcohols, amines, sulfides, carbonyls	Papadopoulou et al. (2013)
Sensor development based on QCM coated with PANI/titanium dioxide (TiO_2). To detect three toxic gases	Spoilage	PANI/Titanium dioxide (TiO_2)	Ammonia, hydrogen sulfide, and trimethylamine	Cui et al. (2015)
QCM data from beef sample were trained by machine learning ensemble to categorize beef as fresh, semifresh, spoil	Spoilage	Polypyrrole polymers	Headspace odor	Mohareb et al. (2016)
Categorize beef into three quality classes (fresh, semifresh, and spoiled) multioutput fuzzy wavelet neural network	Spoilage	Modified metallo-porphyrins	Headspace odor	Kodogiannis (2017)

CDs, Cyclodextrins; MIP, molecularly imprinted polymer; *PAA*, polyacrylic acid; *PANI*, polyaniline; *PE*, polyethylene; *PFBBr*, pentafluorobenzyl bromide; *PVC*, polyvinyl chloride; *RTIL*, room temperature ionic liquid; *β-CD*, β-cyclodextrin.

cup-shaped cyclophanes with a space in between that can capture methane. The double-trimerization process for synthesizing cryptophane-A is as shown in Fig. 4.24. Cryptophane-A was dissolved in tetrahydrofuran and coated onto the QCM, achieving a methane sensing sensitivity of 0.103 Hz/ppm at 20°C (Sun et al., 2009).

Sulfur dioxide (SO_2) and nitrogen dioxide (NO_2), stemming from the combustion of fossil fuels, precipitate the formation of acid rain, smog, and respiratory ailments, rendering their

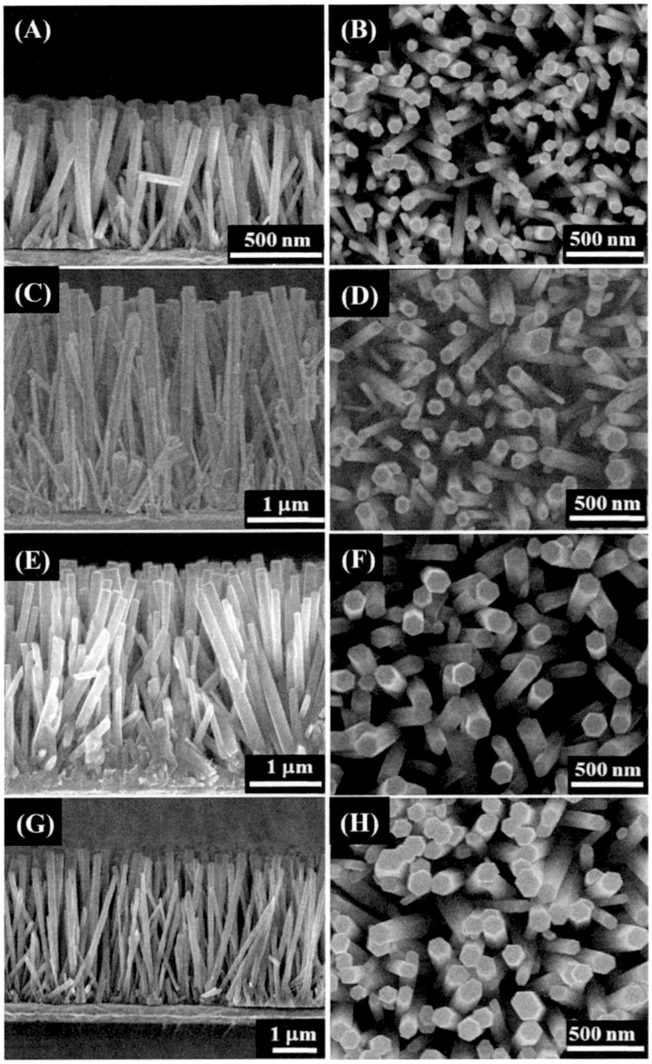

FIGURE 4.22 **FE-SEM images of the as-grown ZnO nanorods on the Au electrode of the QCM**. FE-SEM images of ZnO nanorods on QCM gold electrodes: Different growth stages from 1 to 4 h, showing top and side views of the structural evolution. *QCM*, Quartz crystal microbalance. (A–B) 1 hour: side view, top view. (C–D) 2 hours: side view, top view. (E–F) 3 hours: side view, top view. (G–H) 4 hours: side view, top view. Source: *From Minh, V. A., Anh Tuan, L., Quang Huy, T., Ngoc Hung, V., & Van Quy, N. (2013). Enhanced NH_3 gas sensing properties of a QCM sensor by increasing the length of vertically orientated ZnO nanorods.* Applied Surface Science, *265. https://doi.org/10.1016/j.apsusc.2012.11.028.*

monitoring an imperative component of air quality management. SO_2 can be absorbed well on the oxygenated-defective site of the nanotube surface. The researcher explored the use of multiwall carbon nanotubes (MWNTs) as SO_2 gas sensors (Hoang et al., 2019). The SEM image of MWNTs is shown in Fig. 4.25A. The further purification of the MWNTs caused the Raman spectroscopy to show a higher carbon defect site (higher peak at 1328 cm^{-1} compared to peak at 1582 cm^{-1} of the functional CNT) in Fig. 4.25B, which improved the gas sensing property. It can measure SO_2 down to 15 ppm at a 6.2 Hz frequency shift, compared to sensing NO_2 at around 4.5 Hz.

Due to its ease of coating and nonspecificity, QCM has been exclusively chosen by many materials science researchers as a sensor to

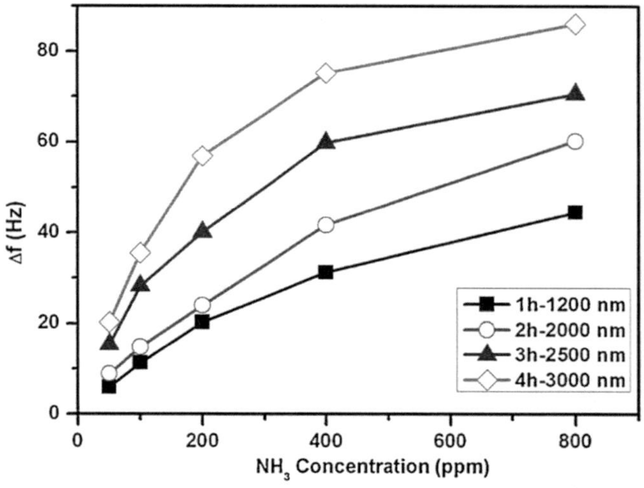

FIGURE 4.23 **The frequency shift of the QCM sensors based on ZnO nanorods.** Frequency shifts of ZnO nanorod-coated QCM at varying NH_3 concentrations for rod lengths grown over 1–4 h. *QCM*, Quartz crystal microbalance. Source: *From Minh, V. A., Anh Tuan, L., Quang Huy, T., Ngoc Hung, V., & Van Quy, N. (2013). Enhanced NH_3 gas sensing properties of a QCM sensor by increasing the length of vertically orientated ZnO nanorods.* Applied Surface Science, *265. https://doi.org/10.1016/j.apsusc.2012.11.028.*

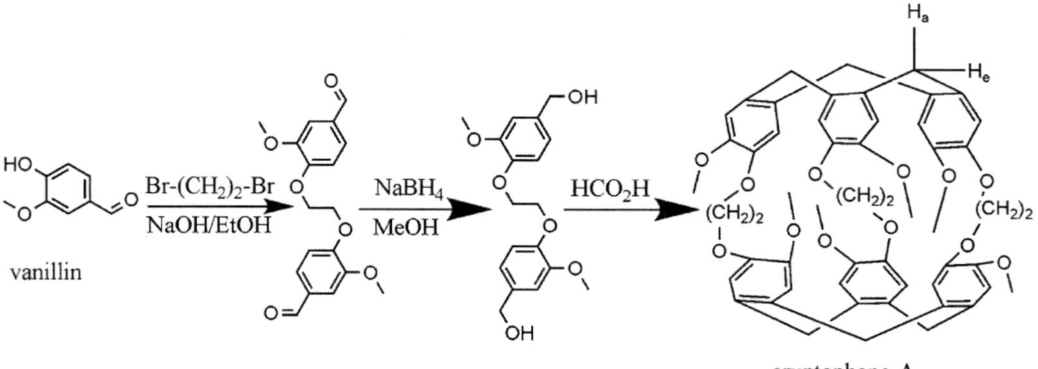

FIGURE 4.24 **Double-trimerization synthesis of cryptophane-A.** Synthesis of cryptophane-A: vanillin is first converted to bis-vanillin derivative via dibromoethane reaction, reduced to bis-vanillyl alcohol by $NaBH_4$, then cyclized into cryptophane-A using formic acid, and purified by silica gel chromatography. Source: *From Sun, P., Jiang, Y., Xie, G., Du, X., & Hu, J. (2009). A room temperature supramolecular-based quartz crystal microbalance (QCM) methane gas sensor.* Sensors and Actuators, B: Chemical, *141(1). https://doi.org/10.1016/j.snb.2009.06.012.*

demonstrate its material sensing properties. QCM is particularly versatile in detecting various gases prevalent in gas monitoring. Table 4.3 provides an overview of publications from the years 2010 to 2023.

4.4.2.2 Volatile organic compounds and other gases

For uncommon environmental gases, the applications of these sensors sometimes were not to detect the gases directly from the known source but to the application of quality control or toxic gas detection, which require the sensors to be able to detect a variety of the concerned gases, especially the family of VOCs. VOCs are a group of organic chemicals that can easily evaporate into the air at room temperature. They are called "volatile" because of their tendency to become vapor or gas. VOCs can be emitted from a variety of sources, both natural

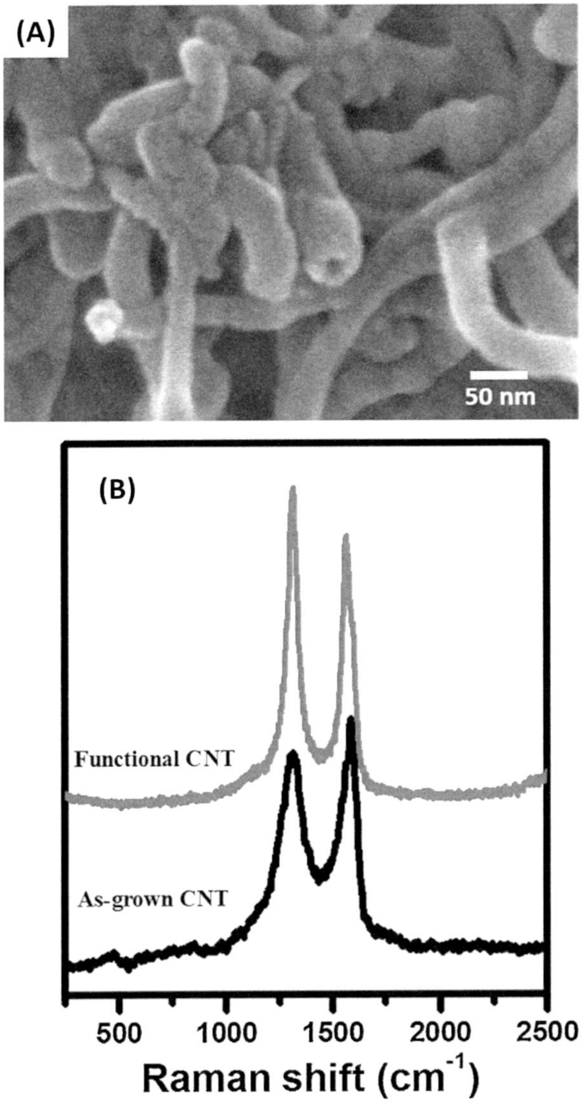

FIGURE 4.25 **FE-SEM image Raman spectroscopy of the functional MWNTs.** (A) presents an FE-SEM image of MWNTs on a gold electrode of a QCM. (B) displays Raman spectroscopic analysis of both as-grown and functionalized MWNTs. *MWNTs*, Multi-wall carbon nanotubes; *QCM*, quartz crystal microbalance. Source: *From Hoang, N. D., Van Cat, V., Hoai Nam, M., Ngoc Phan, V., Tuan Le, A., & Van Quy, N. (2019). Enhanced SO_2 sensing characteristics of multi-wall carbon nanotubes based mass-type sensor using two-step purification process.* Sensors and Actuators, A: Physical, *295. https://doi.org/10.1016/j.sna.2019.06.046.*

and human-made, and they can have various effects on the environment and human health.

A sensing system to detect various VOCs was developed by Speller et al. (2015). With the selection of ionic liquid, the salt that remains a stable liquid at room temperature without vaporing, the sensitivity of [OMIm][Br] and [OMIm][SCN] ionic liquids was tested with *n*-alcohols, chloromethanes, aromatic hydrocarbons, and nitriles. The author measured QCM change at multiple harmonics (1, 3, 5, 7, 9, and 11), making it six frequency measurements for a single QCM. The QCM response for each harmonic concept is shown in Fig. 4.26. The result showed that at different harmonics, the frequency responses are

TABLE 4.3 Overview application of electronic noses (E-nose) based on quartz crystal microbalance (QCM) sensor in common hazardous gases sensing.

Summary	Coating material	Target substance	Ref.
PANI/TiO_2-based gas sensor showed good sensitivity toward NH_3, H_2S, and trimethylamine	(PANI)/Titanium dioxide (TiO_2)	NH_3, H_2S, and trimethylamine	Cui et al. (2015)
The sensor based on graphene/PANI detection limit was improved compared to conventional PANI	Graphene/PANI	NH_3	Wu et al. (2013)
Length of ZnO nanorod has an effect on NH_3 gas sensing sensitivity	ZnO nanorods	NH_3	Minh et al. (2013)
QCM sensor coated with CA nanofibers with layer-by-layer deposition of polyethylenimine and GO for NH_3 sensing application	CA/PEI/GO nanofiber	NH_3	Jia et al. (2016)
Fabrication of a QCM sensor for CO detection using PEDOT:PSS/PVP composite nanofibers as coating material	PEDOT:PSS/PV nanofiber	CO	Zhang et al. (2016)
Study the adsorption property of ferrocene-chitosan with CO using a QCM sensor	Ferrocene chitosan	CO	Bayram et al. (2017)
Study of utilizing the novel material 1-*n*-butyl-2,3-dimethylimidazolium hexafluorophosphate as a chemical vapor sensor	*n*-Butyl-2,3-dimethylimidazolium hexafluorophosphate	CCl_4, $CHCl_3$, toluene, 1-propanol, acetone, ethanol, acetonitrile, methanol	Regmi et al. (2012)
NCD was growth onto QCM using microwave plasma CVD process. The sensitivity to both NH_3 and CO_2 were improved with the NCD coating	NCD	NH_3, CO_2	Varga et al. (2015)
The PEI film modified with starch showed an improvement of CO_2 sensing sensitivity	PEI	CO_2	Xie et al. (2015)
Synthesized COF-5 was used for CO_2 sensing application	COF-5	CO_2	Xu et al. (2023)
The polymer nanocomposites based on GNRs were coated on QCM and exhibit good sensitivity toward NH_3 gas	GNR	CO, NH_3, and N_2O	Trajcheva et al. (2021)
QCM sensors based on graphene composite materials for NO_2 detection	Composite of SnO_2, AgNPs, and graphene (GO)	NO_2	Qi et al. (2016)
Thin zinc oxide (ZnO) was deposited on QCM using the ALD technique for NO_2 sensing	ZnO	NO_2	Boyadjiev et al. (2016)
Polybutadiene-based QCM sensor showed high sensitivity toward O_3 sensing	Polybutadiene	O_3	Guillemot et al. (2019)
SnO_2-coated QCM using noise spectroscopy to extract the feature for O_3 gas sensing	SnO	O_3	Lim et al. (2017)

(*Continued*)

TABLE 4.3 (Continued)

Summary	Coating material	Target substance	Ref.
Polybutadiene-coated QCMs were used, and the study on various flow rates and concentrations of ozone gas was tested	Polybutadiene	O_3	Muller et al. (2011)
The UiO-MOF was applied to both QCM and fluorescent sensors, when exposing to O_3 the frequency change can be observed	UiO-MOF	O_3	Yang et al. (2020)
QCM showed good sensitivity, selectivity, and stable life for 13 days	Polyvinyl pyrrolidone	H_2S	Wu et al. (2014)
The sensitivity of β-Ga_2O_3/rGO material toward H_2S gas was evaluated by EIS technique	β-Ga_2O_3/rGO	H_2S	Balasubramani et al. (2020)
Multiwall carbon nanotubes with improved defect sites were deposited on QCM. The best response was toward SO_2 gas	MWNTs	SO_2	Hoang et al. (2019)
The comparison between polystyrene latex spheres and PSF as coating materials for SO_2 detection	Polystyrene latex spheres	SO_2	Hwang et al. (2013)
Exploration of three different coatings polypyrrole, PEDOT, and polystyrene were tested for its sensitivity for sensing SO_2 and NO_2	Polypyrrole, PEDOT, and polystyrene	SO_2, NO_2	Yang et al. (2011)
In order to fabricate SO_2 sensor gamma-Fe_2O_3 nanoparticles, three samples of different [Fe^{3+}] and [Fe^{2+}] precursors were fabricated	γ-Fe_2O_3 nanoparticles	SO_2	Thanh Vinh et al. (2021)

ALD, Atomic layer deposition; *CA*, cellulose acetate; *COF*, covalent organic frameworks; *GNR*, graphene nanoribbon; *GO*, graphene oxide; *MWNTs*, multiwall carbon nanotubes; *NCD*, nanocrystalline diamond; *PEDOT*, poly(3,4-ethylenedioxythiophene); *PEI*, polyehtyleneimine.

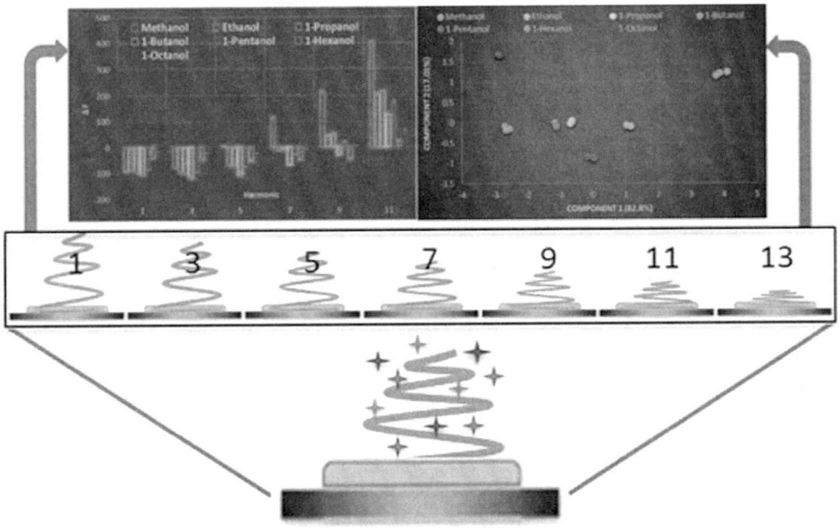

FIGURE 4.26 Diagram of QCM measurement using harmonics. Response to various alcohols: a bar graph displaying sensor responses and a PCA scatter plot showing data clusters, set above a QCM sensor operating at various harmonic frequencies. *PCA*, Principal component analysis; *QCM*, quartz crystal microbalance. Source: *From Speller, N. C., Siraj, N., Regmi, B. P., Marzoughi, H., Neal, C., & Warner, I. M. (2015). Rational design of QCM-D virtual sensor arrays based on film thickness, viscoelasticity, and harmonics for vapor discrimination.* Analytical Chemistry, *87(10). https://doi.org/10.1021/ac5046824.*

different. Applying the PCA and discriminant analysis (DA) techniques for 18 analytes at multiple concentrations for interclass discrimination, the authors achieved accuracies ranging from 97.6% to 99.4%.

One of the applications of VOC sensing found in the publication is children's toys contaminated with potentially harmful VOCs being detected. Oleneva et al. (2020) used an E-nose consisting of eight QCM sensors with different coatings, such as multilayer carbon nanotubes, propolis, and poly(ethylene glycol). A total of 75 toy samples were analyzed, and based on the presence of certain VOCs identified through GC/MS, they were classified as either "toxic" or "clean." The PCA plot is shown in Fig. 4.27. The analysis included the use of regularized discriminant analysis (RDA), which serves as a middle ground between LDA and quadratic discriminant analysis (QDA). While LDA operates under the assumption of a common covariance matrix for all classes, QDA assumes unique covariance matrices for each class. RDA thus offers a compromise, blending the simplicity of LDA with the flexibility of QDA. In the application of this method, along with SVM classifications, the study achieved accuracy rates of 88% and 96%, respectively. This high level of accuracy

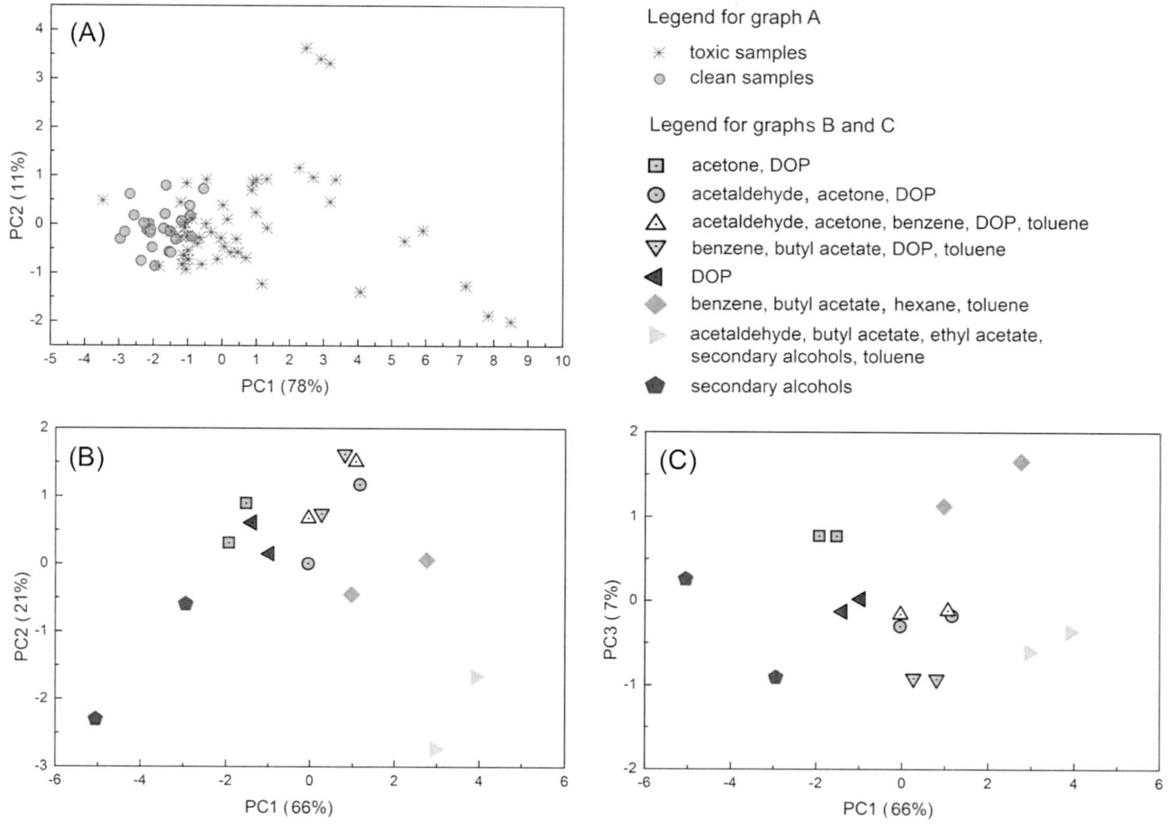

FIGURE 4.27 **PCA scores graph of contaminated toy samples.** (A) PCA scores graph for the entire electronic nose data set. (B and C) PCA score graphs for eight classes of toxic samples with different gas phase content. *PCA*, Principal component analysis. Source: *From Oleneva, E., Kuchmenko, T., Drozdova, E., Legin, A., & Kirsanov, D. (2020). Identification of plastic toys contaminated with volatile organic compounds using QCM gas sensor array.* Talanta, *211. https://doi.org/10.1016/j.talanta.2019.120701.*

TABLE 4.4 Overview applications of electronic noses (E-nose) based on quartz crystal microbalance (QCM) sensor of volatile organic compounds (VOCs) and other gases.

Summary	Coating material	Target substance	Ref.
Electronic nose based on eight QCM sensor array	MCNTs, propolis, propolis modified by $FeCl_3$, various poly(ethylene glycol), polyphenyl ether, trioctylphosphine oxide	VOCs	Oleneva et al. (2020)
Titanate sol–gel layers tested with different alcohols and hydrocarbon molecules	Titanate sol–gel layers imprinted with carbonic acids	Alcohols and engine oils	Latif et al. (2011)
QCM coated with 2 ionic liquids tested with 18 analytes using QDA	Ionic liquid: [OMIm][Br] and [OMIm][SCN]	VOCs	Speller et al. (2015)
Functionalize TiO_2 nanofibers by PEI were deposited on QCM for formaldehyde gas sensing	TiO_2 fibers	Formaldehyde	Wang et al. (2012)
QCM coated with modified MCNT was found to have high sensitivity toward various amines	MCNT modified with poly(diehylene glycol succinate) in acetone and $ZrO(NO_3)_2 \cdot 2\,H_2O$ (Azr)	Alkylamines	Kuchmenko & Umarkhanov (2013)
Exploring the possibility of using polyaniline coating for detecting formic acid gas	Polyaniline	Formic acid gas	Yan et al. (2012)
Complex sensing material of copper and DDS was tested as a formaldehyde sensor	DDS and copper complex	Formaldehyde	Wang et al. (2017)
A wireless QCM working at four harmonics coated with polyvinylpyrrolidone for detecting various VOC gases	Polyvinylpyrrolidone	VOCs	Su et al. (2020)

DDS, 4,4′-Diaminodiphenyl sulfone; *IL*, ionic liquid; *MCNTs*, multilayer carbon nanotubes; *PEI*, polyethyleneimine; *PSF*, polystyrene film; *QDA*, quadratic discriminant analysis; *rGO*, reduced graphene oxide.

demonstrates the effectiveness of combining these analytical techniques in identifying potentially hazardous substances in children's toys.

There are more publications of VOCs and other gases sensing that are shown in Table 4.4.

4.4.2.3 Humidity

This section was very popular among researchers. Humidity sensing has vast applications across many industries. In agriculture, humidity sensing is crucial for monitoring and controlling the moisture levels in soil, air, and plant environments. This information helps farmers optimize irrigation, prevent diseases, and enhance crop yields. In healthcare settings, humidity sensors can be essential for monitoring and controlling the environmental conditions in hospitals, laboratories, and storage areas for sensitive medications and equipment.

ZnO was chosen to be a sensing material for water. The oxygen atom of water is attracted to the positively charged zinc ions (Zn^{2+}) on the ZnO surface. And forming a surface hydroxyl group (OH^-) on the ZnO surface. Ates et al. explore the use of ZnO powder using the sol–gel method. ZnO was doped with Sn with various atomic ratios of 0.1%, 1%, and 2%. The solution of ZnO nanopowder was drop coated onto QCM. Its frequency shift Sn-doped ZnO samples at a ratio of 0.1%, 1%, and 2% were 1980, 1310, 1120, and 1250 Hz, respectively, at the relative humidity of 68%–70%.

Graphene oxide (GO) is a focus of research due to its unique properties when applied to QCM sensors. GO is a type of functionalized graphene with C = O and C–OH groups on its carbon sheets and carboxyl groups at the edges, making it highly hydrophilic. Researchers created three GO-coated QCM samples with varying thicknesses using the spin-coating technique. These sensors showed a maximum sensitivity of 22.1 Hz/% RH and a linear frequency response in a broad relative humidity range of 6.4%–93.5% RH, with a high regression coefficient of 0.9718, demonstrating the potential of GO in enhancing QCM sensor performance in humidity detection.

In summary, humidity sensing plays a vital role in various industries and applications. Accurate and timely humidity measurements are essential for making informed decisions and taking appropriate actions in a wide range of fields. The summary table of humidity sensing is in Table 4.5.

4.4.3 Healthcare

4.4.3.1 *Breath analysis*

The exhaled breath of humans contains more than 3000 VOCs that are associated with internal bodily processes. E-noses have the potential to analyze breath patterns and serve as a noninvasive method for screening and analyzing respiratory and systemic diseases. By utilizing a sensor array and artificial neural network, E-noses can detect specific VOC patterns, offering valuable insights for disease monitoring. Within this section, we elucidate the utilization of breath analysis for nonalcohol-related applications.

Di Natale et al. (2003) explored the use of an E-nose for diagnosing lung cancer based on the volatile compounds present in exhaled air. The researchers collected breath samples from a total of 60 individuals, including 35 lung cancer patients, 18 healthy individuals for reference, and 9 individuals after surgical therapy. They coated a QCM sensor array with various metalloporphyrins, chosen for their similarity to the active site of the olfactory receptor protein. This coating exhibited good sensitivity to previously identified potential breath markers for lung cancer. By employing a technique called "partial least squares-discriminant analysis" (PLS-DA), the E-nose achieved a 100% accuracy rate in classifying lung cancer patients and accurately classified 94% of the reference samples, the PLS-DA plot of the patients as shown in Fig. 4.28. Among the post-surgery patients, the E-nose correctly identified 44% of the cases, with the remaining samples classified as healthy references. The altered breath composition caused by the presence of lung cancer was sufficient to accurately identify affected individuals.

There was a study conducted to explore the detection of blood glucose and HbA1c levels by analyzing exhaled gas (Saraoglu et al., 2013). In diabetes patients, there is a change in the acetone content in breath. However, since the concentration of acetone in exhaled breath is very low, ranging from 0.1 to 10 ppm, a preconcentrator is required. Zeolite material was used to absorb acetone molecules from breath. The frequency change results from a QCM were then compared against glucose and HbA1c levels. The study employed a radial basis function neural network and achieved prediction accuracies of 83.03% for HbA1c parameters and 74.76% for glucose parameters.

4.4.3.2 *Alcohol*

Sensing alcohol has been a significant focus in various fields, including industry, breath analysis, and food spoilage detection. Several

TABLE 4.5 Overview application of electronic noses (E-nose) based on quartz crystal microbalance (QCM) sensor for humidity sensing.

Summary	Coating material	Ref.
ZnO doped with Sn prepared with the sol–gel method for humidity sensing	Sn-doped ZnO nanopowder	Ates et al. (2013)
Drop-casted ZnO nanoparticles on QCM respond sharply to humidity	ZnO nanoparticles	Erol et al. (2010)
QCM was spin coated with graphene oxide, and it retains good linearity in the range of 6.4%–93.5% RH with R^2 of 0.9718	GO	Yao et al. (2011)
Diamine-decorated graphene oxide improved the wettability of the sensor; its sensitivity was greatly improved while retaining high *Q* factor	Amine GO film	Ho and Wu (2020)
Self-assembled PDDAC and GO nanocomposite films their measuring range is 0%–97% RH	PDDAC and GO	Ren et al. (2018)
Complex composite of graphene oxide/tin dioxide/polyaniline (GO/SnO_2/PANI) was spray-coated on QCM; it archived a sensitivity of 29.1 Hz/%RH with a response time of 7 s	Graphene oxide/tin dioxide/polyaniline (GO/SnO_2/PANI)	Zhang et al. (2018)
A novel metal organic framework of Cu^I/Cu^{II} was synthesized and tested with alcohol, acetone, dichloromethane, and water, and it showed great affinity toward water vapor	Metal–Organic framework: $\{[Cu^{II}(oda)_2]_3[Cu^I(4,4'\text{-bipy})]_6 \cdot 2DMF\}_n$ $(1 \cdot 2DMF)$	Jiao et al. (2016)
A drop-coated QCM sensor with C_{60}/GO was fabricated. The sensor showed that the response is nonlinear; as humidity becomes larger, the frequency drastically changed.	Fullerene (C_{60})/GO	Ding et al. (2018)
Drop-coated Sb/WO_3 ink on QCM annealed at 200 C showed the highest response to humidity. The application of respiration monitoring in face masks is also explained here	Sb-doped WO_3 nanocrystals	Li et al. (2017, 2022)
QCM coated with mesoporous silica SBA-15 showed high sensitivity and stability for sensing water vapor	Hexagonal lamelliform monodisperse mesoporous silica SBA-15	Zhu et al. (2010)
Nafion/MoS_2-coated QCM exhibits high sensitivity and small hysteresis when varying the humidity from 11.3% to 97.3%	Nafion/MoS_2	Yu et al. (2022)

GO, Graphene oxide; *PANI*, polyaniline; *PDDAC*, poly(dimethyl diallyl ammonium chloride).

coating materials have been developed and applied for gas sensing purposes. We have chosen a selection of several publications that are particularly interesting in the field of QCM alcohol sensors.

A material called mesoporous silica called KIT-6 was used as a coating material for a variety of VOCs (Ayad et al., 2019). A mesoporous material refers to a nanoporous substance that encompasses pores ranging in diameter from 2 to 50 nm. KIT-6 was studied extensively as a promising adsorbent. The author was interested in developing a gas sensor based on the adsorbing properties of KIT-6 material. The coating procedure was a simple drop-coating method. The results showed that sensor sensitivity is in the order of butanol > propanol > ethanol > methanol, where the sensitivities are 33.25, 24, 9.9, and 5 Hz/mg/L, respectively. Further study for the different coating materials was conducted. By applying the

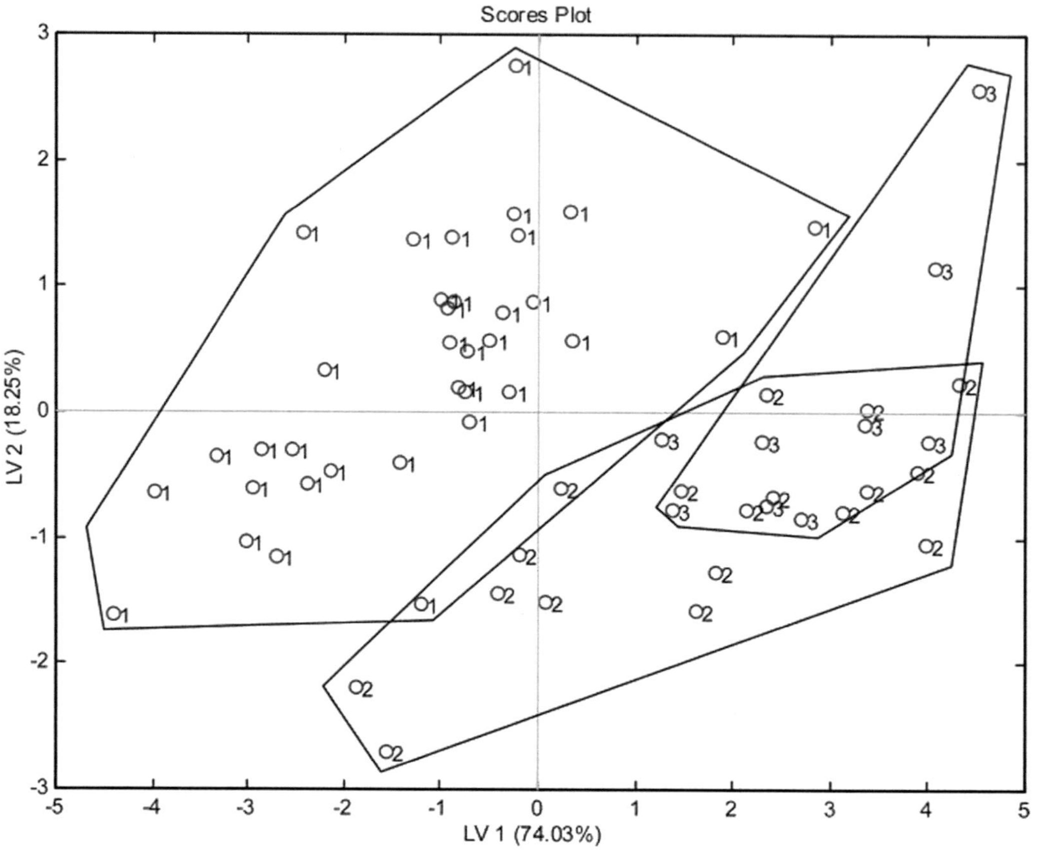

FIGURE 4.28 **Plot of the first two latent variables of the PLS-DA model built to classify the lung cancer data.** Data are labeled (O_1) lung cancer disease, (O_2) reference population, and (O_3) postsurgery samples. *DA*, Discriminant analysis. Source: *From Saevels, S., Lammertyn, J., Berna, A. Z., Veraverbeke, E. A., Natale, C. D., & Nicolai, B. M. (2003). Electronic nose as a non-destructive tool to evaluate the optimal harvest date of apples.* Postharvest Biology and Technology, *30(1). https://doi.org/10.1016/S0925-5214(03)00059-0.*

drop-casting method of colloidal polypyrrole (PPy)/poly(N-vinylpyrrolidone) nanorods/nanotubes onto the QCM (Torad et al., 2021). It was found by Torad et al. that the sensor demonstrated rapid response and effective sensitivity toward ethanol, 1-propanol, and 1-butanol with the sensitivity of 0.50, 0.52, and 0.57 Hz/mg/L, respectively.

The COVID-19 pandemic, which has affected the entire world and caused more than 6 million deaths worldwide, has increased interest in rapid detection methods for COVID-19. Török et al. (2022) developed a breathalyzer based on a QCM sensor with an ethanal-sensitive membrane using Methyl Red, Methyl Red sodium salt, and NaOH. It was observed that infected patients exhibited simultaneous elevations in the levels of ethanal (acetaldehyde) and acetone in their breath. The membrane sensor was capable of detecting a 50 Hz deviation when exposed to 1 ppm of ethanal, resulting in a sensitivity of 0.02 ppm/Hz. The resonance frequency of a healthy individual's breath (negative for COVID-19) remained unchanged.

The figure shows two images of a 3D-printed device: the left image displays an internal view with electronic components and wiring, featuring what appears to be a sensor or a lens at the front; the right image provides a top-down view of the device's interior, revealing a circuit board, a breadboard with electronic components, and additional wiring, possibly indicating a prototype or custom-built electronic instrument.

4.4.3.3 *Drugs and narcotic*

E-noses provide numerous benefits for the detection of drugs and narcotics. They possess the ability to recognize a broad spectrum of substances, encompassing illegal drugs, pharmaceuticals, and precursor chemicals, which renders them very useful to diverse environments like law enforcement agencies, customs checkpoints, and forensic laboratories. Also, E-noses offer real-time outcomes, enabling prompt decision-making and immediate response, especially in time-sensitive situations like border control or drug raids. The rapidity they provide is vital for ensuring timely actions.

Most QCM biosensors operate in the liquid phase, one reason being the convenience in preparing the solutions and the benefit of no humidity interfering with the measurement. The E-nose application is more challenging due to its gas phase measurement. There was an attempt to overcome this issue by Frisk et al. (2006). He fabricated a micromachined system to adopt the flowing channel of liquid as an adsorption site for narcotic vapor. Their micromachined cylindrical pillar designs allow vapor or aerosol to be trapped more easily due to their liquid–air interface, which forms a condensation site. The sample sheet of 100 ng heroin and 50 ng cocaine was prepared. Their design responses to heroin and cocaine with a frequency shift of 50 and 15 Hz, respectively.

Cali and Persaud (2020) modified OBPs from *Anopheles gambiae* (OBP1 and OBP47). OBP1 was adapted and mutated to have high affinity to cannabinol, 3,4-methylenedioxy methamphetamine (Ecstasy), and cocaine hydrochloride. Self-assembled monolayer techniques were employed to immobilize proteins onto 20 MHz QCMs. The saturated analytical grade cocaine hydrochloride vapor was exposed to their OBP-modified QCM sensor (AgamOBP1-S82P, AgamOBP1-K93H) compared with wild-type OPB (WTAgamOBP1). The result showed that it has good specificity toward cocaine HCl vapor when compared to 2-phenylethanol. The stability test was conducted for over 10 months without signs of degradation.

Application of QCM sensors in healthcare may not be as ubiquitous as industrial quality control and environmental monitoring. This may be due to in medical and healthcare fields not being aware of the potential of QCM and E-nose technologies in medical applications, or it may be due to its interdisciplinary nature making it challenging to publish in traditional medical or healthcare journals, leading to fewer healthcare-specific publications. In Table 4.6, we show the extensive application of using QCM in healthcare.

4.4.4 Explosive detection

There was a handful of research about using E-nose for explosive detection. E-noses do not require physical contact with objects or individuals, making them a noninvasive detection method. This is particularly useful in scenarios where manual inspection may be challenging or dangerous. While E-noses are not typically used as the primary method for explosive detection in high-security settings like airports or military installations, they can play a complementary role in certain applications.

A reliable E-nose technology for detecting explosives is challenging. To be able to detect the residue of the small trace amount of explosive vapor is difficult due to the extensively low vapor pressures of explosives at ambient temperature, especially the prevalent explosives

TABLE 4.6 Overview application of electronic noses (E-nose) based on quartz crystal microbalance (QCM) sensor in Healthcare.

Application	Category	Coating material	Target substance	Ref.
Discrimination between lung cancer and healthy individuals using array of metalloporphyrin-coated QCM sensor	Breath analysis	Metalloporphyrins	Alkanes and benzene derivatives	Di Natale et al. (2003)
Discrimination between lung cancer and other lung diseases using array of metalloporphyrin-coated QCM sensor	Breath analysis	Metalloporphyrins	Aniline, *O*-toluidine, and cyclopentane	D'Amico et al. (2010)
Prediction of blood glucose and HbA1c levels by analyzing acetone concentrations in exhaled breath using an RBFNN	Breath analysis	Zeolites, fullerene (C_{60}), chiral materials, polypyrrole, carbon graphites, ITO films, and oligonucleotides	Acetone	Saraoglu et al. (2013)
Electronic nose and GC−MS analyses were performed on the breath of lung cancer patients. The PSL-DA models were used to compare performance between E-nose and GC−MS	Breath analysis	Metalloporphyrins	VOCs	Capuano et al. (2014)
The study also explored skin analysis using a quartz crystal microbalance sensor system for different age groups in diagnosing lung cancer	Breath analysis	MWCNT1, Zr1, DCH-18C6, GA1, and PEGS	VOCs	Gashimova et al. (2022)
Development of a 3D calcium spirobifluorene metal−organic framework for VOC sensing	Alcohol	Metal−organic framework: Ca-SBF	Toluene, ethanol, acetone, benzene	Fang et al. (2018)
Electrospun polyvinyl acetate nanofiber−modified quartz crystal microbalance for the detection of primary alcohol vapor	Alcohol	Electrospun PVA	*n*-Pentanol, *n*-butanol, *n*-propanol, ethanol, methanol	Rianjanu et al. (2020)
QCM sensor coated with colloidal PPy/PVP for detecting alcohol vapor	alcohol	PPy/PVP	Ethanol, 1-propanol, 1-butanol	Torad et al. (2021)
Micromachined microfluidic system to capture airborne sample to liquid and flow to the QCM sensing area	Drugs and narcotic	Antibody, antigen	Cocaine, heroine	Frisk et al. (2006)
Odorant binding proteins from *Anopheles gambiae* (OBP1 and OBP47) immobilized on QCM for detecting drug	Drugs and narcotic	OBP1, OBP47	3,4-Methylenedioxy methamphetamine, cocaine hydrochloride	Cali and Persaud (2020)
Exploring five different structures of phenylene-bridged azaacene molecules as materials for sensing precursors of narcotics	Drugs and narcotic	Phenylene-bridged, TIPS-alkynylated azaacenes	Henylacetone, safrole, and GBL γ-phenylacetone, safrole, and GBL	Prantl et al. (2022)

DA, Discriminant analysis; *GBL*, γ-butyrolactone; *GC−MS*, gas chromatography−mass spectrometry; *OBP*, odorant binding proteins; PPy, polypyrrole; PVP, poly(*N*-vinylpyrrolidone); *RBFNN*, radial basis function neural network; *VOC*, volatile organic compound.

belonging to nitroexplosive classes, such as TNT, RDX, and PETN. Eslami and Alizadeh (2019) developed a QCM sensor modified with PPy doped with bromophenol blue (BPB) in nanosphere and nanorod form (Fig. 4.29). It can detect various nitroexplosive compounds, such as RDX, TNT, PETN, and HMX, with the highest sensitivity toward TNT in its PPy-BPB compound, as shown in Fig. 4.30.

There are only a few publications discussing the use of QCM-based E-nose for explosive detection. The main challenge in this field is the low concentration of the target material, and QCM may not have enough sensitivity to detect the target vapor without the use of a preconcentrator. Additionally, explosives can produce a wide range of VOCs, making it challenging to distinguish between harmless odors and potentially dangerous substances. We have compiled the applications of using QCM for explosive detection in Table 4.7.

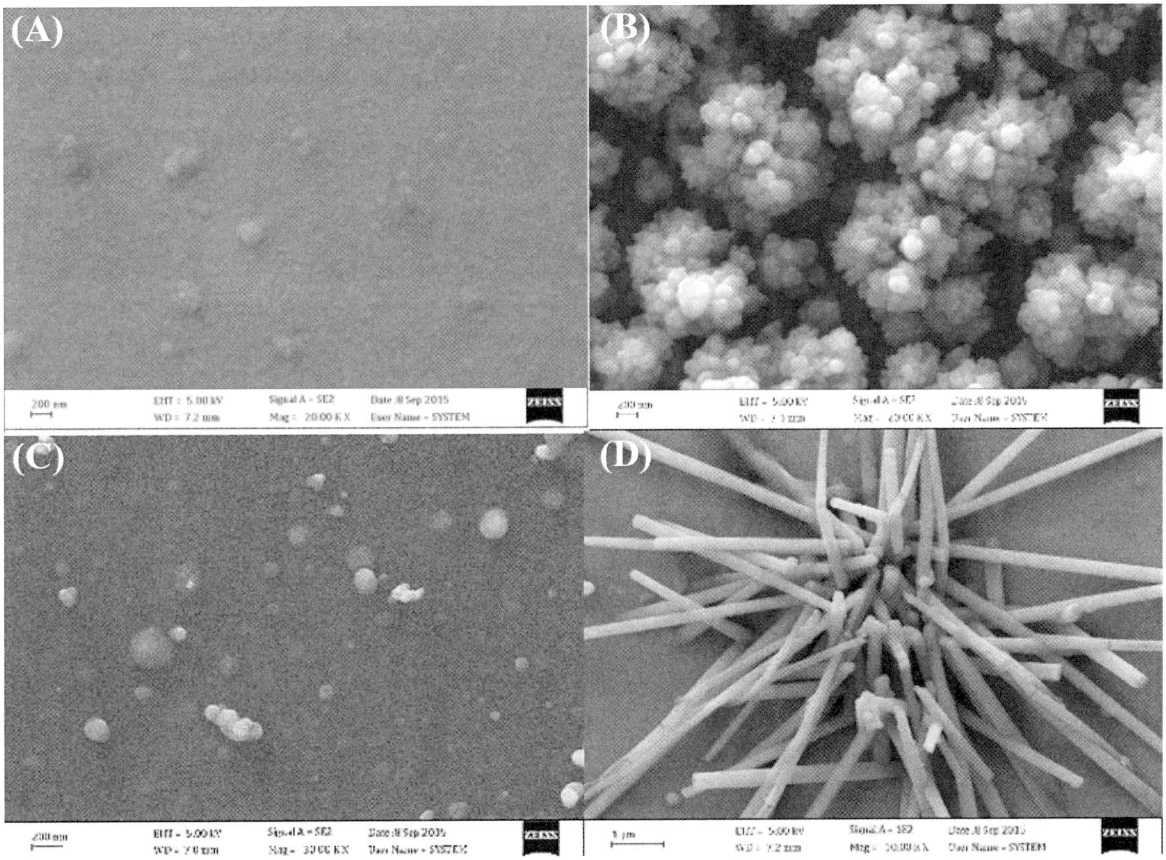

FIGURE 4.29 FE-SEM images of prepared PPy-BPB film deposited on QCM electrode with different conditions. (A) Concentrations of BPB (0.0001 mol/L) after 3 synthesis cycles; (B) 30 synthesis cycles; (C) concentrations of BPB (0.005 mol/L) after 3 synthesis cycles; (D) 30 synthesis cycles. *BPB*, Bromophenol blue; *PPy*, polypyrrole; *QCM*, quartz crystal microbalance. Source: *From Reza Eslami, M., & Alizadeh, N. (2019). Ultrasensitive and selective QCM sensor for detection of trace amounts of nitroexplosive vapors in ambient air based on polypyrrole—Bromophenol blue nanostructure.* Sensors and Actuators, B: Chemical, *278. https://doi.org/10.1016/j.snb.2018.09.060.*

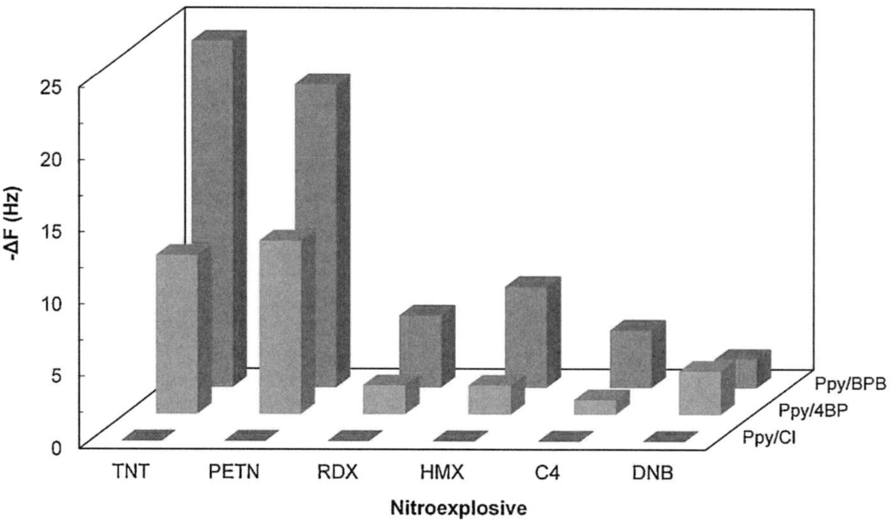

FIGURE 4.30 **Effect of dopant type on sensor response for 10 ppb of various nitroexplosives.** Comparative sensor responses—the QCM sensor with Cl-doped PPy showed no response to various nitroexplosive vapors, while the 4 BP-doped PPy sensor exhibited notable sensitivity, especially to TNT. *PPy*, Polypyrrole; *QCM*, quartz crystal microbalance. Source: *From Reza Eslami, M., & Alizadeh, N. (2019). Ultrasensitive and selective QCM sensor for detection of trace amounts of nitroexplosive vapors in ambient air based on polypyrrole—Bromophenol blue nanostructure.* Sensors and Actuators, B: Chemical, *278. https://doi.org/10.1016/j.snb.2018.09.060.*

TABLE 4.7 Overview application of electronic noses (E-nose) based on quartz crystal microbalance (QCM) sensor in Explosive detection.

Application	Coating material	Target substance	Ref.
TiO_2 nanostructures for the detection of NO_2 using QCM sensor	TiO_2 nanostructures	Nitrogen oxides (NO_x)	Procek et al. (2015)
PPy-BPB in nanosphere and nanorod forms on a QCM gold electrode for detecting various nitroexplosives	PPy-BPB	TNT, RDX, HMX	Eslami and Alizadeh (2019)
Development of a QCM-based TNT sensor using a self-assembled monolayer of OEG-alkyl thiols and TNT-analog	SAM of TNT-analog molecules	TNT	Larsson et al. (2006)
PEGM bonded with peptide that is specific to TNT coated on QCM	TNT peptide receptor/PEGM	TNT, DNT	Cerruti et al. (2009)
Exploring the sensitivity of phenylene dendrimer as a sensing material for detecting TATP	Phenylene dendrimer	TATP	Lubczyk et al. (2010)
EQCM sensor fabrication uses IL $BMIBF_4$ as electrolyte and sorption solvent for detecting ENB	IL $BMIBF_4$	ENB, dinitrotoluene	Yu et al. (2009)

ENB, Ethyl nitrobenzene; *IL*, ionic liquid; *OEG*, oligo(ethylene glycol); *PEGM*, poly(ethylene-co-glycidyl methacrylate); *PPy-BPB*, polypyrrole-bromophenol blue; *TATP*, triacetone triperoxide.

4.5 Summary

In this book chapter, we have delved into the fascinating realms of QCM sensors and E-Nose technology, uncovering their profound impact on sensing, diagnostics, and various applications. We began by explaining the basic principles, physical structures, and the sensor's behavior. Additionally, we derived its equivalent circuit, a critical aspect for utilizing QCM as a sensor, as it enables the design of measurement techniques based on the electrical circuit of the QCM.

Furthermore, we showcased various fields of applications, demonstrating how QCM can function as a single gas sensor or as part of a sensor array to construct an E-Nose system. E-Noses are equipped with sensor arrays and pattern recognition algorithms, enabling the detection and identification of odors and volatile compounds. Their applications span a wide range, including food quality control, environmental monitoring, medical diagnostics, industrial processes, and even security and defense.

In conclusion, QCM sensors and E-nose technology are not just scientific curiosities but practical tools that can transform the way we perceive and interact with our environment. Their versatility, sensitivity, and potential to provide real-time data make them indispensable in fields ranging from healthcare to industry and beyond.

4.6 Artificial Intelligence disclosure

During the preparation of this work, the author(s) used ChatGPT 4 to only assist the author to improve readability and clean up grammar. After using this tool/service, the author(s) reviewed and edited the content as needed and took full responsibility for the content of the publication.

References

Addabbo, T., Bertocci, F., Fort, A., Mugnaini, M., Shahin, L., Vignoli, V., & Rocchia, S. (2014). A DDS-based multiharmonic frequency meter for QCM sensor applications. *Procedia Engineering, 87*. Available from https://doi.org/10.1016/j.proeng.2014.11.664.

Ali, S. B., Ghatak, B., Debabhuti, N., Sharma, P., Ghosh, A., Tudu, B., Bhattacharya, N., & Bandyopadhyay, R. (2018). Detection of β-caryophyllene in mango using a quartz crystal microbalance sensor. *Sensors and Actuators, B: Chemical, 255*. Available from https://doi.org/10.1016/j.snb.2017.09.131.

Ates, T., Tatar, C., & Yakuphanoglu, F. (2013). Preparation of semiconductor ZnO powders by sol-gel method: Humidity sensors. *Sensors and Actuators, A: Physical, 190*. Available from https://doi.org/10.1016/j.sna.2012.11.031.

Ayad, M. M., Torad, N. L., Minisy, I. M., Izriq, R., & Ebeid, E. Z. M. (2019). A wide range sensor of a 3D mesoporous silica coated QCM electrodes for the detection of volatile organic compounds. *Journal of Porous Materials, 26*(6), 1731–1741. Available from https://doi.org/10.1007/s10934-019-00765-3.

Balasubramani, V., Nowshath Ahamed, A., Chandraleka, S., Krishna Kumar, K., Kuppusamy, M. R., & Sridhar, T. M. (2020). Highly sensitive and selective H_2S gas sensor fabricated with β-Ga_2O_3/rGO. *ECS Journal of Solid State Science and Technology, 9*(5). Available from https://doi.org/10.1149/2162-8777/ab9a18.

Bayram, A., Özbek, C., Şenel, M., & Okur, S. (2017). CO gas sorption properties of ferrocene branched chitosan derivatives. *Sensors and Actuators, B: Chemical, 241*. Available from https://doi.org/10.1016/j.snb.2016.08.175.

Boyadjiev, S. I., Georgieva, V., Yordanov, R., Raicheva, Z., & Szilágyi, I. M. (2016). Preparation and characterization of ALD deposited ZnO thin films studied for gas sensors. *Applied Surface Science, 387*. Available from https://doi.org/10.1016/j.apsusc.2016.06.007.

Cali, K., & Persaud, K. C. (2020). Modification of an *Anopheles gambiae* odorant binding protein to create an array of chemical sensors for detection of drugs. *Scientific Reports, 10*(1). Available from https://doi.org/10.1038/s41598-020-60824-7.

Capuano, R., Martinelli, E., Ghezzi, S., Paolesse, R., Di Natale, C., D'Amico, A., Santonico, M., & Pennazza G. (2014). An investigation about the origin of the lung cancer signalling VOCs in breath. Proceedings of IEEE Sensors, Valencia, Spain. 10.1109/ICSENS.2014.6985267.

Cerruti, M., Jaworski, J., Raorane, D., Zueger, C., Varadarajan, J., Carraro, C., Wuk Lee, S., Maboudian, R., & Majumdar, A. (2009). Polymer-oligopeptide composite coating for selective detection of explosives in water. *Analytical Chemistry*, *81*(11). Available from https://doi.org/10.1021/ac8019174.

Cui, S., Wang, J., & Wang, X. (2015). Fabrication and design of a toxic gas sensor based on polyaniline/titanium dioxide nanocomposite film by layer-by-layer self-assembly. *RSC Advances*, *5*(72). Available from https://doi.org/10.1039/c5ra06388d.

D'Amico, A., Pennazza, G., Santonico, M., Martinelli, E., Roscioni, C., Galluccio, G., & Paolesse, R. (2010). C. Di Natale, An investigation on electronic nose diagnosis of lung cancer. *Lung Cancer (Amsterdam, Netherlands)*, *68*(2), 170–176.

Debabhuti, N., Ghatak, B., Parasrampuria, V., Md Rafiqul, S., Agarwal Hassan, P., Dutta Gupta S., Sharma P., Tudu, B., Bandyopadhyay, R., Ali, S.B., & Bhattacharyya, N. (2017). Development of a QCM sensor for detection of trans-2-hexenal in tomatoes. International Conference on Intelligent Control, Power and Instrumentation, ICICPI 2016. 10.1109/ICICPI.2016.7859681

Debabhuti, N., Mukherjee, S., Neogi, S., Sharma, P., Hossain Sk, U., Maiti, S., Poddar Sarkar, M., Tudu, B., Bhattacharyya, N., & Bandyopadhyay, R. (2022). A study of vegetable oil modified QCM sensor to detect β-pinene in Indian cardamom. *Talanta*, *236*. Available from https://doi.org/10.1016/j.talanta.2021.122837.

Di Natale, C., Macagnano, A., Martinelli, E., Paolesse, R., D'Arcangelo, G., Roscioni, C., Finazzi-Agrò, A., & D'Amico, Ò. , A. (2003). Lung cancer identification by the analysis of breath by means of an array of non-selective gas sensors. *Biosensors and Bioelectronics*, *18*(10). Available from https://doi.org/10.1016/S0956-5663(03)00086-1.

Ding, X., Chen, X., Chen, X., Zhao, X., & Li, N. (2018). A QCM humidity sensor based on fullerene/graphene oxide nanocomposites with high quality factor. *Sensors and Actuators, B: Chemical*, *266*. Available from https://doi.org/10.1016/j.snb.2018.03.143.

Ema, K., Yokoyama, M., Nakamoto, T., & Moriizumi, T. (1989). Odour-sensing system using a quartz-resonator sensor array and neural-network pattern recognition. *Sensors and Actuators*, *18*(3–4). Available from https://doi.org/10.1016/0250-6874(89)87036-2.

Erol, A., Okur, S., Comba, B., Mermer, Ö., & Arikan, M. Ç. (2010). Humidity sensing properties of ZnO nanoparticles synthesized by sol-gel process. *Sensors and Actuators, B: Chemical*, *145*(1). Available from https://doi.org/10.1016/j.snb.2009.11.051.

Escuderos, M. E., Sánchez, S., & Jiménez, A. (2010). Virgin olive oil sensory evaluation by an artificial olfactory system, based on Quartz Crystal Microbalance (QCM) sensors. *Sensors and Actuators, B: Chemical*, *147*(1). Available from https://doi.org/10.1016/j.snb.2010.01.071.

Fang, X., Wang, L., He, X., Xu, J., & Duan, Z. (2018). A 3D calcium spirobifluorene metal-organic framework: Single-crystal-to-single-crystal transformation and toluene detection by a quartz crystal microbalance sensor. *Inorganic Chemistry*, *57*(4). Available from https://doi.org/10.1021/acs.inorgchem.7b02671.

Ferrari, M., & Ferrari, V. (2009). An oscillator circuit for dual-harmonic tracking of frequency and resistance in quartz resonator sensors. *Measurement Science and Technology*, *20*(12). Available from https://doi.org/10.1088/0957-0233/20/12/124005.

Frisk, T., Rönnholm, D., Van Der Wijngaart, W., & Stemme, G. (2006). A micromachined interface for airborne sample-to-liquid transfer and its application in a biosensor system. *Lab on a Chip*, *6*(12). Available from https://doi.org/10.1039/b612526n.

Gashimova, E., Temerdashev, A., Porkhanov, V., Polyakov, I., Perunov, D., & Dmitrieva, E. (2022). Non-invasive exhaled breath and skin analysis to diagnose lung cancer: Study of age effect on diagnostic accuracy. *ACS Omega*, *7*(46). Available from https://doi.org/10.1021/acsomega.2c06132.

Ghatak, B., Ali, S. B., Prasad, A., Ghosh, A., Sharma, P., Tudu, B., Pramanik, P., & Bandyopadhyay, R. (2018). Application of polymethacrylic acid imprinted quartz crystal microbalance sensor for detection of 3-carene in mango. *IEEE Sensors Journal*, *18*(7). Available from https://doi.org/10.1109/JSEN.2018.2794607.

Guha, A., Sandström, N., Ostanin, V. P., van der Wijngaart, W., Klenerman, D., & Ghosh, S. K. (2019). Simple and ultrafast resonance frequency and dissipation shift measurements using a fixed frequency drive. *Sensors and Actuators, B: Chemical*, *281*. Available from https://doi.org/10.1016/j.snb.2018.11.052.

Guillemot, M., Ravera, C., & Castel, B. (2019). Development of quartz crystal microbalance based sensor for real-Time ozone monitoring. ISOEN 2019 – 18th International Symposium on Olfaction and Electronic Nose, Proceedings, 10.1109/ISOEN.2019.8823424.

Hardoyono, F., & Windhani, K. (2021). Identification and detection of bioactive compounds in turmeric (*Curcuma longa* L.) using a gas sensor array based on molecularly imprinted polymer quartz crystal microbalance. *New Journal of Chemistry*, *45*(38). Available from https://doi.org/10.1039/d1nj03640h.

Hardoyono, F., Windhani, K., Sambodo, H., & Pudjianto, H. (2019). Identification of bioactive compounds in ginger based on molecularly imprinted polymer quartz crystal microbalance gas sensor. *IOP Conference Series: Materials Science and Engineering*. Available from https://doi.org/10.1088/1757-899X/546/3/032012.

Hayden, O., Lieberzeit, P. A., Blaas, D., & Dickert, F. L. (2006). Artificial antibodies for bioanalyte detection –

Sensing viruses and proteins. *Advanced Functional Materials*, *16*(10). Available from https://doi.org/10.1002/adfm.200500626.

Ho, C. Y., & Wu, Y. S. (2020). Diamine decorated graphene oxide film on quartz crystal microbalance for humidity-sensing analysis. *Applied Surface Science*, *510*. Available from https://doi.org/10.1016/j.apsusc.2020.145257.

Hoang, N. D., Van Cat, V., Hoai Nam, M., Ngoc Phan, V., Tuan Le, A., & Van Quy, N. (2019). Enhanced SO_2 sensing characteristics of multi-wall carbon nanotubes based mass-type sensor using two-step purification process. *Sensors and Actuators, A: Physical*, *295*. Available from https://doi.org/10.1016/j.sna.2019.06.046.

Huang, X., Bai, Q., Hu, J., & Hou, D. (2017). A practical model of quartz crystal microbalance in actual applications. *Sensors (Switzerland)*, *17*(8). Available from https://doi.org/10.3390/s17081785.

Hwang, M. J., Geun Shim, W., Seok Choi, W., & Moon, H. (2013). Quartz crystal microbalance sensor coated with nano-sized polystyrene latex spheres for SO_2 detection. *Journal of Nanoscience and Nanotechnology*, *13*(6). Available from https://doi.org/10.1166/jnn.2013.7000.

Iqbal, N., Mustafa, G., Rehman, A., Biedermann, A., Najafi, B., Lieberzeit, P. A., & Dickert, F. L. (2010). QCM-Arrays for sensing terpenes in fresh and dried herbs via bio-mimetic MIP layers. *Sensors*, *10*(7). Available from https://doi.org/10.3390/s100706361.

Janshoff, A., Galla, H. J., & Steinem, C. (2000). Piezoelectric mass-sensing devices as biosensors – An alternative to optical biosensors? *Angewandte Chemie – International Edition*, *22*(39). Available from https://doi.org/10.1002/1521-3773(20001117)39:22 < 4004::AID-ANIE4004 > 3.0.CO;2-2.

Jia, Y., Yu, H., Zhang, Y., Dong, F., & Li, Z. (2016). Cellulose acetate nanofibers coated layer-by-layer with polyethylenimine and graphene oxide on a quartz crystal microbalance for use as a highly sensitive ammonia sensor. *Colloids and Surfaces B: Biointerfaces*, *148*. Available from https://doi.org/10.1016/j.colsurfb.2016.09.007.

Jiao, C., Jiang, X., Chu, H., Jiang, H., & Sun, L. (2016). A mixed-valent CuI/CuII metal-organic framework with selective chemical sensing properties. *CrystEngComm*, *18* (44). Available from https://doi.org/10.1039/c6ce01827k.

Juncai, H., Yaohua, H., & Kangquan, G. (2014). Application of modified quartz crystal sensors: Detection of isobutyl alcohol, ethyl acetate and ethylene. *International Journal of Agricultural and Biological Engineering*, *7*(5).

Kanazawa, K. K., & Gordon, J. G. (1985). The oscillation frequency of a quartz resonator in contact with liquid. *Analytica Chimica Acta*, *175*(C). Available from https://doi.org/10.1016/S0003-2670(00)82721-X.

Kodogiannis, V. S. (2017). Application of an electronic nose coupled with fuzzy-wavelet network for the detection of meat spoilage. *Food and Bioprocess Technology*, *10*(4). Available from https://doi.org/10.1007/s11947-016-1851-6.

Kuchmenko, T. A., & Umarkhanov, R. U. (2013). Peculiarities of microweighing of trace quantities of alkylamines on polymer and solid-state thin films. *Journal of Analytical Chemistry*, *68*(4). Available from https://doi.org/10.1134/S1061934813040059.

Larsson, A., Angbrant, J., Ekeroth, J., Månsson, P., & Liedberg, B. (2006). A novel biochip technology for detection of explosives – TNT: Synthesis, characterisation and application. *Sensors and Actuators, B: Chemical*, *113*(2). Available from https://doi.org/10.1016/j.snb.2005.07.025.

Latif, U., Rohrer, A., Lieberzeit, P. A., & Dickert, F. L. (2011). QCM gas phase detection with ceramic materials-VOCs and oil vapors. *Analytical and Bioanalytical Chemistry*, *400*(8). Available from https://doi.org/10.1007/s00216-011-4684-1.

Li, Q., Gu, Y., & Fei Wang, N. (2017). Application of random forest classifier by means of a QCM-based E-nose in the identification of Chinese liquor flavors. *IEEE Sensors Journal*, *17*(6). Available from https://doi.org/10.1109/JSEN.2017.2657653.

Li, Z., Teng, M., Yang, R., Lin, F., Fu, Y., Lin, W., Zheng, J., Zhong, X., Chen, X., Yang, B., & Liao, Y. (2022). Sb-doped WO3 based QCM humidity sensor with self-recovery ability for real-time monitoring of respiration and wound. *Sensors and Actuators B: Chemical*, *361*. Available from https://doi.org/10.1016/j.snb.2022.131691.

Lim, M., Malhotra, A., Mills Steven Muth, J., & Lee Bongmook Misra, V. (2017). Metal oxide gas sensing characterization by low frequency noise spectroscopy. Proceedings of IEEE Sensors, Orlando, FL, USA. Available from https://doi.org/10.1109/ICSENS.2016.7808835.

Lu, C. S., & Lewis, O. (1972). Investigation of film-thickness determination by oscillating quartz resonators with large mass load. *Journal of Applied Physics*, *43*(11). Available from https://doi.org/10.1063/1.1660931.

Lubczyk, D., Siering, C., Lörgen, J., Shifrina, Z. B., Müllen, K., & Waldvogel, S. R. (2010). Simple and sensitive online detection of triacetone triperoxide explosive. *Sensors and Actuators, B: Chemical*, *143*(2). Available from https://doi.org/10.1016/j.snb.2009.09.061.

Minh, V. A., Anh Tuan, L., Quang Huy, T., Ngoc Hung, V., & Van Quy, N. (2013). Enhanced NH 3 gas sensing properties of a QCM sensor by increasing the length of vertically orientated ZnO nanorods. *Applied Surface Science*, *265*. Available from https://doi.org/10.1016/j.apsusc.2012.11.028.

Mohareb, F., Papadopoulou, O., Panagou, E., John Nychas, G., & Bessant, C. (2016). Ensemble-based support vector machine classifiers as an efficient tool for quality assessment of beef fillets from electronic nose data. *Analytical Methods*, *8*(18). Available from https://doi.org/10.1039/c6ay00147e.

Mujahid, A. A., & Adeel Dickert, F. L. (2019). An overview of high frequency acoustic sensors—QCMs, SAWs and FBARs—Chemical and biochemical applications. *Sensors (Switzerland), 19*(20). Available from https://doi.org/10.3390/s19204395.

Muller, J. B. A., Smith, C. E., Newton, M. I., & Percival, C. J. (2011). Evaluation of coated QCM for the detection of atmospheric ozone. *Analyst, 136*(14). Available from https://doi.org/10.1039/c1an15106a.

Nakamoto, T., Fukuda, A., & Moriizumi, T. (1993). Perfume and flavour identification by odour-sensing system using quartz-resonator sensor array and neural-network pattern recognition. *Sensors and Actuators: B. Chemical, 10*(2). Available from https://doi.org/10.1016/0925-4005(93)80030-F.

Nakamoto, T., Nakamura, K., & Moriizumi, T. (1996). Study of oscillator-circuit behavior for QCM gas sensor. Proceedings of the IEEE Ultrasonics Symposium, San Antonio, TX, USA. Available from https://doi.org/10.1109/ultsym.1996.583990.

Nakamoto, T., & Kobayashi, T. (1994). Development of circuit for measuring both Q variation and resonant frequency shift of quartz crystal microbalance. *IEEE Transactions on Ultrasonics, Ferroelectrics, and Frequency Control, 41*(6). Available from https://doi.org/10.1109/58.330261.

Nakamoto, T., & Moriizumi, T. (1990). A theory of a quartz crystal microbalance based upon a mason equivalent circuit. *Japanese Journal of Applied Physics, 29*(5 R). Available from https://doi.org/10.1143/JJAP.29.963.

Okur, S., Sarheed, M., Huber, R., Zhang, Z., Heinke, L., Kanbar, A., Wöll, C., Nick, P., & Lemmer, U. (2021). Identification of mint scents using a QCM based e-nose. *Chemosensors, 9*(2). Available from https://doi.org/10.3390/chemosensors9020031.

Oleneva, E., Kuchmenko, T., Drozdova, E., Legin, A., & Kirsanov, D. (2020). Identification of plastic toys contaminated with volatile organic compounds using QCM gas sensor array. *Talanta, 211*. Available from https://doi.org/10.1016/j.talanta.2019.120701.

Papadopoulou, O. S., Panagou, E. Z., Mohareb, F. R., John, G., & Nychas, E. (2013). Sensory and microbiological quality assessment of beef fillets using a portable electronic nose in tandem with support vector machine analysis. *Food Research International, 50*(1). Available from https://doi.org/10.1016/j.foodres.2012.10.020.

Prantl, E., Hahn, S., Bunz, U. H. F., & Waldvogel, S. R. (2022). Experimental and computational studies of phenylene-bridged azaacenes as affinity materials for sensing using quartz crystal microbalances. *Organic Materials, 4*(02). Available from https://doi.org/10.1055/a-1873-5186.

Procek, M., Stolarczyk, A., Pustelny, T., & Maciak, E. (2015). A study of a QCM sensor based on TiO_2 nanostructures for the detection of NO_2 and explosives vapours in air. *Sensors (Switzerland), 15*(4). Available from https://doi.org/10.3390/s150409563.

Qi, P., Wang, Z., Wang, R., Xu, Y., & Zhang, T. (2016). Studies on QCM-type NO_2 gas sensor based on graphene composites at room temperature. *Chemical Research in Chinese Universities, 32*(6). Available from https://doi.org/10.1007/s40242-016-6129-z.

Regmi, B. P., Monk, J., El-Zahab, B., Das, S., Hung, F. R., Hayes, D. J., & Warner, I. M. (2012). A novel composite film for detection and molecular weight determination of organic vapors. *Journal of Materials Chemistry, 22*(27). Available from https://doi.org/10.1039/c2jm31623d.

Ren, X., Zhang, D., Wang, D., Li, Z., & Liu, S. (2018). Quartz crystal microbalance sensor for humidity sensing based on layer-by-layer self-assembled PDDAC/graphene oxide film. *IEEE Sensors Journal, 18*(23). Available from https://doi.org/10.1109/JSEN.2018.2872854.

Reza Eslami, M., & Alizadeh, N. (2019). Ultrasensitive and selective QCM sensor for detection of trace amounts of nitroexplosive vapors in ambient air based on polypyrrole—Bromophenol blue nanostructure. *Sensors and Actuators, B: Chemical, 278*. Available from https://doi.org/10.1016/j.snb.2018.09.060.

Rianjanu, A., Triyana, K., Nugroho, D. B., Kusumaatmaja, A., & Roto, R. (2020). Electrospun polyvinyl acetate nanofiber modified quartz crystal microbalance for detection of primary alcohol vapor. *Sensors and Actuators, A: Physical, 301*. Available from https://doi.org/10.1016/j.sna.2019.111742.

Rodahl, M., & Kasemo, B. (1996). Frequency and dissipation-factor responses to localized liquid deposits on a QCM electrode. *Sensors and Actuators, B: Chemical, 37*(1−2). Available from https://doi.org/10.1016/S0925-4005(97)80077-9.

Saevels, S., Lammertyn, J., Berna, A. Z., Veraverbeke, E. A., Di Natale, C., & Nicolaï, B. M. (2003). Electronic nose as a non-destructive tool to evaluate the optimal harvest date of apples. *Postharvest Biology and Technology, 30*(1). Available from https://doi.org/10.1016/S0925-5214(03)00059-0.

Sankaran, S., Panigrahi, S., & Mallik, S. (2011). Odorant binding protein based biomimetic sensors for detection of alcohols associated with *Salmonella* contamination in packaged beef. *Biosensors and Bioelectronics, 26*(7). Available from https://doi.org/10.1016/j.bios.2010.07.122.

Melih Saraoglu, H., Osman Selvi, A., Ali Ebeoglu, M., & Tasaltin, C. (2013). Electronic nose system based on quartz crystal microbalance sensor for blood glucose and hba1c levels from exhaled breath odor. *IEEE Sensors Journal, 13*(11). Available from https://doi.org/10.1109/JSEN.2013.2265233.

Sasaki, K., Furusawa, H., Nagamine, K., & Tokito, S. (2018). Detection of odorant molecules in the gaseous phase using α-, β-, and γ-cyclodextrin films on a quartz crystal microbalance. *Technologies, 6*(3). Available from https://doi.org/10.3390/technologies6030063.

Sauerbrey, G. (1959). Verwendung von Schwingquarzen zur Wägung dünner Schichten und zur Mikrowägung. *Zeitschrift für Physik, 155*(2). Available from https://doi.org/10.1007/BF01337937.

Sharma, P., Ghosh, A., Tudu, B., Prasad Bhuyan, L., Tamuly, P., Bhattacharyya, N., Bandyopadhyay, R., & Chatterjee, A. (2014). Detection of linalool in black tea using a quartz crystal microbalance sensor. *Sensors and Actuators, B: Chemical, 190*. Available from https://doi.org/10.1016/j.snb.2013.08.088.

Sharma, P., Ghosh, A., Tudu, B., Prasad Bhuyan, L., Tamuly, P., Bhattacharyya, N., Bandyopadhyay, R., & Das, U. (2015). A quartz crystal microbalance sensor for detection of geraniol in black tea. *IEEE Sensors Journal, 15*(2). Available from https://doi.org/10.1109/JSEN.2014.2359741.

Sharma, P., Tudu, B., Prasad Bhuyan, L., Tamuly, P., Bhattacharyya, N., & Bandyopadhyay, R. (2016). Detection of methyl salicylate in black tea using a quartz crystal microbalance sensor. *IEEE Sensors Journal, 16*(13). Available from https://doi.org/10.1109/JSEN.2016.2564978.

Songkhla, S. N., & Nakamoto, T. (2021). Interpretation of quartz crystal microbalance behavior with viscous film using a mason equivalent circuit. *Chemosensors, 9*(1). Available from https://doi.org/10.3390/chemosensors9010009.

Speller, N. C., Siraj, N., Regmi, B. P., Marzoughi, H., Neal, C., & Warner, I. M. (2015). Rational design of QCM-D virtual sensor arrays based on film thickness, viscoelasticity, and harmonics for vapor discrimination. *Analytical Chemistry, 87*(10). Available from https://doi.org/10.1021/ac5046824.

Speller, N. C., Siraj, N., Vaughan, S., Speller, L. N., & Warner, I. M. (2016). Assessment of QCM array schemes for mixture identification: Citrus scented odors. *RSC Advances, 6*(98). Available from https://doi.org/10.1039/c6ra16988k.

Su, X., Chen, D., Li, N., Carl Stevenson, A., Li, G., & Hu, R. (2020). A wireless electrode-free QCM-D in a multiresonance mode for volatile organic compounds discrimination. *Sensors and Actuators, A: Physical, 305*. Available from https://doi.org/10.1016/j.sna.2020.111938.

Sun, P., Jiang, Y., Xie, G., Du, X., & Hu, J. (2009). A room temperature supramolecular-based quartz crystal microbalance (QCM) methane gas sensor. *Sensors and Actuators, B: Chemical, 141*(1). Available from https://doi.org/10.1016/j.snb.2009.06.012.

Thanh Vinh, N., Van Dang, T., Thi Hang, B., Tuan Le, A., Tri Tuan, N., Khanh Vinh, L., & Van Quy, N. (2021). Effect of ferric ion [Fe^{3+}] and [Fe^{2+}] on SO_2 adsorption ability of γ-Fe_2O_3 nanoparticles for mass-type gas sensors. *Sensors and Actuators, A: Physical, 331*. Available from https://doi.org/10.1016/j.sna.2021.112981.

Tolentino, M. A. K. P., Albano, D. R. B., & Sevilla, F. B. (2018). Piezoelectric sensor for ethylene based on silver (I)/polymer composite. *Sensors and Actuators, B: Chemical, 254*. Available from https://doi.org/10.1016/j.snb.2017.07.015.

Toniolo, R., Pizzariello, A., Dossi, N., Lorenzon, S., Abollino, O., & Bontempelli, G. (2013). Room temperature ionic liquids as useful overlayers for estimating food quality from their odor analysis by quartz crystal microbalance measurements. *Analytical Chemistry, 85*(15). Available from https://doi.org/10.1021/ac401151m.

Torad, N. L., Minisy, I. M., Sharaf, H. M., Stejskal, J., Yamauchi, Y., & Ayad, M. M. (2021). Gas sensing properties of polypyrrole/poly(N-vinylpyrrolidone) nanorods/nanotubes-coated quartz-crystal microbalance sensor. *Synthetic Metals, 282*, 116935. Available from https://doi.org/10.1016/J.SYNTHMET.2021.116935.

Török, Z. M., Frederic Blaser, A., Kavianynejad, K., de Torrella, C. G. M. G., Nsubuga, L., Kumar Mishra, Y., Rubahn, H. G., & de Oliveira Hansen, R. (2022). Breath biomarkers as disease indicators: Sensing techniques approach for detecting breath gas and COVID-19. *Chemosensors, 10*(5). Available from https://doi.org/10.3390/chemosensors10050167.

Trajcheva, A., Politakos, N., Pérez, B. T., Joseph, Y., Blazevska Gilev, J., & Tomovska, R. (2021). QCM nanocomposite gas sensors — Expanding the application of waterborne polymer composites based on graphene nanoribbon. *Polymer, 213*. Available from https://doi.org/10.1016/j.polymer.2020.123335.

Varga, M., Laposa, A., Kulha, P., Kroutil, J., Husak, M., & Kromka, A. (2015). Quartz crystal microbalance gas sensor with nanocrystalline diamond sensitive layer. *Physica Status Solidi (B) Basic Research, 252*(11). Available from https://doi.org/10.1002/pssb.201552229.

Wang, L., Wang, Z., Xiang, Q., Chen, Y., Duan, Z., & Xu, J. (2017). High performance formaldehyde detection based on a novel copper (II) complex functionalized QCM gas sensor. *Sensors and Actuators, B: Chemical, 248*. Available from https://doi.org/10.1016/j.snb.2016.12.015.

Wang, X., Cui, F., Lin, J., Ding, B., Yu, J., & Al-Deyab, S. S. (2012). Functionalized nanoporous TiO_2 fibers on quartz crystal microbalance platform for formaldehyde sensor. *Sensors and Actuators, B: Chemical, 171–172*. Available from https://doi.org/10.1016/j.snb.2012.05.050.

Ward, M. D., & Delawski, E. J. (1991). Radial mass sensitivity of the quartz crystal microbalance in liquid media. *Analytical Chemistry, 63*(9). Available from https://doi.org/10.1021/ac00009a009.

Wu, M., Xie, G. Z., Zhou, Y., & Tai, H. L. (2014). A PVP-based quartz crystal microbalance sensor for H_2S detection. *Applied Mechanics and Materials, 651–653*. Available from https://doi.org/10.4028/http://www.scientific.net/AMM.651-653.191.

Wu, Z., Chen, X., Zhu, S., Zhou, Z., Yao, Y., Quan, W., & Liu, B. (2013). Enhanced sensitivity of ammonia sensor using graphene/polyaniline nanocomposite. *Sensors and Actuators, B: Chemical, 178*. Available from https://doi.org/10.1016/j.snb.2013.01.014.

Xie, G. Z., Kang, T., Zhou, Y., Xie, T., Ling Tai, H., & Dong Jiang, Y. (2015). QCM sensors based on PEI films for CO_2 detection. *Journal of Electronic Science and Technology, 13*(2). Available from https://doi.org/10.3969/j.issn.1674-862X.2015.02.017.

Xu, X., He, L., Cao, S., Bing, Y., Sui, N., Zhang, D., Zhou, T., & Zhang, T. (2023). Humidity-mediated QCM sensor based on COF-5 for carbon dioxide detection. *IEEE Electron Device Letters*. Available from https://doi.org/10.1109/LED.2023.3277894.

Yan, Y., Lu, D., Zhou, H., Hou, H., Zhang, T., Wu, L., & Cai, L. (2012). Polyaniline-modified quartz crystal microbalance sensor for detection of formic acid gas. *Water, Air, and Soil Pollution, 223*(3). Available from https://doi.org/10.1007/s11270-011-0943-1.

Yang, C. Y., Hwang, M. J., Ryu, D. W., Park, J. H., Ryu, M. S., & Moon, H. (2011). A quartz crystal microbalance-based sensor system coated with functional polymers for SO_2 and NO_2 detection. *Journal of Nanoscience and Nanotechnology, 11*(8). Available from https://doi.org/10.1166/jnn.2011.4790.

Yang, X. L., Hua Xie, M., Cai, W., Shao, R., Bin Zang, R., Feng Guan, R., & Feng, Y. (2020). Postmodified dual functional UiO sensor for selective detection of ozone and tandemly derived sensing of Al^{3+}. *Analytical Chemistry, 92*(17). Available from https://doi.org/10.1021/acs.analchem.0c01082.

Yang, Y., Zhang, W., Guo, Z., Zhang, Z., Zhu, H., Yan, R., & Zhou, L. (2017). Stability enhanced, repeatability improved parylene-C passivated on QCM sensor for aPTT measurement. *Biosensors and Bioelectronics, 98*. Available from https://doi.org/10.1016/j.bios.2017.06.021.

Yao, Y., Chen, X., Guo, H., & Wu, Z. (2011). Graphene oxide thin film coated quartz crystal microbalance for humidity detection. *Applied Surface Science, 257*(17). Available from https://doi.org/10.1016/j.apsusc.2011.04.028.

Yu, L., Huang, Y., Jin, X., Mason, A. J., & Zeng, X. (2009). Ionic liquid thin layer EQCM explosives sensor. *Sensors and Actuators, B: Chemical, 140*(2). Available from https://doi.org/10.1016/j.snb.2009.04.038.

Yu, X., Chen, X., Yu, X., Chen, X., Ding, X., Zhao, X., & Tang, K. (2022). Ultrahighly sensitive QCM humidity sensor based on Nafion/MoS_2 hybrid thin film. *IEEE Transactions on Electron Devices, 69*(3). Available from https://doi.org/10.1109/TED.2021.3137088.

Zainuddin, A. A., Nurashikin Nordin, A., Ab Rahim, R., Aini Md Ralib, A., Khan, S., Guines, C., Chatras, M., & Pothier, A. (2018). Verification of quartz crystal microbalance array using vector network analyzer and open QCM. *Indonesian Journal of Electrical Engineering and Computer Science, 10*(1). Available from https://doi.org/10.11591/ijeecs.v10.i1.pp84-93.

Zhang, D., Wang, D., Zong, X., Dong, G., & Zhang, Y. (2018). High-performance QCM humidity sensor based on graphene oxide/tin oxide/polyaniline ternary nanocomposite prepared by in-situ oxidative polymerization method. *Sensors and Actuators, B: Chemical, 262*. Available from https://doi.org/10.1016/j.snb.2018.02.012.

Zhang, H. D., Yan, X., Hua Zhang, Z., Feng Yu, G., Peng Han, W., Cheng Zhang, J., & Ze Long, Y. (2016). Electrospun PEDOT:PSS/PVP nanofibers for CO gas sensing with quartz crystal microbalance technique. *International Journal of Polymer Science, 2016*. Available from https://doi.org/10.1155/2016/3021353.

Zhu, Y., Yuan, H., Xu, J., Xu, P., & Pan, Q. (2010). Highly stable and sensitive humidity sensors based on quartz crystal microbalance coated with hexagonal lamelliform monodisperse mesoporous silica SBA-15 thin film. *Sensors and Actuators, B: Chemical, 144*(1). Available from https://doi.org/10.1016/j.snb.2009.10.053.

CHAPTER

5

Odor biosensor

Hongchao Deng[1,2]

[1]Zhejiang Huanmao Auto-Control Technology Co., Ltd., Hangzhou, Zhejiang, P.R. China [2]Institute of Integrated Research, Institute of Science Tokyo, Yokohama, Japan

5.1 Odor biosensor

Before the invention of current odor sensors, humans have trained animals such as dogs and pigs for food hunting for thousands of years. Moreover, the behaviors of animals can indicate the appearance of some specific odorants (Hansson & Stensmyr, 2011; McBride et al., 2014; Pelletier et al., 2015). Hence, researchers utilized various creatures for odor detection that could be regarded as an origin of odor biosensors (Bodyak & Slotnick, 1999; Furton & Myers, 2001; Kiddy et al., 1978; Lambin et al., 2005; Otto et al., 2002; Pham-Delègue et al., 1997). Considering that the animal behaviors are not always stable and only can transmit limited information, scientists proposed diverse odor biosensors that inherit the outstanding detection performance from creature's olfaction. In this chapter, we first introduce several kinds of odor biosensors. Then the odor biosensors based on cell-expressing olfactory receptors (ORs) are explained in detail owing to their good development prospects, stable properties, and ease of operation.

5.2 Conception of odor biosensor

The sense organs enable us to obtain the information from the external environment. Based on these feelings, humans are able to make correct decisions. However, it is far from enough to only rely on sensory organs in our daily activities because of limitations of detection range, sensitivity, safety, and so on. To enhance the detection ability, numerous sensors are developed. Sensor is a kind of detection device that can sense some specific substance and transform the information into electrical or other required forms of signal. According to the measured object, sensors can be divided into different categories such as light, acoustic, odor, pressure, and temperature.

Odor sensor is used to detect the odor concentration and component. The first step of detection is the combination between sensing material and target odorant molecule, then transducer transforms this combination into electrical, light, or acoustic signal, and the signal magnitude usually has a dose-dependency property (Fig. 5.1). Therefore the characteristics

Digital Technologies in Olfaction
DOI: https://doi.org/10.1016/B978-0-443-15721-9.00005-1

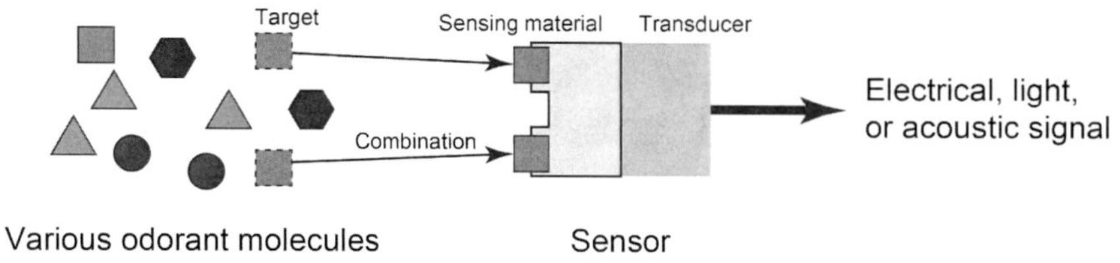

FIGURE 5.1 The schematic representation of odor sensor.

of sensing material directly determine the performance of sensor, and the main effort in improving the sensor detection ability is focused on discovering or creating better sensing materials. Apart from the ordinary materials, some biological materials have ultra-high sensitivity and selectivity toward their ligands, thereby being utilized in odor sensor that is usually named odor biosensor.

An odor biosensor consists of a biological sensing material and a transducer. Here we briefly list several odor biosensors with different measurement methods (Fig. 5.2). The combination between odorant molecules and biological materials leads to various changes in mass, fluorescence, potential, conformation, and so on. The transducers can transform these changes into desired signals thus completing the detection process.

5.3 Current odor biosensors

Attracted by the superb odor detection ability of the animal olfactory system, scholars combined the biological materials with the conventional transducers to create odor biosensors. The applied biological materials allow biosensors to have better sensitivity and selectivity compared with normal odor sensors. In Fig. 5.2, we present several typical transducers adopted in odor biosensors. In addition, odor biosensors also can be separated into different categories based on the sensitive materials such as olfactory bulb (OB), olfactory epithelium (OE), olfactory sensory neuron (OSN), cell-expressing OR, nanovesicle, OR protein, and odorant binding protein (OBP). Now we introduce the mechanism of those odor biosensors in sequence.

5.3.1 Olfactory bulb

In a mammalian olfactory system, one mature OSN only expresses one type of OR, while one OR can combine with one or multiple odorant molecules. The odor perception starts from the odorant molecules captured by ORs in OE. This combination between ORs and odor molecules leads to an ion influx thus changing the membrane potential of OSNs. Then this olfactory information is sent from OSN via its axon to OB. Inside the OB, the odor-specific response pattern is coded and processed by mitral/tufted cells (M/Ts) (Davison & Katz, 2007). After that, the encoded signal is transmitted to the olfactory cortex for further processing to generate a sense of smell. In this whole odor perception procedure, the OB plays a role of primary processing unit and its output is closely related to the applied odors. Therefore OB is utilized for building the odor biosensors.

In odor biosensors based on OB, the experiment animals were mainly mice or rats (Dasgupta et al., 2022; Gao et al., 2018; Liu et al., 2020; Zhu et al., 2022; Zhuang et al., 2013, 2014, 2015). After the animal was anesthetized, a microelectrode array (MEA) was implanted to the OB for obtaining the electrophysiological signal (Fig. 5.3A). Before applying the odorant stimulation, M/Ts already had some spontaneous signals (Fig. 5.3B). After the odor stimulation was applied, the frequency of spike also

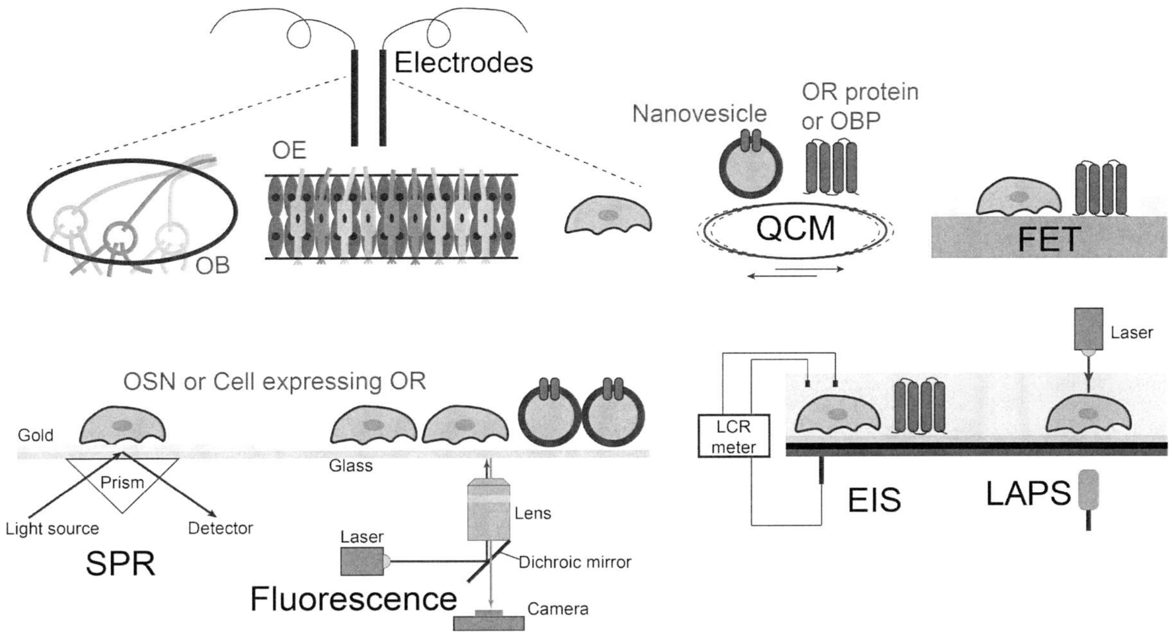

FIGURE 5.2 **Examples of odor biosensors.** The odorant to be measured is first combined with the biological materials, then the variation induced by this combination is captured by the ordinary transducers. The abbreviations and some detection principles are as follows: *OB*, olfactory bulb; *OE*, olfactory epithelium; *OBP*, odorant binding protein; *OSN*, olfactory sensory neuron; *QCM*, quartz crystal microbalance, QCM utilizes the piezoelectric effect of the quartz crystal to convert the surface mass change of the quartz crystal electrode into the frequency change of the output electrical signal of the oscillating circuit; *FET*, field effect transistor, the combination between target substance and biological sensing material changes the surface charge density that is equivalent to the gate potential change, and the channel current between the source and drain also changes accordingly; *SPR*, surface plasmon resonance, under the total internal reflection condition, the incident light causes the resonance of the thin gold layer plasma, resulting in the reflected light energy being almost zero at a certain angle (SPR angle), the SPR angle is extremely sensitive to any change of the gold layer surface; Fluorescence, when OR captures its target odorant it opens an ion channel resulting in the calcium ion influx, calcium ion sensitive fluorescent protein or fluorescent dye inside the cell or nanovesicle change the ion concentration increase into fluorescent intensity increase; *EIS*, electrochemical impedance spectroscopy, apply a small-amplitude alternating current signal with different frequencies to the electrochemical system, the combination between target substance and biological sensing materials influences the system impedance or impedance phase angle; *LAPS*, light-addressable potentiometric sensors, it can sense the surface potential change based on photo current induced by modulated light beam. It also provides two-dimensional surface potential change distribution by scanning the light beam.

increased (Fig. 5.3C). The spike pattern varied from odor to odor; thus the classification of four odors could be realized (Fig. 5.3D, E). There are several ways to further improve the performance of odor biosensor based on OB. For example, signals captured in OB may come from different glomeruli. To improve the specificity of odor detection, Gao et al. (2018) handled the experiment animal with transgenic technique so one specific area in OB was highlighted by fluorescence. This area had the glomeruli that connected with M72 OSN, and the electrophysiological signal from this area could detect 10^{-5} M liquid sample trinitrotoluene and distinguish trinitrotoluene from other similar structure chemicals. Another example is the bioengineer method to overexpress OR3 on the rat OE and record the electrophysiological signals from OB (Zhu et al., 2022). The detection limit could reach around 10^{-5} M toward four ligand odorants.

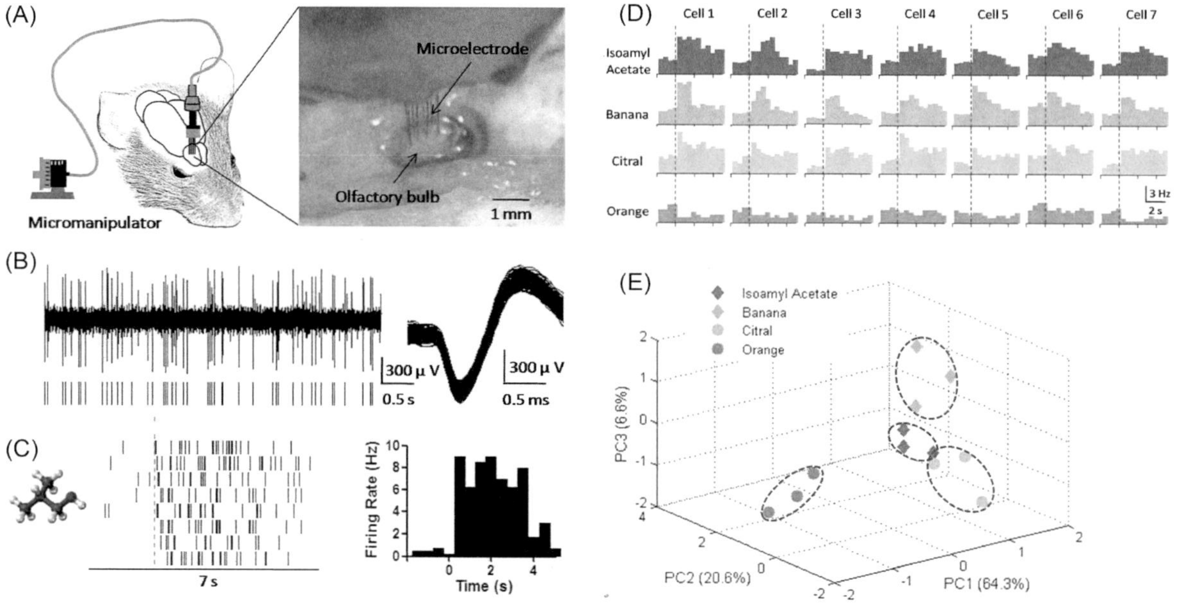

FIGURE 5.3 **Odor biosensor based on olfactory bulb (OB).** (A) MEA formed by eight electrode pairs is implanted into rats' OB for obtaining the electrophysiological signal. (B) Spontaneous signal of M/Ts, the left top is one typical raw signal, the left bottom is its raster plot, and the right is an enlarged spike waveform. (C) Activities from one M/T cell before and after isoamyl acetate applied, the left is raster and right is peristimulus time histograms, the dashed line indicates the timing of odorant application, and the tests repeated eight times thus one row in the left figure represents for one test. (D) Peristimulus time histograms of seven M/Ts under different target odorant stimulations. (E) The response patterns are handled by PCA for extracting information of isoamyl acetate, banana, orange, and citral. MEA, Microelectrode array; *M/Ts*, mitral/tufted cells; *PCA*, principal component analysis. *Source: Modified from Zhuang, L., Guo, T., Cao, D., Ling, L., Su, K., Hu, N., & Wang, P. (2015). Detection and classification of natural odors with an in vivo bioelectronic nose.* Biosensors and Bioelectronics, *67, 694–699. https://doi.org/10.1016/j.bios.2014.09.102 with permission from Elsevier.*

5.3.2 Olfactory epithelium

The odor biosensor based on OB is an in vivo detection method. Moreover, the OE can be stripped off from the nasal cavity for in vitro odor biosensor. The olfactory tissue consists of three types of cells: OSN, support, and basal. The specific binding between ORs and their ligands happens in the cilia of OSN, and then an electric signal is transmitted through OSN axons. MEA and light-addressable potentiometric sensors (LAPS) were suitable to investigate this electric signal (Liu et al., 2011; Liu, Ye, Xiao, et al., 2010; Liu, Ye, Yu, et al., 2010). An intact rat OE was isolated and fixed on the surface of MEA or LAPS (Fig. 5.4A–C). To receive the odorant stimulation, the cilia were exposed. Similar to the OB, the olfactory tissue also has spontaneous potential signal. After the target odorants were applied, the cell burst discharge pattern also changed (Fig. 5.4D, E). These specific response patterns can be exploited for odor discrimination.

5.3.3 Olfactory sensory neuron

OE comprises OSN, support cell, and basal cell. Among these cells, only OSN can sense the target odorant, thereby using the OSN to form an odor biosensor seems more efficient. The OE tissue was isolated from animal and handled with papain or other protein enzymes for

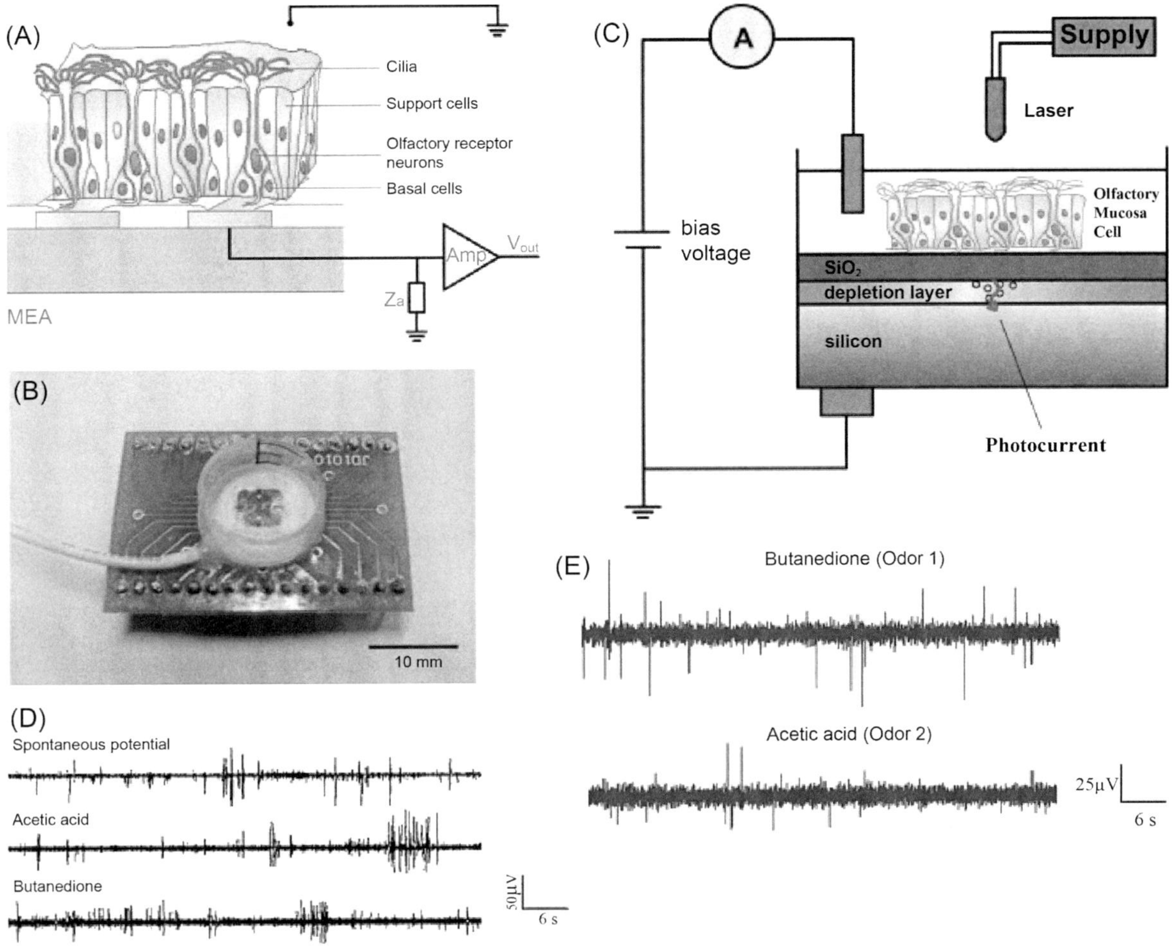

FIGURE 5.4 Odor biosensors based on olfactory epithelium (OE). (A, B) OE combined with MEA. (C) OE combined with LAPS. (D) Electric signals captured by MEA under different odorant stimulations. and (E) Electric signals captured by LAPS under different odorant stimulations. *LAPS*, Light-addressable potentiometric sensors; *MEA*, microelectrode array. *Source: (A, B, D) are from Liu, Q., Ye, W., Xiao, L., Du, L., Hu, N., & Wang, P. (2010). Extracellular potentials recording in intact olfactory epithelium by microelectrode array for a bioelectronic nose.* Biosensors and Bioelectronics, 25*(10), 2212–2217. https://doi.org/10.1016/j.bios.2010.02.024 with permission from Elsevier; (C, E) are from Liu, Q., Ye, W., Yu, H., Hu, N., Du, L., Wang, P., & Yang, M. (2010). Olfactory mucosa tissue-based biosensor: A bioelectronic nose with receptor cells in intact olfactory epithelium.* Sensors and Actuators B: Chemical, 146*(2), 527–533. https://doi.org/10.1016/j.snb.2009.12.032 with permission from Elsevier.*

obtaining isolated cells. These cells could be combined with electrodes or fluorescence (Datta-Chaudhuri et al., 2016; Suzuki et al., 2016). When the OSN cells were loaded near the electrodes, their electrophysiological signal was captured (Fig. 5.5A). The selectivity of this voltage signal was determined by the OR type expressed on the corresponding OSN (Fig. 5.5B). Also, the OSN cells could be labeled with fluorescent dye such as Fluo-4AM (Fig. 5.5C), and then the odor stimulation response was presented as fluorescent intensity change (Fig. 5.5D). Both sensing methods are suitable for forming a big sensor array. In addition, some researchers cultured

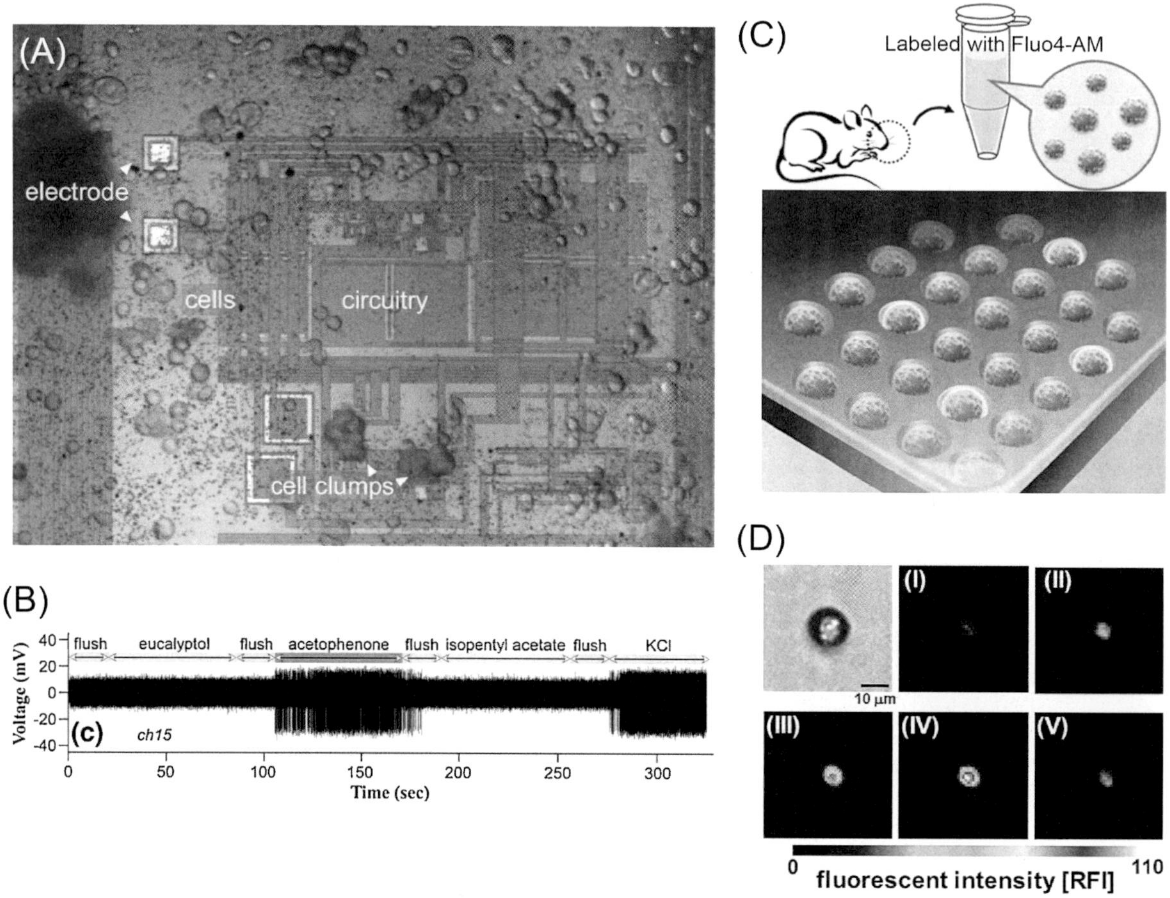

FIGURE 5.5 **Odor biosensors based on olfactory sensory neuron (OSN).** (A) OSN combined with electrodes. (B) Electric signals captured by electrodes under different odorant stimulations. (C) OSN combined with fluorescence. (D) Fluorescence captured by camera under odorant stimulations. *Source: (A, B) are from Datta-Chaudhuri, T., Araneda, R.C., Abshire, P. & Smela, E. (2016). Olfaction on a chip.* Sensors and Actuators, B: Chemical, 235, *74–78. https://doi.org/10.1016/j.snb.2016.05.048 with permission from Elsevier; (C, D) are from Suzuki, M., Yoshimoto, N., Shimono, K., & Kuroda, S. (2016). Deciphering the receptor repertoire encoding specific odorants by time-lapse single-cell array cytometry.* Scientific Reports, *6(1), https://doi.org/10.1038/srep19934.*

OSNs and OB neurons together in vitro for better mimicking the animal olfactory system (Gao et al., 2021).

5.3.4 Cell-expressing olfactory receptor

The OR and OR coreceptor (Orco) can be heterologously expressed on another type of cell such as HEK-293 or Sf21. This cell has similar function to OSN cell while being easier to culture. The odor biosensors based on cell-expressing OR have two main advantages compared with the OSN-based biosensors: First, researchers can freely select the OR type to be expressed; second, the newly established cell line can be passaged for a long time while maintaining stable sensing performance. Therefore the odor biosensors based on cell-expressing OR were well developed in recent decades.

The odor biosensors based on cell-expressing OR are mainly combined with electrodes or fluorescence (Lee, Oh, et al., 2015; Misawa et al., 2010; Mitsuno et al., 2015; Sato & Takeuchi, 2014). In the electrodes method, two electrodes formed a voltage clamp and inserted into the sensing cell—for example, *Xenopus laevis* oocyte—that expresses insect ORs to detect the ion concentration change (Fig. 5.6A) (Misawa et al., 2010). When the ligand was applied, the OR captured it and opened an ion channel. The influx of calcium ions resulted in a negative charge from the current electrode then reached a new stable state. Therefore the temporal current

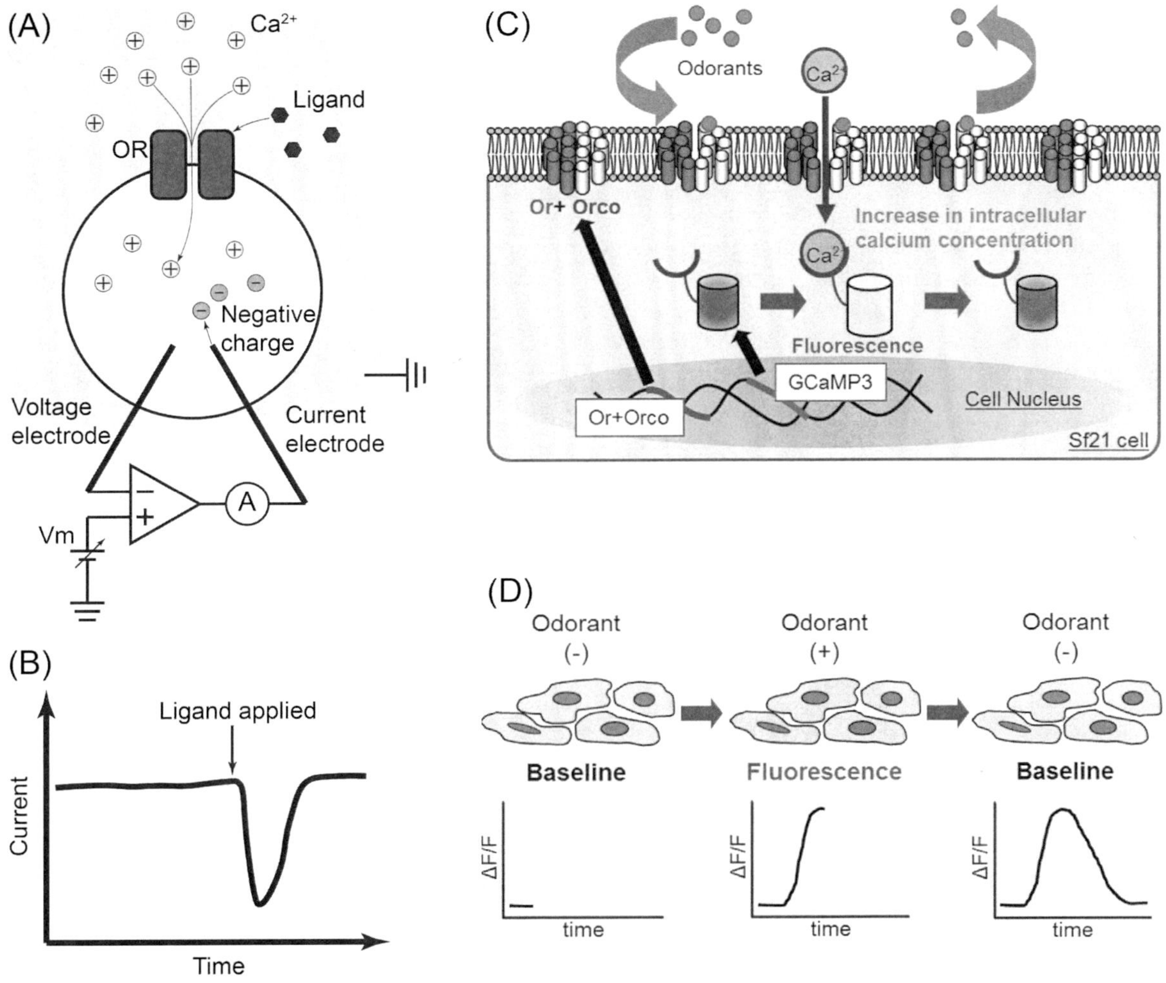

FIGURE 5.6 Odor biosensors based on cell-expressing olfactory receptor (OR). (A) Cell combined with electrodes, the voltage electrode can sense the membrane potential change. (B) Current variation induced by odorant stimulation. (C) Cell combined with fluorescence, besides the OR, a calcium sensitive fluorescent protein GCaMP3 also exists. (D) Fluorescence variation induced by odorant stimulation. *Source: (C, D) are from Mitsuno, H., Sakurai, T., Namiki, S., Mitsuhashi, H., & Kanzaki, R. (2015). Novel cell-based odorant sensor elements based on insect odorant receptors.* Biosensors and Bioelectronics, 65, *287–294, https://doi.org/10.1016/j.bios.2014.10.026.*

change had odor-dose-dependency property (Fig. 5.6B). This biosensor can detect liquid-phase chemicals at a few ppb concentrations and distinguish different types of odorants. In the fluorescence method, the increase of calcium ion concentration induced by combination of OR and ligands is presented as an increase of fluorescence intensity. Thence, a calcium indicator dye such as Fluo-4 AM or calcium sensitive fluorescent protein such as GCaMP3s is necessary (Lee, Oh, et al., 2015; Mitsuno et al., 2015). After the target ligand was applied, the cell brightness first increased then decreased (Fig. 5.6C, D) and this variation was exploited for computing the biosensor response. The cell number in the fluorescence method is much more than the electrode method. Obtaining the response from a batch of cells can mitigate the influence due to individual differences. Cells expressing different ORs could be arranged to constitute a sensing array thus further enhancing the detection ability (Termtanasombat et al., 2016).

Compared with using OB, OE, and OSN, researchers can freely determine the OR type in cell-expressing OR-based odor biosensors. When a panel of ORs was employed to form an OR array, it was able to present distinct response patterns toward odorants of different types and concentrations (Fig. 5.7) (Kida et al., 2018). Meanwhile, some ORs have a narrowly tuned response pattern, which means they can perform qualitative or quantitative detections of specific odorants (Stensmyr et al., 2012). This property is also guaranteed in nanovesicle or OR-based odor biosensors.

5.3.5 Nanovesicle

Cells secrete extracellular vesicles for intercellular communication. Those vesicles contain several substances such as protein, deoxyribonucleic acid (DNA), and ribonucleic acid (RNA). Therefore the nanovesicle from

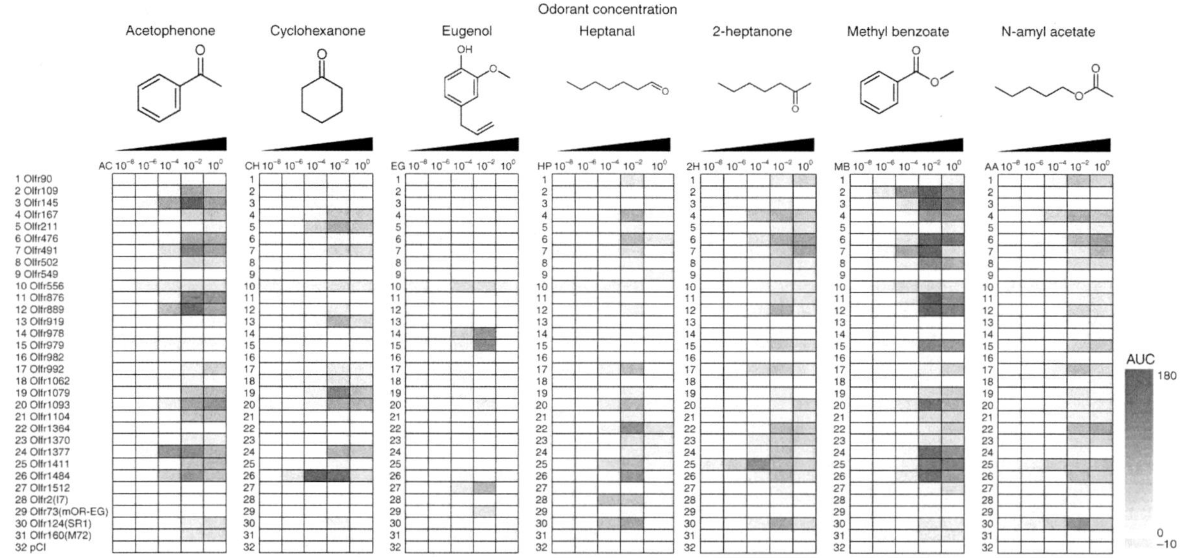

FIGURE 5.7 The response pattern of a panel of olfactory receptors (ORs) toward odorants of different types and concentrations. *AUC*, Area under the curve. *Source: From Kida, H., Fukutani, Y., Mainland, J. D., de March, C. A., Vihani, A., Li, Y. R., Chi, Q., Toyama, A., Liu, L., Kameda, M., Yohda, M., & Matsunami, H. (2018). Vapor detection and discrimination with a panel of odorant receptors.* Nature Communications, *9(1), https://doi.org/10.1038/s41467-018-06806-w.*

the cell-expressing OR also has OR on its membrane and could be used for odor detection (Kalluri & LeBleu, 2020). Fig. 5.8A illustrates the procedure of obtaining the desired nanovesicle (Jin et al., 2012): First transfected the target OR gene into cells, then the OR protein was expressed; second applied some agitation to accelerate the emission secretion of nanovesicle; third separated the cell and nanovesicle through centrifugation. These collected nanovesicles have the similar detection ability with the original cells. The size of nanovesicles varies from tens to hundreds of nanometers that are smaller than normal cells (around tens micrometers) (Calò et al., 2012; Cho et al., 2021; Khadka et al., 2019). Therefore, besides measuring the fluorescent intensity change, the nanovesicles also could be combined with field effect transistor (FET), electrochemical impedance spectroscopy (EIS), quartz crystal microbalance (QCM), and so on (Ahn et al., 2015; Calò et al., 2012; Cho

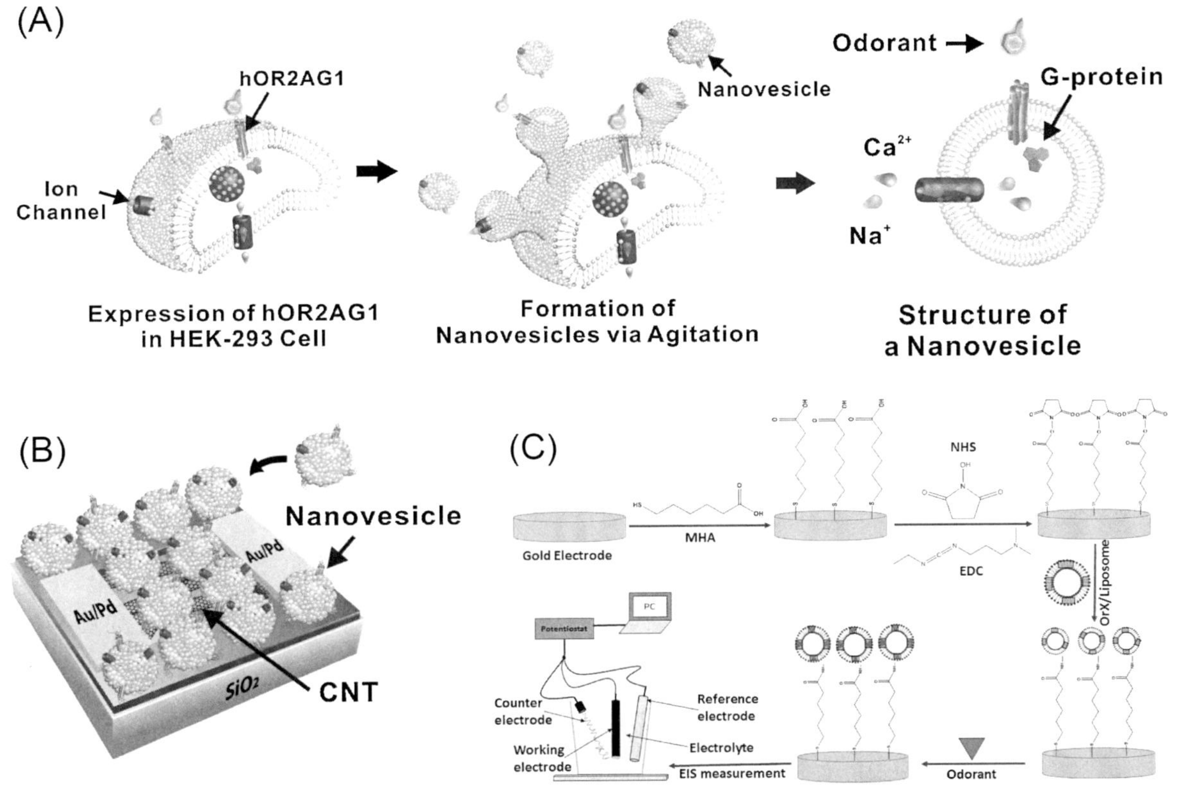

FIGURE 5.8 Odor biosensors based on nanovesicle. (A) The procedure of producing nanovesicle from a cell, the generated nanovesicle have the same membrane proteins and cytosolic components with the original cell while its size is much smaller. (B) Nanovesicle combined with FET, the FET here is swCNT-based FET. (C) Nanovesicle combined with EIS. *FET*, Field effect transistor; *swCNT*, single-walled carbon nanotube. *Source: (A, B) are adapted from Jin, H. J., Lee, S. H., Kim, T. H., Park, J., Song, H. S., Park, T. H., & Hong, S. (2012). Nanovesicle-based bioelectronic nose platform mimicking human olfactory signal transduction.* Biosensors and Bioelectronics, *35(1), 335–341, https://doi.org/10.1016/j.bios.2012.03.012 with permission from Elsevier; (C) is from Khadka, R., Aydemir, N., Carraher, C., Hamiaux, C., Colbert, D., Cheema, J., Malmström, J., Kralicek, A., & Travas-Sejdic J. (2019). An ultrasensitive electrochemical impedance-based biosensor using insect odorant receptors to detect odorants.* Biosensors and Bioelectronics, 126, *207–213, https://doi.org/10.1016/j.bios.2018.10.043 with permission from Elsevier.*

et al., 2021; Jin et al., 2012; Khadka et al., 2019; Park et al., 2012; Son et al., 2015).

In the fluorescence method, if the host cell already expressed fluorescent protein, the nanovesicles can be directly employed for experiment. Otherwise, a fluorescent dye was necessary. In reference (Jin et al., 2012), the cell was incubated in the buffer solution containing fluorescent dye Fura-2AM, then the generated nanovesicle also had the Fura-2AM inside. The nanovesicle was transferred into a 96-well plate, and the odorant was added, the fluorescent brightness increased, while the light intensity from no OR protein nanovesicle (reference group) showed no change. In the FET method, the gate electrode in the FET sensor was handled with poly-D-lysine solution for enhancing the attachment of nanovesicle. The FET was incubated in the solution with nanovesicle, and the nanovesicles in the buffer solution were attracted to the poly-D-lysine (Fig. 5.8B). After this, the FET with nanovesicle could be used as an ordinary FET, and its conductance varied with the odorant concentrations (Ahn et al., 2015; Calò et al., 2012; Son et al., 2015). In the EIS method, the gold electrodes were first handled with 6-Mercaptohexanoic acid then 1-ethyl-3-((3-dimethylaminopropyl)-carbodiimide) for better capturing the nanovesicle (Fig. 5.8C). The Nyquist plot of EIS exhibited odor-dose dependency, and the detection limit could reach 1 fM (Khadka et al., 2019). In addition, the nanovesicle could also be utilized as sensing materials of other sensors such as QCM (Khadka et al., 2019), the detection mechanism would not change, also the sensitivity and selectivity were ensured.

5.3.6 Olfactory receptor protein

The OR protein can specially bind with its ligands. Such selectivity is the basis of odor detection. There are vast types of ORs that exist in various animals such as pig, honeybee, fruit fly, human, mosquito, and so on (Jones et al., 2011; Leal, 2013). For a determined target odorant, we always can find one or several ORs that are capable of sensing it. Meanwhile, the size of OR protein is much smaller than cell-expressing OR or nanovesicle-containing OR, which means it is easier to combine with various transducers. Therefore the biosensors based on OR protein might be the most commonly used odor biosensors.

There are two methods to obtain the OR protein: First is obtained from the cell membrane. The plasmid containing OR DNA was transfected into the cell, for example, HEK-293 for expressing the desired OR on the cell plasma membrane (Du et al., 2013). The cell was maintained in the culture medium for a period. At this moment, this cell is already suitable for biosensor based on cell-expressing OR. However, for getting the OR protein, researchers added the membrane protein extraction kit to the cell, then sonicated the cell and centrifuged it to acquire the membrane protein. Second is cell-free protein synthesis (Kaiser et al., 2008; Ritz et al., 2013). Because the production of OR protein in heterologous cell systems might encounter some difficulties such as protein aggregation and cytotoxicity while the cell-free protein synthesis can avoid those issues. Some commercial available cell-free protein synthesis systems could be chosen for ensuring the successful rate (Chen et al., 2019). The generated OR protein was mixed with other components thus requiring centrifugation and purification operations to obtain the final product.

There are mainly three ways to utilize the obtained OR protein: first is directly using the OR protein as the sensing element (Du et al., 2013; Kwon et al., 2015; Lee, Jin, et al., 2012; Lee, Kwon, et al., 2012). For example, in reference (Lee, Jin, et al., 2012), the hOR2AG1 (a kind of human OR) was expressed in *Escherichia coli* and collected. After this, a single-walled carbon nanotube-based FET sensor was handled by 1,5-diaminonaphthalene and glutaraldehyde. Then the hOR2AG1 protein was deposited on the swCNT to improve the sensitivity and selectivity of this odor biosensor (Fig. 5.9A). The OR protein also could be used as sensitive materials for other sensors such as EIS (Chen et al., 2019; Hou et al., 2007), QCM (Ko & Park, 2005; Sung et al., 1981), and surface acoustic wave (SAW) (Wu et al., 2011). Besides the OR protein, the OR peptides are also capable of recognizing the odorant molecules. The ligand binding portion is part of the whole OR protein. Researchers insisted OR-peptide-based odor biosensor would be more compact. Also, a shorter amino acid chain is more stable. Hence, several OR peptide–based odor biosensors were manufactured (Fig. 5.9B) (Lee, Lim, et al., 2015; Sankaran et al., 2011).

The second method is inserting OR into a nanodisc (Lee et al., 2018; Murugathas et al., 2019; Oh et al., 2019; Yang et al., 2017). Compared with the bare OR protein or peptide, the OR protein structure in nanodisc is believed to be stable even in the dried condition (Oh et al., 2019). One example of nanodisc containing OR protein is shown in Fig. 5.9C. The lipids were mixed and solubilized in buffer solution, then the purified OR protein was added. After incubation with agitation overnight, the nanodisc was fabricated.

The third method is OR protein embedded into bilayer lipid membrane (Misawa et al., 2019). Its structure is somewhat like a cell. But the formation method is totally different: n-decane, 1,2-dioleoyl-sn-glycero-3-phosphocholine, 1,2-dioleoyl-sn-glycero-3-phosphoethanolamine, agarose gel, and buffer solution were subsequently added to form the lipid membrane. Then OR/Orco complex was added to the wells to finish the formation procedure. When the OR combined with its ligand, the ion channel formed by the OR/Orco opened and an ion flow emerged. The current recorded by two Ag/AgCl electrodes at the bottom of the wells changed (Fig. 5.9D). Since the amount of OR/Orco pairs on the bilayer lipid membrane was not stable, the target odorant sometimes could not trigger a visible current change. To overcome this problem, an odor biosensor with 16 double wells was fabricated (Yamada et al., 2021). The target odorant was introduced into each well through several symmetrically distributed flow channels. The increase of wells enhanced the robustness of this odor biosensor.

5.3.7 Odorant binding protein

The insect OSNs exist in the olfactory sensillum and are submerged in the sensillum lymph. The vertebrate OSNs exist in the OE and are covered by a layer of mucus. Both the sensillum lymph and mucus are aqueous solutions, while the majority of odorant molecules are hydrophobic. Therefore OBP is secreted to carry the gas-phase hydrophobic odorants through lymph or mucus to reach OSNs (Matarazzo et al., 2002; Pevsner et al., 1990; Steinbrecht, 1998). Similar to OR protein, OBP is also used as the sensing materials of different sensors such as QCM, FET, SAW, and interdigitated electrodes (Di Pietrantonio et al., 2013; Ko & Park, 2008; Larisika et al., 2015; Lu et al., 2015; Son et al., 2016). Furthermore, the molecular transporting function of OBP helps to enhance the response magnitude in normal odor biosensors. In reference (Choi et al., 2022), Rat I7 OR protein was assembled in a nanodisc then immobilized on the floating electrode of CNT-FET (Fig. 5.10A). The FET was placed in a case filled with buffer solution. The target odorant—octanal—was applied to the FET in the liquid phase or gas phase. The existence of OBP could make the conductance change after odor stimulation became larger (Fig. 5.10B). The combination of OR and OBP can better imitate the human mucosa in gas-phase odor sensing.

In addition to OBP, pheromone binding protein (PBP) also exists in the sensillum lymph or

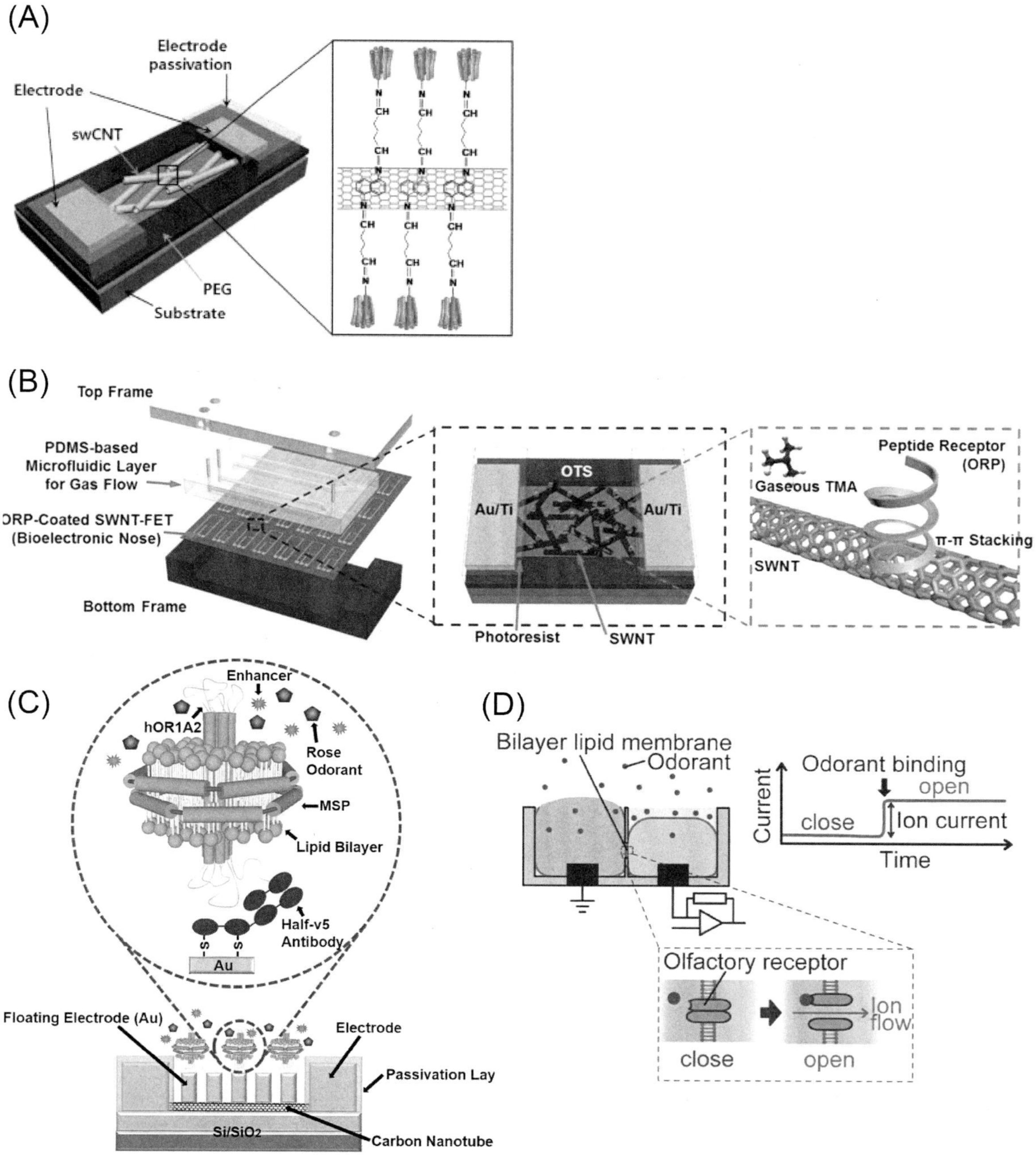

FIGURE 5.9 Odor biosensors based on olfactory receptor (OR) protein. (A) OR protein combined with swCNT-FET, the swCNT-FET is handled with 1,5-diaminonaphthalene and glutaraldehyde thus making the deposition of OR protein easier. (B) OR peptide combined with swCNT-FET, only the ligand binding portion of OR is generated and utilized. (C) OR protein fixed in a nanodisc then combined with floating electrode (FE)-based CNT-FET. (D) OR protein combined with bilayer lipid membrane, the response signal is recorded by two Ag/AgCl electrodes in the bottom of wells. *FET*, Field

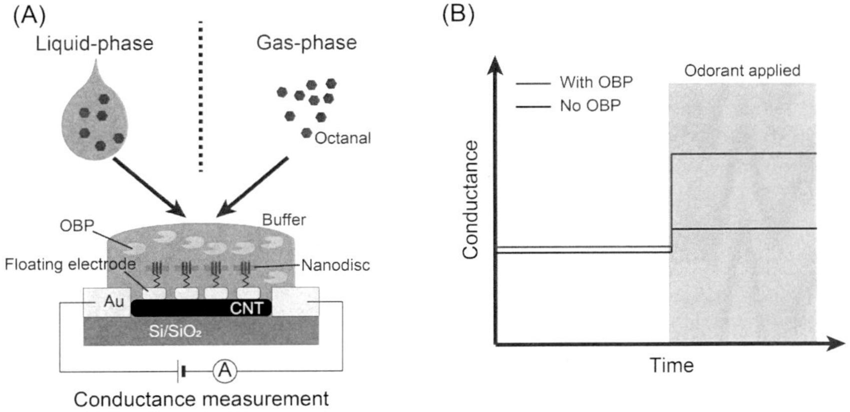

FIGURE 5.10 **Odorant binding protein acting as a molecular transporter to improve the odor biosensor response.** (A) The structure of odor biosensor, the nanodisc was immobilized on the floating electrode, the odor stimulation could be conducted in gas or liquid phase, the current between source and drain was measured during the experiment. (B) With the existence of OBP, the conductance change became more significant that meant the biosensor is more sensitive. *OBP*, Odorant binding protein.

mucus. The PBPs only combine with conspecific animal-released pheromones; thus the sensitivity and specificity are very high (Campanacci et al., 2001; Forstner et al., 2009; Leal et al., 2005; Wojtasek & Leal, 1999). Based on this characteristic, plenty of PBP-based odor biosensors are explored in recent decades (Bonazza et al., 2021; Katre et al., 2009; Lu et al., 2019).

5.3.8 Other sensing materials for odor biosensors

The sensing materials in the aforementioned odor biosensors have a close relationship with the animal olfactory system. However, any other biological substance that can specifically combine with odorant molecules is possible for applying in odor biosensors. Here, we provide two instances of those biosensors.

First is enzyme-based biosensor. Ye et al. (2015) developed a fiber-optic biochemical gas sensing system. In this system, a flow-cell with nicotinamide adenine dinucleotide (NADH)-dependent secondary alcohol dehydrogenase (S-ADH) immobilized membrane was attached onto a fiber-optic NADH measurement structure (Fig. 5.11A). When NADH was consumed by the enzymatic reaction of S-ADH, the consumption was proportional to the acetone vapor. Hence, the change of fluorescence emitted from NADH represented the change of acetone concentration.

effect transistor; *swCNT*, single-walled carbon nanotube. *Source: (A) is from Lee, S. H., Jin, H. J., Song, H. S., Hong, S., & Park, T. H. (2012). Bioelectronic nose with high sensitivity and selectivity using chemically functionalized carbon nanotube combined with human olfactory receptor.* Journal of Biotechnology, *157(4), 467–472, https://doi.org/10.1016/j.jbiotec.2011.09.011 with permission from Elsevier; (B) is from Lee, S. H., Lim, J. H., Park, J., Hong, S. & Park, T. H. (2015). Bioelectronic nose combined with a microfluidic system for the detection of gaseous trimethylamine.* Biosensors and Bioelectronics, *71, 179–185. https://doi.org/10.1016/j.bios.2015.04.033 with permission from Elsevier; (C) is from Lee, M., Yang, H., Kim, D., Yang, M., Park, T. H., & Hong, S. (2018). Human-like smelling of a rose scent using an olfactory receptor nanodisc-based bioelectronic nose.* Scientific Reports, *8(1), https://doi.org/10.1038/s41598-018-32155-1; (D) is reproduced with the kind permission of ACS Publications From Misawa, N., Fujii, S., Kamiya, K., Osaki, T., Takaku, T., Takahashi, Y., & Takeuchi, S. (2019). Construction of a biohybrid odorant sensor using biological olfactory receptors embedded into bilayer lipid membrane on a chip.* ACS Sensors, *4(3), 711–716, https://doi.org/10.1021/acssensors.8b01615.*

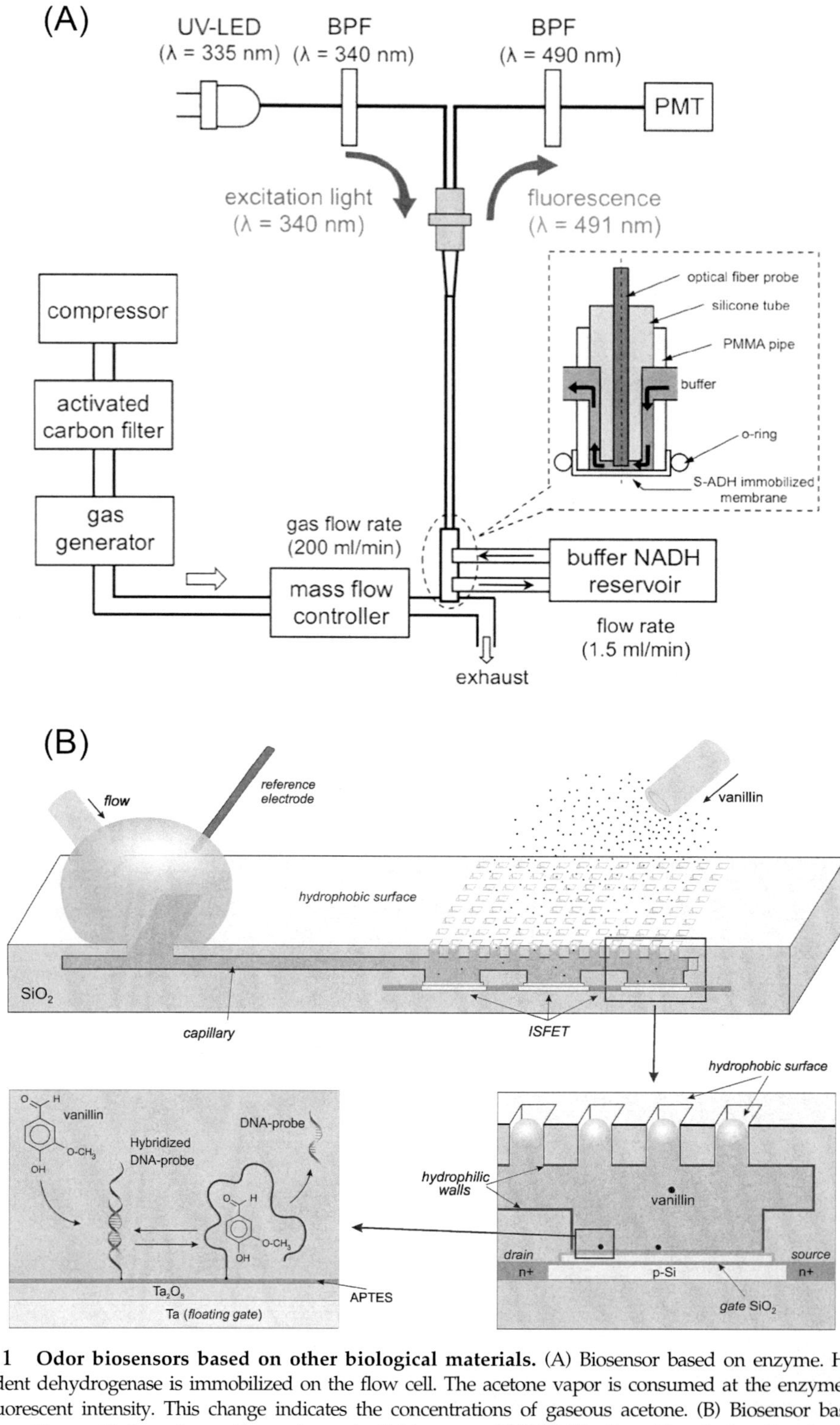

FIGURE 5.11 Odor biosensors based on other biological materials. (A) Biosensor based on enzyme. Here the enzyme NADH dependent dehydrogenase is immobilized on the flow cell. The acetone vapor is consumed at the enzyme membrane and changes the fluorescent intensity. This change indicates the concentrations of gaseous acetone. (B) Biosensor based on aptamer. When vanillin combines with the aptamer, the surface potential increases. *NADH*, Nicotinamide adenine dinucleotide. *Source: (A) is from Ye, M., Chien, P. J., Toma, K., Arakawa, T., & Mitsubayashi, K. (2015). An acetone bio-sniffer (gas phase biosensor) enabling assessment of lipid metabolism from exhaled breath.* Biosensors and Bioelectronics, *73, 208–213, http://www.elsevier.com/locate/bios. https://doi.org/10.1016/j.bios.2015.04.023 with permission from Elsevier; (B) is from Kuznetsov, A. E., Komarova, N. V., Kuznetsov, E. V., Andrianova, M. S., Grudtsov, V. P., Rybachek, E. N., Puchnin, K. V., Ryazantsev, D. V. & Saurov, A. N. (2019). Integration of a field effect transistor-based aptasensor under a hydrophobic membrane for bioelectronic nose applications.* Biosensors and Bioelectronics, *129, 29–35. https://doi.org/10.1016/j.bios.2019.01.013 with permission from Elsevier.*

Second is aptamer-based biosensor. Aptamer can be RNA, single-stranded DNA or double-stranded DNA. It can bind to target molecules with extremely high affinity and specificity. The combination of aptamer and target molecule is similar to that of antigen-antibody, so it can be used as a sensing element in biosensors. In reference (Kuznetsov et al., 2019), aptamer was immobilized onto the sensitive surface of an ion-selective FET (ISFET). When the target odorant (here was vanillin) was applied, it diffused through the pores and dissolved into the buffer solution, then captured by the Van74 DNA aptamer (Fig. 5.11B). Vanillin molecules replaced the hybridized probes on the sensitive surface resulting in an increase of surface potential. Therefore the surface potential value detected by ISFET increased with the vanillin concentration.

Besides enzyme and aptamer, other materials such as taste receptor protein (Son & Park, 2018; Son et al., 2017) and major urinary protein (Scorsone et al., 2021) were also employed for odor biosensors. Due to the space limitation, we cannot enumerate all types of odor biosensors here. The biological materials that are employed in odor biosensors vary from one to another. So the common advantages or disadvantages of those materials are hard to conclude. In Table 5.1, we make a simple comparison of different odor biosensors.

Among all the materials listed in the Table 5.1, OB, OE, and OSN are obtained from experimental animals, and their response signals are close to the original olfactory signal. Hence, the biosensors based on these materials also could also be exploited for exploring the

TABLE 5.1 Comparison of different materials.

Material	Advantages	Disadvantages	Transducers
OB	Easy to obtain Similar sensitivity with animal olfaction	Low selectivity Low reproducibility High operation complexity	MEA
OE	Easy to obtain	Low selectivity Low reproducibility Short lifetime	MEA and LAPS
OSN	Easy to obtain	Low reproducibility Short lifetime	Electrode, LAPS, fluorescence
Cell-expressing OR	Stable characteristics Low operation complexity Good selectivity	Hard to obtain a favorable cell line Individual differences between cells	Electrodes, fluorescence, FET, and LAPS
Nanovesicle	Good sensitivity Good selectivity	High preparation complexity	FET, EIS, and QCM
OR protein	High sensitivity High selectivity Easy to combine with transducers	High preparation complexity	FET, QCM, SAW, and EIS
OBP	Good sensitivity Easy to combine with transducers	High preparation complexity	QCM, FET, SAW, and IDE

EIS, Electrochemical impedance spectroscopy, *FET*, field effect transistor; *IDE*, interdigitated electrodes; *LAPS*, light-addressable potentiometric sensors; *MEA*, microelectrode array; *OB*, olfactory bulb; *OBP*, odorant binding protein; *OE*, olfactory epithelium; *OR*, olfactory receptor; *OSN*, olfactory sensory neuron; *QCM*, quartz crystal microbalance; *SAW*, surface acoustic wave.

mechanism of the biological olfactory system (Gao et al., 2021). Nanovesicle, OR protein, and OBP have small size as well as outstanding sensitivity and selectivity. Those materials are widely integrated with various types of transducers (Khadka et al., 2019; Oh et al., 2019; Yamada et al., 2021). Cell-expressing OR can be treated as isolated OSN, so the biosensors based on cell-expressing OR can imitate animal olfactory systems (Lee, Oh, et al., 2015). Also, the ORs to be expressed are determined by the researchers, thereby the sensitivity and selectivity of those biosensors are guaranteed. Meanwhile, once a stable cell line is obtained, its characteristics remain stable after several generations (Mitsuno et al., 2015). Considering those aforesaid advantages, we will further introduce odor biosensors based on cell-expressing OR in addition to the description in Section 5.3.4.

5.4 Odor biosensors based on cell-expressing olfactory receptor

5.4.1 Discovery and heterologous expression of olfactory receptor

In 1989 David and Randall proposed that an olfactory neuron-specific G protein has a close relationship with the odorant signal transduction (Jones & Reed, 1989). Previous studies from other researchers suggested that olfactory signal transduction is mediated by an adenylate cyclase cascade coupled through a stimulatory G protein (Pace et al., 1985). To find out this G protein, David and Randall screened the rat olfactory cDNA library and discovered a protein that can activate adenylate cyclase, also the RNA of this protein only appeared in olfactory neuroepithelium instead of other tissues such as brain (OB removed), kidney, liver, lung, heart, and intestine. In 1991 Buck and Axel cloned and characterized 18 different members of a multigene family that encodes seven transmembrane domain proteins (now we know they are OR proteins) whose expression is restricted to the OE (Buck & Axel, 1991). Those members are believed to encode a diverse family of odorant receptors. The information pathway and information coding in different animal olfactory systems were later studied by other scientists (Buck, 2000; Duchamp-Viret et al., 1999; Parmentier et al., 1992; Reed, 1992; Ressler et al., 1994; Sato et al., 2008; Wang et al., 1993).

Heterologous expression of OR is a method to better investigate and utilize the molecular mechanism of olfaction. Dahmen et al. (1992) injected the messenger RNA (mRNA) isolated from 4-week-old rat or adult catfish olfactory tissues into the *Xenopus* Oocytes. Two to six days after injection, the oocytes were put in an experimental chamber and the ligands at desired concentrations were applied. The membrane current of the oocyte sometimes changed with odorant stimulation. But the responses were not stable because the injected mRNA types were unknown. In 1993 Raming et al. transfected the *Spodoptera frugiperda* (Sf9) cell with rat OR DNA (OR5, OR12) via recombinant virus. Two days after infection, membranes from Sf9 cells were assayed for odorant-induced generation of cyclic adenosine monophosphate or inositol trisphosphate (IP_3). The IP_3 concentration of OR5-expressing Sf9 cell increased under some odorant stimulations. Similar responses could also be found in isolated OSN. This result reveals the function of OR and indicates the success of heterologous OR expression in host cells. Zhuang and Matsunami (2008) used the plasmid to transfer the OR DNA into the host cell. They also mentioned that the expression of receptor-transporting protein can promote proper targeting of OR proteins to the cell surface (Dwyer et al., 1998; Gimelbrant et al., 2001). With those techniques, the expression of insect or vertebrate ORs on insect cell, vertebrate cell, and bacterium (yeast, *E. coli*) was achieved

(Gimelbrant et al., 1999; Kiefer et al., 1996; Krautwurst et al., 1998; Radhika et al., 2007; Zhang et al., 1997). The ligand libraries of corresponding ORs could be established on the basis of the transfected cells.

5.4.2 Liquid-phase odor biosensor

Although OR protein can be expressed in bacteria, insect cells, or vertebrate cells, considering that the protein modification in prokaryotic cells is worse than in eukaryotic cells, scientists mainly utilized the insect or vertebrate cell for heterologous expression of OR. These heterologously expressed OR maintained the similar detection ability as the OR on the OSN, thereby can be utilized for exploring the target odorants of relevant OR as well as forming the odor biosensor. The cell that expresses OR requires an aqueous extracellular environment to keep its viability, so the odor biosensors based on cell-expressing OR were first proposed in the liquid-phase condition.

The liquid-phase odor biosensors based on cell-expressing OR have been researched for more than two decades (Kajiya et al., 2001; Radhika et al., 2007; Touhara et al., 1999). Some brief introduction about those odor biosensors is provided in Section 5.3.4; here we provide more detailed descriptions.

5.4.2.1 Sensing methods

The electrode and fluorescence methods have been illustrated in Section 5.3.4. Besides them, the cell-expressing OR can also be combined with surface plasmon resonance (SPR), LAPS, and FET. In reference (Lee, Ko, et al., 2009), HEK-293 cell-expressing rat OR (OR I7) was fixed on the SPR sensor chip (Fig. 5.12A). To ensure the stable attachment, the chip was treated with poly-D-lysine, the cell was cultured for a period before experiment. SPR is sensitive to the tiny refractive index changes happening at the surface of the sensor chip. The binding between OR and odorant leads to a calcium ion influx that can be captured by the SPR. The reflectance of SPR changed with the application of the target odorant (octanal). Furthermore, the magnitude of reflectance change had the dose-dependent characteristic with the target odorant.

Fig. 5.12B illustrates an LAPS sensor measured the OR response from the HEK-293 cell-expressing *Caenorhabditis elegans* OR (ODR-10) (Du et al., 2014). The specific binding of OR and odorants leads to the calcium ion influx thus increasing the intracellular Ca^{2+} concentration. It also equals the decrease of extracellular Ca^{2+} concentration. Therefore the LAPS sensor executed the extracellular proton concentration measurement. The OR response presented as the decreased LAPS output potential. The magnitude of change had linear dose dependence in some odorant concentration range.

Terutsuki et al. (2017) integrated the Sf21 cell-expressing silkmoth OR (BmOR3) or *Drosophila melanogaster* OR (Or13a) with FET (Fig. 5.12C). The gate electrodes were extended from the FET gates, and the sensing area sizes were $50 \times 50\ \mu m^2$ and $100 \times 100\ \mu m^2$. The Sf21 cells were loaded on the extended gate electrodes. The drain current varied with the supply of target odorant. For the sake of continuous measurement, a perfusion system was developed. A peristaltic pump provided the assay buffer solution to wash away the odorant from previous stimulations. They also explored the Sf21 cell state on aluminum-based layers (Terutsuki, 2018). The calcium imaging and cell-growth observations showed the cell could function as normal odorant sensor elements, which means the FET can utilize aluminum-based electrode materials for cost-effective sensor systems.

When comparing the biosensors based on the electrode, fluorescence, SPR, LAPS, and FET methods, the electrode method always requires inserting electrodes into the cell, this invasive operation leads to a short lifespan.

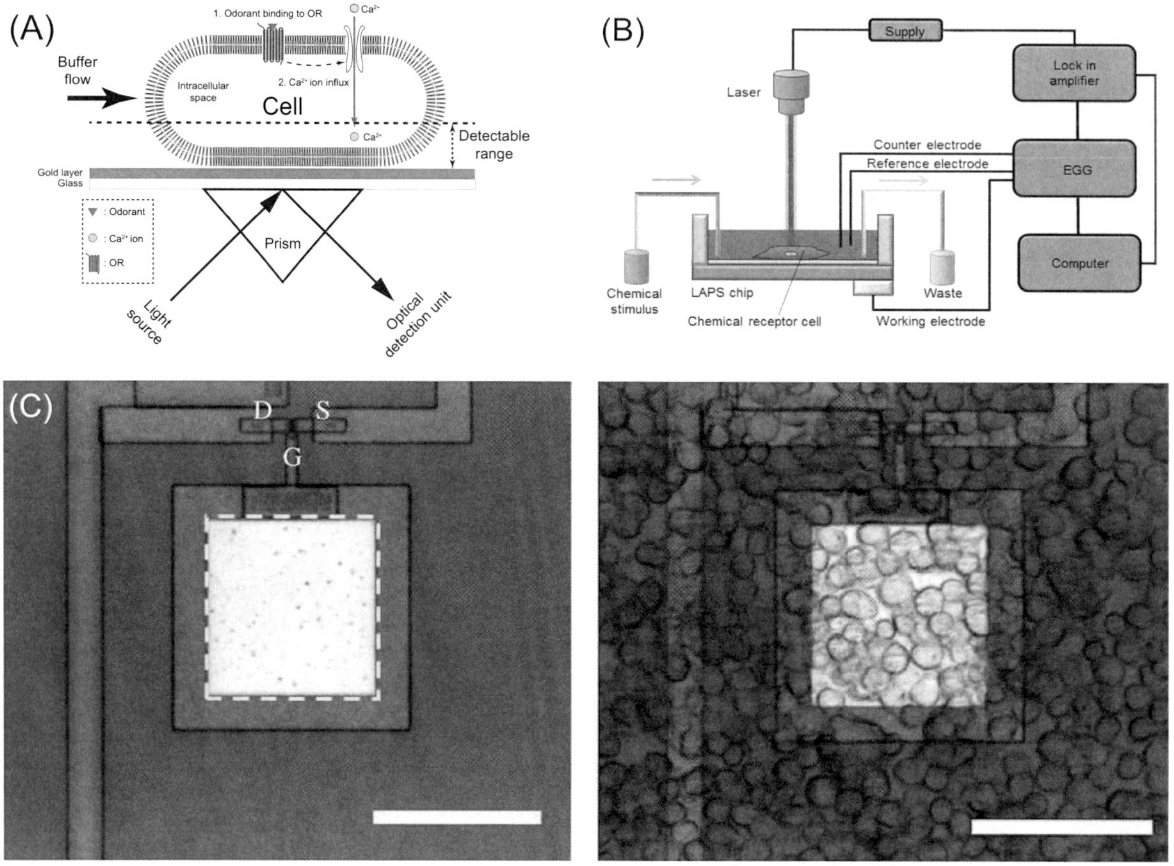

FIGURE 5.12 **Other methods used in odor biosensors based on cell-expressing olfactory receptor (OR).** (A) Cell combined with SPR, the refractive index change induced by the combination of OR and odorant is captured by the SPR sensor. (B) Cell combined with LAPS, the LAPS sensor records the extracellular proton concentration change. (C) Cell combined with FET. The electrophysiological signal influences the FET drain current. Left figure: no cell; Right figure: cell attached. The white scale bars equal 100 μm. *FET*, Field effect transistor; *LAPS*, light-addressable potentiometric sensors; *SPR*, surface plasmon resonance. *Source: (A) is adapted from Lee, S. H., Ko, H. J., & Park, T. H. (2009). Real-time monitoring of odorant-induced cellular reactions using surface plasmon resonance.* Biosensors and Bioelectronics, *25(1), 55–60, https://doi.org/10.1016/j.bios.2009.06.007 with permission from Elsevier; (B) is from Du, L., Zou, L., Zhao, L., Huang, L., Wang, P., & Wu C. (2014). Label-free functional assays of chemical receptors using a bioengineered cell-based biosensor with localized extracellular acidification measurement.* Biosensors and Bioelectronics, *54, 623–627, https://doi.org/10.1016/j.bios.2013.11.049 with permission from Elsevier; (C) is from Terutsuki, D. (2018). Increasing cell–device adherence using cultured insect cells for receptor-based biosensors.* Royal Society Open Science, 5. *(Original work published 2018).*

Although some scholars designed a planar electrode (noninvasive) to measure the OR response (Lee, Jun, et al., 2009), its detection threshold was not favorable (~mM). The SPR, LAPS, and FET methods were noninvasive, but the cell number in the sensing area was limited. The fluorescence method is also a noninvasive method. The experiment could start immediately after several minutes of cell adhesion. In general, hundreds of cells are used to calculate the biosensor response thus lightening the impact of individual difference.

Therefore the odor biosensor based on the fluorescence method has been well developed in recent years (Cho et al., 2021; Hirata et al., 2019; Park et al., 2013; Yokoshiki et al., 2022).

5.4.2.2 Odor biosensor improvement

In the very beginning, the biosensors based on cell-expressing OR only can detect limited types of odorants and the sensor responses were quite unstable (Dahmen et al., 1992; Raming et al., 1993). With the effort of researchers, various OR ligand libraries were established (Münch & Galizia, 2016) and improvements were achieved to enhance the odor biosensor detection performance.

Increasing the OR types in the sensing area can improve the odor biosensor efficiency. Misawa et al. (2010) fabricated a fluidic system and fixed *Xenopus* oocytes expressing different types of ORs in it (Fig. 5.13A). When different odorant solutions were applied, the oocytes only responded to their relevant ligands. Termtanasombat et al. (2016) used a

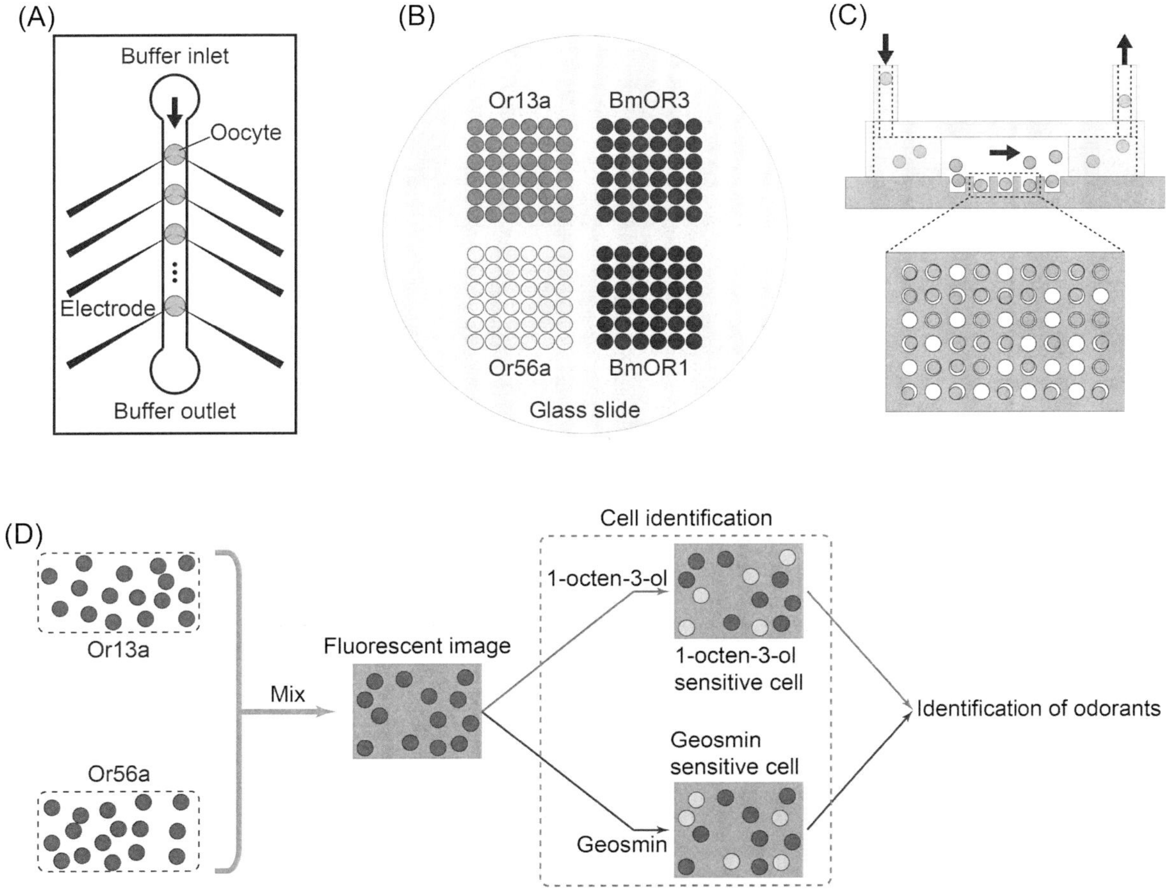

FIGURE 5.13 Methods to increase the olfactory receptor (OR) types in the sensing area. (A) Design a fluidic system and fix multiple cells. (B) Create a cell pattern with different ORs. (C) Manufacture a microfluidic microwell array to trap isolated cells. (D) Use cell image processing to identify the cell type in the cell mixture.

patterned polydimethylsiloxane (PDMS) and cell-immobilizing reagents named biocompatible anchor for membrane to create a clear pattern of Sf21 cells. This pattern consisted of four separated areas, and the cell lines in each area were Or13a, Or56a, BmOR1, and BmOR3, respectively (Fig. 5.13B). This sensor array could yield detectable responses toward target 1-octen-3-ol, geosmin, bombykol, and bombykal mixtures. Figueroa et al. (2010) manufactured a microfluidic microwell array to trap the isolated OSN cells (Fig. 5.13C). The OSNs were incubated with fluorescent dye Fluo4-AM. When the odorants were added, the fluorescence of some OSNs altered. In addition to the OSNs, the cells expressing OR are also suitable for microchamber array chip (Suzuki et al., 2016). The aforementioned three methods require fabrication of electrode voltage clamp or cell pattern or microarray thereby are relatively time-consuming. To further enhance the efficiency, Sukekawa et al. (2019) put forward an image processing method to handle OR mixture without arranging the cell position (Fig. 5.13D). Sf21 cells expressing Or13a or Or56a were mixed together then loaded to a fluidic channel. The target odorants for Or13a and Or56a are 1-octen-3-ol and geosmin, respectively (Münch & Galizia, 2016). The cells were randomly distributed in the experiment area. The region of the individual cell was identified using the Circle Hough Transform function in MATLAB®. The individual cell brightness change was computed through cell tracking technique (Fig. 5.14). The liquid-phase 1-octen-3-ol and geosmin samples were added in sequence. According to the responses toward different odorant stimulations, the OR labels were added to the cells. The labeled cells then were utilized to discriminate the type of later odorant stimulations (Table 5.2).

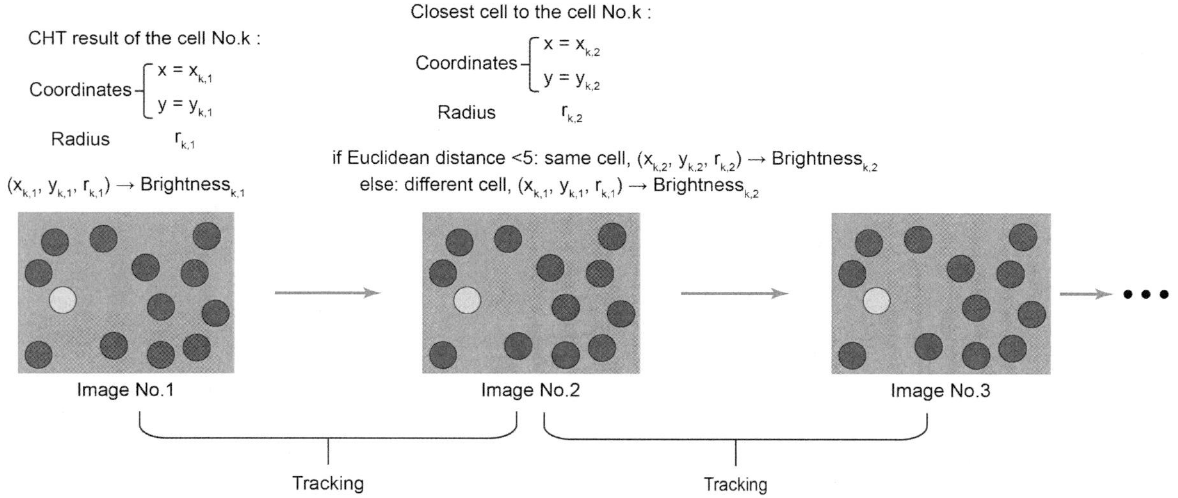

FIGURE 5.14 **Illustration of cell tracking technique.** We first obtain the coordinates and radii of all identified cells in the first image. Among all these cells, suppose we want to track the cell No.k (the bright cell in the figure), and we record its coordinates and radius then calculate its brightness. These data are saved, then we move to the second image. The CHT operation in the second image also identifies many cells. Among those cells, we find a cell that has closest coordinates with the cell No.k in the first image. If their Euclidean distance is less than five pixels, we update the coordinates; otherwise, we copy the coordinates from the first image. The new brightness is computed and saved. Repeat this procedure for all subsequent images until the final one, and the brightness of each cell is computed. *CHT*, Circle Hough Transform.

TABLE 5.2 Odor discrimination results with Mahalanobis distance to centroid of each category. A sample is classified as a category with shorter distance.

Experiment number		1	2	3	4	5	6	7	8	9	10
Result	OCT sensitive	0.02	0.83	0.08	0.86	0.07	1.19	0.15	1.35	0.12	0.93
	GEO sensitive	0.99	0.01	1.63	0.01	0.85	0.05	0.80	0.05	1.71	0.04
Identification result		OCT	GEO	OCT	GEO	OCT	GEO	OCT	GEO	OCT	GEO

OCT: 1-Octen-3-ol, *GEO*: Geosmin.

To overcome the problem that cell fluorescent intensity was not bright enough, a lock-in method was employed by Mujiono et al. (2016, 2017) to mitigate the influence from bright ambient light. The excitation light from a laser diode was an on-off modulation signal. The lock-in technique enables the detection of light synchronous with the modulated excitation light thus improving the signal-to-noise ratio. Hirata et al. (2019) manufactured collagen micropillars with HEK-293 cells expressing OR. The micropillars were made by pouring the collagen solution with cells into a PDMS mold. Therefore the cells not only existed in a single plane but also in all spaces of the entire micropillar. The brightness captured in the micropillar was much brighter compared with in two-dimensional culture.

Unlike the electrical signal, the fluorescence image was two-dimensional information so the biosensor responses were usually handled offline (Deng et al., 2021; Hirata et al., 2019; Lee, Oh, et al., 2015; Mitsuno et al., 2015; Termtanasombat et al., 2016). Sukekawa et al. (2021) executed the online image processing through a commercially available board that consisted of a field-programmable gate array (FPGA) with a hard processor system (HPS). The cell images were recorded by a camera and transferred to the FPGA part. The HPS part accessed the images stored in the FPGA memory and processed them online for individual cell brightness. Based on the OR responses obtained, the mixture composition of known odor could be iteratively adjusted closer to that of unknown sample. This procedure is also named active sensing.

Normally, the target odorant was dissolved in an organic solvent, for example, dimethyl sulfoxide, then diluted by buffer solution before applied to the cells (Sukekawa et al., 2017). To extend the detection range, some gas-phase volatile organic compounds (VOCs) dissolution methods were necessary. Bubbling VOCs through an organic solvent or capturing VOCs in a water/oil system were two suitable ways (Bouchard & Hunkeler, 2014; Sekiguchi et al., 2020). In addition, Terutsuki et al. (2020) established a highly effective VOC dissolution method based on mist atomization. Fig. 5.15 is the concept of this method. The atomizer generated plenty of mist to increase the surface area of the liquid exposed to the gas. When the mist impacted the water surface, the generated fine bubbles further enhanced VOC dissolution speed (Thoroddsen et al., 2012).

5.4.3 Gas-phase odor biosensor

Considering that most odorants exist in the atmosphere, the gas-phase odor biosensors have broader application scenarios compared with the liquid-phase conditions. Although several VOC dissolution strategies have been proposed in Section 5.4.2.2, the direct gas-phase odorant detection is simpler and more efficient.

In the gas-phase odor biosensors based on cell-expressing OR, the cells should be

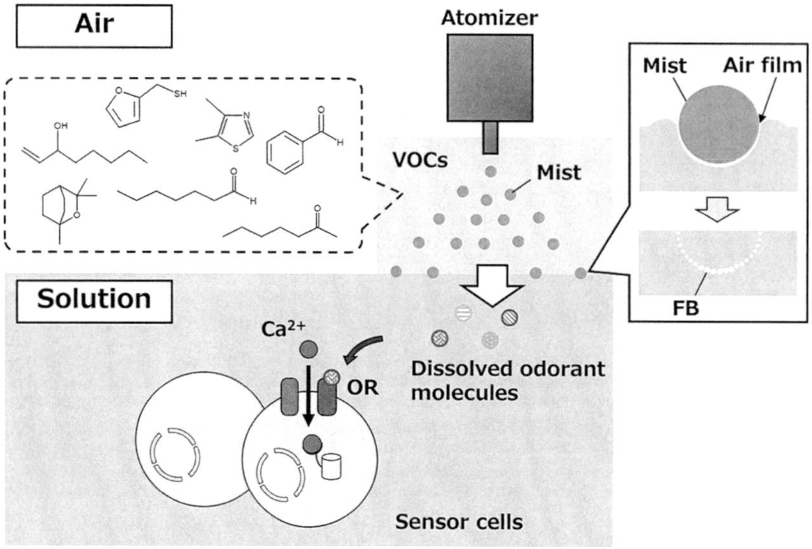

FIGURE 5.15 **Illustration of dissolving volatile organic compounds (VOCs) with poor water solubility into aqueous solution.** The mist and fine bubbles generated by the atomizer can accelerate the VOCs dissolution. *FB*, Fine bubble. *Source: From Terutsuki, D., Mitsuno, H., Sato, K., Sakurai, T., Mase, N., & Kanzaki, R. (2020). Highly effective volatile organic compound dissolving strategy based on mist atomization for odorant biosensors.* Analytica Chimica Acta, 1139, *178–188, https://doi.org/10.1016/j.aca.2020.09.043 with permission from Elsevier.*

submerged in an aqueous buffer medium. Most odorant molecules, however, are hydrophobic. Therefore the major challenge in gas-phase odor biosensor is how to deliver the odor to the cells while maintaining their viability. Scholars attempted multiple approaches to solve this puzzle.

5.4.3.1 *Odor natural dissolution*

Benefiting from the superb sensitivity of ORs, the odor biosensors can detect even a tiny amount of target ligand dissolved in the liquid phase. Kida et al. (2018) recorded the cell brightness change during the odorant vapor dissolution. The pure odorant was diluted with mineral oil, and the odorant solution was put into a sealed luminescence measurement chamber. After incubating for 5 minutes, the odorant vapor in the luminometer chamber reached an equilibration state. Then they plated Hana3A cells expressing ORs on a 96-well plate and added 25 μl of odorant between the wells of the 96-well plate. The plate was put into a luminescence measurement chamber. With the natural dissolution of odorant vapor into the cell culture medium, the ORs on the cell membrane captured the liquid-phase ligands and the cell fluorescent intensity increased.

5.4.3.2 *Special device*

Some specially designed devices were proposed for gas-phase odor measurement. Lee, Oh, et al. (2015) designed a microfluidic platform to imitate the human olfactory system. The cells were loaded on a polycarbonate porous membrane for 48 hours. Then the membrane was flipped over, and some cell culture medium was added into the hole in the PDMS bottom layer. After this, all layers were assembled together as Fig. 5.16A. The cells were surrounded by the medium, which guaranteed their viability. The target odorant was injected into the assembled device using a syringe pump, and the stopcock in the inlet was closed. The channel in the PDMS top layer was filled with odorant, the odorant molecules in the gas phase penetrated the porous membrane to reach the cells in the liquid phase, and then a fluorescence intensity increase could be observed.

Hirata et al. (2018) utilized the cell-laden micropillars to detect gaseous odorant on a liquid–air interface. A pedestal mold was made by cutting a silicone sheet about one square. The collagen was poured into the pedestal mold and incubated for 15 minutes. A PDMS mold with a microwell array was fabricated by standard photolithography method. The cells were suspended in the collagen solution and dropped on PDMS microwell mold. The PDMS microwell mold with collagen pillars was stamped on the pedestal for fixing them together (Fig. 5.16B). The collagen pedestal absorbed the medium and supplied to the pillars thus preventing the pillars from dry damage. Therefore the cells could maintain the wet condition and responded normally toward gaseous odorant stimulations. Furthermore, the pillars can generate a wider gas–liquid interface as compared with a whole block of collagen. It accelerated the reaction of cells to odorants.

5.4.3.3 Special cell

Since air exposure is the main problem of gas-phase odor biosensors, choosing a type of cell that can tolerate dry condition would be a proper path for maintaining the stable detection ability. Nam and Takeuchi (2017) chose the corneal epithelial cells for volatile odorant detection because this cell can survive longer at continuous exposure of air in contrast to other parenchymal cells. An acryl device consisting of medium storage and odorant detection area was fabricated. The detection part has a dorm-shape to prevent the odorant from diffusing to the outside of

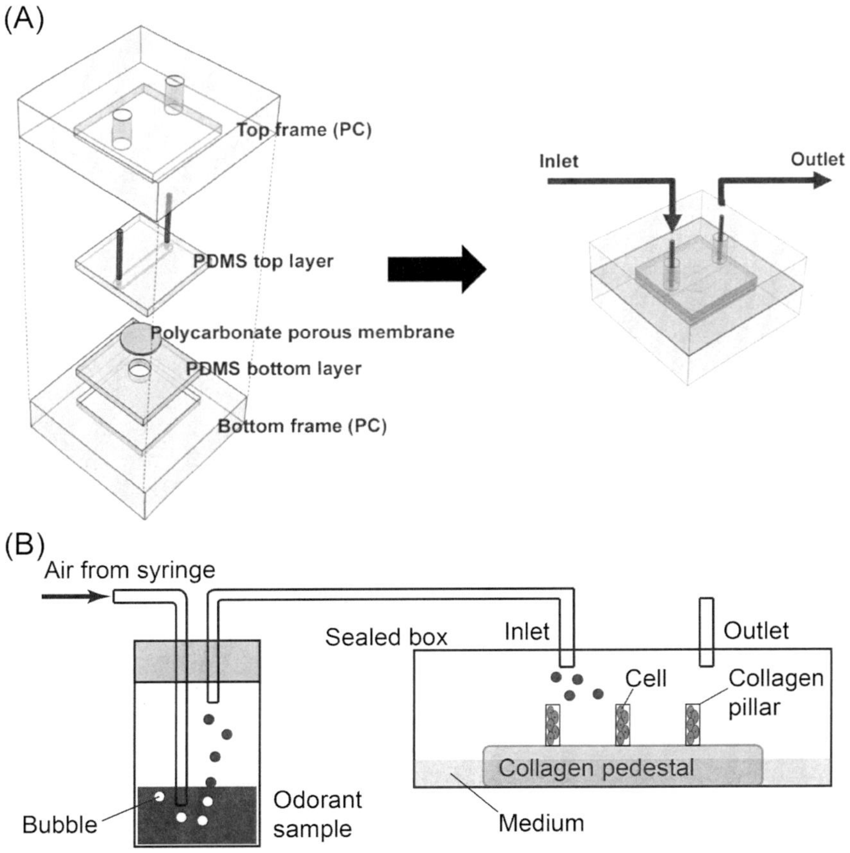

FIGURE 5.16 **Special devices for gas-phase measurement.** (A) Schematic diagram of the microfluidic system. The cells are attached to the polycarbonate porous membrane. The low side of the membrane is attached with the cell culture medium in the PDMS bottom layer hole. The upper side of the membrane is exposed to the target odorant. PC means polycarbonate in this figure. (B) Schematic diagram of cell-laden pillars. The medium is absorbed by the pedestal then compensated for the evaporated liquid in pillars. The cells in the pillars can stay alive for a long time. *PDMS*, polydimethylsiloxane. *Source: (A) is from Lee, S. H., Oh, E. H., & Park, T. H. (2015). Cell-based microfluidic platform for mimicking human olfactory system.* Biosensors and Bioelectronics, 74, *554–561, http://www.elsevier.com/locate/bios. https://doi.org/10.1016/j.bios.2015.06.072 with permission from Elsevier.*

the device. Channels existed at the bottom of the device for adjusting the height of the medium in the storage. The cells were loaded in a collagen gel at the detection area. During no-odor stimulation period, the cells were submerged in the buffer solution. Before the supply of target odorant, the medium was removed then the cells could be exposed to volatile odorant.

5.4.3.4 *Thin liquid film*

The human OE cells are covered by a layer of mucus ranging between approximately 0.5 and 5 μm (Johnson, 2011). This mucus keeps the epithelial cells moist for stable cell viability. Hence, covering a thin-liquid film on the cells expressing ORs can maintain the detection function, while allowing hydrophobic odorant molecules to penetrate it and combine with the ORs expressed on the cell membrane. Sato and Takeuchi (2014) covered the cell spheroid with a thin buffer solution layer to gain multiple OR responses. They first used the microfluidic technique to produce a cell spheroid (Kim et al., 2012), then loaded the spheroid on a hydrogel microchamber and covered it with an extracellular buffer solution, including 1.5% agarose (w/v). The presence of buffer solution was checked through confocal laser Raman microscopy. Although the thickness of the buffer solution was only around 100 μm, the Raman scattering results indicated the buffer medium might exist for more than 1 hour. The OR response was obtained by measuring the spheroid extracellular field potential. The spheroid could respond to odor stimulation for longer than half an hour.

Deng et al. (2021) loaded buffer solution with cells into a small well and applied the odor stimulations. A piece of tape with a 5 mm × 5 mm hole for holding the buffer solution was pasted on a glass slide. The cells were resuspended in buffer solution and loaded into the square hole. After cell adhesion for a few minutes, the additional buffer medium was removed and the target odorant was applied to the square hole. Since the odor was directly applied to the experiment area, the cell responded to the stimulation rapidly. The fluorescent intensity peak appeared around 30 seconds after odor stimulation, and gas volume for single stimulation was less than 1 mL. The lifetime of this biosensor was around 11 minutes.

5.4.3.5 *Liquid thickness control and liquid exchange*

In the liquid-phase odorant detection, the liquid film is quite thick thus no dry out problem (Fig. 5.17A). For direct gas-phase odorant detection, the thin-liquid film method is easy to operate and has fast OR response (Fig. 5.17B). But it also has one significant drawback that the biosensor lifetime is relatively short owing to the dry out of liquid film covering the cell. To overcome this problem, Deng et al. (2022a) came up with a liquid thickness control and liquid exchange method (Fig. 5.17C). In the experiment chamber, two electrodes contacted the bottom of the experiment area (Fig. 5.18). When the water evaporated from the experiment area, the impedance between two electrodes also increased. Therefore the impedance value represented the liquid layer thickness. A feedback control structure was introduced to maintain the liquid film thickness (Fig. 5.19). The set point of this feedback control was obtained in the beginning of the experiment. During the experiment, the difference between set point and current impedance value was input into the microcontroller for adjusting the syringe pump output. The syringe pump infused or withdrew liquid from the experiment area thus kept the liquid film thickness same with the initial state.

With the stable liquid film, the cell lifetime was considerably extended. However, the biosensor lifetime was still similar to no liquid thickness control condition. The reason might be two: First was multiple chemical substances existing in the buffer medium while only water evaporating from the cell area resulting in the

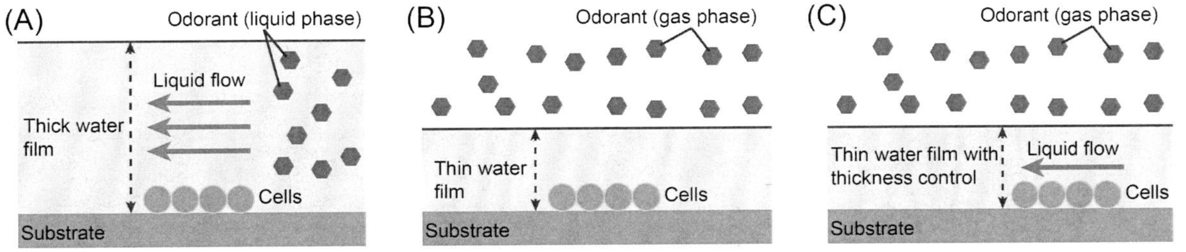

FIGURE 5.17 Different odorant detection conditions. (A) Liquid phase odorant detection, odorant is dissolved into liquid phase then applied to cell. (B) Odorant is directly detected in gas phase under thin-liquid film. (C) Odorant is directly detected in gas phase under thin-liquid film, liquid thickness control and liquid exchange methods are utilized to extend the biosensor lifetime. *Source: From Deng, H., Mitsuno, H., Kanzaki, R., & Nakamoto, T. (2022a). Extending lifetime of gas-phase odor biosensor using liquid thickness control and liquid exchange.* Biosensors and Bioelectronics, 199, *https://doi.org/10.1016/j.bios.2021.113887.*

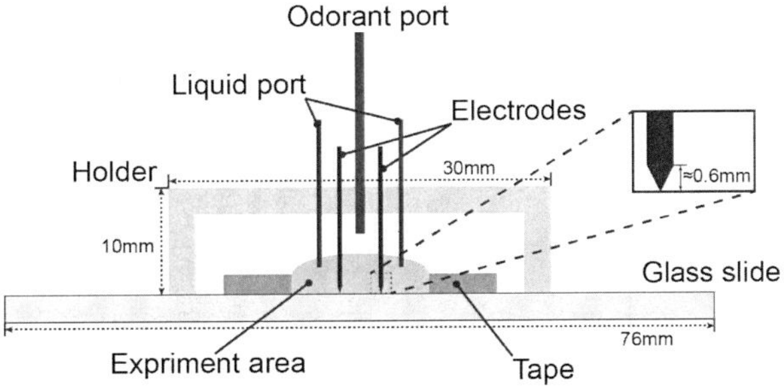

FIGURE 5.18 Schematic of the experiment chamber. Two electrodes contact the bottom of the experiment area. The impedance between two electrodes represents the liquid film thickness. *Source: Adapted from Deng, H., Mitsuno, H., Kanzaki, R., & Nakamoto, T. (2022a). Extending lifetime of gas-phase odor biosensor using liquid thickness control and liquid exchange.* Biosensors and Bioelectronics, 199, *https://doi.org/10.1016/j.bios.2021.113887.*

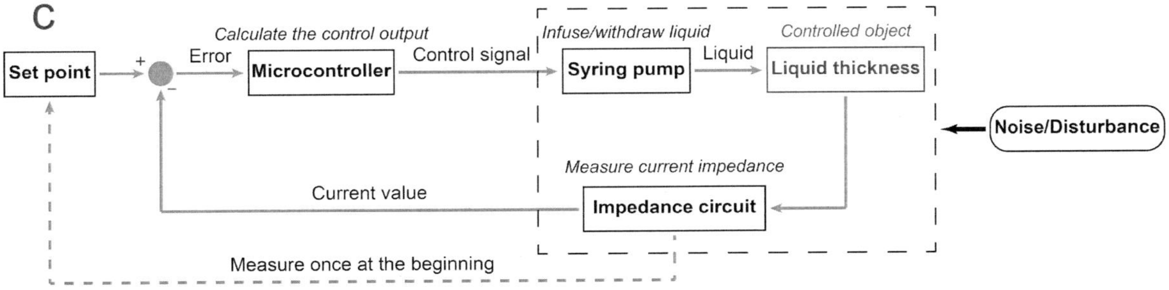

FIGURE 5.19 Mechanism of feedback control method. The blue dashed line from "Impedance circuit" to "Set point" means the impedance measurement only executes once at the beginning of the experiment, whereas the blue solid line represents that they continue the entire experiment. The black dotted box indicates the components affected by noise and disturbance. *Source: From Deng, H., Mitsuno, H., Kanzaki, R., & Nakamoto, T. (2022a). Extending lifetime of gas-phase odor biosensor using liquid thickness control and liquid exchange.* Biosensors and Bioelectronics, 199, *https://doi.org/10.1016/j.bios.2021.113887.*

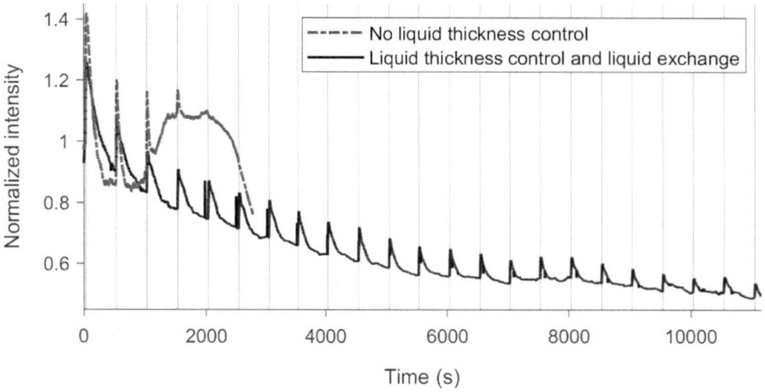

FIGURE 5.20 Extending the lifetime of gas-phase odor biosensor using liquid thickness control and liquid exchange. Red vertical line indicates the timing of vapor exposure, and the fluorescence curves are normalized on the basis of the intensity when first stimulation is applied. In no control condition, the OR response decreases to zero within half an hour (magenta dash line). In liquid thickness control and liquid exchange condition, the OR response can last more than 3 h (blue solid line). *OR*, Olfactory receptor.

ion density increase; second was the residual odorant molecules accumulating in the experiment area. To settle this issue, another syringe pump was introduced to form the liquid exchange thus continuously refreshing the buffer solution in the experiment area. The minimum liquid exchange speed was capable of extending the biosensor lifespan. Faster liquid exchange speed removed more residual odorant molecules thus bringing better OR response. The combination of liquid thickness control and liquid exchange prolonged the biosensor lifetime from half an hour to more than 3 hours (Fig. 5.20). It is also longer than only covering a thin-liquid film in Section 5.4.3.4.

In addition, Deng et al. (2022b) adjusted the liquid layer thickness by changing the impedance set points. The OR response was faster and larger under a thinner liquid layer. However, a too thin-liquid layer also brought about the instability in liquid thickness control. Hence, a suitable liquid film is essential for favorable biosensor performance.

5.5 Summary

The development of odor biosensors is still in the early stage. With the deepening of research, numerous odor biosensors with higher sensitivity and selectivity, faster response time, and wider detection range will be manufactured. Those odor biosensors can help us to explore the secret of animal olfaction as well as benefit our daily lives.

References

Ahn, J. H., Lim, J. H., Park, J., Oh, E. H., Son, M., Hong, S., & Park, T. H. (2015). Screening of target-specific olfactory receptor and development of olfactory biosensor for the assessment of fungal contamination in grain. *Sensors and Actuators, B: Chemical, 210*, 9–16. Available from https://doi.org/10.1016/j.snb.2014.12.060.

Bodyak, N., & Slotnick, B. (1999). Performance of mice in an automated olfactometer: Odor detection, discrimination and odor memory. *Chemical Senses, 24*(6), 637–645. Available from https://doi.org/10.1093/chemse/24.6.637, http://chemse.oxfordjournals.org/.

Bonazza, C., Zhu, J., Hasler, R., Mastrogiacomo, R., Pelosi, P., & Knoll, W. (2021). Responses of the pheromone-binding protein of the silk moth *Bombyx mori* on a graphene biosensor match binding constants in solution. *Sensors (Switzerland)., 21*(2), 1–12. Available from https://doi.org/10.3390/s21020499, https://www.mdpi.com/1424-8220/21/2/499/pdf.

Bouchard, D., & Hunkeler, D. (2014). Solvent-based dissolution method to sample gas-phase volatile organic compounds for compound-specific isotope analysis. *Journal of Chromatography. A, 1325*, 16–22. Available from https://doi.org/10.1016/j.chroma.2013.11.055.

Buck, L., & Axel, R. (1991). A novel multigene family may encode odorant receptors: A molecular basis for odor recognition. *Cell., 65*(1), 175–187. Available from https://doi.org/10.1016/0092-8674(91)90418-x.

Buck, L. B. (2000). The molecular architecture of odor and pheromone sensing in mammals. *Cell.*, *100*(6), 611–618. Available from https://doi.org/10.1016/S0092-8674(00)80698-4, https://www.sciencedirect.com/journal/cell.

Calò, A., Sanmartí-Espinal, M., Iavicoli, P., Persuy, M. A., Pajot-Augy, E., Gomila, G., & Samitier, J. (2012). Diffusion-controlled deposition of natural nanovesicles containing G-protein coupled receptors for biosensing platforms. *Soft Matter.*, *8*(46), 11632–11643. Available from https://doi.org/10.1039/c2sm25893e, http://www.rsc.org/Publishing/Journals/sm/Article.asp?Type = CurrentIssue.

Campanacci, V., Krieger, J., Bette, S., Sturgis, J. N., Lartigue, A., Cambillau, C., Breer, H., & Tegoni, M. (2001). Revisiting the specificity of *Mamestra brassicae* and *Antheraea polyphemus* pheromone-binding proteins with a fluorescence binding assay. *Journal of Biological Chemistry.*, *276*(23), 20078–20084. Available from https://doi.org/10.1074/jbc.m100713200.

Chen, F., Wang, J., Du, L., Zhang, X., Zhang, F., Chen, W., Cai, W., Wu, C., & Wang, P. (2019). Functional expression of olfactory receptors using cell-free expression system for biomimetic sensors towards odorant detection. *Biosensors and Bioelectronics.*, *130*, 382–388. Available from https://doi.org/10.1016/j.bios.2018.09.032.

Cho, S. W., Ko, H. J., & Park, T. H. (2021). Identification of a lung cancer biomarker using a cancer cell line and screening of olfactory receptors for biomarker detection. *Biotechnology and Bioprocess Engineering.*, *26*(1), 55–62. Available from https://doi.org/10.1007/s12257-020-0132-4, http://www.springerlink.com/content/1226-8372.

Choi, D., Lee, S. J., Baek, D., Kim, S.-O., Shin, J., Choi, Y., Cho, Y., Bang, S., Park, J. Y., Lee, S. H., Park, T. H., & Hong, S. (2022). Bioelectrical nose platform using odorant-binding protein as a molecular transporter mimicking human mucosa for direct gas sensing. *ACS Sensors.*, *7*(11), 3399–3408. Available from https://doi.org/10.1021/acssensors.2c01507.

Dahmen, N., Wang, H., & Margolis, F. L. (1992). Expression of olfactory receptors in Xenopus oo*cytes. Journal of Neurochemistry*, *58*(3), 1176–1179. Available from https://doi.org/10.1111/j.1471-4159.1992.tb09379.x.

Dasgupta, D., Warner, T. P. A., Erskine, A., & Schaefer, A. T. (2022). Coupling of mouse olfactory bulb projection neurons to fluctuating odor pulses. *Journal of Neuroscience.*, *42*(21), 4278–4296. Available from https://doi.org/10.1523/JNEUROSCI.1422-21.2022, https://www.jneurosci.org/content/jneuro/42/21/4278.full.pdf.

Datta-Chaudhuri, T., Araneda, R. C., Abshire, P., & Smela, E. (2016). Olfaction on a chip. *Sensors and Actuators, B: Chemical*, *235*, 74–78. Available from https://doi.org/10.1016/j.snb.2016.05.048.

Davison, I. G., & Katz, L. C. (2007). Sparse and selective odor coding by mitral/tufted neurons in the main olfactory bulb. *Journal of Neuroscience.*, *27*(8), 2091–2101. Available from https://doi.org/10.1523/JNEUROSCI.3779-06.2007, http://www.jneurosci.org/cgi/reprint/27/8/2091, United States.

Deng, H., Mitsuno, H., Kanzaki, R., & Nakamoto, T. (2022a). Extending lifetime of gas-phase odor biosensor using liquid thickness control and liquid exchange. *Biosensors and Bioelectronics.*, *199*. Available from https://doi.org/10.1016/j.bios.2021.113887.

Deng, H., Mitsuno, H., Kanzaki, R., & Nakamoto, T. (2021). Gas phase odorant detection by insect olfactory receptor. *IEEE Sensors Journal*, *21*(19), 21184–21191. Available from https://doi.org/10.1109/jsen.2021.3103286.

Deng, H., Mitsuno, H., Kanzaki, R., & Nakamoto, T. (2022b). Study of liquid film thickness for gas phase odor biosensor. *IEEE Sensors Journal*, *22*(17), 16785–16793. Available from https://doi.org/10.1109/jsen.2022.3192567.

Di Pietrantonio, F., Cannatà, D., Benetti, M., Verona, E., Varriale, A., Staiano, M., & D'Auria, S. (2013). Detection of odorant molecules via surface acoustic wave biosensor array based on odorant-binding proteins. *Biosensors and Bioelectronics.*, *41*(1), 328–334. Available from https://doi.org/10.1016/j.bios.2012.08.046.

Du, L., Wu, C., Peng, H., Zou, L., Zhao, L., Huang, L., & Wang, P. (2013). Piezoelectric olfactory receptor biosensor prepared by aptamer-assisted immobilization. *Sensors and Actuators B: Chemical.*, *187*, 481–487. Available from https://doi.org/10.1016/j.snb.2013.02.009.

Du, L., Zou, L., Zhao, L., Huang, L., Wang, P., & Wu, C. (2014). Label-free functional assays of chemical receptors using a bioengineered cell-based biosensor with localized extracellular acidification measurement. *Biosensors and Bioelectronics.*, *54*, 623–627. Available from https://doi.org/10.1016/j.bios.2013.11.049.

Duchamp-Viret, P., Chaput, M. A., & Duchamp, A. (1999). Odor response properties of rat olfactory receptor neurons. *Science (New York, N.Y.)*, *284*(5423), 2171–2179. Available from https://doi.org/10.1126/science.284.5423.2171.

Dwyer, N. D., Troemel, E. R., Sengupta, P., & Bargmann, C. I. (1998). Odorant receptor localization to olfactory cilia is mediated by ODR-4, a novel membrane-associated protein. *Cell.*, *93*(3), 455–466. Available from https://doi.org/10.1016/S0092-8674(00)81173-3, https://www.sciencedirect.com/journal/cell.

Figueroa, X. A., Cooksey, G. A., Votaw, S. V., Horowitz, L. F., & Folch, A. (2010). Large-scale investigation of the olfactory receptor space using a microfluidic microwell array. *Lab on a Chip*, *10*(9), 1120–1127. Available from https://doi.org/10.1039/b920585c, http://pubs.rsc.org/en/journals/journal/lc.

Forstner, M., Breer, H., & Krieger, J. (2009). A receptor and binding protein interplay in the detection of a distinct pheromone component in the silkmoth *Antheraea polyphemus*. *International Journal of Biological Sciences*, *5*(7), 745–757. Available from https://doi.org/10.7150/ijbs.5.745.

Furton, K. G., & Myers, L. J. (2001). The scientific foundation and efficacy of the use of canines as chemical detectors for explosives. *Talanta*, *54*(3), 487–500. Available from https://doi.org/10.1016/S0039-9140(00)00546-4.

Gao, K., Gao, F., Li, J., He, C., Liu, M., Zhu, Q., Qian, Z., Ma, T., & Wang, P. (2021). Biomimetic integrated olfactory sensory and olfactory bulb systems in vitro based on a chip. *Biosensors and Bioelectronics.*, *171*. Available from https://doi.org/10.1016/j.bios.2020.112739.

Gao, K., Li, S., Zhuang, L., Qin, Z., Zhang, B., Huang, L., & Wang, P. (2018). In vivo bioelectronic nose using transgenic mice for specific odor detection. *Biosensors and Bioelectronics.*, *102*, 150–156. Available from https://doi.org/10.1016/j.bios.2017.08.055.

Gimelbrant, A. A., Haley, S. L., & McClintock, T. S. (2001). Olfactory receptor trafficking involves conserved regulatory steps. *Journal of Biological Chemistry.*, *276*(10), 7285–7290. Available from https://doi.org/10.1074/jbc.M005433200.

Gimelbrant, A. A., Stoss, T. D., Landers, T. M., & McClintock, T. S. (1999). Truncation releases olfactory receptors from the endoplasmic reticulum of heterologous cells. *Journal of Neurochemistry*, *72*(6), 2301–2311. Available from https://doi.org/10.1046/j.1471-4159.1999.0722301.x.

Hansson, B. S., & Stensmyr, M. C. (2011). Evolution of insect olfaction. *Neuron*, *72*(5), 698–711. Available from https://doi.org/10.1016/j.neuron.2011.11.003.

Hirata, Y., Morimoto, Y., Nam, E., & Takeuchi, S. (2019). Portable biohybrid odorant sensors using cell-laden collagen micropillars. *Lab on a Chip*, *19*(11), 1971–1976. Available from https://doi.org/10.1039/c9lc00131j.

Hirata, Y., Morimoto, Y., & Takeuchi, S. (2018). *Cell-laden micropillars detect gaseous odorants on a liquid-air interface*. Proceedings of the IEEE International Conference on Micro Electro Mechanical Systems (MEMS) (pp. 304–307), 10.1109/MEMSYS.2018.8346546.

Hou, Y., Jaffrezic-Renault, N., Martelet, C., Zhang, A., Minic-Vidic, J., Gorojankina, T., Persuy, M.-A., Pajot-Augy, E., Salesse, R., Akimov, V., Reggiani, L., Pennetta, C., Alfinito, E., Ruiz, O., Gomila, G., Samitier, J., & Errachid, A. (2007). A novel detection strategy for odorant molecules based on controlled bioengineering of rat olfactory receptor I7. *Biosensors and Bioelectronics.*, *22*(7), 1550–1555. Available from https://doi.org/10.1016/j.bios.2006.06.018.

Jin, H. J., Lee, S. H., Kim, T. H., Park, J., Song, H. S., Park, T. H., & Hong, S. (2012). Nanovesicle-based bioelectronic nose platform mimicking human olfactory signal transduction. *Biosensors and Bioelectronics.*, *35*(1), 335–341. Available from https://doi.org/10.1016/j.bios.2012.03.012.

Johnson, D. C. (2011). Airway mucus function and dysfunction. *Journal of Medicine*, *364*(10). Available from https://doi.org/10.1056/NEJMc1014719, http://www.nejm.org/doi/pdf/10.1056/NEJMc1014719.

Jones, D. T., & Reed, R. R. (1989). Golf: An olfactory neuron specific-G protein involved in odorant signal transduction. *Science (New York, N.Y.)*, *244*(4906), 790–795. Available from https://doi.org/10.1126/science.2499043.

Jones, P. L., Pask, G. M., Rinker, D. C., & Zwiebel, L. J. (2011). Functional agonism of insect odorant receptor ion channels. *Proceedings of the National Academy of Sciences of the United States of America*, *108*(21), 8821–8825. Available from https://doi.org/10.1073/pnas.1102425108, http://www.pnas.org/content/108/21/8821.full.pdf + html.

Kaiser, L., Graveland-Bikker, J., Steuerwald, D., Vanberghem, M., Herlihy, K., & Zhang, S. (2008). Efficient cell-free production of olfactory receptors: Detergent optimization, structure, and ligand binding analyses. *Proceedings of the National Academy of Sciences of the United States of America*, *105*(41), 15726–15731. Available from https://doi.org/10.1073/pnas.0804766105, http://www.pnas.org/content/105/41/15726.full.pdf, United States.

Kajiya, K., Inaki, K., Tanaka, M., Haga, T., Kataoka, H., & Touhara, K. (2001). Molecular bases of odor discrimination: Reconstitution of olfactory receptors that recognize overlapping sets of odorants. *Journal of Neuroscience.*, *21* (16), 6018–6025. Available from https://doi.org/10.1523/jneurosci.21-16-06018.2001, http://www.jneurosci.org.

Kalluri, R., & LeBleu, V. S. (2020). The biology, function, and biomedical applications of exosomes. *Science (New York, N.Y.)*, *367*(6478). Available from https://doi.org/10.1126/science.aau6977, https://science.sciencemag.org/content/367/6478/eaau6977/tab-pdf.

Katre, U. V., Mazumder, S., Prusti, R. K., & Mohanty, S. (2009). Ligand binding turns moth pheromone-binding protein into a pH sensor. *Journal of Biological Chemistry.*, *284*(46), 32167–32177. Available from https://doi.org/10.1074/jbc.m109.013383.

Khadka, R., Aydemir, N., Carraher, C., Hamiaux, C., Colbert, D., Cheema, J., Malmström, J., Kralicek, A., & Travas-Sejdic, J. (2019). An ultrasensitive electrochemical impedance-based biosensor using insect odorant receptors to detect odorants. *Biosensors and Bioelectronics.*, *126*, 207–213. Available from https://doi.org/10.1016/j.bios.2018.10.043.

Kida, H., Fukutani, Y., Mainland, J. D., de March, C. A., Vihani, A., Li, Y. R., Chi, Q., Toyama, A., Liu, L., Kameda, M., Yohda, M., & Matsunami, H. (2018). Vapor detection and discrimination with a panel of odorant receptors. *Nature Communications.*, *9*(1). Available from https://doi.org/10.1038/s41467-018-06806-w.

Kiddy, C. A., Mitchell, D. S., Bolt, D. J., & Hawk, H. W. (1978). Detection of estrus-related odors in cows by trained dogs. *Biology of Reproduction*, *19*(2), 389–395. Available from https://doi.org/10.1095/biolreprod19.2.389.

Kiefer, H., Krieger, J., Olszewski, J. D., von Heijne, G., Prestwich, G. D., & Breer, H. (1996). Expression of an olfactory receptor in *Escherichia coli*: Purification, reconstitution, and ligand binding. *Biochemistry*, *35*(50), 16077–16084. Available from https://doi.org/10.1021/bi9612069.

Kim, T., Doh, I., & Cho, Y. H. (2012). On-chip three-dimensional tumor spheroid formation and pump-less perfusion culture using gravity-driven cell aggregation and balanced droplet dispensing. *Biomicrofluidics.*, *6*(3). Available from https://doi.org/10.1063/1.4739460, http://scitation.aip.org/content/aip/journal/bmf.

Ko, H. J., & Park, T. H. (2005). Piezoelectric olfactory biosensor: Ligand specificity and dose-dependence of an olfactory receptor expressed in a heterologous cell system. *Biosensors and Bioelectronics.*, *20*(7), 1327–1332. Available from https://doi.org/10.1016/j.bios.2004.05.002.

Ko, H. J., & Park, T. H. (2008). Enhancement of odorant detection sensitivity by the expression of odorant-binding protein. *Biosensors and Bioelectronics.*, *23*(7), 1017–1023. Available from https://doi.org/10.1016/j.bios.2007.10.008.

Krautwurst, D., Yau, K. W., & Reed, R. R. (1998). Identification of ligands for olfactory receptors by functional expression of a receptor library. *Cell.*, *95*(7), 917–926. Available from https://doi.org/10.1016/S0092-8674(00)81716-X, https://www.sciencedirect.com/journal/cell.

Kuznetsov, A. E., Komarova, N. V., Kuznetsov, E. V., Andrianova, M. S., Grudtsov, V. P., Rybachek, E. N., Puchnin, K. V., Ryazantsev, D. V., & Saurov, A. N. (2019). Integration of a field effect transistor-based aptasensor under a hydrophobic membrane for bioelectronic nose applications. *Biosensors and Bioelectronics*, *129*, 29–35. Available from https://doi.org/10.1016/j.bios.2019.01.013, http://www.elsevier.com/locate/bios.

Kwon, O. S., Song, H. S., Park, S. J., Lee, S. H., An, J. H., Park, J. W., Yang, H., Yoon, H., Bae, J., Park, T. H., & Jang, J. (2015). An ultrasensitive, selective, multiplexed superbioelectronic nose that mimics the human sense of smell. *Nano Letters*, *15*(10), 6559–6567. Available from https://doi.org/10.1021/acs.nanolett.5b02286, http://pubs.acs.org/journal/nalefd.

Lambin, M., Déglise, P., & Gauthier, M. (2005). Antennal movements as indicators of odor detection by worker honeybees. *Apidologie.*, *36*(1), 119–126. Available from https://doi.org/10.1051/apido:2004076.

Larisika, M., Kotlowski, C., Steininger, C., Mastrogiacomo, R., Pelosi, P., Schütz, S., Peteu, S. F., Kleber, C., Reiner-Rozman, C., Nowak, C., & Knoll, W. (2015). Electronic olfactory sensor based on *A. mellifera* odorant-binding protein 14 on a reduced graphene oxide field-effect transistor. *Angewandte Chemie International Edition*, *54*(45), 13245–13248. Available from https://doi.org/10.1002/anie.201505712.

Leal, W. S. (2013). Odorant reception in insects: Roles of receptors, binding proteins, and degrading enzymes. *Annual Review of Entomology*, *58*, 373–391. Available from https://doi.org/10.1146/annurev-ento-120811-153635.

Leal, W. S., Chen, A. M., Ishida, Y., Chiang, V. P., Erickson, M. L., Morgan, T. I., & Tsuruda, J. M. (2005). Kinetics and molecular properties of pheromone binding and release. *Proceedings of the National Academy of Sciences*, *102*(15), 5386–5391. Available from https://doi.org/10.1073/pnas.0501447102.

Lee, M., Yang, H., Kim, D., Yang, M., Park, T. H., & Hong, S. (2018). Human-like smelling of a rose scent using an olfactory receptor nanodisc-based bioelectronic nose. *Scientific Reports.*, *8*(1). Available from https://doi.org/10.1038/s41598-018-32155-1.

Lee, S. H., Jin, H. J., Song, H. S., Hong, S., & Park, T. H. (2012). Bioelectronic nose with high sensitivity and selectivity using chemically functionalized carbon nanotube combined with human olfactory receptor. *Journal of Biotechnology*, *157*(4), 467–472. Available from https://doi.org/10.1016/j.jbiotec.2011.09.011.

Lee, S. H., Jun, S. B., Ko, H. J., Kim, S. J., & Park, T. H. (2009). Cell-based olfactory biosensor using microfabricated planar electrode. *Biosensors and Bioelectronics.*, *24*(8), 2659–2664. Available from https://doi.org/10.1016/j.bios.2009.01.035.

Lee, S. H., Ko, H. J., & Park, T. H. (2009). Real-time monitoring of odorant-induced cellular reactions using surface plasmon resonance. *Biosensors and Bioelectronics.*, *25*(1), 55–60. Available from https://doi.org/10.1016/j.bios.2009.06.007.

Lee, S. H., Kwon, O. S., Song, H. S., Park, S. J., Sung, J. H., Jang, J., & Park, T. H. (2012). Mimicking the human smell sensing mechanism with an artificial nose platform. *Biomaterials*, *33*(6), 1722–1729. Available from https://doi.org/10.1016/j.biomaterials.2011.11.044.

Lee, S. H., Lim, J. H., Park, J., Hong, S., & Park, T. H. (2015). Bioelectronic nose combined with a microfluidic system for the detection of gaseous trimethylamine. *Biosensors and Bioelectronics*, *71*, 179–185. Available from https://doi.org/10.1016/j.bios.2015.04.033, http://www.elsevier.com/locate/bios.

Lee, S. H., Oh, E. H., & Park, T. H. (2015). Cell-based microfluidic platform for mimicking human olfactory system.

Biosensors and Bioelectronics, 74, 554–561. Available from https://doi.org/10.1016/j.bios.2015.06.072, http://www.elsevier.com/locate/bios.

Liu, P., Cao, T., Xu, J., Mao, X., Wang, D., & Li, A. (2020). Plasticity of sniffing pattern and neural activity in the olfactory bulb of behaving mice during odor sampling, anticipation, and reward. *Neuroscience Bulletin, 36*(6), 598–610. Available from https://doi.org/10.1007/s12264-019-00463-9.

Liu, Q., Hu, N., Ye, W., Cai, H., Zhang, F., & Wang, P. (2011). Extracellular recording of spatiotemporal patterning in response to odors in the olfactory epithelium by microelectrode arrays. *Biosensors and Bioelectronics., 27*(1), 12–17. Available from https://doi.org/10.1016/j.bios.2011.03.010.

Liu, Q., Ye, W., Xiao, L., Du, L., Hu, N., & Wang, P. (2010). Extracellular potentials recording in intact olfactory epithelium by microelectrode array for a bioelectronic nose. *Biosensors and Bioelectronics., 25*(10), 2212–2217. Available from https://doi.org/10.1016/j.bios.2010.02.024.

Liu, Q., Ye, W., Yu, H., Hu, N., Du, L., Wang, P., & Yang, M. (2010). Olfactory mucosa tissue-based biosensor: A bioelectronic nose with receptor cells in intact olfactory epithelium. *Sensors and Actuators B: Chemical., 146*(2), 527–533. Available from https://doi.org/10.1016/j.snb.2009.12.032.

Lu, D., Xu, Q., & Pang, G. (2019). A bombykol electrochemical receptor sensor and its kinetics. *Bioelectrochemistry (Amsterdam, Netherlands), 128*, 263–273. Available from https://doi.org/10.1016/j.bioelechem.2019.04.009.

Lu, Y., Yao, Y., Zhang, Q., Zhang, D., Zhuang, S., Li, H., & Liu, Q. (2015). Olfactory biosensor for insect semiochemicals analysis by impedance sensing of odorant-binding proteins on interdigitated electrodes. *Biosensors and Bioelectronics., 67*, 662–669. Available from https://doi.org/10.1016/j.bios.2014.09.098.

Matarazzo, V., Zsürger, N., Guillemot, J. C., Clot-Faybesse, O., Botto, J. M., Dal Farra, C., Crowe, M., Demaille, J., Vincent, J. P., Mazella, J., & Ronin, C. (2002). Porcine odorant-binding protein selectively binds to a human olfactory receptor. *Chemical Senses, 27*(8), 691–701. Available from https://doi.org/10.1093/chemse/27.8.691, http://chemse.oxfordjournals.org/.

McBride, C. S., Baier, F., Omondi, A. B., Spitzer, S. A., Lutomiah, J., Sang, R., Ignell, R., & Vosshall, L. B. (2014). Evolution of mosquito preference for humans linked to an odorant receptor. *Nature, 515*(7526), 222–227. Available from https://doi.org/10.1038/nature13964, http://www.nature.com/nature/index.html.

Misawa, N., Fujii, S., Kamiya, K., Osaki, T., Takaku, T., Takahashi, Y., & Takeuchi, S. (2019). Construction of a biohybrid odorant sensor using biological olfactory receptors embedded into bilayer lipid membrane on a chip. *ACS Sensors., 4*(3), 711–716. Available from https://doi.org/10.1021/acssensors.8b01615.

Misawa, N., Mitsuno, H., Kanzaki, R., & Takeuchi, S. (2010). Highly sensitive and selective odorant sensor using living cells expressing insect olfactory receptors. *Proceedings of the National Academy of Sciences of the United States of America, 107*(35), 15340–15344. Available from https://doi.org/10.1073/pnas.1004334107, http://www.pnas.org/content/107/35/15340.full.pdf, Japan.

Mitsuno, H., Sakurai, T., Namiki, S., Mitsuhashi, H., & Kanzaki, R. (2015). Novel cell-based odorant sensor elements based on insect odorant receptors. *Biosensors and Bioelectronics., 65*, 287–294. Available from https://doi.org/10.1016/j.bios.2014.10.026.

Mujiono, T., Sukekawa, Y., Nakamoto, T., Mitsuno, H., Nakajima, Y., Kanzaki, R., & Misawa, N. (2016). Lock-in measurement technique in fluorescent instrumentation system for cell-based odor sensor. *IEEJ Transactions on Sensors and Micromachines., 136*(3), 83–89. Available from https://doi.org/10.1541/ieejsmas.136.83.

Mujiono, T., Sukekawa, Y., Nakamoto, T., Mitsuno, H., Termtanasombat, M., Kanzaki, R., & Misawa, N. (2017). Sensitivity improvement by applying lock-in technique to fluorescent instrumentation for cell-based odor sensor. *Sensors and Materials., 29*(1), 65–76. Available from https://doi.org/10.18494/SAM.2017.1378, http://myukk.org/sm_pdf/SM1301.pdf.

Münch, D., & Galizia, C. G. (2016). DoOR 2.0—Comprehensive mapping of *Drosophila melanogaster* odorant responses. Nature Publishing Group, Germany. *Scientific Reports., 6*. Available from https://doi.org/10.1038/srep21841, http://www.nature.com/srep/index.html.

Murugathas, T., Zheng, H. Y., Colbert, D., Kralicek, A. V., Carraher, C., & Plank, N. O. V. (2019). Biosensing with insect odorant receptor nanodiscs and carbon nanotube field-effect transistors. *ACS Applied Materials & Interfaces, 11*(9), 9530–9538. Available from https://doi.org/10.1021/acsami.8b19433.

Nam, E., & Takeuchi, S. (2017). *Volatile odorant detection by corneal epithelial cells using a perfusable fluidic chamber*. Proceedings of the IEEE International Conference on Micro Electro Mechanical Systems (MEMS) (pp. 412–413), Institute of Electrical and Electronics Engineers Inc. Japan. 10.1109/MEMSYS.2017.7863429

Oh, J., Yang, H., Jeong, G. E., Moon, D., Kwon, O. S., Phyo, S., Lee, J., Song, H. S., Park, T. H., & Jang, J. (2019). Ultrasensitive, selective, and highly stable bioelectronic nose that detects the liquid and gaseous cadaverine. *Analytical Chemistry, 91*(19), 12181–12190. Available from https://doi.org/10.1021/acs.analchem.9b01068.

Otto, J., Brown, M. F., & Long, W. (2002). Training rats to search and alert on contraband odors. *Applied Animal Behaviour Science, 77*(3), 217–232. Available from https://doi.org/10.1016/S0168-1591(02)00052-7.

Pace, U., Hanski, E., Salomon, Y., & Lancet, D. (1985). Odorant-sensitive adenylate cyclase may mediate olfactory reception. *Nature, 316*(6025), 255–258. Available from https://doi.org/10.1038/316255a0.

Park, J., Lim, J. H., Jin, H. J., Namgung, S., Lee, S. H., Park, T. H., & Hong, S. (2012). A bioelectronic sensor based on canine olfactory nanovesicle-carbon nanotube hybrid structures for the fast assessment of food quality. *Analyst., 137*(14), 3249–3254. Available from https://doi.org/10.1039/c2an16274a, http://pubs.rsc.org/en/journals/journal/an.

Park, T. H., Lee, S. H., Oh, E. H., Lee, S. H., & Ko, H. J. (2013). Cell-based high-throughput odorant screening system through visualization on a microwell array. *Biosensors and Bioelectronics., 53*, 18–25.

Parmentier, M., Libert, F., Schurmans, S., Schiffmann, S., Lefort, A., Eggerickx, D., Ledent, C., Mollereau, C., Gérard, C., Perret, J., Grootegoed, A., & Vassart, G. (1992). Expression of members of the putative olfactory receptor gene family in mammalian germ cells. *Nature, 355*(6359), 453–455. Available from https://doi.org/10.1038/355453a0.

Pelletier, J., Xu, P., Yoon, K. S., Clark, J. M., & Leal, W. S. (2015). Odorant receptor-based discovery of natural repellents of human lice. *Insect Biochemistry and Molecular Biology, 66*, 103–109. Available from https://doi.org/10.1016/j.ibmb.2015.10.009, http://www.journals.elsevier.com/insect-biochemistry-and-molecular-biology/.

Pevsner, J., Hou, V., Snowman, A. M., & Snyder, S. H. (1990). Odorant-binding protein. Characterization of ligand binding. *Journal of Biological Chemistry., 265*(11), 6118–6125. Available from https://doi.org/10.1016/s0021-9258(19)39300-7.

Pham-Delègue, M. H., Blight, M. M., Kerguelen, V., Le Métayer, M., Marion-Poll, F., Sandoz, J. C., & Wadhams, L. J. (1997). Discrimination of oilseed rape volatiles by the honeybee: Combined chemical and biological approaches. *Entomologia Experimentalis et Applicata., 83*(1), 87–92. Available from https://doi.org/10.1046/j.1570-7458.1997.00160.x, http://onlinelibrary.wiley.com/journal/10.1111/(ISSN)1570-7458.

Radhika, V., Proikas-Cezanne, T., Jayaraman, M., Onesime, D., Ha, J. H., & Dhanasekaran, D. N. (2007). Chemical sensing of DNT by engineered olfactory yeast strain. *Nature Chemical Biology, 3*(6), 325–330. Available from https://doi.org/10.1038/nchembio882.

Raming, K., Krieger, J., Strotmann, J., Boekhoff, I., Kubick, S., Baumstark, C., & Breer, H. (1993). Cloning and expression of odorant receptors. *Nature, 361*(6410), 353–356. Available from https://doi.org/10.1038/361353a0.

Reed, R. R. (1992). Signaling pathways in odorant detection. *Neuron, 8*(2), 205–209. Available from https://doi.org/10.1016/0896-6273(92)90287-N.

Ressler, K. J., Sullivan, S. L., & Buck, L. B. (1994). Information coding in the olfactory system: Evidence for a stereotyped and highly organized epitope map in the olfactory bulb. *Cell., 79*(7), 1245–1255. Available from https://doi.org/10.1016/0092-8674(94)90015-9.

Ritz, S., Hulko, M., Zerfaß, C., May, S., Hospach, I., Krasteva, N., Nelles, G., & Sinner, E. K. (2013). Cell-free expression of a mammalian olfactory receptor and unidirectional insertion into small unilamellar vesicles (SUVs). *Biochimie, 95*(10), 1909–1916. Available from https://doi.org/10.1016/j.biochi.2013.06.021.

Sankaran, S., Panigrahi, S., & Mallik, S. (2011). Olfactory receptor based piezoelectric biosensors for detection of alcohols related to food safety applications. *Sensors and Actuators B: Chemical., 155*(1), 8–18. Available from https://doi.org/10.1016/j.snb.2010.08.003.

Sato, K., Pellegrino, M., Nakagawa, T., Nakagawa, T., Vosshall, L. B., & Touhara, K. (2008). Insect olfactory receptors are heteromeric ligand-gated ion channels. *Nature, 452*(7190), 1002–1006. Available from https://doi.org/10.1038/nature06850.

Sato, K., & Takeuchi, S. (2014). Chemical vapor detection using a reconstituted insect olfactory receptor complex. *Angewandte Chemie International Edition., 53*(44), 11798–11802. Available from https://doi.org/10.1002/anie.201404720.

Scorsone, E., Manai, R., Cali, K., Ricatti, M. J., Farno, S., Persaud, K., & Mucignat, C. (2021). Biosensor array based on ligand binding proteins for narcotics and explosives detection. *Sensors and Actuators B: Chemical., 334*. Available from https://doi.org/10.1016/j.snb.2021.129587.

Sekiguchi, K., Yasui, F., & Fujii, E. (2020). Capturing of gaseous and particulate pollutants into liquid phase by a water/oil column using microbubbles. *Chemosphere, 256*. Available from https://doi.org/10.1016/j.chemosphere.2020.126996.

Son, M., Cho, D.-G., Lim, J. H., Park, J., Hong, S., Ko, H. J., & Park, T. H. (2015). Real-time monitoring of geosmin and 2-methylisoborneol, representative odor compounds in water pollution using bioelectronic nose with human-like performance. *Biosensors and Bioelectronics., 74*, 199–206. Available from https://doi.org/10.1016/j.bios.2015.06.053.

Son, M., Kim, D., Kang, J., Lim, J. H., Lee, S. H., Ko, H. J., Hong, S., & Park, T. H. (2016). Bioelectronic nose using odorant binding protein-derived peptide and carbon nanotube field-effect transistor for the assessment of salmonella contamination in food. *Analytical Chemistry, 88*

(23), 11283–11287. Available from https://doi.org/10.1021/acs.analchem.6b03284.

Son, M., Kim, D., Ko, H. J., Hong, S., & Park, T. H. (2017). A portable and multiplexed bioelectronic sensor using human olfactory and taste receptors. *Biosensors and Bioelectronics*, *87*, 901–907. Available from https://doi.org/10.1016/j.bios.2016.09.040, http://www.elsevier.com/locate/bios.

Son, M., & Park, T. H. (2018). The bioelectronic nose and tongue using olfactory and taste receptors: Analytical tools for food quality and safety assessment. *Biotechnology Advances*, *36*(2), 371–379. Available from https://doi.org/10.1016/j.biotechadv.2017.12.017, http://www.elsevier.com/inca/publications/store/5/2/5/4/5/5/index.htt.

Steinbrecht, R. A. (1998). Odorant-binding proteins: Expression and function. *Annals of the New York Academy of Sciences*, *855*, 323–332. Available from https://doi.org/10.1111/j.1749-6632.1998.tb10591.x, http://www.blackwellpublishing.com/0077-8923.

Stensmyr, M. C., Dweck, H. K. M., Farhan, A., Ibba, I., Strutz, A., Mukunda, L., Linz, J., Grabe, V., Steck, K., Lavista-Llanos, S., Wicher, D., Sachse, S., Knaden, M., Becher, P. G., Seki, Y., & Hansson, B. S. (2012). A conserved dedicated olfactory circuit for detecting harmful microbes in drosophila. *Cell.*, *151*(6), 1345–1357. Available from https://doi.org/10.1016/j.cell.2012.09.046, https://www.sciencedirect.com/journal/cell.

Sukekawa, Y., Mitsuno, H., Kanzaki, R., & Nakamoto, T. (2021). Binary mixture quantification using cell-based odor biosensor system with active sensing. *Biosensors and Bioelectronics.*, *179*. Available from https://doi.org/10.1016/j.bios.2021.113053.

Sukekawa, Y., Mitsuno, H., Kanzaki, R., & Nakamoto, T. (2019). Odor discrimination using cell-based odor biosensor system with fluorescent image processing. *IEEE Sensors Journal*, *19*(17), 7192–7200. Available from https://doi.org/10.1109/jsen.2019.2916377.

Sukekawa, Y., Mujiono, T., Nakamoto, T., Mitsuno, H., Nakajima, Y., Kanzaki, R., & Misawa, N. (2017). Development of automated flow measurement system for cell-based odor sensor. *Electronics and Communications in Japan.*, *100*(9), 41–49. Available from https://doi.org/10.1002/ecj.11981, http://onlinelibrary.wiley.com/journal/10.1002/(ISSN)1942-9541.

Sung, J. H., Ko, H. J., & Park, T. H. (1981). Piezoelectric biosensor using olfactory receptor protein expressed in *Escherichia coli*. *Biosensors and Bioelectronics.*, *21*.

Suzuki, M., Yoshimoto, N., Shimono, K., & Kuroda, S. (2016). Deciphering the receptor repertoire encoding specific odorants by time-lapse single-cell array cytometry. *Scientific Reports.*, *6*(1). Available from https://doi.org/10.1038/srep19934.

Termtanasombat, M., Mitsuno, H., Misawa, N., Yamahira, S., Sakurai, T., Yamaguchi, S., Nagamune, T., & Kanzaki, R. (2016). Cell-based odorant sensor array for odor discrimination based on insect odorant receptors. *Journal of Chemical Ecology*, *42*(7), 716–724. Available from https://doi.org/10.1007/s10886-016-0726-7.

Terutsuki, D., Mitsuno, H., Sato, K., Sakurai, T., Mase, N., & Kanzaki, R. (2020). Highly effective volatile organic compound dissolving strategy based on mist atomization for odorant biosensors. *Analytica Chimica Acta*, *1139*, 178–188. Available from https://doi.org/10.1016/j.aca.2020.09.043.

Terutsuki, D. (2018). Increasing cell–device adherence using cultured insect cells for receptor-based biosensors. *Royal Society Open Science.*, *5*.

>Terutsuki, D., Mitsuno, H., Okamoto, Y., Sakurai, T., Tixier-Mita, A., Toshiyoshi, H., Mita, Y., & Kanzaki R. (2017). *Odor-sensitive field effect transistor (OSFET) based on insect cells expressing insect odorant receptors*. Proceedings of the IEEE International Conference on Micro Electro Mechanical Systems (MEMS) (pp. 394–397), Available from https://doi.org/10.1109/MEMSYS.2017.7863424.

Thoroddsen, S. T., Thoraval, M.-J., Takehara, K., & Etoh, T. G. (2012). Micro-bubble morphologies following drop impacts onto a pool surface. *Journal of Fluid Mechanics*, *708*, 469–479. Available from https://doi.org/10.1017/jfm.2012.319.

Touhara, K., Sengoku, S., Inaki, K., Tsuboi, A., Hirono, J., Sato, T., Sakano, H., & Haga, T. (1999). Functional identification and reconstitution of an odorant receptor in single olfactory neurons. *Proceedings of the National Academy of Sciences of the United States of America*, *96*(7), 4040–4045. Available from https://doi.org/10.1073/pnas.96.7.4040.

Wang, H. W., Wysocki, C. J., & Gold, G. H. (1993). Induction of olfactory receptor sensitivity in mice. *Science (New York, N.Y.)*, *260*(5110), 998–1000. Available from https://doi.org/10.1126/science.8493539.

Wojtasek, H., & Leal, W. S. (1999). Conformational change in the pheromone-binding protein from *Bombyx mori* induced by pH and by interaction with membranes. *Journal of Biological Chemistry.*, *274*(4 3), 30950–30956. Available from https://doi.org/10.1074/jbc.274.43.30950.

Wu, C., Du, L., Wang, D., Wang, L., Zhao, L., & Wang, P. (2011). A novel surface acoustic wave-based biosensor for highly sensitive functional assays of olfactory receptors. *Biochemical and Biophysical Research Communications*, *407*(1), 18–22. Available from https://doi.org/10.1016/j.bbrc.2011.02.073.

Yamada, T., Sugiura, H., Mimura, H., Kamiya, K., Osaki, T., & Takeuchi, S. (2021). Highly sensitive VOC detectors using insect olfactory receptors reconstituted into lipid bilayers. *Science Advances.*, *7*(3). Available from https://doi.org/10.1126/sciadv.abd2013.

Yang, H., Kim, D., Kim, J., Moon, D., Song, H. S., Lee, M., Hong, S., & Park, T. H. (2017). Nanodisc-based bioelectronic nose using olfactory receptor produced in *Escherichia coli* for the assessment of the death-associated odor cadaverine. *ACS Nano.*, *11*(12), 11847–11855. Available from https://doi.org/10.1021/acsnano.7b04992.

Ye, M., Chien, P. J., Toma, K., Arakawa, T., & Mitsubayashi, K. (2015). An acetone bio-sniffer (gas phase biosensor) enabling assessment of lipid metabolism from exhaled breath. *Biosensors and Bioelectronics*, *73*, 208–213. Available from https://doi.org/10.1016/j.bios.2015.04.023, http://www.elsevier.com/locate/bios.

Yokoshiki, Y., Nagayoshi, K., Mitsuno, H., Niki, S., Kanzaki, R., Tokuda, T., & Nakamoto, T. (2022). Improving performance of odor biosensor system using olfactory receptor. *Sensors and Materials.*, *34*(4), 1617–1626. Available from https://doi.org/10.18494/SAM3765, https://sensors.myu-group.co.jp/article.php?ss = 3765.

Zhang, Y., Chou, J. H., Bradley, J., Bargmann, C. I., & Zinn, K. (1997). The *Caenorhabditis elegans* seven-transmembrane protein ODR-10 functions as an odorant receptor in mammalian cells. *Proceedings of the National Academy of Sciences of the United States of America.*, *94* (22), 12162–12167. Available from https://doi.org/10.1073/pnas.94.22.12162.

Zhu, P., Liu, S., Tian, Y., Chen, Y., Chen, W., Wang, P., Du, L., & Wu, C. (2022). In vivo bioelectronic nose based on a bioengineered rat realizes the detection and classification of multiodorants. *ACS Chemical Neuroscience.*, *13* (12), 1727–1737. Available from https://doi.org/10.1021/acschemneuro.2c00015.

Zhuang, H., & Matsunami, H. (2008). Evaluating cell-surface expression and measuring activation of mammalian odorant receptors in heterologous cells. *Nature Protocols*, *3*(9), 1402–1413. Available from https://doi.org/10.1038/nprot.2008.120.

Zhuang, L., Guo, T., Cao, D., Ling, L., Su, K., Hu, N., & Wang, P. (2015). Detection and classification of natural odors with an in vivo bioelectronic nose. *Biosensors and Bioelectronics.*, *67*, 694–699. Available from https://doi.org/10.1016/j.bios.2014.09.102.

Zhuang, L., Hu, N., Dong, Q., Liu, Q., & Wang, P. (2013). A high sensitive in vivo biosensing detection for odors by multiunit in rat olfactory bulb. *Sensors and Actuators B: Chemical.*, *186*, 308–314. Available from https://doi.org/10.1016/j.snb.2013.06.032.

Zhuang, L., Hu, N., Tian, F., Dong, Q., Hu, L., Li, R., & Wang, P. (2014). A high-sensitive detection method for carvone odor by implanted electrodes in rat olfactory bulb. *Chinese Science Bulletin.*, *59*(1), 29–37. Available from https://doi.org/10.1007/s11434-013-0044-1.

CHAPTER

6

Active odor sensing system

Manuel Aleixandre

Institute of Integrated Research, Institute of Science Tokyo, Yokohama, Japan

6.1 Active sensing

When a sensor system adapts in synchrony with an ever-evolving world, refining its observations to gather more information or to enhance the process, we are talking about active sensing. Active sensing can be defined as a sensing procedure in which measurement conditions are modified on the basis of the specific measurement being made, with the aim of enhancing the measurement's accuracy or acquiring additional information (Nakamoto, 2016). In certain studies, active sensing involves emitting energy, as seen in echolocation used by bats and dolphins or in man-made radar (Nelson & MacIver, 2006). However, this study exclusively focuses on the former interpretation, which is allocative sensing (involving the modification of the sensor system), and does not encompass homeoactive sensing (involving modifications to the external environment) (Zweifel & Hartmann, 2020).

Fig. 6.1 illustrates the procedure of an active sensing process. The top row denotes the sequential steps (from 1 to 4). The second and third rows offer a visual representation and a corresponding explanation of each step. In the first step, the system (depicted as a blue box) begins in its initial state ($\mathbf{P}_i$). In the second step, a measure of the environment ($\mathbf{E}_i$) (depicted as a green box) is acquired ($\mathbf{S}_i$). This measurement is a function (*f*) of both the environment and the internal state of the sensor system. Moving to the third step, the system evaluates the measurement and the internal state and then calculates adjustments to the parameters of the sensor system ($\mathbf{P}_{i+1}$) to improve the measurement quality or gather additional information. Finally, the internal state of the sensor is modified, and the procedure returns to step 1. The bottom row illustrates an example of an active measurement process that involves the localization of an object by modifying the sensor system's position. The system progressively moves closer to the object in each iteration. In detail, in the first step, the robot (depicted as a small blue square) interacts with the environment to locate an object (depicted as a green spiked star) using two sensors (depicted as red dots). In the second step, a measurement is obtained, which is a function of the environment (the location of the object to be detected) and the state of the robot (its position and orientation). In the next step, the active sensing algorithm calculates the modification of the position and

Digital Technologies in Olfaction
DOI: https://doi.org/10.1016/B978-0-443-15721-9.00006-3

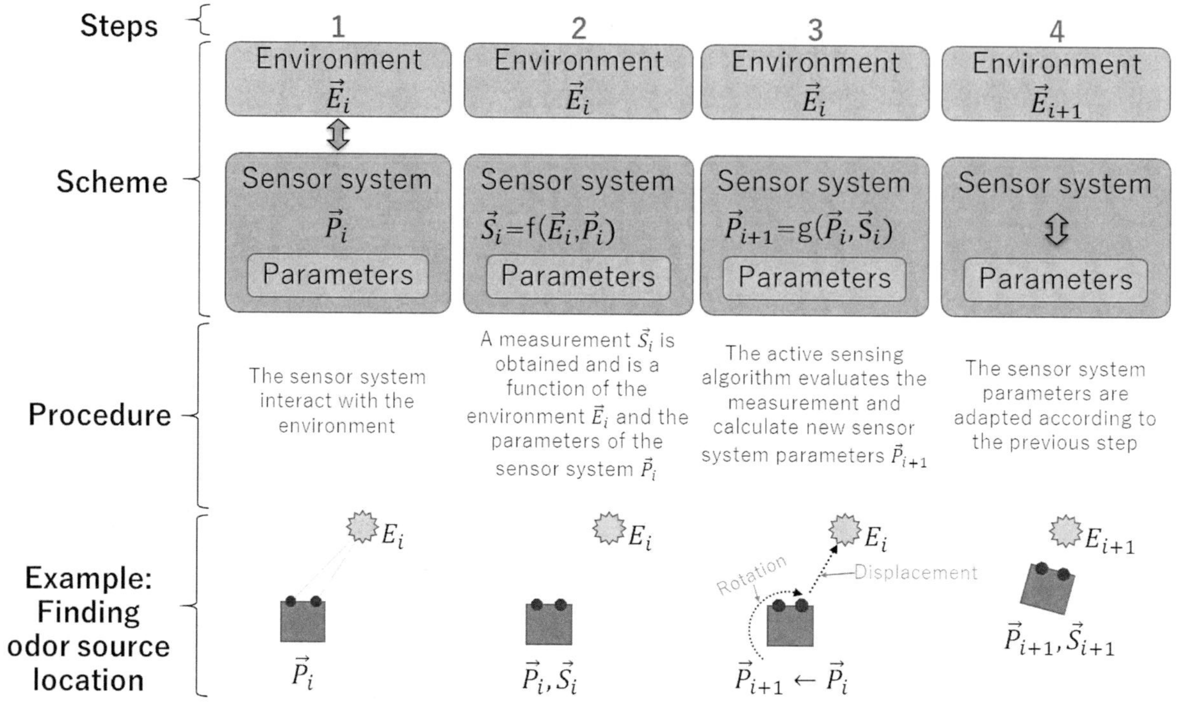

FIGURE 6.1 Scheme of active sensing. Active sensing is illustrated here. Columns depict the different steps, and rows depict different aspects of each step: step number, scheme of the action, comment, and illustrated example.

orientation of the robot to, finally, approach the object in the last step. These steps are repeated until the object is determined to have been found.

Nature offers several instances of active sensing that provide valuable insights into its potential advantages. Very clear examples of active sensing include the precise eye movements employed to center an object within the more sensitive part of the field of vision (Ballard, 1991), as well as the tactile movements of whiskers in certain mammals to detect objects (O'Connor et al., 2010). Active sense also happens in olfaction, where even simple organisms like larvae exhibit active sensing behaviors, as demonstrated by Gomez-Marin and Louis (2012). In their research, it was described how *Drosophila melanogaster* larvae, with a nervous system comprising fewer than 10,000 neurons, employ active sensing to locate odor sources. The study revealed that the larvae tilt their heads laterally to sense odor gradients, and their behavior cannot be explained by random movements or stereo-olfaction.

In particular, olfaction in mammals is inherently linked to active sensing due to the initial inhalation of air as the primary step in the olfactory process, which makes active sensing and mammal olfaction closely related. This close relationship is substantiated by investigations into sniffing patterns (Jordan et al., 2018; Wachowiak, 2011) that demonstrate how these patterns modulate neural activity, enhancing the neural encoding of odors and facilitating odor identification (Jordan et al., 2018). Another fundamental example of the active sensing in nature is the location of odor sources (Thesen et al., 1993) that has inspired significant advancements in the field of active sensing related to odor and pollution source localization.

Since gas sensor systems have a wide variety of configurations and belong to a big number of technologies (Hunter et al., 2020; Nazemi et al., 2019), the characteristics that can be tuned during the active measurement are very diverse. Among the parameters subject to modification during the measurement procedure encompass location (Deveza et al., 1994; Ishida et al., 1994; Jing et al., 2021; Rozas et al., 1991), carrier gas flow rate (Wang et al., 2017), operation temperature (Gosangi & Gutierrez-Osuna, 2009; Gosangi & Gutierrez-Osuna, 2010; Herrero-Carrón et al., 2015; Martinelli et al., 2012), gas mixture composition (Nakamoto et al., 1994; Sukekawa et al., 2021; Yokoshiki & Nakamoto, 2019b), or selection of light frequencies in optical measurement (Huang & Gutierrez-Osuna, 2016; Huang et al., 2012). While many studies explore the offline impact of these parameters or optimize system performance for specific applications (Lee & Reedy, 1999; Ziyatdinov et al., 2015a), their real-time adjustment during the measurement process remains relatively underexplored. Nevertheless, they serve as a valuable starting point when implementing active sensing algorithms.

The additional information obtained by the active sensing procedure can be leveraged to improve the discrimination between gases or mixtures, quantification of gas concentrations, improve measurement stability, reduce measurement duration, or obtain odor source location. In the following sections, we study the main research lines currently in development, categorizing them based on the parameters they modify.

6.2 Temperature of operation

While the modulation of the sensor temperature of metal oxide resistive sensors has been systematically studied with different heating patterns or profiles (Lee & Reedy, 1999), the closed-loop control of the temperature during the measurement has been not widely explored.

An initial work (Nakamoto et al., 1991) investigated the adjustment of the operating temperature in metal oxide resistive gas sensors for the optimization of gas discrimination. In this experiment, two commercial resistive sensors were employed to discriminate among three distinct gases: acetone, ammonia, and hexane. The algorithm conducted measurements at several sensor temperatures and calculated the separation index among the three classes based on multiple metrics (Euclidean distance, vector angle, Mahalanobis distance, and Wilks' lambda). Subsequently, the method calculated the gradient of this separation respect to the temperature and used this gradient to optimize sensor temperatures and improve the separation index. It repeated the procedure in iterations till it found the optimal sensor temperature profile (Fig. 6.2).

In subsequent studies (Gosangi & Gutierrez-Osuna, 2009; Gosangi & Gutierrez-Osuna, 2010), the fact that the resistive metal oxide sensor responses depend on the operation temperature was also used. This opened the door to extracting various temperature-dependent sensor features, enabling the acquisition of additional information. Instead of scanning all the possible temperature settings, an algorithm based on a Markov model that mapped sensor resistance to operation temperatures was designed. Subsequently, an optimal temperature sequence was determined using a partially observable Markov decision process (POMDP) that maximized classification accuracy while minimized sensing cost.

The procedure is as follows: Initially, all possible classes for the gas being measured were assigned equal probabilities. Then, an iterative loop began. Based on the information available, a sensor and a temperature setting were selected, leading to a measurement. This measurement updated the model, the

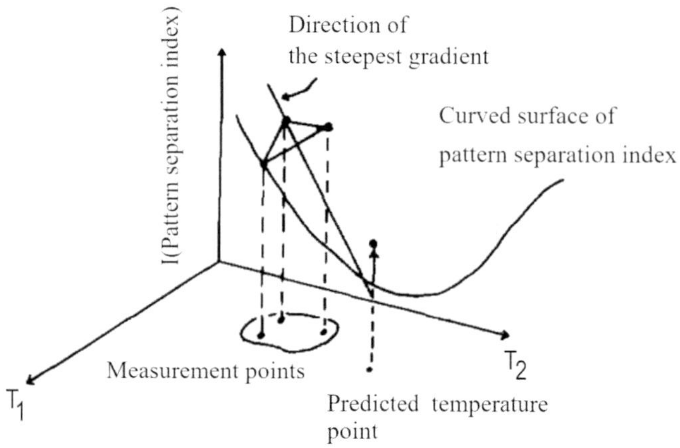

FIGURE 6.2 Illustration of the gradient descent procedure. In the plot, the horizontal axis represents the temperatures of the sensors, while the vertical axis represents the separation index among the measured gases. Three points are measured, and then the gradient is calculated resulting in a direction to modify the temperature of the sensors. Source: *From Nakamoto, T., Fukuda, T., & Moriizumi, T. (1991). Gas identification system using plural sensors with characteristics of plasticity.* Sensors and Actuators B: Chemical, *3(1), 1–6. https://doi.org/10.1016/0925-4005(91)85001-y.*

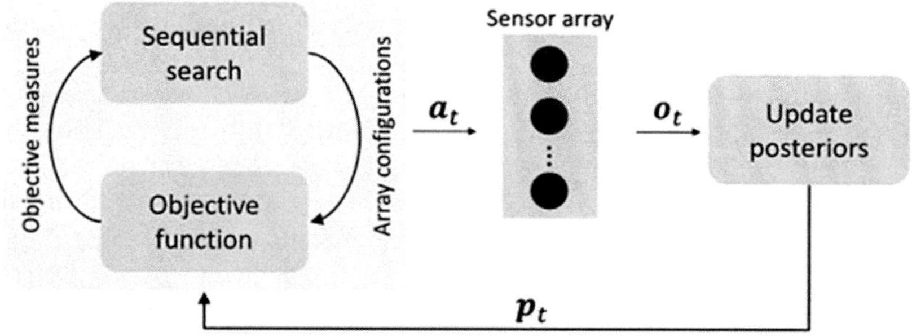

FIGURE 6.3 Scheme of the Posterior-Weighed Active Search for temperature modulated active sensing. Scheme of the Posterior-Weighed Active Search, it uses a sequential search to produce several array configurations that are then evaluated by the objective function. The optimal configuration is selected to operate the sensor array that in turn is used to update the algorithm. Source: *From Gosangi, R., & Gutierrez-Osuna, R. (2014). Active classification with arrays of tunable chemical sensors.* Chemometrics and Intelligent Laboratory Systems, *132, 91–102. https://doi.org/10.1016/j.chemolab.2014.01.001.*

probability distributions, and the measurement cost. This procedure was iterated till the cost exceeded the classification improvement measured by Bayes risk. This iterative process reduced the measurement steps with a constrained accuracy loss (Gosangi & Gutierrez-Osuna, 2009). In a separate study (Gosangi & Gutierrez-Osuna, 2010), it was successfully experimentally applied to a single sensor (a commercial metal oxide resistive sensor), eight operation temperatures, and three gases (ammonia, acetone, and isopropyl alcohol).

In more recent works, the same group improved the methods by studying several algorithms, and in Gosangi and Gutierrez-Osuna (2014), a posterior-weighted active search method was used as illustrated by Fig. 6.3. This method integrated feature subset selection and sequential Bayesian filtering, resulting in superior classification accuracy and robustness when compared to passive forward selection algorithms.

Other studies have explored temperature adjustment based on the sensor measurement although the modulation in these works is not dependent on an evaluation of the performance, but it is done in an automatic, predetermined pattern. For instance, in Martinelli et al.

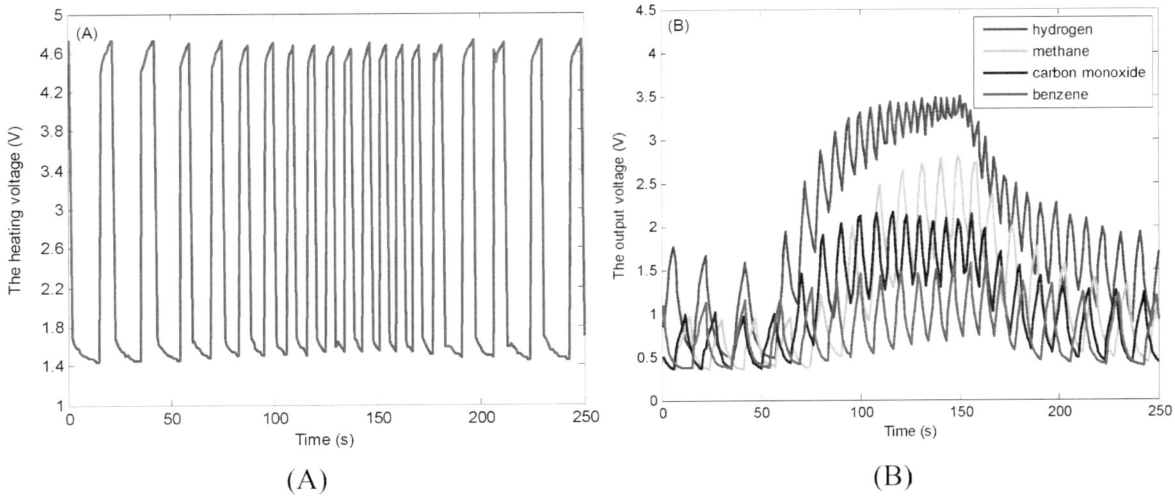

FIGURE 6.4 Temperature oscillation modulation for active sensing. Active modulation of temperature oscillation for active sensing. (A) Heater voltage profile that is produced for 30 ppm of benzene and (B) resistance profiles for the four gases measured. Source: *From He, A., Wei, G., Yu, J., Li, M., Li, Z., & Tang, Z. (2019). A novel sparse representation classification method for gas identification using self-adapted temperature modulated gas sensors.* Sensors, 19(9). *https://doi.org/10.3390/s19092173.*

(2012), a resistive gas sensor was modulated with an oscillating heater signal. A self-adaptive strategy modified the oscillation with the sensor resistance. This approach improved the sensor system performance compared to a static approach. The effectiveness of this approach was demonstrated on a single commercial resistive metal oxide sensor, which was used to differentiate among nitrobenzene, carbon monoxide, nitric oxide, and a mixture of nitric oxide and nitrobenzene across concentrations ranging from 10 to 50 ppm.

In He et al. (2019), a similar method coupled with more advanced machine learning algorithms based on sparse representation and convolutional neural networks was employed for classifying four gases at different concentrations (hydrogen, methane, carbon monoxide, and benzene). Three commercial resistive metal oxide sensors were employed, and the heating oscillation frequency was modified. Fig. 6.4 illustrates how the heater voltage modulation frequency was adjusted in relation to the output voltage (related to sensor resistance), with higher frequencies corresponding to higher sensor signals.

While the previous approaches employed a basic feedback loop, Herrero-Carrón et al. (2015) explored the possibility of modulating the heaters of commercial metal oxide resistive sensors to achieve a predefined sensor resistance target. The closed-loop was implemented using a proportional, integrative, and derivative control with the heater voltage serving as the control signal itself. They demonstrated how this approach improved measurement stability and accuracy compared to the conventional sensor measurement of gases. A single commercial metal oxide resistive sensor was able to discriminate among methanol, ethanol, and butanol at several concentrations.

Likewise, in Dominguez-Pumar et al. (2016), an on-off feedback loop was employed for controlling the temperature of metal oxide resistive nanosensors (nanoneedles of WO_3 decorated with Au nanoparticles fabricated by the authors). In this case as well, the controlled variable (heater voltage) was adopted as the new measurement. As a result, the measurement response time was reduced from over 1000 seconds to approximately 50 seconds.

Another parameter related to the sensor operational temperature is the temperature of a preconcentrator. Positioned between the odor source and the gas sensors, a preconcentrator selectively absorbs the target gases for a determined time duration, followed by heating to release the gases rapidly. While preconcentrators are typically used to enhance sensitivity, selectivity, or remove interferents, Nakamoto et al. (2000) explored the iterative adjustment of a Tenax preconcentrator's desorption temperature to optimize discrimination capabilities during measurements, similar to the approach described earlier (Nakamoto et al., 1991). In this study, the preconcentrator was exposed to the target gases flow for 13 seconds. Subsequently, desorption started at 80°C, and three rapid temperature increases were applied to the preconcentrator heater. Seven quartz crystal microbalance (QCM) sensors coated with different sensing materials were used to differentiate between two gases (pure propyl acetate and a mixture of propyl acetate and hexyl acetate). The two intermediate preconcentrator temperatures were adjusted by gradient descent so they maximized the separation index.

Unlike in nature, where controlling or modulating temperature is challenging, electronic devices offer the convenience of easily adjusting temperature as an operating sensor parameter. Moreover, temperature has a significant influence on the sensory attributes of various chemical sensors, such as oxide resistive sensors. As has been shown in this section, these properties make temperature modulation one of the most straightforward strategies in active sensing. With ongoing sensor miniaturization, the speed or the temperature modulation can be faster and power consumption could decrease. Although achieving precise temperature control may present challenges due to the difficulty in measuring temperature with precision, voltage can serve as a reliable proxy for temperature.

6.3 Gas flow rate

The gas rate flow is one very important part of the active sensing in nature (Jordan et al., 2018), and the impact of carrier gas flow has been a subject of investigation in artificial sensor systems on some occasions. A study (Ziyatdinov et al., 2015b) focused on the modulation of gas flow at a frequency of 0.08 Hz applied to a metal oxide resistive sensor inspired by the biological olfaction. This setup yielded additional information when frequencies higher than 0.7 Hz were used. However, the modulation of flow rate in active sensing remains moslty an underexplored area.

In Wang et al. (2017), the approach was to employ a Deep *Q* Network (DQN) to modulate the carrier gas flow rate. DQN is an algorithm that combines deep convolutional neural network architectures with reinforcement learning and aims to estimate the action-value function *Q* by the convolutional network, while the reinforcement algorithm updates the parameters of such network (Mnih et al., 2015). In this work, a convolutional neural network was used as guide for the reinforce learning (by approximating the rewards), while the DQN selected the best flow rate at each iteration. Utilizing 12 commercial resistive metal oxide sensors, the trained system was able to classify five gases in less than six steps in most of the cases. The resulting system exhibited a good classification performance and proved adaptable to scenarios characterized by limited data samples and labeled information.

A variation to the gas rate flow methods was employed in Nakamoto and Sumitimo (2003) where a preconcentrator was incorporated as illustrated in Fig. 6.5. In this study, the absorption times of the preconcentrator were adjusted within a closed-loop to achieve a response independent of the gas concentration, effectively mitigating saturation issues. The design of the preconcentration had to be adjusted to improve the absorption and desorption times and avoid changes in the response patterns. The system, equipped with three QCM gas sensors, successfully discriminated among *trans*-2-hexenyl acetate, *trans*-2-hexenal, ethyl valerate, and isobutyric acid at different concentrations. Moreover, the preconcentrator's inclusion allowed for the removal of humidity-related interference, enabling discrimination between propionic and isobutyric acids at different concentrations and humidity ranges.

In nature, gas flow rate modulation is very common, breath patterns, and movements of antennae are easy to realize. In electronic devices, however, manipulating gas flow rate demands more complex, larger, and costlier experimental systems. This additional complexity is reflected in the limited number of studies that explore this parameter, both as an active sensing strategy and in terms of its influence on sensory characteristics.

6.4 Frequency selection

Optical sensors often rely on the spectral characteristics of the target gases, and active sensing can be employed to select what frequencies to sample using spectroscopy methods. In Dinakarababu et al. (2011), an adaptive feature specific spectrometer algorithm was introduced. The method is based on features that consist of linear combinations of several frequencies realized experimentally by an active mirror array. The algorithm dynamically adapted the features during the measurement process by a selection made by sequential hypothesis testing implemented with principal components and probabilistic weighting of a library of spectrographs of chemical compounds. For identifying chemical substances with accuracies below 2%, the method reduced the number of steps required.

Furthermore, in Huang et al. (2012), a system incorporating a tunable infrared interferometer was developed. The algorithm selected the optimal measuring frequencies based on POMDPs that minimized sensing cost in a similar way to their work on Gosangi and Gutierrez-Osuna (2014). Nonnegative matrix factorization was employed to establish concentration-independent absorption profiles and identify single components regardless of their concentrations. By implementing this method with a tunable Fabry–Perot interferometer that operated at 66

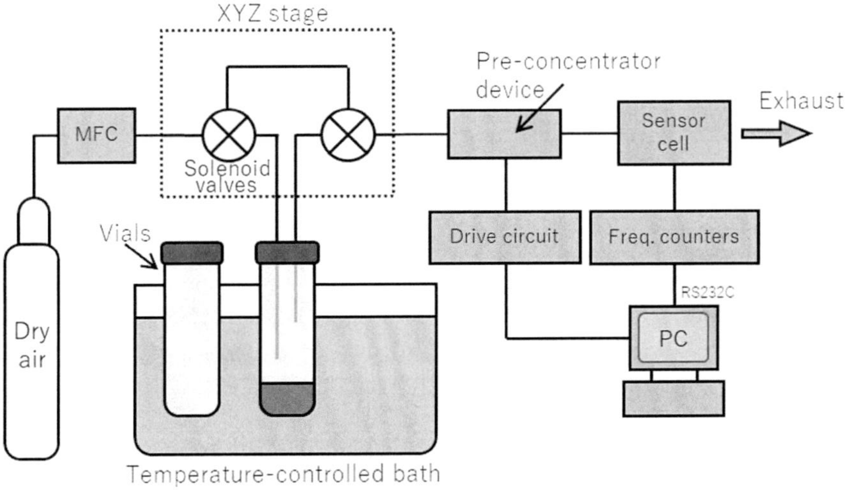

FIGURE 6.5 Scheme of flow modulation for active sensing with preconcentration. Scheme of time modulated preconcentrator for active sensing where the closed loop from the measurement of the preconcentrator driving circuits can be seen. Source: *Re-drawn from Nakamoto, T., & Sumitimo, E. (2003). Study of robust odor sensing system with auto-sensitivity control.* Sensors and Actuators, B: Chemical, 89 *(3), 285–291. https://doi.org/10.1016/S0925-4005(03)00003-0.*

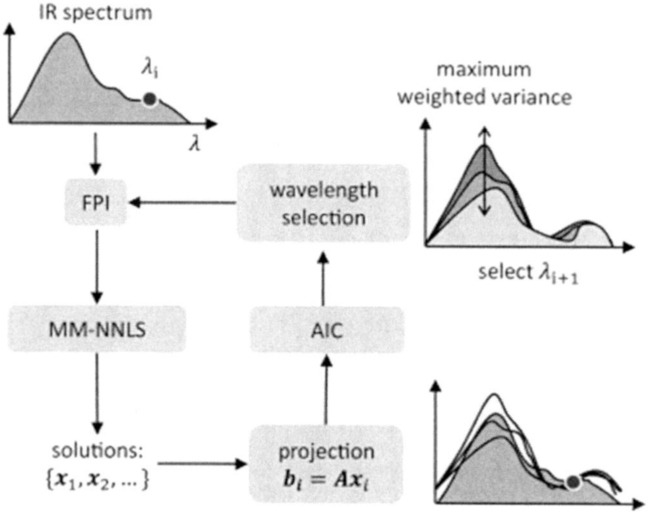

FIGURE 6.6 **Scheme of multimodal wavelength selection algorithm for active sensing based on optical measurement.** Scheme of multimodal wavelength selection algorithm. FPI: Fabry–Perot interferometer used to realize the measurement at the selected frequency; MM-NNLS: multimodal nonnegative least squares used to evaluate the possible solutions; AIC: Akaike information criterion for selecting the optimal interrogation frequency. Source: *From Huang, J., & Gutierrez-Osuna, R. (2015). Active wavelength selection for mixture analysis with tunable infrared detectors.* Sensors and Actuators B: Chemical, 208, 245–257. *https://doi.org/10.1016/j.snb.2014.10.094.*

wavelengths ranging from 3 μm to 4.3 μm in 0.02 μm increments, the system successfully classified eight gases (air, brush cleaner, lacquer thinner, denatured alcohol, acetone, xylene, isopropyl alcohol, and propanol) at concentration between 20% and 80%.

Subsequent studies by the same group improved the system by incorporating multimodal nonnegative matrix factorization alongside sparsity regularization for mixtures. The multimodal algorithm solver maintained multiple candidate solutions, which led to increased sensing efficiency and stability compared to static methods in simulated measurements (Huang & Gutierrez-Osuna, 2013).

Akaike information criterion was added to the algorithm in their more recent work on this subject (Huang & Gutierrez-Osuna, 2015) as illustrated by Fig. 6.6. In experimental testing involving binary mixtures of vapor from dilutions of 2.5% isopropyl and 5% isopropyl alcohol, the system successfully determined the correct mixture at the fifth iteration. Additionally, in theoretically simulated mixtures with more components, the method exhibited robustness in the face of ill-conditioned problems and low signal-to-noise ratios.

Although the optical method lacks direct inspiration from the natural world, the active sensing approach improves the system accuracy and robustness. There is still ample space for innovation in this family of algorithms; however, the experimental implementation of such modifications in optical systems remains challenging.

6.5 Modification of location

The most common application of active odor sensing is the modification of the sensor placement to extract information about the location of odor sources. One of the earliest works was done by Rozas et al. (1991). In this work, they mounted four commercial gas sensors on a mobile platform separated a few centimeters. That platform could move on the horizontal plane and could rotate horizontally. When the platform was performing the active sense, the platform determined the gas gradient in each position by rotating four time 90 degrees and then moved in the gradient direction, searching for odor sources.

But since the odor distribution follows the air stream and is distributed in very complex plumes (Pannunzi & Nowotny, 2019), more sophisticated algorithms are being explored. In this direction, another of the earliest works was done by Ishida et al. (1994). In their work, they combined gas sensors with anemometers to locate odor sources, and they studied two algorithms to move the sensors under several wind conditions as illustrated in Fig. 6.7. One consisted of direct movement toward the computed odor source, and the other one implemented a zigzag movement. In most of the wind speed conditions, they were able to find the odor source.

These first works were based on gradient-based algorithms. Even if they are very intuitive, they struggle with turbulent plumes. So more advanced methods have been developed. According to comprehensive recent reviews such as Chen and Huang (2019) and Jing et al. (2021), they can classify be in four categories: gradient-based algorithms, bioinspired algorithms, probabilistic and map-based algorithms, and multirobot algorithms. Fig. 6.8 shows a Venn diagram of the four categories from Chen and Huang (2019).

Bioinspired algorithms mimic behaviors of organisms such as moths, bacteria, and lobsters. They use preset movement rules rather than gradients (Chen & Huang, 2019). One of the earliest works on bioinspired algorithms for odor source localization is based on the silkworm moth (Kuwana & Shimoyama, 1998), which mimics the plume tracking strategies used to find mates. This pioneering bioinspired approach paved the way for other organism-influenced algorithms. It consisted on the

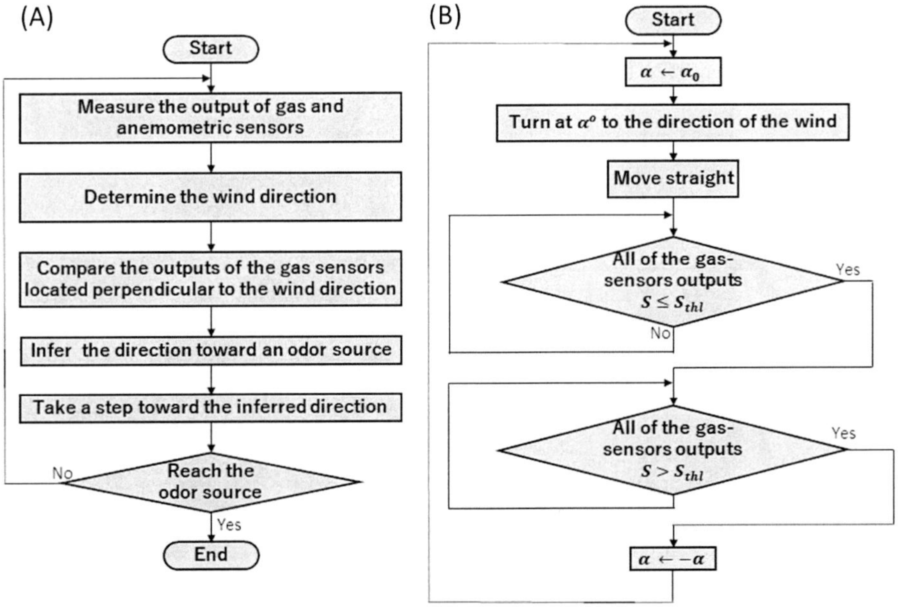

FIGURE 6.7 Two early algorithms for odor source location. Two early algorithms for odor source location. The two algorithms for the odor source localization by active sensing with a robotic platform with gas sensors and anemometers. (A) step-by-step and (B) zigzag approach turning the robot and in which a threshold sensor output (S_{thl}) was used to determine whether the system was inside an odor plume. Source: *Re-drawn from Ishida, H., Suetsugu, K., Nakamoto, T., & Moriizumi T. (1994), Study of autonomous mobile sensing system for localization of odor source using gas sensors and anemometric sensors.* Sensors and Actuators A: Physical, 45*(2), 153–157. https://doi.org/10.1016/0924-4247(94)00829-9.*

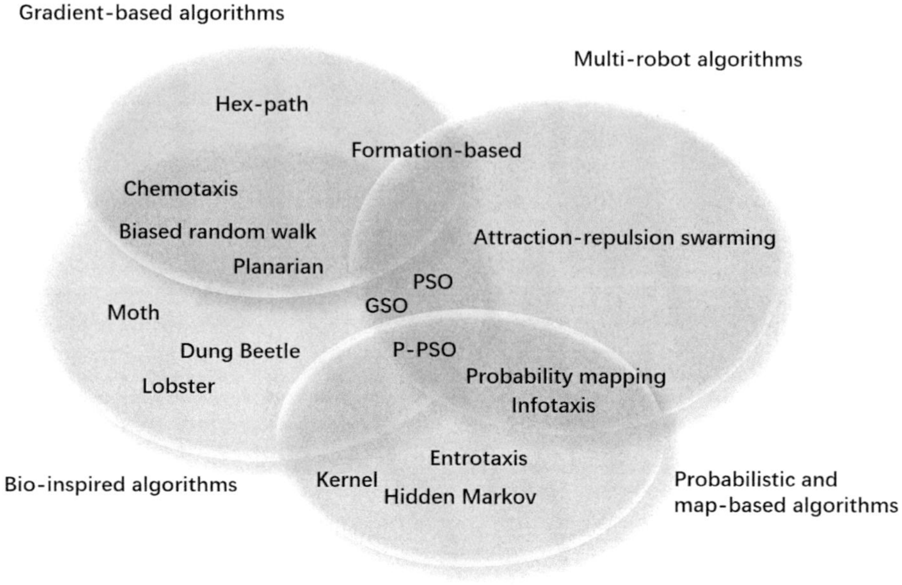

FIGURE 6.8 **Venn diagram of algorithms for active sensing for odor location according to their algorithm.** As the illustration shows, some of the algorithms could be classified in more than one category. Source: *From Chen, Xin-xing., & Huang, Jian. (2019). Odor source localization algorithms on mobile robots: A review and future outlook.* Robotics and Autonomous Systems, 112, *123–136. https://doi.org/10.1016/j.robot.2018.11.014.*

application of two preset movement patterns, surging and casting. During casting, the robot moved in a zigzag movement to relocate the plume, and during the surging, the robot moved directly toward the odor source following the plume. Other significant algorithms include a lobster-inspired one that moved the robot toward higher concentrations and move backward when no odor is sensed (Grasso & Atema, 2002), or a dung beetle inspired "zigzagging" across a plume by turning when the edge is reached (Russell et al., 2003). Overall, these algorithms provide a simple yet robust approach to odor source localization.

Multirobot algorithms use collaboration between robots to share information and speed up the search for odor sources. Hayes et al. (2002) proposed in 2011 a first application of swarm intelligence or multirobot algorithms for odor source localization. In their study, they did an offline optimization of the search parameters to locate the odor source, and the experimental results proved that multiple robots perform more efficiently than a single autonomous robot. Other common swarm algorithms studied include aggregation, pattern formation, collective movement, task allocation, source search, and collective mapping (Faria Dias et al., 2021). These algorithms greatly improve efficiency over single robots, but if not well designed the efficiency can be lower when robot interfere with each other. Moreover, the cost of having multiple robots can be much higher than other solutions.

In the probabilistic algorithms, a probability distribution of the locations of the odor source is first estimated and then improved in iterations till the probability of the odor source is well defined. The Infotaxis (Vergassola et al., 2007) was one of the first works using this approach. In the Infotaxis method, instead of measuring the gradient (a common and straightforward approach), the robot makes an estimation of the probability distribution and then move trying to maximize the expected rate of information acquisition. The information accumulates faster as the robot gets closer to the source making the algorithm converge. It showed its applicability in very diluted limits or weak odor sources. Later variations used kernel functions or hidden Markov models to further improve modeling and computational efficiency. Overall, these probabilistic algorithms are powerful, but they rely on accurate concentration data and maps.

Recent advancements are being done with machine learning (Chen et al., 2021), swarms (Junhan Wang et al., 2022), 3D environments (Ercolani & Martinoli, 2020), problems with multiple sources of odors (Jain et al., 2019), and fusion of multimodal sensing information (Ishida et al., 2012; Shen et al., 2021), among others.

6.6 Chemical gas blending feedback systems

In certain applications, active sensing procedures offer the potential to provide additional information that improves the measurement characteristics, compensating some of the most challenging and undesired characteristics of chemical sensor, such as sensor drift, repeatability, or sensor reproducibility. In this context, Nakamoto et al. (1994) introduced an active sensing procedure in 1994 that coupled a gas mixture blender system with a gas sensor system in a closed loop. The underlying concept of this procedure is depicted in Fig. 6.9. Initially, in the first step, a gas sensor array (represented as a grid of four sensors of different colors) measures a target gas (in this example, the head space of a vial with an unknown volatile liquid), and the gas follows the back arrow. In a second step, the same gas sensor array measures a blended gas that is internal to the active sensing system. In the figure, this is represented as a smaller vial sending the head

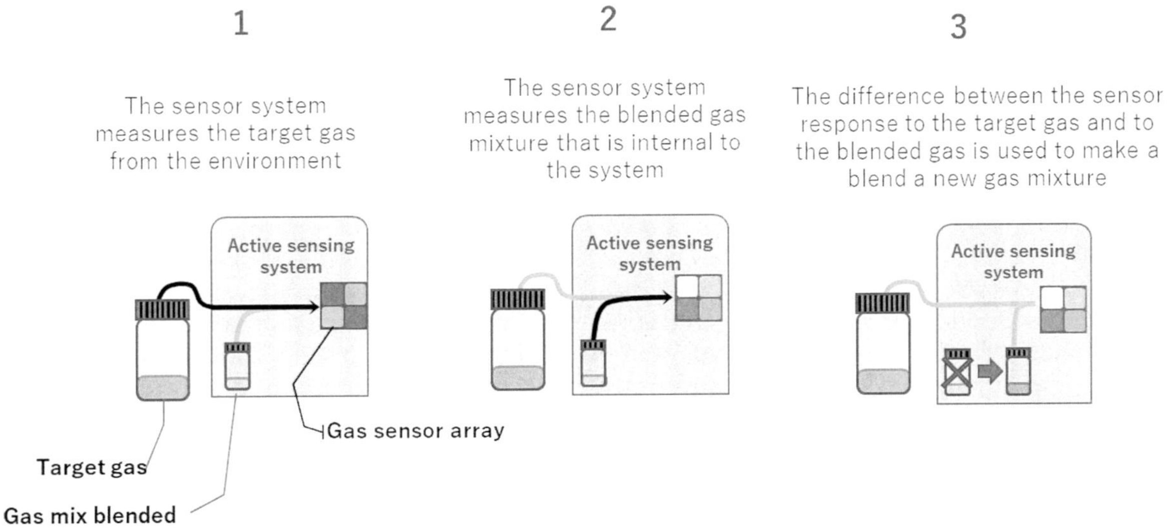

FIGURE 6.9 Scheme of the active sensing procedure based on chemical gas blending feedback. Scheme of the active sensing procedure based on chemical gas blending feedback. Detailed step-by-step illustration of the active sensing procedure with gas blending system feedback. Each column is one step and each row one detail of the step, number of the step, explanation of the step, and scheme of the action taking place in the step.

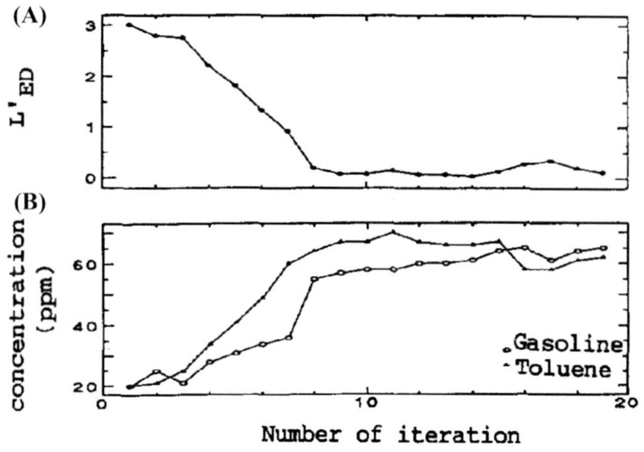

FIGURE 6.10 **Result of the gradient descent search for the quantification gasoline toluene.** Experimental dynamic curves of the gradient descent search for the quantification of 60 ppm of gasoline and 60 ppm of toluene. (A) Weighted distance in the sensor space and (B) quantification at each iteration step. Source: *From Nakamoto, T., Ustumi, S., Yamashita, N., Moriizumi, T., & Sonoda, Y. (1994). Active gas/odor sensing system using automatically controlled gas blender and numerical optimization technique.* Sensors and Actuators B: Chemical, 20*(2–3), 131–137. https://doi.org/10.1016/0925-4005(93)01193-8.*

space to the sensor array along the dark arrow. Then, in a third step, an algorithm evaluates and compares the sensor response of the target gas and to blended gas and calculate a new composition of the blended gas. The process is repeated till the responses match, and the composition of the blended gas is considered the response of the system.

Applying this framework, the procedure outlined in Nakamoto et al. (1994) involved two steps. First, a gas sensor system comprising three commercial tin oxide resistive sensors measured a target gas mixture. This target gas was composed of different gases at different ratios. Then, in the second step, a gas blender delivered several concentrations and compared them to the initial gas sensor measurements. Employing two search algorithms, a simplex optimization and a gradient descent algorithm, adjustments were made to the blended gas to replicate the original sensor response to the target gas. Fig. 6.10 shows the result of the search for the case of the gradient descent.

Subsequent refinement of this procedure was presented in a later study (Nakamoto et al., 1995), which involved computing the gradient through least-squares analysis over multiple measurements to mitigate errors and applying singular value decomposition (SVD) to reduce sensor array collinearity when the algorithm was applied to the mixture quantification.

After these initial works, a substantial advancement was achieved by designing an optimal control-loop system in Nakamoto et al. (1997). In this study, the gas sensors (QCM coated in different sensing films) were modeled by a discrete state equation of order one. The model parameters were estimated through a calibration process, which involved conducting measurements of the sensor responses when exposed to gases generated by the gas blender. These calibrations were calculated using the least-squares method. The estimated parameters of the sensor model response were used to design an optimal control by classical control theory. The improvement allowed a faster response with a convergence within two minutes and the accurate reproduction of a mixture comprising three distinct odorant gases.

Thanks to the introduction of more advanced gas blender systems, the same research group further improved the active sensing algorithms by increasing their complexity and sophistication of their active sensing algorithm.

Building upon the previous method, which had already demonstrated resilience against environmental variables such as temperature and humidity, enhancements were made by

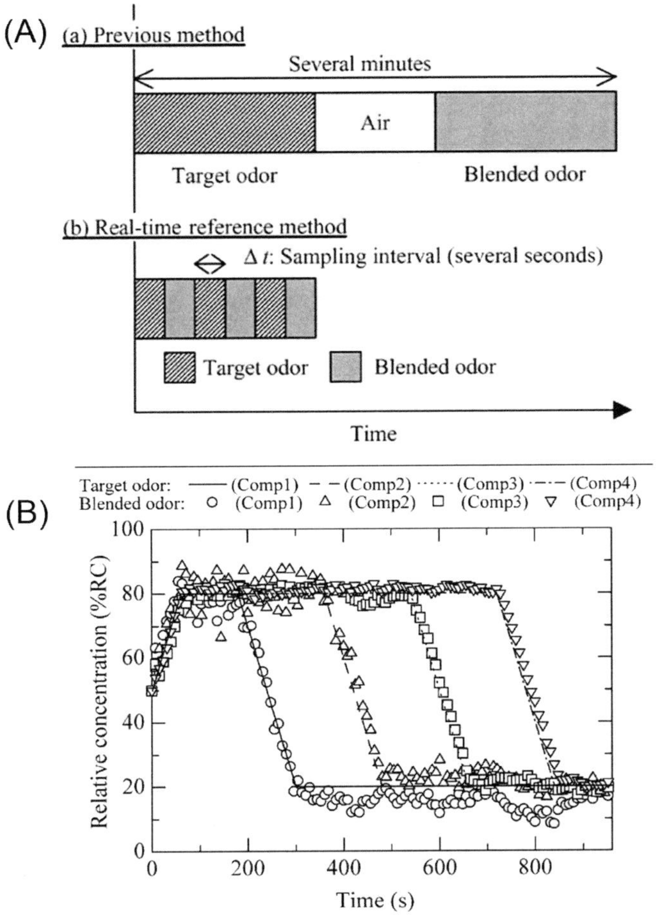

FIGURE 6.11 Comparison between previous and real-time reference method. (A) The comparison of the measurement steps in the original method and the improved version where the alternating blended and target measurements can be seen. (B) The result of the improved version on a real-time experiment. Source: *From Yamanaka, T., & Nakamoto, T. (2003). Real-time reference method in odor recorder under environmental change.* Sensors and Actuators B: Chemical, *93(1-3), 51–56. https://doi.org/10.1016/s0925-4005(03)00202-8.*

alternating the measurement of the reference odor with that of the blended odor (Yamanaka & Nakamoto, 2003), as depicted in Fig. 6.11A. This procedure facilitated the real-time tracking of evolving target odors, as illustrated in Fig. 6.11B. It also allowed the comparison of the target odor and blended odors measurements under the same environment conditions, increasing the robustness of the system against fluctuations of humidity and temperature. This practical application found utility in the real-time recording, transmission, and reproduction of odor mixtures comprising four compounds, even under humidity and temperature changes up to 13%.

One of the most prevalent challenges in artificial olfactory systems is the inherent collinearity of the responses of the gas sensors that compose the sensor arrays. This collinearity deteriorates the performance of the algorithms for classification or quantification. Moreover, in the context of active sensing, it restricts the number of gas components that can be effectively employed in chemical gas blending feedback systems. Aiming to solve this problem Yamanaka et al. (2003) proposed applying SVD to the active sensing algorithms based on the optimal control loop. Rather than utilizing individually the entire array of sensors (composed by eight QCM sensors coated in different sensing materials) and complete array of odor-reproducing channels, SVD was used to reduce their dimensionality. Thanks to this approach, they were able to reproduce the odor of apple based in eight different odor components while using only four dimensions in the control loop thanks to the SVD dimensional reduction.

In more recent works (Aleixandre & Nakamoto, 2022), the procedure was modified to incorporate online learning control, increasing the robustness of the control loop against humidity variations and sensor drift. Unlike the previous approach outlined above (Yamanaka & Nakamoto, 2003), this work omitted the alternation between reference and target odor measurements. Instead, it introduced a recalibration of the sensor model during active measurements that was used to recalculate the control-loop gains. The procedure's detailed scheme is presented in Fig. 6.12A. An initial calibration provided the initial estimation of the sensor response model to the gases from the odor blender. Then, the active sensing was initialized by a setup step by measuring the target odor and recording the sensor measurement. Following this setup, the active sensing algorithm started generating and measuring a blended odor. The most recent measurement served a dual purpose. First, it was used as calibration data by revising the sensor model parameter estimates through recursive least squares. These updated parameters were then employed to redesign the control-loop system. The second use of the most recent measurement was as part of the active sensing procedure. It was compared to the sensor response to the target odor, and if this difference was small enough, the procedure was considered finish. Otherwise, the difference was employed to adjust the odor blender using the closed-loop control system.

The system was tested in three different scenarios, no perturbation, sensor drift (consisting in an interval of 3 weeks between calibration and operation), and humidity change (consisting on a change on humidity of 15% between the calibration and the operation). Fig. 6.12B illustrates the differences in performance across these scenarios. In the first one, the difference is negligible as both systems, with fixed sensor model parameters and with online learning, could measure the odor mixtures correctly. However, in the remaining two scenarios, sensor drift and humidity change, the active sensing system with fixed sensor model parameters displayed significantly higher error rates compared to the system employing online learning. This outcome showed the increased stability offered by online learning when confronting challenges such as a 15% humidity fluctuation and sensor drift extending over 3 weeks.

Further refinement has been done by Gong et al. (2022) that optimized the operation of the QCM gas sensors by increasing the measured sensor features. In that work, they showed that by monitoring additional harmonic resonant frequencies and corresponding resistances, the active sensing was able to improve the procedure due to the additional gathered information.

The active sensing has also been used with field asymmetric ion mobility spectrometry (FAIMS) for tracking changes of gas mixture concentrations using an optimal controller (Yokoshiki & Nakamoto, 2019a) as shown by Fig. 6.13. Also, measuring with FAIMS was also combined with gradient descent to quantify mixtures of the components and improved the algorithm with adaptive learning rate (Yokoshiki & Nakamoto, 2019b).

The robustness of the methods was further shown in some recent works (Sukekawa et al., 2021) where the application of the active sensing was extended to sensors based on insect olfactory receptor-expressing cells. In this case, instead of a control loop, they used an iterative triangle grid search. Thanks to the comparative measurement, the prototype was able to

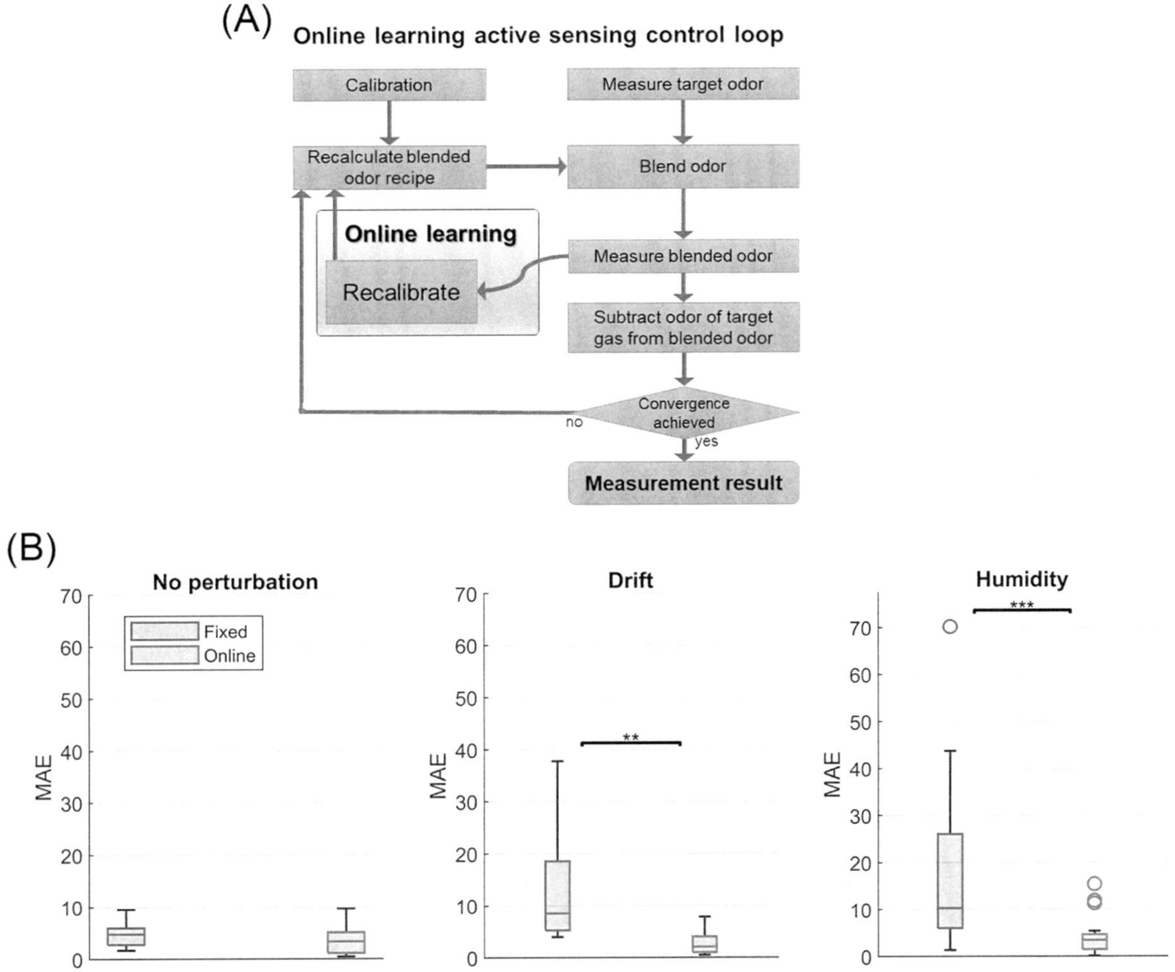

FIGURE 6.12 **Adaptive control of active sensing.** Adaptive control of active sensing. (A) Scheme of the steps taken place in the online learning of the active sensing, (B) mean absolute error on the final prediction of the measurement on the different scenarios: without disturbance, with sensor drift, and ambient humidity. The middle of the boxes are the medians, the box comprises the 75th and 25th percentiles of the data, the whiskers represent the maximum and minimum values and the circles represent outliers (1.5 times above or below the box). The asterisks represent significance of the medians, two $P \leq .01$ and three $P \leq .001$.

overcome the high noise of the cell biological responses. After this initial work in this field, a paper (Deng et al., 2023) from the same laboratory, significant improved the active algorithm in this biological application. This was achieved by obtaining a reference response model through curve fitting the cell response to known gas intensities. This advancement allowed for quantification within 400 seconds, faster than previous response times above 1500 seconds. They also proved the possibility of applying a variation of the method to quantify mixtures (Deng et al., 2024). Furthermore, the active sensing method effectively

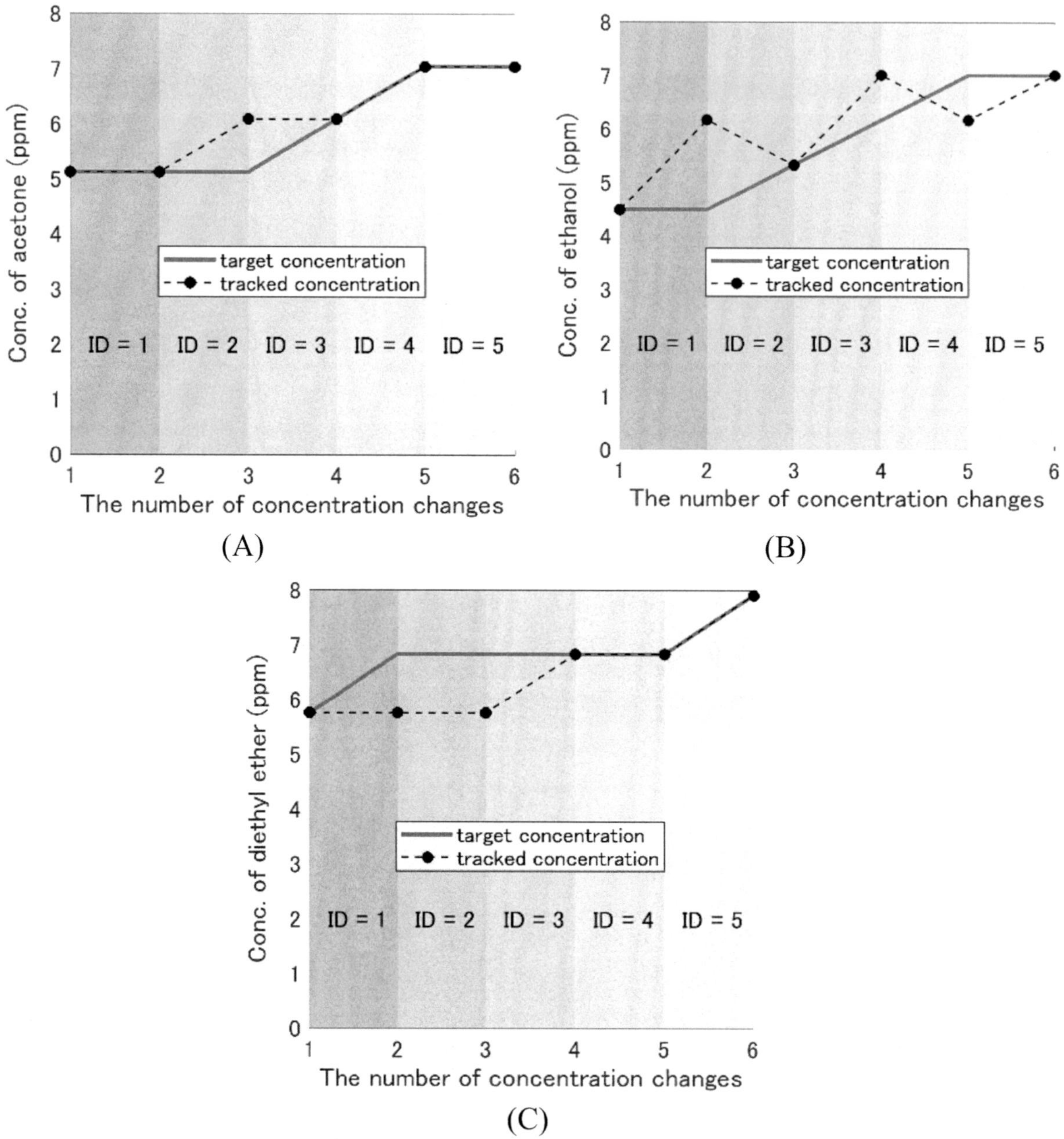

FIGURE 6.13 **Tracking of the concentration changes with the gradient descent active sensing of a ternary mixture.** In the three plots, the red line is the evolving targets and the black dots the realized quantification: (A) acetone, (B) ethanol, and (C) diethyl ether. Source: *From Yokoshiki, Y., & Nakamoto, T. (2019a). On-line mixture quantification to track temporal change of composition using FAIMS.* Sensors, 19(24), *https://doi.org/10.3390/s19245442.*

addressed issues commonly associated with biological odor sensors, such as sensor nonlinearity, drift, and aging.

6.7 Future studies

As illustrated in this chapter, active sensing can significantly improve measurements, particularly concerning stability. By adapting measurement conditions to specific problems and environmental factors, the system can achieve heightened robustness, even if the measurement procedure has to proceed for longer times. The active sensing is specially well suited to the location of gas or odors sources, as shown by the numerous works in this field with the intersection of gas sensors and robotics. As seen in the chapter in many examples, other improvements are shown also in classification, discrimination, and time of measurement compared to other measurement methods.

However, due to the increased complexity of experimental setups, that have to implement means to modify the sensing parameters, the more elaborated algorithms needed, and the good integration of both into experimental systems, active sensing remains underutilized and static methods are still mainly used.

There exists ample potential for advancing this research area by drawing inspiration from nature or exploring other factors that may influence measurements in artificial gas sensor systems, such as voltage operation for electrochemical sensors or geometry of the sensor chamber. Additionally, the algorithms offer opportunities for improvement, including insights from neurological processing in nature or the integration of machine learning techniques.

Abbreviations used in this chapter:

- Singular Value Decomposition (SVD)
- Markov Decision Process (POMDP)
- Quartz Crystal Microbalance (QCM)
- Field Asymmetric Ion Mobility Spectrometry (FAIMS)

References

Aleixandre, M., & Nakamoto, T. (2022). Online learning for active odor sensing based on a QCM gas sensor array and an odor blender. *IEEE Sensors Journal, 1*, 22310–22318. Available from https://doi.org/10.1109/JSEN.2022.3215127.

Ballard, D. H. (1991). Animate vision. *Artificial Intelligence, 48*(1), 57–86. Available from https://doi.org/10.1016/0004-3702(91)90080-4.

Chen, X., Fu, C., & Huang, J. (2021). A deep Q-network for robotic odor/gas source localization: Modeling, measurement and comparative study. *Measurement, 183*. Available from https://doi.org/10.1016/j.measurement.2021.109725.

Chen, X.-x, & Huang, J. (2019). Odor source localization algorithms on mobile robots: A review and future outlook. *Robotics and Autonomous Systems, 112*, 123–136. Available from https://doi.org/10.1016/j.robot.2018.11.014.

Deng, H., Mitsuno, H., Kanzaki, R., Nomoto, S., & Nakamoto, T. (2023). Gas-phase odorant fast quantification by odor biosensor based on reference response model. *IEEE Sensors Journal, 23*(20), 24169–24178. Available from https://doi.org/10.1109/jsen.2023.3309968.

Deng, H., Mitsuno, H., Kuroda, E., Niki, S., Kanzaki, R., & Nakamoto, T. (2024). Gas-phase odor mixture quantification based on relative comparison method using multiple olfactory receptors. *Sensors and Actuators B: Chemical, 401*. Available from https://doi.org/10.1016/j.snb.2023.134995.

Deveza, R., Thiel, D., Russell, A., & Mackay-Sim, A. (1994). Odor sensing for robot guidance. *International Journal of Robotics Research., 13*(3), 232–239. Available from https://doi.org/10.1177/027836499401300305.

Dinakarababu, D. V., Golish, D. R., & Gehm, M. E. (2011). Adaptive feature specific spectroscopy for rapid chemical identification. *Optics Express, 19*(5), 4595–4610. Available from https://doi.org/10.1364/OE.19.004595.

Dominguez-Pumar, M., Kowalski, L., Calavia, R., & Llobet, E. (2016). Smart control of chemical gas sensors for the reduction of their time response. *Sensors and Actuators B: Chemical., 229*, 1–6. Available from https://doi.org/10.1016/j.snb.2016.01.081.

Ercolani, C., & Martinoli, A. (2020). 3D odor source localization using a micro aerial vehicle: *System design and performance evaluation*. IEEE International Conference on Intelligent Robots and Systems, 6194-6200. Available from https://doi.org/10.1109/IROS45743.2020.9341501.

Faria Dias, P. G., Silva, M. C., Rocha Filho, G. P., Vargas, P. A., Cota, L. P., & Pessin, G. (2021). Swarm robotics: A perspective on the latest reviewed concepts and applications. *Sensors, 21*(6), 1–30. Available from https://doi.org/10.3390/s21062062, https://www.mdpi.com/1424-8220/21/6/2062/pdf.

Gomez-Marin, A., & Louis, M. (2012). Active sensation during orientation behavior in the *Drosophila* larva: More sense than luck. *Current Opinion in Neurobiology, 22*(2), 208–215. Available from https://doi.org/10.1016/j.conb.2011.11.008.

Gosangi, R., & Gutierrez-Osuna, R. (2014). Active classification with arrays of tunable chemical sensors. *Chemometrics and Intelligent Laboratory Systems, 132*, 91–102. Available from https://doi.org/10.1016/j.chemolab.2014.01.001.

Gong, N., Aleixandre, M., & Nakamoto, T. (2022). Odor recorder based on an array of QCM sensors using frequency shifts and resistance changes of multiple harmonics. *Proceedings of IEEE Sensors*, 1–4. Available from https://doi.org/10.1109/SENSORS52175.2022.9967066.

Gosangi, R., & Gutierrez-Osuna, R. (2009). Active chemical sensing with partially observable Markov decision processes. *AIP Conference Proceedings, 1137*, 562–565. Available from https://doi.org/10.1063/1.3156617.

Gosangi, R., & Gutierrez-Osuna, R. (2010). Active temperature programming for metal-oxide chemoresistors. *IEEE Sensors Journal, 10*(6), 1075–1082. Available from https://doi.org/10.1109/JSEN.2010.2042165.

Grasso, F. W., & Atema, J. (2002). Integration of flow and chemical sensing for guidance of autonomous marine robots in turbulent flows. *Environmental Fluid Mechanics*, 2(1–2), 95–114. Available from https://doi.org/10.1023/A:1016275516949.

Hayes, A. T., Martinoli, A., & Goodman, R. M. (2002). Distributed odor source localization. *IEEE Sensors Journal, 2*(3), 260–271. Available from https://doi.org/10.1109/JSEN.2002.800682.

He, A., Wei, G., Yu, J., Li, M., Li, Z., & Tang, Z. (2019). A novel sparse representation classification method for gas identification using self-adapted temperature modulated gas sensors. *Sensors., 19*(9). Available from https://doi.org/10.3390/s19092173.

Herrero-Carrón, F., Yáñez, D. J., Rodríguez, F. de B., & Varona, P. (2015). An active, inverse temperature modulation strategy for single sensor odorant classification. *Sensors and Actuators B: Chemical, 206*, 555–563. Available from https://doi.org/10.1016/j.snb.2014.09.085.

Huang, J., & Gutierrez-Osuna, R. (2013). Active analysis of chemical mixtures with multi-modal sparse non-negative least squares. *ICASSP, IEEE International Conference on Acoustics, Speech and Signal Processing*, 8756–8760. https://doi.org/10.1109/ICASSP.2013.6639376.

Huang, J., & Gutierrez-Osuna, R. (2015). Active wavelength selection for mixture analysis with tunable infrared detectors. *Sensors and Actuators B: Chemical., 208*, 245–257. Available from https://doi.org/10.1016/j.snb.2014.10.094.

Huang, J., & Gutierrez-Osuna, R. (2016). Active wavelength selection for mixture identification with tunable mid-infrared detectors. *Analytica Chimica Acta, 937*, 11–20. Available from https://doi.org/10.1016/j.aca.2016.08.005.

Huang, J., Gosangi, R., & Gutierrez-Osuna, R. (2012). Active concentration-independent chemical identification with a tunable infrared sensor. *IEEE Sensors Journal, 12*(11), 3135–3142. Available from https://doi.org/10.1109/jsen.2012.2212186.

Hunter, G. W., Akbar, S., Bhansali, S., Daniele, M., Erb, P. D., Johnson, K., Liu, C.-C., Miller, D., Oralkan, O., & Hesketh, P. J. (2020). Editors' choice—critical review—a critical review of solid sate gas sensors. *Journal of The Electrochemical Society, 167*(3), 037570. Available from https://doi.org/10.1149/1945-7111/ab729c.

Ishida, H., Suetsugu, K., Nakamoto, T., & Moriizumi, T. (1994). Study of autonomous mobile sensing system for localization of odor source using gas sensors and anemometric sensors. *Sensors and Actuators A: Physical, 45* (2), 153–157. Available from https://doi.org/10.1016/0924-4247(94)00829-9.

Ishida, H., Wada, Y., & Matsukura, H. (2012). Chemical sensing in robotic applications: A review. *IEEE Sensors Journal, 12*(11), 3163–3173. Available from https://doi.org/10.1109/JSEN.2012.2208740.

Jain, U., Tiwari, R., & Godfrey, W. W. (2019). Multiple odor source localization using diverse-PSO and group-based strategies in an unknown environment. *Journal of Computational Science, 34*, 33–47. Available from https://doi.org/10.1016/j.jocs.2019.04.008.

Jing, T., Meng, Q. H., & Ishida, H. (2021). Recent progress and trend of robot odor source localization. *IEEJ Transactions on Electrical and Electronic Engineering, 16*(7), 938–953. Available from https://doi.org/10.1002/tee.23364, http://onlinelibrary.wiley.com/journal/10.1002/(ISSN)1931-4981.

Jordan, R., Fukunaga, I., Kollo, M., & Schaefer, A. T. (2018). Active sampling state dynamically enhances olfactory bulb odor representation. *Neuron, 98*(6), 1214–1228. Available from https://doi.org/10.1016/j.neuron.2018.05.016, http://www.cell.com/neuron/home, e5.

Kuwana, Y., & Shimoyama, I. (1998). A pheromone-guided mobile robot that behaves like a Silkworm Moth with living antennae as pheromone sensors. *The International Journal of Robotics Research, 17*(9), 924–933. Available from https://doi.org/10.1177/027836499801700902.

Lee, A. P., & Reedy, B. J. (1999). Temperature modulation in semiconductor gas sensing. *Sensors and Actuators, B: Chemical, 60*(1), 35–42. Available from https://doi.org/10.1016/S0925-4005(99)00241-5.

Martinelli, E., Polese, D., Catini, A., D'Amico, A., & Natale, C. D. (2012). Self-adapted temperature modulation in metal-oxide semiconductor gas sensors. *Sensors and Actuators B: Chemical, 161*(1), 534–541. Available from https://doi.org/10.1016/j.snb.2011.10.072.

Mnih, V., Kavukcuoglu, K., Silver, D., Rusu, A. A., Veness, J., Bellemare, M. G., Graves, A., Riedmiller, M., Fidjeland, A. K., Ostrovski, G., Petersen, S., Beattie, C., Sadik, A., Antonoglou, I., King, H., Kumaran, D., Wierstra, D., Legg, S., & Hassabis, D. (2015). Human-level control through deep reinforcement learning. *Nature, 518*(7540), 529–533. Available from https://doi.org/10.1038/nature14236, http://www.nature.com/nature/index.html.

Nakamoto, T. (2016). Olfactory display and odor recorder. *Essentials of machine olfaction and taste* (pp. 247–314). Wiley. Available from http://onlinelibrary.wiley.com/book/10.1002/9781118768495, 10.1002/9781118768495.ch7.

Nakamoto, T., & Sumitimo, E. (2003). Study of robust odor sensing system with auto-sensitivity control. *Sensors and Actuators, B: Chemical, 89*(3), 285–291. Available from https://doi.org/10.1016/S0925-4005(03)00003-0.

Nakamoto, T., Fukuda, T., & Moriizumi, T. (1991). Gas identification system using plural sensors with characteristics of plasticity. *Sensors and Actuators B: Chemical, 3*(1), 1–6. Available from https://doi.org/10.1016/0925-4005(91)85001-y.

Nakamoto, T., Ustumi, S., Yamashita, N., Moriizumi, T., & Sonoda, Y. (1994). Active gas/odor sensing system using automatically controlled gas blender and numerical optimization technique. *Sensors and Actuators B: Chemical, 20*(2-3), 131–137. Available from https://doi.org/10.1016/0925-4005(93)01193-8.

Nakamoto, T., Matsushita, H., & Okazaki, N. (1995). Improvement of optimization algorithm in active gas/odor sensing system. *Sensors and Actuators A: Physical, 50*(3), 191–196. Available from https://doi.org/10.1016/0924-4247(95)01039-4.

Nakamoto, T., Okazaki, N., & Mornzumi, T. (1997). High speed active gas/odor sensing system using adaptive control theory. *Sensors and Actuators B: Chemical, 41*(1-3), 183–188. Available from https://doi.org/10.1016/s0925-4005(97)80293-6.

Nakamoto, T., Isaka, Y., Ishige, T., & Moriizumi, T. (2000). Odor-sensing system using preconcentrator with variable temperature. *Sensors and Actuators B: Chemical, 69*(1-2), 58–62. Available from https://doi.org/10.1016/s0925-4005(00)00388-9.

Nazemi, H., Joseph, A., Park, J., & Emadi, A. (2019). Advanced micro- and nano-gas sensor technology: A review. *Sensors, 19*(6). Available from https://doi.org/10.3390/s19061285.

Nelson, M. E., & MacIver, M. A. (2006). Sensory acquisition in active sensing systems. *Journal of Comparative Physiology A, 192*(6), 573–586. Available from https://doi.org/10.1007/s00359-006-0099-4.

O'Connor, D. H., Clack, N. G., Huber, D., Komiyama, T., Myers, E. W., & Svoboda, K. (2010). Vibrissa-based object localization in head-fixed mice. *Journal of Neuroscience, 30*(5), 1947–1967. Available from https://doi.org/10.1523/JNEUROSCI.3762-09.2010, http://www.jneurosci.org/cgi/reprint/30/5/1947.

Pannunzi, M., & Nowotny, T. (2019). Odor stimuli: Not just chemical identity. *Frontiers in Physiology, 10*. Available from https://doi.org/10.3389/fphys.2019.01428.

Rozas, R., Morales, J., & Vega, D. (1991). *Artificial smell detection for robotic navigation, Fifth International Conference on Advanced Robotics 'Robots in Unstructured Environments* (2). IEEE. Available from http://doi.org/10.1109/ICAR.1991.240354.

Russell, R. A., Bab-Hadiashar, A., Shepherd, R. L., & Wallace, G. G. (2003). A comparison of reactive robot chemotaxis algorithms. *Robotics and Autonomous Systems, 45*(2), 83–97. Available from https://doi.org/10.1016/s0921-8890(03)00120-9.

Shen, X., Yuan, J., & Shan, Y. (2021). A novel plume tracking method in partial 3D diffusive environments using multi-sensor fusion. *Expert Systems with Applications, 178*, 114993. Available from https://doi.org/10.1016/j.eswa.2021.114993.

Sukekawa, Y., Mitsuno, H., Kanzaki, R., & Nakamoto, T. (2021). Binary mixture quantification using cell-based odor biosensor system with active sensing. *Biosensors and Bioelectronics, 179*. Available from https://doi.org/10.1016/j.bios.2021.113053.

Thesen, A., Steen, J. B., & Døving, K. B. (1993). Behaviour of dogs during olfactory tracking. *Journal of Experimental Biology, 180*(1), 247–251. Available from https://doi.org/10.1242/jeb.180.1.247.

Vergassola, M., Villermaux, E., & Shraiman, B. I. (2007). 'Infotaxis' as a strategy for searching without gradients. *Nature, 445*, 406–409. Available from https://doi.org/10.1038/nature05464.

Wachowiak, M. (2011). All in a Sniff: Olfaction as a model for active sensing. *Neuron, 71*(6), 962–973. Available from https://doi.org/10.1016/j.neuron.2011.08.030.

J. Wang, Y. Lin, R. Liu, J. Fu, Odor source localization of multi-robots with swarm intelligence algorithms: A review. Frontiers in Neurorobotics. **16** (2022), doi:10.3389/fnbot.2022.949888.

Wang, Y., Xing, J., & Qian, S. (2017). Selectivity enhancement in electronic nose based on an optimized DQN. *Sensors, 17*(10). Available from https://doi.org/10.3390/s17102356.

Yamanaka, T., & Nakamoto, T. (2003). Real-time reference method in odor recorder under environmental change.

Sensors and Actuators B: Chemical, *93*(1–3), 51–56. Available from https://doi.org/10.1016/s0925-4005(03)00202-8.

Yamanaka, T., Matsumoto, R., & Nakamoto, T. (2003). Fundamental study of odor recorder for multicomponent odor using recipe exploration method based on singular value decomposition. *IEEE Sensors Journal*, *3* (4), 468–474. Available from https://doi.org/10.1109/JSEN.2003.815778.

Yokoshiki, Y., & Nakamoto, T. (2019a). On-line mixture quantification to track temporal change of composition using FAIMS. *Sensors*, *19*(24). Available from https://doi.org/10.3390/s19245442.

Yokoshiki, Y., & Nakamoto, T. (2019b). Ternary gas mixture quantification using field asymmetric ion mobility spectrometry (FAIMS). *Sensors*, *19*(13). Available from https://doi.org/10.3390/s19133007.

Ziyatdinov, A., Fonollosa, J., Fernández, L., Gutierrez-Gálvez, A., Marco, S., & Perera, A. (2015a). Bioinspired early detection through gas flow modulation in chemo-sensory systems. *Sensors and Actuators B: Chemical*, *206*, 538–547. Available from https://doi.org/10.1016/j.snb.2014.09.001.

Ziyatdinov, A., Fonollosa, J., Fernández, L., Gutierrez-Gálvez, A., Marco, S., & Perera, A. (2015b). Data set from gas sensor array under flow modulation. *Data in Brief*, *3*, 131–136. Available from https://doi.org/10.1016/j.dib.2015.02.016.

Zweifel, N. O., & Hartmann, M. J. Z. (2020). Defining "active sensing" through an analysis of sensing energetics: Homeoactive and alloactive sensing. *Journal of Neurophysiology*, *124*(1), 40–48. Available from https://doi.org/10.1152/jn.00608.2019.

SECTION 3

Cheminformatics in machine olfaction

CHAPTER

7

Fundamentals of chemometrics techniques

Takamichi Nakamoto

Institute of Integrated Research, Institute of Science Tokyo, Yokohama, Japan

7.1 Classification of chemometrics techniques

When we use a set of variables for data analysis, the multivariate data analysis techniques are useful. Especially, the techniques oriented to chemistry are called chemometrics (Beebe et al., 1998; Muhammad et al., 1986). Recently, chemoinformatics is also used in the similar situation. We use a variety of multivariate analysis methods for sensory test and sensor-array data (William & Goldstein, 1984). We should select an appropriate method among them according to the aim of the analysis.

The multivariate analysis techniques are classified as dependence method and independence method as are shown in Fig. 7.1. Dependence methods seek to explain one or more criterion measures based upon a set of predictor variables. On the other hand, independence methods attempt to provide insights into the underlying structure of the data by simplifying the complexities, primarily through data reduction.

In the case of neural network, nonlinear models can be constructed. They are classified into supervised and unsupervised methods.

Supervised method uses training data with labels, and the neural network is trained according to target inputs in the manner similar to the situation that a supervisor (teacher) guides students to obtain the correct answer to a question. Unsupervised method uses training data without labels, and the neural network is trained without target signals.

There are a variety of chemometrics techniques. In this chapter, we focus on feature extraction technique, classification method, regression method, and clustering techniques. They frequently appear in cheminformatics. Moreover, several points to be considered when we use a neural network are explained. We do not explain a neural network itself since there are many textbooks of neural network. However, it is necessary to be careful about several points when we use it in chemometrics domain.

7.2 Feature extraction techniques

There are several techniques to extract features. In pattern recognition, the curse of dimensionality often occurs. It means that pattern classifier capability decreases if the data dimension is too high. On the other hand, data cannot be expressed well if the data

Digital Technologies in Olfaction
DOI: https://doi.org/10.1016/B978-0-443-15721-9.00007-5

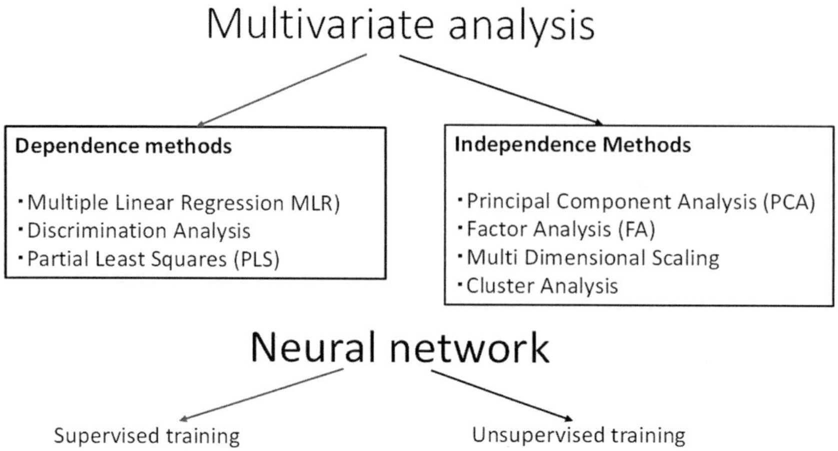

FIGURE 7.1 Classification of chemometrics techniques. They are classified as dependence and independence methods.

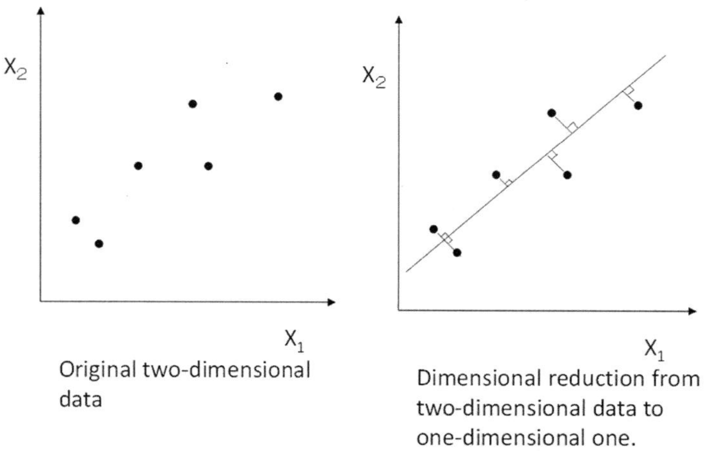

FIGURE 7.2 Method to reduce dimension. Two-dimensional data is reduced to one-dimensional one.

dimension is too low. It is essential to select appropriate dimension to obtain a high capability of pattern recognition. That is why feature extraction is often used.

Another reason for extracting feature is to visualize data. It is hard to understand the relationship between the highly dimensional data. For example, we cannot understand immediately which data point is close to the particular point. However, we can understand the closeness between the data points immediately when the data is visualized after data dimension is reduced to two or three one. We often do the feature extraction when we want to observe the data structure.

There are several techniques to extract the feature. The well-known technique is principal component analysis (PCA). Similar technique is factor analysis. The technique relevant to neural network is an auto encoder. We first learn PCA and then auto encoder a little.

7.2.1 Principal component analysis

PCA is widely used dimensional reduction technique to visualize a data point in two- or three-dimensional space accompanied with the least information loss. The example shown in Fig. 7.2 is the dimensional reduction from two dimension to one dimension. The axis is determined so that the sum of the distances from the data points to the axis can be minimized. Principal axis is set so that data variance can be included as much as possible. This method can be extended to m-dimensional data.

7.2.1.1 *Principal axes*

PCA has m principal axes when the dimension of the data is m. The first principal axis has the largest variance. The second principal axis is orthogonal to the first principal axis, which has the second largest variance. The third principal axis is orthogonal to the first and second principal axes, and it has the third largest variance. The mth principal axis is orthogonal to the first, second, ..., $m-1$th principal axes, and it has the mth largest (the least) variance.

7.2.1.2 *Derivation of principal axis*

Data matrix structure (n samples $\times$ p variables) is shown in Fig. 7.3.

The principal component scores, $Z_1, \ldots, Z_m$, are obtained after the dimensional reduction from p dimension to m dimension is performed:

$$
\begin{aligned}
Z_1 &= \ell_{11}x_1 + \ell_{12}x_2 + \cdots + \ell_{1p}x_p = \sum_{i=1}^{p} \ell_{1i}x_i \\
&\vdots \\
Z_k &= \ell_{k1}x_1 + \ell_{k2}x_2 + \cdots + \ell_{kp}x_p = \sum_{i=1}^{p} \ell_{ki}x_i \\
&\vdots \\
Z_m &= \ell_{m1}x_1 + \ell_{m2}x_2 + \cdots + \ell_{mp}x_p = \sum_{i=1}^{p} \ell_{mi}x_i
\end{aligned}
\tag{7.1}
$$

under the constraint below,

$$\ell_{k1}^2 + \ell_{k2}^2 + \cdots + \ell_{kp}^2 = \sum_{i=1}^{p} \ell_{ki}^2 = 1. \tag{7.2}$$

Coefficient ℓ_{kj} is determined so that the variance of Z_k can be maximized.

The coefficients $\{\ell_{1i}\}$ related to Z_1 (score of first principal component) should maximize $V(Z_1)$, the variance of Z_1 under the constraint of Eq. (7.2):

$$V(Z_1) = \frac{1}{n-1}\sum_{\alpha=1}^{n}\left(Z_{\alpha 1} - \overline{Z_1}\right)^2 = \frac{1}{n-1}\sum_{\alpha=1}^{n}\left(\sum_{i=1}^{p} \ell_{1i}(x_{\alpha 1} - \overline{x_i})\right)^2.$$

The equation above is converted to the following equation:

$$
\begin{aligned}
V(Z_1) &= \sum_{i=1}^{p}\sum_{i'=1}^{p} \ell_{1i}\ell_{1i'} \frac{1}{n-1}\sum_{\alpha=1}^{n}(x_{\alpha i} - \overline{x_i})(x_{\alpha i'} - \overline{x_{i'}}) \\
&= \sum_{i=1}^{p}\sum_{i'=1}^{p} \ell_{1i}\ell_{1i'} V_{ii'},
\end{aligned}
\tag{7.3}
$$

where

$$V_{ii'} = \frac{1}{n-1}\sum_{\alpha=1}^{n}(x_{\alpha i} - \overline{x_i})(x_{\alpha i'} - \overline{x_{i'}}) \tag{7.4}$$

and

$$\overline{x_i} = \frac{1}{n}\sum_{j=1}^{n} x_{ji}. \tag{7.5}$$

Using the method of Lagrange multiplier, $\{\ell_{1i}\}$ $(i = 1, 2, \ldots, p)$ to maximize Q,

$$Q = \sum_{i=1}^{p}\sum_{i'=1}^{p} \ell_{1i}\ell_{1i'} V_{ii'} - \lambda\left(\sum_{i=1}^{p} \ell_{1i}^2 - 1\right), \tag{7.6}$$

should be obtained. λ is a constant. Thus,

$$\frac{\partial Q}{\partial \ell_{1i}} = \sum_{i'}^{p} \ell_{1i'} V_{ii'} - \lambda \ell_{1i} = 0 \quad (i = 1, 2, \ldots, p), \tag{7.7}$$

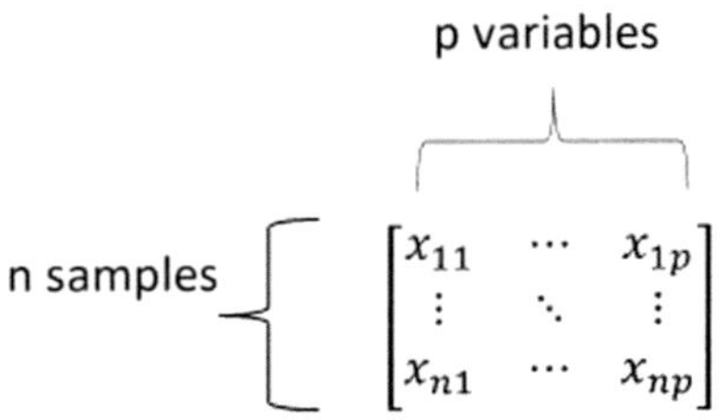

FIGURE 7.3 **Data structure for PCA.** Data matrix consists of n samples and p variables.

and, then, equations

$$\begin{aligned} (V_{11}-\lambda)\ell_{11}+V_{12}\ell_{12}+\cdots+V_{1p}\ell_{1p} &= 0 \\ V_{12}\ell_{11}+(V_{22}-\lambda)\ell_{12}+\cdots+V_{2p}\ell_{1p} &= 0 \\ &\vdots \\ V_{1p}\ell_{11}+V_{2p}\ell_{12}+\cdots\left(V_{pp}-\lambda\right)\ell_{1p} &= 0 \end{aligned} \tag{7.8}$$

are obtained. The determinant $|V-\lambda I|$ should be zero to obtain nontrivial solution:

$$|V-\lambda I| = \begin{vmatrix} V_{11}-\lambda & \cdots & V_{1p} \\ \vdots & \ddots & \vdots \\ V_{1p} & \cdots & V_{pp}-\lambda \end{vmatrix} = 0 \tag{7.9}$$

This is the pth-order polynomial equation of λ. Let the solutions of Eq. (7.9) be $\lambda_1, \lambda_2, \ldots, \lambda_p \quad (\lambda_1 > \lambda_2 > \cdots > \lambda_p)$.

Using Eqs. (7.2) and (7.7),

$$\begin{aligned} V(Z_1) &= \sum_{i=1}^{p}\sum_{i'=1}^{p}\ell_{1i}\ell_{1i'}V_{ii'} \\ &= \sum_{i=1}^{p}\ell_{1i}(\lambda_1\ell_{1i}) = \lambda_1\sum_{i=1}^{p}\ell_{1i}{}^2 = \lambda_1. \end{aligned} \tag{7.10}$$

Thus the variance of Z_1 is equal to the largest eigenvalue λ_1 and $\{\ell_{1i}\}$ $(i = 1, 2, \ldots, p)$ should be the eigenvector associated with the eigenvalue λ_1.

It is possible to obtain other principal components $(Z_2, Z_3, \ldots, Z_m)$ using the same procedure.

7.2.1.3 *Summary of method to obtain principal components*

You should obtain the eigenvalues of variance–covariance matrix V in the descent order. Then, let a set of coefficients $\{\ell_{1i}\}, \{\ell_{2i}\}, \ldots, \{\ell_{mi}\}\left(i = 1, 2, \ldots, p\right)$ be the eigenvectors with unity magnitudes associated with the eigenvalues.

The proportion of the kth principal component is

$$\frac{\lambda_k}{\sum_{k=1}^{p}\lambda_k}.$$

This value is an indicator of the amount of the information of the kth principal component. The cumulative proportion,

$$\frac{\sum_{i=1}^{m}\lambda_i}{\sum_{k=1}^{p}\lambda_k}, \tag{7.11}$$

is often used as an indicator of the amount of the information when we use the first m principal components.

7.2.2 Auto encoder

Although PCA is a linear dimensional reduction technique, an auto encoder is nonlinear dimensional reduction technique using a neural network (Goodfelllow et al., 2016). The auto encoder is one type of multilayer perceptron (MLP). Its structure is shown in Fig. 7.4.

The number of output layer units is set to be the same as the number of input layer units. Target inputs for output-layer units are the same as those of input layer units. Thus auto encoder behaves like a neural network with unsupervised learning although it is a neural network with supervised learning. The number of middle hidden layer units is set to be small so that the feature of the data can appear there. Thus the outputs of the middle hidden layer can be regarded as the auto encoder outputs.

The cost function used here consists of the sum of the square of the errors at the output layer and the regularization term as is shown in the following equation:

$$E = \frac{1}{2}\sum_{k}\left(y_k - \widehat{y_k}\right)^2 + \frac{\lambda}{2}\sum_{k}\sum_{j}\left|w_{kj}\right|, \tag{7.12}$$

where λ is a positive constant to control the regularization term. y_k, $\widehat{y_k}$, and w_{kj} are the output and predicted values of neuron k at the output layer and weight between neuron k and neuron j, respectively. Regularization is used to prevent overfitting. It uses an additional term in an error function that is typically chosen to impose a penalty on the complexity of the model. The usual

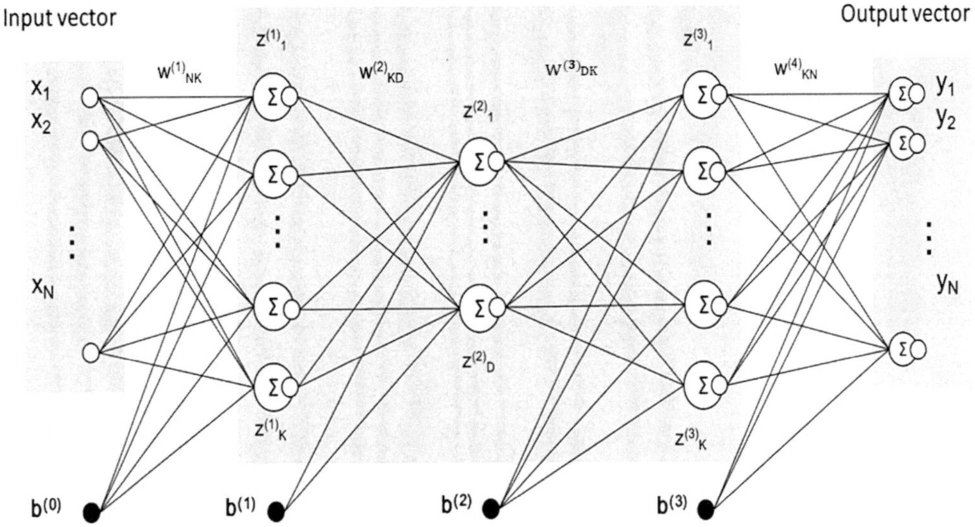

FIGURE 7.4 **Structure of auto encoder.** x_{ni} corresponds to the ith element of the nth sample. $\boldsymbol{b}^{(j)}$ is the bias in the jth layer. $y_n(x_n;W) = f(W^{(4)}f(W^{(3)}f(W^{(2)}f(W^{(1)}x_n + \boldsymbol{b}^{(0)}) + \boldsymbol{b}^{(1)}) + \boldsymbol{b}^{(2)}) + \boldsymbol{b}^{(3)})$ is the output of an autoencoder for a given x_n, where $\boldsymbol{W} = \{W^{(4)}, W^{(3)}, W^{(2)}, W^{(1)}, \boldsymbol{b}^{(0)}, \boldsymbol{b}^{(1)}, \boldsymbol{b}^{(2)}, \boldsymbol{b}^{(3)}\}$. f is a p-dimensional sigmoid function, $f(\boldsymbol{a}) = [1/(1 + \exp(-a_1)), \ldots 1/(1 + \exp(-a_p))]$. Source: *From Nozaki, Yuji, Nakamoto, Takamichi, Schaefer, & Andreas. (2016). Odor impression prediction from mass spectra.* PLoS One, *11(6). https://doi.org/10.1371/journal.pone.0157030.*

expression of L1 norm penalty term, that is, the sum of the absolute values of the weights, is used in the cost function.

Since the auto encoder has nonlinear property, it is expected to have a high capability of dimensional reduction. Fig. 7.5 shows the comparison of PCA with auto encoder. In both cases, the reconstruction error decreases as the number of feature increases. The features are principal components in the case of PCA, whereas they are the outputs of middle hidden layer units in the case of the auto encoder. The reconstruction error of the auto encoder is smaller than that of PCA at the same number of features, but the difference between the two becomes smaller as the number of the features increases.

Although a high capability of dimensional reduction for the auto encoder is expected, it takes some time to train the neural network. Moreover, the optimization of the neural-network hyperparameters is required before the actual use. Since PCA is very simple, it is often used to visualize the data. The auto encoder is used when the high capability of the dimensional reduction is required. Recently other technique called t-distributed Stochastic neighbor embedding is sometimes used to visualize the data.

7.3 Classification method

Although PCA can be used to visualize the data, it does not have the capability of classification. For classification, supervised technique is typically used. Since discrimination analysis is the fundamental technique for the classification, we explain it first.

7.3.1 Discrimination analysis

Discrimination analysis provides the method to construct a pattern classifier (Duda et al., 2001). The pattern classifier assigns a vector $\mathbf{x}$ to class ω_i if

$$g_i(\mathbf{x}) > g_j(\mathbf{x}) \quad \forall j \neq i, \tag{7.13}$$

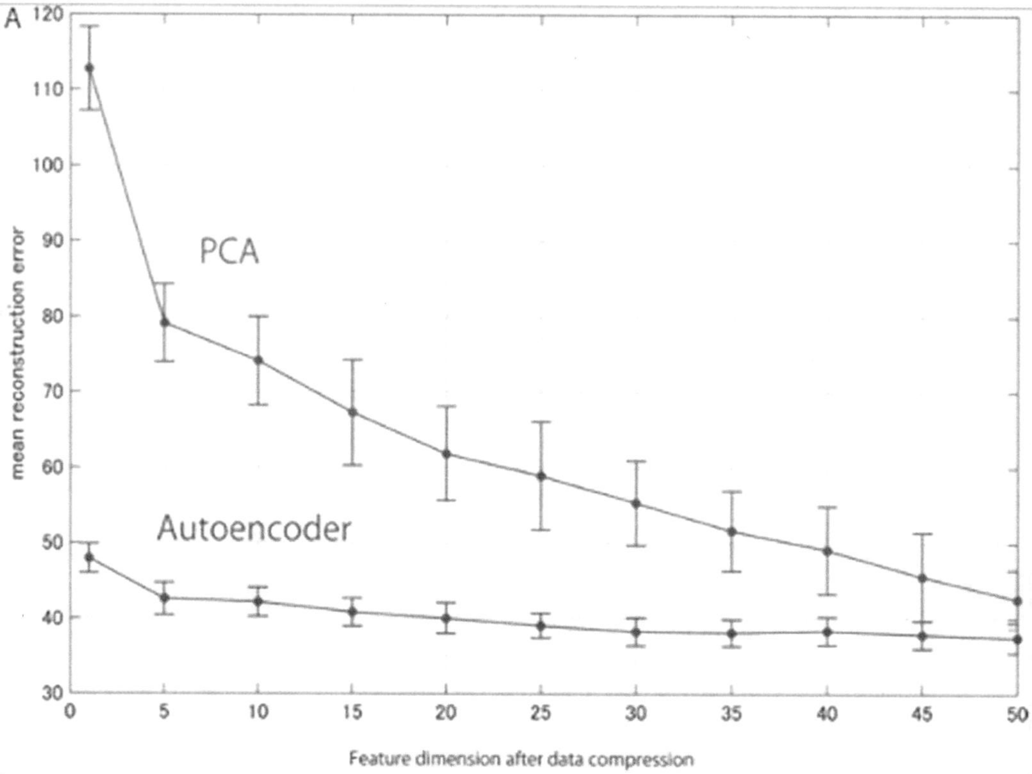

FIGURE 7.5 **Relationship between number of features and reconstruction error.** The data used here is sensory data with 144 original dimension. The number of second- and fourth-layer neurons in five-layer auto encoder was 65. Source: *From Nozaki, Yuji, Nakamoto, Takamichi, Schaefer, & Andreas. (2016). Odor Impression prediction from mass spectra.* PLoS One, *11 (6). https://doi.org/10.1371/journal.pone.0157030.*

where $g_i(\mathbf{x})$ is the ith discriminant function of $\mathbf{x}$. Thus the classifier is viewed as a machine that computes c discriminant functions and select the category with the largest discriminant.

Let us think about the discrimination between categories A and B in two-dimensional space as is shown in Fig. 7.6. The multiple data of categories A and B is given in advance to obtain their gravities. Then, the vector with unknown category is input to the system, and its distance to the gravity A is compared with its distance to the gravity B. The category with the smallest distance is the output of the discrimination analysis.

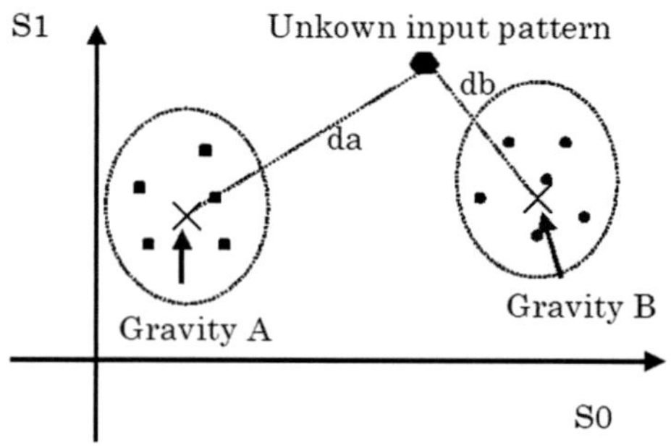

FIGURE 7.6 **Concept of discrimination between groups A and B.** Distance of input vector to gravity of group A is compared with that to gravity of group B.

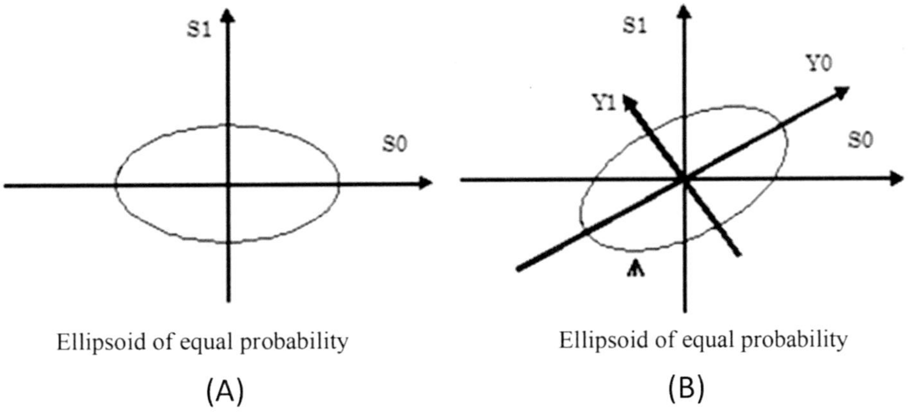

FIGURE 7.7 Data distribution of two variable case: (A) case when two variables are independent, (B) case when two variables are dependent.

Although the distance above is the Euclidean distance, Mahalanobis distance is often used. Since each dimension has different variation, we cannot use the simplest Euclidean distance as is shown in Fig. 7.7A. Thus the technique of autoscaling, which makes every dimension with the zero mean and with the unit standard deviation, is used. The autoscaling is sufficient if each dimension is independent of any other dimensions. However, the autoscaling is not sufficient when each dimension correlates with other dimensions as is shown in Fig. 7.7B. In this case, the coordinate axes should be rotated first so that each axis of the ellipsoid can match the system coordinate. Then, the autoscaling should be performed.

When is d-dimensional vector and $p()$ its probability density distribution, d-dimensional mean vector is

$$\vec{\mu} = E\left[\vec{x}\right] = \int \vec{x}\, p\left(\vec{x}\right) d\,\vec{x}, \tag{7.14}$$

and $d \times d$ variance–covariance matrix Σ is Σ variance–covariance matrix Σ is

$$\sum = E\left[\left(\vec{x} - \vec{\mu}\right)\left(\vec{x} - \vec{\mu}\right)^t\right] = \int \left(\vec{x} - \vec{\mu}\right)\left(\vec{x} - \vec{\mu}\right)^t p\left(\vec{x}\right) d\,\vec{x}, \tag{7.15}$$

where the function $E[\ ]$ is used to obtain expected value and the superscript t means the operation of transposing the vector.

Let the gravity of the kth category $\mu^{(k)}$, the variance–covariance matrix Σ $\Sigma^{(k)}$ and the input vector $\mathbf{x}$. The mahalanobis distance $\Delta(k)$ Δ. The mahalanobis distance $\Delta(k)^2$ between $\mathbf{x}$ and $\mu^{(k)}$ is

$$\Delta(k)^2 = \left(\boldsymbol{x} - \mu^{(k)}\right)^T \Sigma^{(k)-1}\left(\boldsymbol{x} - \mu^{(k)}\right),\ (k = 1, 2), \tag{7.16}$$

where T denotes the transpose operation. The input vector $\mathbf{x}$ is classified as the category k with the smallest mahalanobis distance above. Here, let us assume that $\Sigma^{(1)} = \Sigma^{(2)} = \Sigma$. Since we do not know the true $\mu^{(1)}$, $\mu^{(2)}$, and Σ, their estimates are Σ and Σ, their estimates are $\overline{\boldsymbol{x}}^{(1)}, \overline{\boldsymbol{x}}^{(2)}$, and S, respectively. $\overline{\boldsymbol{x}}^{(1)}$, $\overline{\boldsymbol{x}}^{(2)}$, and S are calculated using the observed data.

The estimated value of $\Delta(k)^2$, that is, $D(k)^2$, is

$$D(k)^2 = \left(\mathbf{x} - \overline{\mathbf{x}}^{(k)}\right)^T S^{-1}\left(\mathbf{x} - \overline{\mathbf{x}}^{(k)}\right), \tag{7.17}$$

where $\mathbf{x}$ is classified as category 1 if $D(2)^2 - D(1)^2 > 0$, whereas the same is classified as category 2 if $D(2)^2 - D(1)^2 < 0$.

$$\begin{aligned} z &= D(2)^2 - D(1)^2 \\ &= (\mathbf{x} - (\overline{\mathbf{x}}^{(1)} + \overline{\mathbf{x}}^{(2)})/2)^T S^{-1}(\overline{\mathbf{x}}^{(1)} - \overline{\mathbf{x}}^{(2)}). \end{aligned} \tag{7.18}$$

The equation above is the linear equation of $\mathbf{x}$ although the Mahalanobis distance itself is in the quadratic form of $\mathbf{x}$. If the assumption that $\Sigma^{(1)} = \Sigma^{(2)} = \Sigma$ is not valid, z has the quadratic form of $\mathbf{x}$. However, the linear form is used in most of cases.

The derivation shown above is the case of binary classification. It can be extended to multiple-class classification (William & Goldstein, 1984).

Finally, the difference between PCA and discrimination analysis is discussed as is shown in Fig. 7.8. In PCA, the principal axes are taken so that their variances can become the largest. However, the overlapping between the categories should be minimized in discrimination analysis, whereas the projections of two groups onto PC1 are almost overlapping in PCA. Thus the axes taken in the two methods are much different as is shown in the figure.

7.3.2 Pattern separation index

The ability of discrimination among groups can be evaluated by a statistical index called Wilks' lambda. This index provides a measure of pattern separation among target groups. We can evaluate the quality of sensor array using Wilks' lambda (Nakamoto et al., 1992).

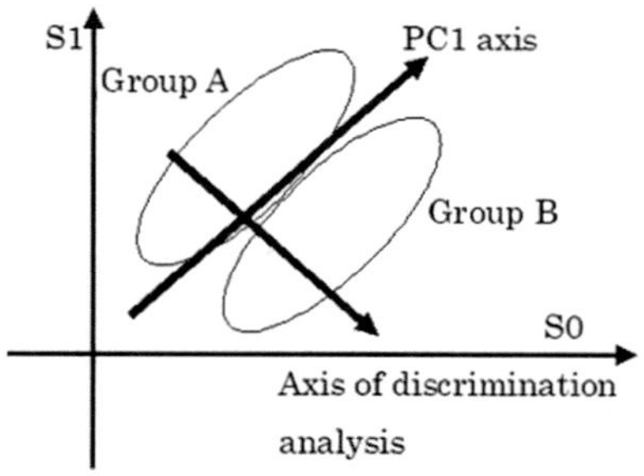

FIGURE 7.8 Difference between principal component analysis and discrimination analysis. Axis of principal component analysis is completely different from that of discrimination analysis.

The pattern separation can be qualitatively evaluated by the ratio of the data variation within a group to that among groups. The good pattern separation is obtained when the data variation among groups is large and the variation within a group is small, as is illustrated in Fig. 7.9. Thus the smaller Wilks' lambda, the better the pattern separation is.

Let q the number of variables, that is, the number of sensors. The matrix of the total mean corrected sums-of-squares of cross-products for the scores on all n observations, that is, the number of total data is

$$T = \sum_{k=1}^{g} \sum_{i=1}^{n_k} (\mathbf{x}_i^k - \mathbf{x}_{av})(\mathbf{x}_i^k - \mathbf{x}_{av})^T. \quad (7.19)$$

The within-group sum-of-squares matrix is

$$W = \sum_{k=1}^{g} \sum_{i=1}^{n_k} (\mathbf{x}_i^k - \mathbf{x}_{av}^k)(\mathbf{x}_i^k - \mathbf{x}_{av}^k)^T, \quad (7.20)$$

where g is the number of groups; n_k, the number of data within group k; $\mathbf{x}_i^k$, ith data vector in group k; $\mathbf{x}_{av}$, the total average vector; and $\mathbf{x}_{av}^k$, the average vector of group k.

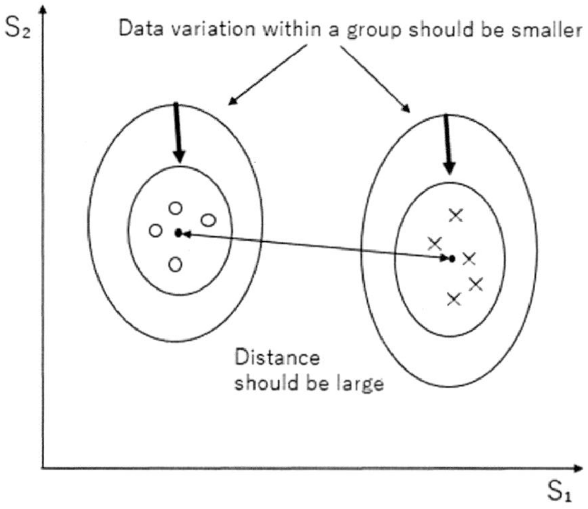

FIGURE 7.9 Concept of pattern separation index. Pattern separation is good when distance between groups is large and data variation within group is small.

Wilks' lambda is the ratio of the determinant W to the determinant T, that is,

$$\Lambda = \frac{W}{T}. \tag{7.21}$$

When we deal with Wilks' lambda mathematically, plural axes, that is, the eigenvectors of the canonical analysis are determined, followed by the data projection onto those axes. The ratio described previously is calculated with respect to each axis and the product of those ratios is Wilks' lambda. This lambda takes value between zero and one. The discrimination accuracy is high if Wilks' lambda is close to zero.

An example of Wilks' lambda is shown in Fig. 7.10A and B. Three electrochemical sensors were used and the result of linear discrimination analysis is plotted there. Electrochemical sensors are explained in Chapter 3. Three analytes, that is, butanol isomers, were used. Wilks' lambda in Fig. 7.10A is 2.0×10^{-1}, indicating small pattern separation. On the other hand, Wilks' lambda in Fig. 7.10B is 1.1×10^{-3}, which is larger than that in Fig. 7.10A. The pattern separation in Fig. 7.10B is obviously better than that in Fig. 7.10A when we see those figures.

When we would like to evaluate the contribution of each variable (sensor) to the pattern separation, we use the partial lambda given by

$$\Lambda\left(x_j|\mathbf{x}^*\right) = \frac{\mathbf{\Lambda}\left(\mathbf{x}^*, x_j\right)}{\mathbf{\Lambda}(\mathbf{x}^*)}, \tag{7.22}$$

where $\mathbf{\Lambda}\left(\mathbf{x}^*, x_j\right)$ is Wilks' lambda with a set of variables with x_j, and $\mathbf{\Lambda}(\mathbf{x}^*)$ is the Wilks lambda of a set of variables without x_j. We can evaluate each variable's contribution using partial lambda.

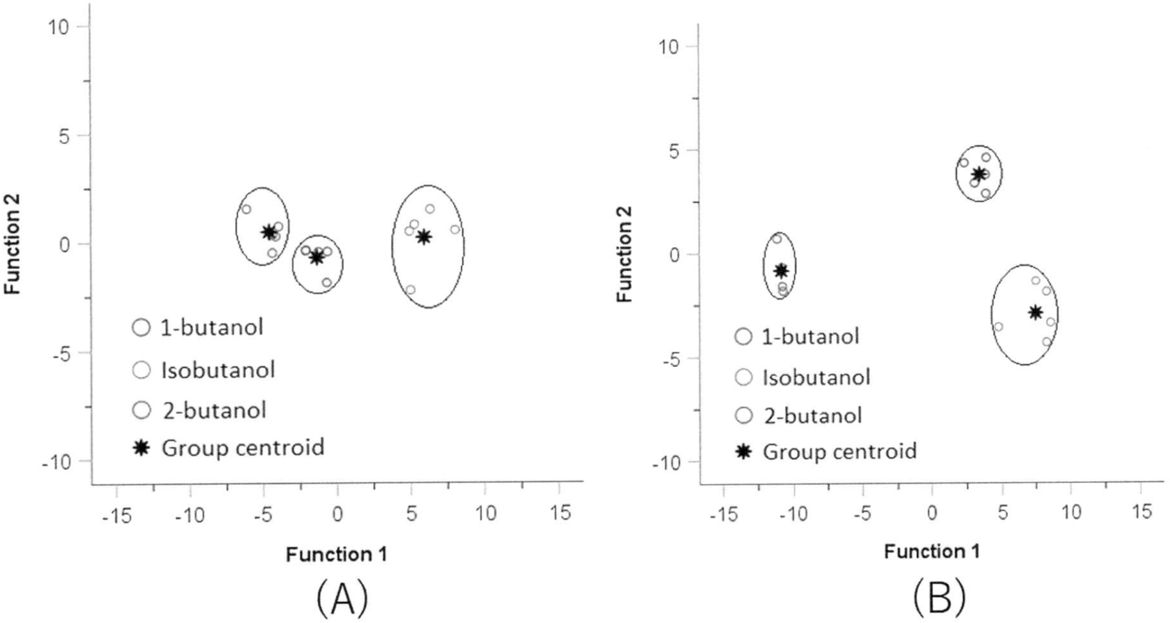

FIGURE 7.10 LDA result of three analytes: (A) same material with different working electrode voltages. Each sample was measured five times. Electrolyte: [EMIM]{Ac}, working electrode voltage:0.25, 0.5, and 0.9 V; (B) different materials with the same working electrode voltage. Electrolytes: [EMIM][Ac], [EMIM] [Otf], and [EMIM][Cl], Working electrode voltage: 0.5 V (Faricha et al., 2023). Source: *From A. Faricha, S. Yoshida, P. Chakraborty, K. Okamoto, T.-F. Mark Chang, M. Sone, & T. Nakamoto. Array of miniaturized amperometric gas sensors using atomic gold decorated Pt/PANI electrodes in room temperature ionic liquid films. Sensors, 23(8)(2023), https://doi.org/10.3390/s23084132.*

7.4 Regression method

7.4.1 Multiple linear regression

Regression analysis is used when quantification, not classification, is required. Most typical problem is to quantify mixture composition when a chemical mixture is provided. Simple linear regression is first explained.

The p variable linear multiple regression model involving the dependent variable y and p independent variables, numbered $x_2, x_3, \ldots, x_p$, can be written as

$$y = \beta_1 + \beta_2 x_2 + \beta_3 x_3 + \cdots + \beta_p x_p + \varepsilon, \tag{7.23}$$

where β_1 is the intercept, $\beta_2, \ldots, \beta_p$ are the partial regression slope coefficients, and ε is the residual term. Multiple linear regression (MLR) technique determines those coefficients.

If we have n observations, we can obtain the following equation:

$$\begin{bmatrix} y_1 \\ y_2 \\ \vdots \\ y_n \end{bmatrix} = \begin{bmatrix} 1 & x_{21} & \cdots & x_{p1} \\ 1 & \vdots & & \vdots \\ \vdots & \vdots & & \vdots \\ 1 & x_{2n} & \cdots & x_{pn} \end{bmatrix} \begin{bmatrix} \beta_1 \\ \beta_2 \\ \vdots \\ \beta_p \end{bmatrix} + \begin{bmatrix} \varepsilon_1 \\ \varepsilon_2 \\ \vdots \\ \varepsilon_n \end{bmatrix}, \tag{7.24}$$

where y_i and ε_i are the ith observation of y and its residual, respectively, and x_{ji} is the ith value of x_j.

Eq. (7.24) can be written as

$$\vec{y} = X\vec{\beta} + \vec{\varepsilon}, \tag{7.25}$$

where

$$X = \begin{bmatrix} 1 & x_{21} & \cdots & x_{p1} \\ 1 & \vdots & & \vdots \\ \vdots & \vdots & & \vdots \\ 1 & x_{2n} & \cdots & x_{pn} \end{bmatrix}, \tag{7.26}$$

$$\vec{\beta} = \begin{bmatrix} \beta_1 \\ \beta_2 \\ \vdots \\ \beta_p \end{bmatrix}, \tag{7.27}$$

and

$$\vec{\varepsilon} = \begin{bmatrix} \varepsilon_1 \\ \varepsilon_2 \\ \vdots \\ \varepsilon_n \end{bmatrix}. \tag{7.28}$$

SSE, error sums-of-squares, is

$$\begin{aligned} \text{SSE} &= \vec{\varepsilon}^T\vec{\varepsilon} = \left(\vec{y} - X\vec{\beta}\right)^T\left(\vec{y} - X\vec{\beta}\right) \\ &= \vec{y}^T\vec{y} - 2\vec{y}^T X\vec{\beta} + \left(X\vec{\beta}\right)^T\left(X\vec{\beta}\right), \end{aligned} \tag{7.29}$$

where SSE is differentiated with respect to $\vec{\beta}$ and we let the derivative be zero to obtain $\vec{\beta}$ for minimizing SSE:

$$\frac{\partial}{\partial\vec{\beta}}(\text{SSE}) = -2X^T\vec{y} + 2X^T X\vec{\beta} = 0. \tag{7.30}$$

Then,

$$(X^T X)\vec{\beta} = X^T\vec{y}, \tag{7.31}$$

and finally we obtain the partial regression slope coefficients in Eq. (7.23):

$$\vec{\beta} = (X^T X)^{-1}X^T\vec{y}. \tag{7.32}$$

We often see the equation above in textbooks.

For the derivation of Eq. (7.30), Eqs. (7.33)–(7.35) are used:

$$\frac{\partial}{\partial\vec{\beta}}(\text{SSE}) = \begin{bmatrix} \dfrac{\partial \text{SSE}}{\partial\beta_1} \\ \dfrac{\partial \text{SSE}}{\partial\beta_2} \\ \vdots \\ \dfrac{\partial \text{SSE}}{\partial\beta_p} \end{bmatrix} \tag{7.33}$$

$$\frac{d}{d\vec{x}}\left(\vec{a}^T\vec{x}\right) = \vec{a} \tag{7.34}$$

$$\frac{d}{d\vec{x}}\vec{x}^T Q\vec{x} = 2Q\vec{x}, \tag{7.35}$$

where Q is a symmetric matrix.

7.4.2 Fitting of multiple linear regression

The equation of MLR does not guarantee its accuracy even if it is optimal in given situation. Thus it is necessary to evaluate its fitting to the data.

SST, the total sums-of-squares, is

$$\mathrm{SST} = \sum_{i=1}^{n} (y_i - \overline{y})^2, \tag{7.36}$$

where $\overline{y}$ is the average of n observations.

SSR, the regression sums-of-squares, is

$$\mathrm{SSR} = \sum_{i=1}^{n} (\hat{y}_i - \overline{y})^2, \tag{7.37}$$

where $\hat{y}_i$ is the ith estimated y.

$$\mathrm{SST} = \mathrm{SSR} + \mathrm{SSE} \tag{7.38}$$

The coefficient of determination, R^2, is defined as

$$R^2 = \frac{\mathrm{SSR}}{\mathrm{SST}} = 1 - \frac{\mathrm{SSE}}{\mathrm{SST}}. \tag{7.39}$$

The fitting is good if R^2 is large whereas the fitting is poor when R^2 is small. R, the square root of coefficient of determination, is called the coefficient of multiple correlation.

7.4.3 Collinearity problem

MLR is the most fundamental technique to obtain the mixture composition of gas/liquid based upon responses of arrayed chemical sensors. The important problem to be solved in the mixture quantification is the collinearity problem.

Let us assume the two-sensor case to obtain the composition of the binary mixture as is illustrated in Fig. 7.11. Each line shows the sensor characteristic and the intersection point of the two lines expresses the composition. Stable solution can be obtained if two lines, that is, two sensor characteristics, are completely different. However, these two lines become almost parallel if two sensor

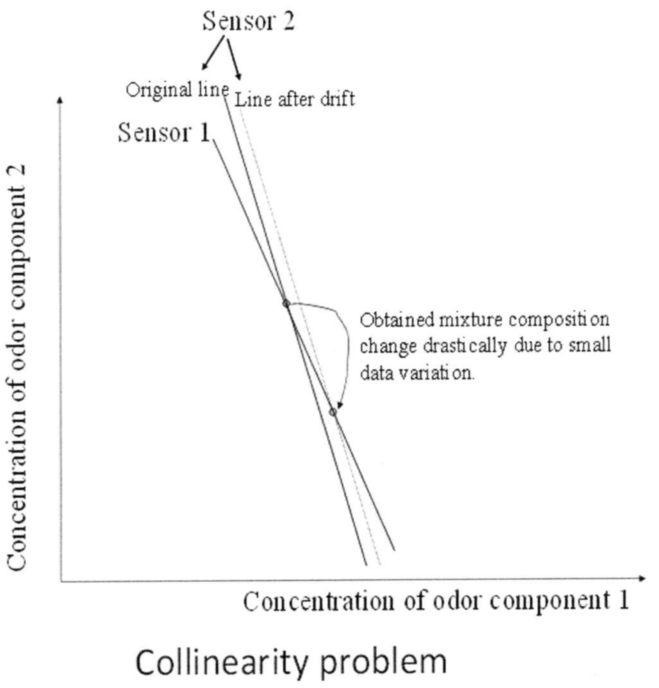

FIGURE 7.11 **Collinearity problem.** Collinearity problem causes unstable solution.

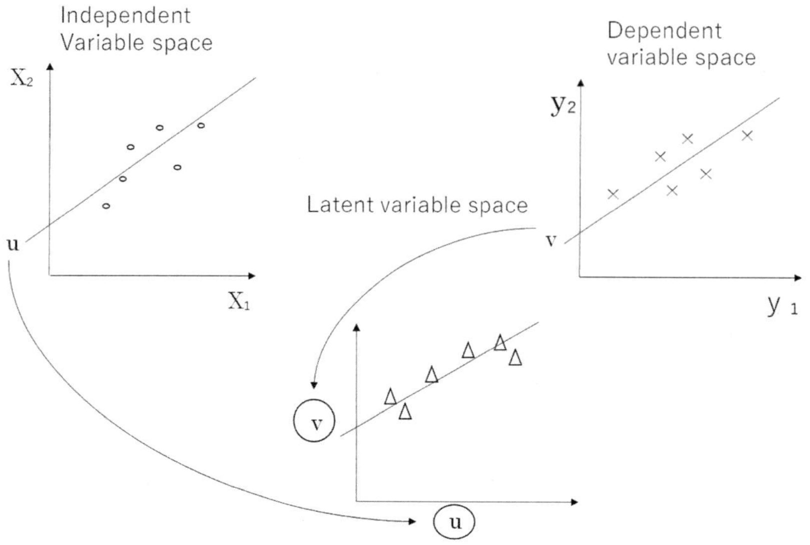

FIGURE 7.12 Principle of PLS method. Latent variables are extracted both for independent and dependent variables. *PLS,* Partial least squares.

characteristics are very similar. Then, the obtained composition becomes instable even if the measurement data fluctuate just a little. In this figure, the location of the intersection point drastically changes with small noise. This problem is called collinearity problem and is the same as that of the calculation of the inverse matrix with almost zero determinant.

Although the best way to solve the problem is to obtain a set of good variables (sensors) with their characteristics orthogonal to each other, there are several methods to reduce the collinearity in chemometrics where statistical and mathematical techniques are used to analyze chemical data. Principal component regression (PCR) and partial least squares (PLS) are often adopted. PCR is the combination of PCA with MLR. PCA is often used to reduce the dimension with minimum information loss as is described in Section 7.2.1. Since the principal component with its small contribution can be regarded as noise and can be ignored, the instability due to the collinearity is reduced. Thus a small number of principal components obtained by PCA are projected onto dependent variables using the MLR. These principal components exhibit the feature of the sensor data and are called latent variables.

PLS method picks up latent variables both for independent and dependent variables (Fig. 7.12) (Geladi & Kowalski, 1986), whereas PCR extracts latent variables only for independent variables of MLR. Latent variables in space of independent variables are projected onto space of dependent variables using regression equation. Number of latent variables is an important parameter for raising prediction accuracy. Thus the cross validation technique is used to determine the appropriate number of the latent variables. PLS typically has higher accuracy than ordinal MLR when it is applied to mixture quantification problem (Carey et al., 1987).

7.4.4 Nonlinear regression method

MLR in Section 7.4.1 can be extended to general polynomial equation if the relationship between independent variables and dependent variables is nonlinear. However, it is necessary to know in advance which polynomial term should be used. In most of cases, we do not have such a priori knowledge. Thus several nonlinear regression techniques without a priori knowledge are introduced here.

Sundgren et al. (1991) applied an artificial neural network to the mixture quantification problem. They used metal oxide semiconductor field effect transistor gas sensor array with nonlinear characteristic. Three-layer MLP was used to quantify the binary mixture. It is straightforward approach at the early stage.

Wang et al. (1995) developed their own model called Chemnet. They focused on metal oxide gas sensor. They expressed the ith sensor output X_i by

$$X_i = R_i \left[1 + \sum_{j=1}^{J} \left(A_{ij} Y_j \right)^{m_{ij}} \right]^{-B_i}, \tag{7.40}$$

where R_i is the sensor resistance in the air; m_{ij} is the gas and sensor-specific parameter; B_i ($B_i > 0$), the parameter specific to the sensor element; Y_j, the jth gas concentration; A_{ij}, the coefficient; J, the number of gases. Although the model above is roughly valid, there is still residual when the curve fitting is performed. Then, they tried to express its residual using standard neural network. Totally Chemnet consists of four layers, and they successfully quantified the binary mixture. It is interesting to introduce the sensor response model into the neural network.

Other interesting technique is locally weighted regression (LWR) (Wang et al., 1994). The nonlinear function can be locally approximated with piecewise linear function. Thus LWR is a piecewise linear method. Its concept is illustrated in Fig. 7.13. Since data point can be interpolated using the data within the local window, the local model at the point A is different from that at the point B. Thus nonlinear function can be expressed even if only the linear model is available. Although Fig. 7.13 is a simple one-dimensional model, it can be extended to multidimensional one.

The important step in LWR is to select the local calibration samples by some distance measure. PCA is used to reduce the dimension of independent variable space, and then

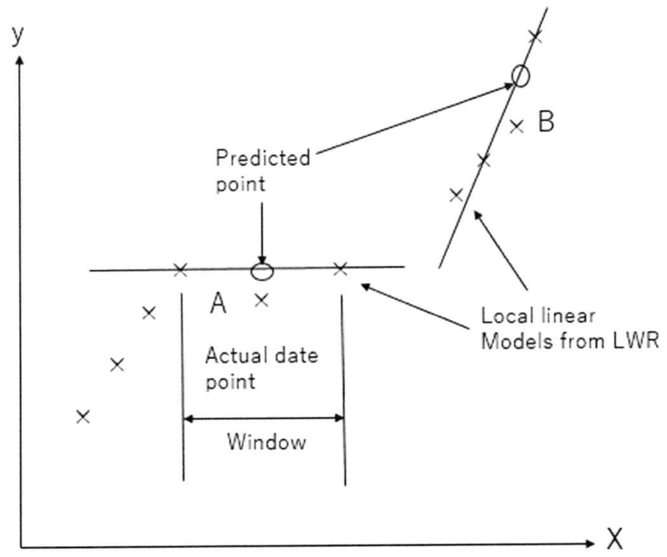

FIGURE 7.13 Illustration of locally weighted regression. Nonlinear function is approximated with piecewise linear functions.

Mahalanobis distance is used as distance measure. Another method (LWR1) is to weight the principal components according to its prediction ability before obtaining Mahalanobis distance measure.

Although those methods only consider independent variable space, it is better to take dependent variable space when distance measure is calculated. The distance measure is obtained by balancing the information in both independent and dependent variable spaces. That method where calibration samples are chosen on the basis of the closeness in two spaces is called LWR2.

The advantage of LWR is just simple even linear function can be used to express whole complicated function. On the other hand, it requires many data points to obtain high prediction accuracy.

7.5 Clustering technique

Cluster analysis is an unsupervised technique to group data close to each other. Since it reveals the similarities among the data points, it can be used for the same purpose of PCA, where the multidimensional data distribution is visualized. Cluster analysis (hierarchical cluster analysis (HCA)) reveals the data structure, whereas PCA provides the information of the direction from data point to data point.

7.5.1 *k*-Means clustering

k-Means clustering is a technique of vector quantization. *k*-Means clustering aims to partition n observations into k clusters in which each observation belongs to the cluster with the nearest mean. This is an unsupervised classification technique and nonHCA. The number of clusters k should be given in advance.

Given an initial set of k cluster centers $\mathbf{m}_1$, ..., $\mathbf{m}_k$, the observation x_p is assigned to cluster j when the distance of x_p to cluster center $\mathbf{m}_j$ is minimum among k cluster centers.

Step 1: $l = 1$

Step 2: $\mathbf{x}_p$ is assigned to cluster j, where $\mathbf{m}_j^{(l)} = \| \min_i x_p - \mathbf{m}_i^{(l)} \|$.

Step 3: Each cluster center $\mathbf{m}_j^{(l+1)}$ is updated according to the following equation:

$$\mathbf{m}_j^{(l+1)} = \frac{1}{N_j} \sum_{x_p \in S_j^{(l)}} x_p, \tag{7.41}$$

where N_j is the number of observations in cluster j and $S_j^{(l)}$ is a set of the observations assigned to the cluster j at the lth update.

Step 4: If $\mathbf{m}_j^{(l+1)} = \mathbf{m}_j^{(l)}$ for every cluster, the assignment of each observation is fixed.

Otherwise, $l \leftarrow l + 1$ and go to Step 2.

Fig. 7.14 shows an example of clustering using *k*-means method

Since the classification result typically depends on k, this algorithm should be tried for various k values.

7.5.2 Hierarchical cluster analysis

HCA is an unsupervised technique to represent the similarity between the data points and the data structure using a dendrogram shown below. The dendrogram shows the order where points are grouped together and shows the distance where they are linked. It is a tree structure that starts at grouping all data points and branches into smaller and smaller groupings until all individual data points are reached.

All data points can be classified as similar and dissimilar groups by ruling a dashed vertical line (cutoff) through the tree structure. Any grouping that occurs to the left side of the line indicates similar values and any which occurs to the right indicates dissimilar groupings. The researcher decides where to rule the line. In Fig. 7.15, the groups based on HCA agree with the actual ones such as ammonia, butanol, ethanol, and isoamyl acetate.

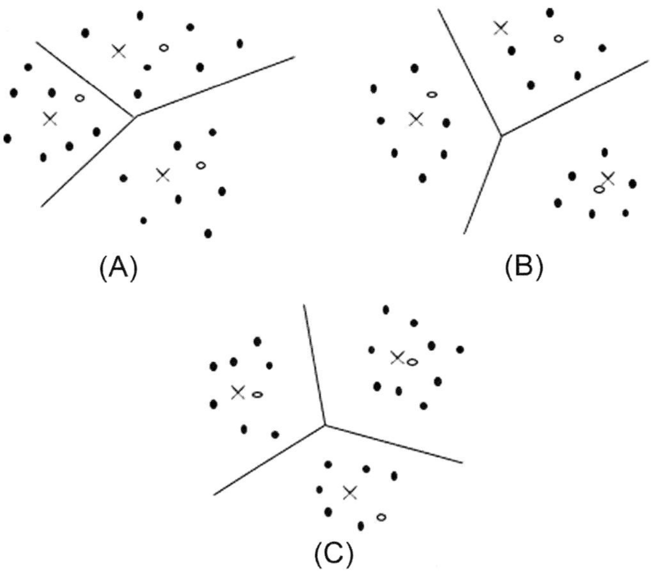

FIGURE 7.14 Clustering using *k*-means method: (A) previous cluster center (x), the updated cluster center (○); (B) clustering result at the second time; (C) clustering result at the third time. Every cluster center agrees with the previous one.

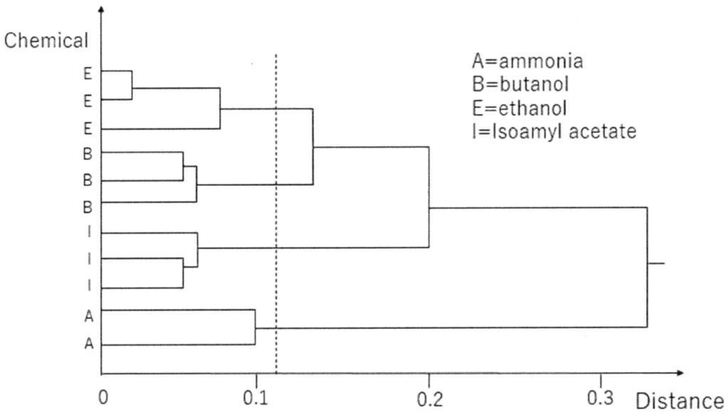

FIGURE 7.15 **Hierarchical cluster analysis of data from gas sensor array.** It shows the structure of multidimensional data using dendrogram.

When you perform HCA, the definition of the distance from one group to other one and the type of distance (dissimilarity) to be measured should be determined.

There are several definitions of distance from group to group. We call that definition of fusion method and several fusion methods are as follows:

Fusion method:

1. Nearest neighbor method (single linkage method)
2. Furthest neighbor method (complete linkage method)
3. Centroid method
4. Ward method

First three methods are illustrated in Fig. 7.16. In nearest neighbor method, that is, single linkage method, the shortest distance among all combination of distance between points of two different groups is taken. In furthest neighbor method, the longest distance among all combination of distances between points of two different groups is

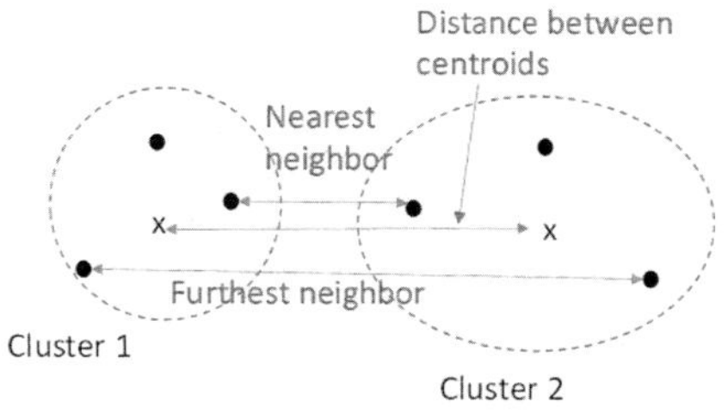

FIGURE 7.16 Several fusion methods used in hierarchical clustering.

taken. In centroid method, the distance between the centroids of two groups is used.

In Ward method, the total sum of squared deviations of every observation from the mean of the cluster where it belongs is considered. The group assignment is done on the basis of the increase in that sum induced from combining ever possible pair of clusters.

Then, several distance types are explained.

Type of distance

1. Euclidean distance

$$D = |\mathbf{x} - \mathbf{m}| \tag{7.42}$$

2. Vector angle

$$\cos\theta = \frac{\mathbf{x}^T\mathbf{m}}{|\mathbf{x}||\mathbf{m}|} \tag{7.43}$$

3. Mahalanobis distance

$$D = (\mathbf{x}-\mathbf{m})^T\Sigma^{-1}(\mathbf{x}-\mathbf{m}) \quad \Sigma\text{: Variance} - \text{covariance matrix} \tag{7.44}$$

In addition to the distances above, correlation coefficient and Tanimoto similarity are sometimes used as distance. Since there are many fusion methods and data types, it is necessary to select those when HCA is applied to data.

7.6 Application of neural network

Neural network is often used in digital olfaction technology. It is a standard pattern recognition technology in odor sensing system. Moreover, it is also used for predicting odor impression (Chapters 9 and 10). Since detail in neural network is described elsewhere (Goodfelllow et al., 2016; Aggarwal, 2018; Alpaydin, 2010), tips to use neural network in digital olfaction technology are explained here.

7.6.1 Neural network structure

There are two ways to make connections between neurons. The first one is a multilayer structure as shown in Fig. 7.17A. The signals propagate from input layer to intermediate layer (hidden layer) and from intermediate layer (hidden layer) to output layer. It is possible to have more than one intermediate layer.

This signal flow is called forward propagation and that flow in the opposite direction is not allowed. Moreover, the connection between the neurons in the same layer is not allowed. The neural network with this structure is called MLP.

Fig. 7.17B shows a mutually coupled neural network. The feedback connection is allowed in this type of the neural network. The most popular network of this type was proposed by Hopfield (Hopfield & Tank, 1985). The Hopfield neural network is often used to solve the optimization problem rapidly. The famous problem is traveling salesman problem, which is NP complete. The abbreviation NP refers to "nondeterministic polynomial time."

In the field of digital olfaction technology, MLP is typically used. The advantage of the neural network over other classifiers is nonlinearity. Moreover, it is not necessary to fix shape of the function. Any function can be flexibly obtained after learning. Fig. 7.18 shows the problem of pattern classification. A complicated boundary between categories can be formed using a neural network, whereas it cannot be done by multivariate analysis such as discrimination analysis. Arbitrary function can

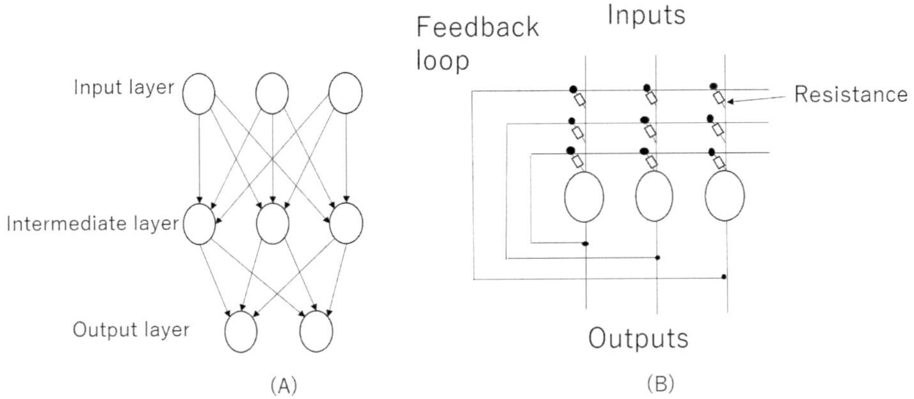

FIGURE 7.17 Neural network structure: (A) multilayer perceptron and (B) structure with feedback.

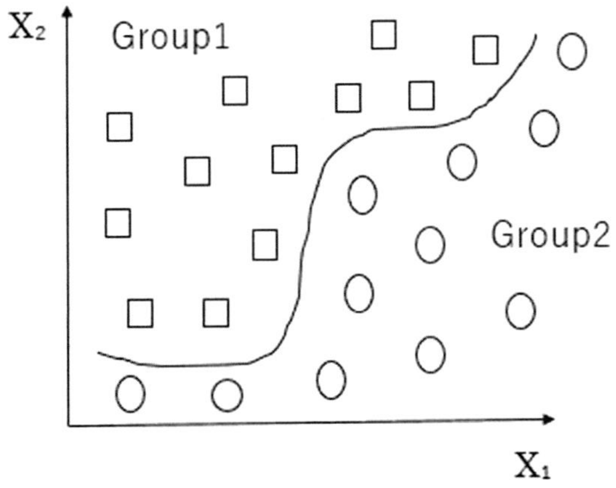

FIGURE 7.18 Data with complicated discrimination boundary. Nonlinear technique can form complicated discrimination boundary.

be approximately realized with arbitrary accuracy if a sufficient number of hidden-layer neurons are available (Funabashi, 1989).

7.6.2 Backpropagation algorithm

There are a huge number of weights in MLP and they should be determined appropriately for pattern classification. The weights are iteratively updated during its training since the whole system is nonlinear. It takes typically long time until the convergence. The supervised learning is performed using target signals for the neurons in the output layer. The signals propagate from the input layer to the output layer during the feedforward phase. In the training phase, the errors between the outputs of the neurons in the output layer and target signals are evaluated, and the weights between intermediate layer and output layer are first updated. Then the weights between input layer and intermediate layer are updated. This training process is called backpropagation (Rumelhart et al., 1986). The feedforward and backward processes are iteratively performed during the training.

The concept of weight optimization is shown in Fig. 7.19. The vertical axis means the cost function in the following equation:

$$E = \sum_{i=1}^{N} (o_i - d_i)^2 + \lambda \sum_{n,m} |W_{nm}|, \tag{7.45}$$

where o_i and d_i are the output and the target signal of the ith neuron in the output layer; λ, constant; W_{nm}, weight; N, the number of output-layer neurons, respectively. The cost function consists of error term and regularization term. The second term is the L1-norm regularization term that suppresses enlarging weight values. If the weight value becomes too

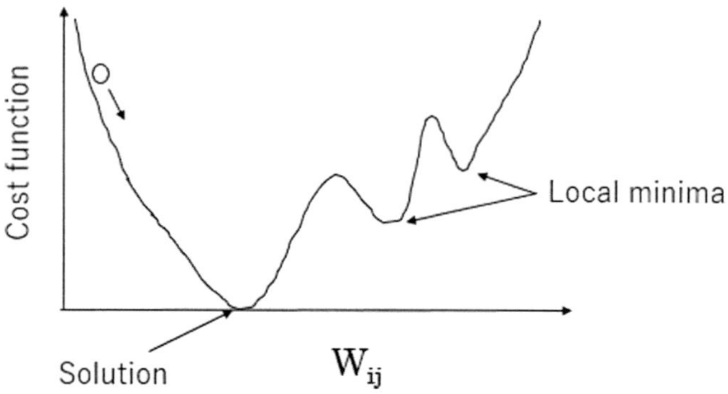

FIGURE 7.19 **Gradient method to update weights.** Solution is searched according to gradient.

large, overlearning problem discussed later appears.

The weight is iteratively updated until the convergence so that E can be minimized. Although multidimensional weight space should be considered in the actual problem, only one weight is shown here for simplicity. The number of weights to be optimized is typically from a few tens to a few tens of thousands. The solution, that is, an optimized set of weights, is not unique and depends upon their initial values. The initial values of the weights are often generated using the random number.

The relationship between W_{ij} and E is not simple. There are several small hills and hollows in Fig. 7.19. The updated weight might be trapped at the local minimum points. Although it is very difficult to solve the local-minima problem, the noise is intentionally added to the weight to escape from the trap. This method is adopted in simulated annealing. Although Fig. 7.19 has only one degree of freedom for the weight change, many degrees of freedom are available in the actual case. In that case, the probability of the trap at the local minima decreases even if there is no guarantee to obtain global minimum.

Another problem we frequently encounter is gradient vanishing problem. Since the activation function of each neuron is a saturated function such as sigmoid or tanh function, learning proceeds very slowly when the data point is located at saturation region. At that time, the shape of the gradient function can be a little modified to have nonzero value.

The training procedure is not described here. The weight update is performed in the following way:

$$\Delta W_{ij}^{(N,N-1)}(n+1) = \eta \delta_i^{(N)} x_j^{(N-1)} + \alpha \Delta W_{ij}^{(N,N-1)}(n), \quad (7.46)$$

where n is the number of learning, η and α, constants; δ_i^N, delta signal; N, layer number, respectively. The first term in Eq. (7.46) decreases the error, and the second term is used to accelerate the convergence.

There are two ways to update weights for multiple observed patterns. The first way is to update the weights for each observed pattern. This method is called stochastic gradient descent method. The second way is to update the weights for all patterns once the weight changes for all patterns are calculated. The latter one is called batch process.

7.6.3 Application of neural network to odor sensing system

Fig. 7.20 shows the application of MLP to an odor sensing system as a pattern classifier. In an odor sensing system, an output pattern from an array of sensors with partially overlapping specificities is recognized by a neural network. The number of neurons in the input layer is the same

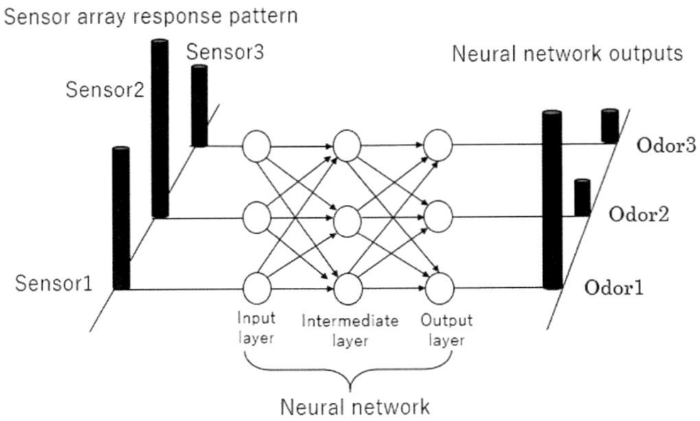

FIGURE 7.20 **MLP used in odor sensing.** Sensor signals are given to an input layer and classification result appears on the output layer. *MLP*, Multilayer perceptron.

as that of the sensors, and the number of neurons in the output layer is the same as that of odor categories. The number of neurons in the intermediate layer is empirically determined in most of cases. When we use MLP with a sufficient number of the neurons in the intermediate layer, arbitrary mapping with arbitrary accuracy can be achieved (Funabashi, 1989).

When the output pattern of odor1 is input to the MLP, the target signal for the output of odor1 neuron is set to 1, whereas the ones for the outputs of odor 2 and odor 3 are set to zeros. When the output patterns of odor 2 and 3 are input to the neural network, the target signals are given in the same way. Although the target signals are one or zero for the classifier, the continuous values are used to express sensory information such as scores of semantic differential method.

7.6.4 Cross-validation method and overlearning

The data used for training cannot be used for measuring pattern recognition capability. The recognition rate, that is, accuracy, for the data not used for training is much lower than that of the data for the training due to the generalization capability. The test data should be separated from the training data. The evaluation method to use the data different from the training data is called cross validation.

The generalization capability is the prediction ability for an unseen data. Since it is always necessary for a pattern classifier to do the prediction for the unseen data, it is very important. Although the accuracy for the training data should be of course high, it is not enough.

Especially, the difference between the two is larger when the problem of the overlearning occurs as is shown in Fig. 7.21. The classification is completely all right for the data points that are exactly the same as those for training. However, the classification result is completely wrong at the data points that are slightly different from the training data points. The cross validation is essential for avoiding the overestimation of the classification performance. Moreover, the magnitude of the weight value tends to be large when over learning occurs.

However, we often encounter the case when the number of observations is insufficient. In this case, the numbers of trained data and test data becomes further smaller when we perform the cross validation. The leave-one-out method illustrated in Fig. 7.22 is useful when the number of the observations is smaller. When we have N data points, $N-1$ data points are used for training and just one data point is used for testing. Then, we exchange a test data point with another point

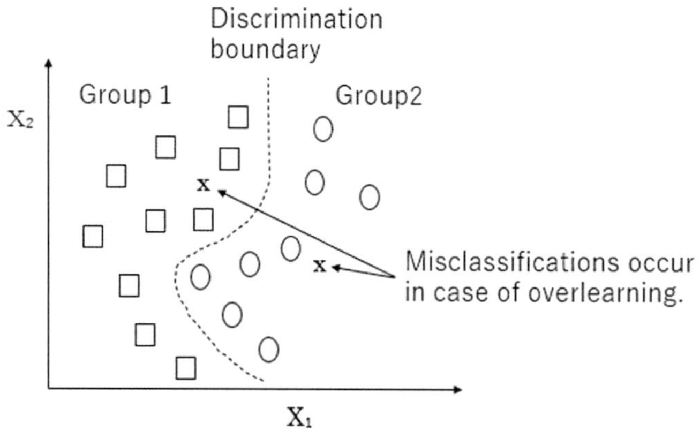

FIGURE 7.21 **Overlearning.** Missclassification occurs when the data point for the test is slightly away from that for training.

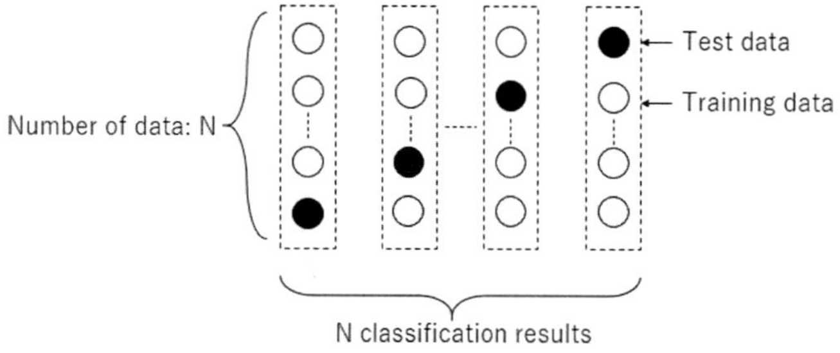

FIGURE 7.22 **Leave-one-out method.** Black circle: test data; white circle: training data.

and rebuild a classifier model again using $N-1$ data points. In the same manner, we can test N data points although we should rebuild a model every time.

7.7 Summary

We learn chemometrics techniques in this chapter. Only fundamental techniques we frequently encounter are picked up, although many other multivariate techniques are available. However, the techniques described here are common in various situations when digital olfaction technology is used.

References

Aggarwal, C. C. (2018). Neural networks and deep learning. Springer.

Alpaydin, E. (2010). Introduction to machine learning. MIT Press.

Beebe, K. R., Pell, R. J., & Seasholtz, M. B. (1998). Chemometrics—A practical guide. Wiley.

Carey, W. P., Beebe, K. R., & Kowalski, B. R. (1987). Multicomponent analysis using an array of piezoelectric crystal sensors. *Analytical Chemistry*, *59*(11), 1529–1534. Available from https://doi.org/10.1021/ac00138a010.

Duda, R. O., Hart, P. E., & Stork, D. G. (2001). *Pattern classification* (pp. 215–281). Wiley-Interscience.

Faricha, A., Yoshida, S., Chakraborty, P., Okamoto, K., Mark Chang, T.-F., Sone, M., & Nakamoto, T. (2023). Array of miniaturized amperometric gas sensors using atomic gold decorated Pt/PANI electrodes in room temperature ionic liquid films. *Sensors*, *23*(8). Available from https://doi.org/10.3390/s23084132.

Funabashi, K.-I. (1989). On the approximate realization of identity mappings by three-layer neural networks. *Neural Network*, *2*(3), 183–192. Available from https://doi.org/10.1016/0893-6080(89)90003-8.

Geladi, P., & Kowalski, B. R. (1986). Partial least-squares regression: A tutorial. *Analytica Chimica Acta*, *185*(C), 1–17. Available from https://doi.org/10.1016/0003-2670(86)80028-9.

Goodfelllow, I., Bengio, Y., & Courville, A. (2016). *Deep learning* (pp. 493–516). The MIT Press.

Hopfield, J. J., & Tank, D. W. (1985). "Neural" computation of decisions in optimization problems. *Biological Cybernetics*, *52*(3), 141–152. Available from https://doi.org/10.1007/bf00339943.

Muhammad, A., Sharaf, L. D., & Kowalski, B. R. (1986). *Chemometrics*. Wiley.

Nakamoto, T., Sasaki, S., Fukuda, A., & Moriizumi, T. (1992). Selection method of sensing membranes in odor-sensing system. *Sensors and Materials*, *4*, 111–120.

Rumelhart, D. E., Hinton, G. E., & Williams, R. J. (1986). Learning representations by back-propagating errors. *Nature*, *323*, 533–536. Available from https://doi.org/10.1038/323533a0.

Sundgren, H., Winquist, F., Lukkari, I., & Lundstrom, I. (1991). Artificial neural networks and gas sensor arrays: Quantification of individual components in a gas mixture. *Measurement Science and Technology*, *2*(5), 464–469. Available from https://doi.org/10.1088/0957-0233/2/5/008.

Wang, Z., Isaksson, T., & Kowalski, B. R. (1994). New approach for distance measurement in locally weighted regression. *Analytical Chemistry*, *66*(2), 249–260. Available from https://doi.org/10.1021/ac00074a012.

Wang, Z., Hwang, J.-N., & Kowalski, B. R. (1995). ChemNets: Theory and application. *Analytical Chemistry*, *67*(9), 1497–1504. Available from https://doi.org/10.1021/ac00105a003.

William, R. D., & Goldstein, M. (1984). *Multivariate Analysis*. Wiley.

CHAPTER

8

Concept of odor space

Dani Prasetyawan and Takamichi Nakamoto

Institute of Integrated Research, Institute of Science Tokyo, Yokohama, Japan

The sense of smell is considered less dominant compared to other senses such as vision and auditory; yet it profoundly influences our subconscious emotions. While it is known that humans can differentiate up to 1–2 million colors using the three basic colors (e.g., the concept of RGB: red, green, and blue), this knowledge has propelled advancements in vision-based technologies, such as displays replicating true-to-life colors. It is easy to say that we can span whole color space using only three-dimensional space (Mamlouk et al., 2017; Secundo et al., 2014). In contrast, our comprehension of the sense of smell lags behind that of vision. The concept of odor space envisions a theoretical framework where various odors are depicted as points within a space, allowing for the evaluation of their similarities and differences based on their spatial positioning (Cleland, 2010; Grabe & Sachse, 2018; Meister, 2015). Unfortunately, there is no universally agreed-upon number for the dimensionality of odor space. Unlike vision, the sense of smell lacks an exact set of primary odors or a standard method for reproducing smells. Several studies proposed different concepts of odor space. The estimated range of odor space fluctuates on the basis of the research methods employed and the particular odors examined. Some studies propose that the dimensionality of odor space could be as minimal as 6–30 dimensions or as extensive as several hundred. Unfortunately, the exact number remains an unresolved issue in the field of olfactory research, given the remarkable complexity and variety of the olfactory system. This chapter will explain several studies that explored the concept of odor space.

8.1 Introduction

The sense of smell is a complex sense without a basic framework for its perceptual dimensions (Castro et al., 2013). In contrast, color perception revolves around three primary colors, establishing a three-dimensional space sufficient to encompass the entire spectrum of color perception (Mamlouk & Martinetz, 2004). Such a foundational structure has not yet been established for the sense of odor. Initial attempts to methodically define the odor space have concentrated on pinpointing a limited number of primary odors (Amoore, 1974; Schiffman, 1974). The idea is that these primary odors, when combined,

Digital Technologies in Olfaction

DOI: https://doi.org/10.1016/B978-0-443-15721-9.00008-7

should encompass the entire gamut of olfactory experiences.

Several researchers were keen on developing a more thorough understanding of how we perceive smells. Individual human responses to odorant stimuli can vary greatly. Factors such as gender, cultural practices, association with specific groups, or belonging to particular social classes can affect how different people perceive odors (Sarafoleanu et al., 2009). Furthermore, the perception of odor is influenced by the stimulus generated by the odorant, which can vary depending on the odorant's structure (Snitz et al., 2019), the human response to it (Secundo et al., 2015), or its chemical and physical properties (Genva et al., 2019).

A widely adopted method in odor space research involves examining the structure−activity relationship between odor molecules and human olfactory perception (Auffarth, 2013; Brookes et al., 2009). The structure−activity relationship concerns the link between physical properties and physiological activity. Meanwhile, the structure−odor relationship delves into the connection between physical properties and the perception of odors. Both these relationships explore how the physicochemical properties of odorants influence physiological reactions and the perception of smell. Example of odorant structures with different functional groups with their odorous alike are shown in Fig. 8.1.

Furthermore, odorants sharing similar structural and functional groups usually exhibit comparable odor profiles. Yet, there are cases where odorants with structural similarities produce different olfactory stimuli. These compounds are known as enantiomers (Brookes et al., 2009). Additionally, there are instances where structurally identical odorants emit distinct scents, referred to as optical isomers. Despite their identical structures, making them challenging to differentiate with instruments such as mass spectrometry, the human nose can easily distinguish between them. Despite these variations, recent research on odor space has increasingly leaned toward a structure-based approach in describing odors (Genva et al., 2019).

On the other hand, it is commonly known that humans are capable of distinguishing around 10,000 different odors thanks to their 400 types of olfactory receptors (ORs), in which each OR varies in sensitivity to specific odorants (Malnic et al., 2004; Sarafoleanu et al., 2009). This suggests the possibility of about 400 unique odor dimensions corresponding to the number of ORs. However, the actual dimensionality of the odor space might be less, considering the overlapping characteristics of some ORs. Additionally, there is growing interest in determining the precise number of odorants that humans can identify.

Despite the commonly accepted theory that humans can distinguish up to 10,000 different odors, some researchers believe this number could be higher. For instance, Mori and colleagues suggested that humans might be able to discern as many as 400,000 odors with molecular weights ranging from 26 to 300 Da. This figure was derived from studies examining the relationship between the structure of odor molecules and human olfactory perception (Mori & Yoshihara, 1995). Moreover, a recent study by Bushdid et al. (2014) indicates that humans may be able to identify over a trillion different odors.

Bushdid et al. conducted a sensory test using various mixtures that contained overlapping sets of odorants. This test aimed to determine the precise number of distinct odors humans perceive. Rather than testing an immense number of individual odorants, Bushdid et al. created mixtures that effectively served as representative odor samples. Through these representative mixtures, they analyzed a mathematical model to estimate the potential number of distinct odors discernible to humans. Here, their study involved sensory tests where participants sniffed a variety of mixtures created from 128 distinct odorous components. These 128 odorous

4,4-dimethyl-2-octeno-δ-lactone	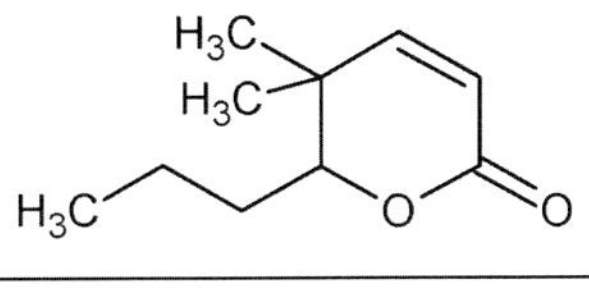	Minty
8-methyl-2-noneno-δ-lactone	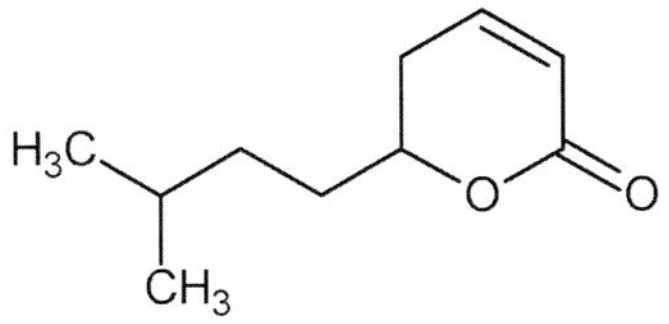	Buttery
5,6,6-trimethyl-2-hepteno-δ-lactone	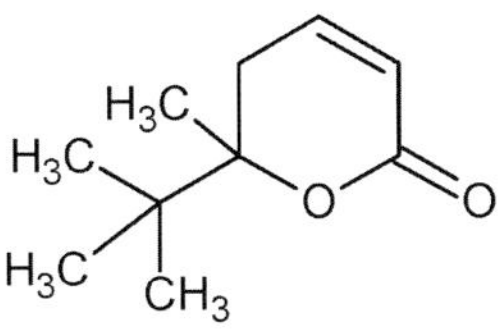	Terpene-like, campherous

FIGURE 8.1 **Example of odorants with common functional groups and their odor alike.** Minty, buttery, and terpene-like molecules. *Source: From Genva, M., Kemene, T., Kenne, D., Magali, L., Laurence, F., & Marie, L. (2019). Is it possible to predict the odor of a molecule on the basis of its structure?* International Journal of Molecular Sciences, *20(12). https://doi.org/10.3390/ijms20123018.*

compounds can cover most of the perceptual and physicochemical of odorous molecules (Haddad et al., 2008; Snitz et al., 2013). They calculated that over 1.72×10^{12} unique mixtures could be produced using a combination of 30 components. Therefore they concluded that the 128-dimensional data or independent odor variables could potentially represent over a trillion different olfactory stimuli. However, Meister (2015) contested this study's claim. Meister argued that the estimation model performed by Bushdid et al. was inadequate and should include dimensionality of physical, neural, and perceptual spaces of olfactory stimuli for a more accurate estimation number.

The possibilities of the number of odorants that human can discriminate and its dimensionalities might vary widely if odor perception is included. Odor perception can be linked to the stimulus given by the odorant samples. This stimulus can be different based on odorant structure (Snitz et al., 2019), human response (Secundo et al., 2015), or chemical and physical features (Genva et al., 2019). For example, odorants with structurally similar functional groups tend to have similar odor characteristics. However, there is a case when the exact same odorant can give different odorant stimuli despite the structural similarity. These are called enantiomers (Brookes et al., 2009). Moreover, structurally identical odorants can have dissimilar odors that are called optical isomers. Due to their "perfectly similar" structure, it is hard to distinguish between them using sensors such as mass spectrometry. But interestingly, the human sense of smell can discriminate between them very well.

This chapter briefly explains the study regarding the concept of odor space. Several studies that are explained in this chapter might overlap with Chapter 11, which is explaining the odor reproduction using odor components. The main difference is that this chapter focused on the approach to the concept and span of the odor space, whereas Chapter 11 focused on the concept of odor reproduction.

8.2 Odor space and its dimensionality

8.2.1 Concept of odor space

Odor space is a hypothetical space in which different odors can be represented as points, and their similarities and differences can be assessed on the basis of their positions through the space (Prasetyawan & Nakamoto, 2023). The dimensionality of odor space can help us in understanding the olfactory system's complexity. In the case of vision sense, there are universally agreed primary colors such as a set of RGB (red, green, and blue). Here, we can say that three-dimensional space is enough to cover whole color space. However, since we do not have such "primary odors" in sense of olfaction, there are no universally agreed upon the number for the dimensionality of odor space. The estimation of dimensionality of odor space varies depending on the methodology and specific set of odors used in the studies. Moreover, some studies are not directly looking for the exact number of dimensionalities of odor space, but they can provide us with the insight of dimensionality of odor space.

An example of understanding odor space based on physicochemical molecule properties of odorant by Bierling et al. is shown in Fig. 8.2. Here, principal component analysis (PCA) was applied to 1389 odor molecules. The first two principal components (PCs) were used to map the entire odor molecules in the two-dimensional spaces (Bierling et al., 2021). Here, each position of odor molecules represents its own odorant characteristics. It is interesting that the figure shows us that several odor molecules might have different properties than others which span further on right side of the map. Most of the odor molecules are gathering together in bottom left part of the map. It can be said the odor molecules gathering together mostly have similar properties, which can be divided further to higher dimension to explain their different properties.

In the 1970s, Berglund et al. (1973) were at the forefront of implementing multidimensional scaling (MDS) in the study of odors. Their research aimed to uncover the basic physicochemical dimensions that define odorants. Their findings suggested that the odor space might be of a lower dimensionality, which was somewhat in contrast to prevailing studies of that era, which hypothesized that the sense of smell involves too many dimensions to accurately represent all odors in an odor space. Berglund's team hypothesized that as few as 21 odors could potentially represent the entire range of odorant space. However, it is important to note that their approach primarily focused on the psychophysical dimensions of odorants, emphasizing those with similar perceived intensities but differing in perceived qualities.

Interesting approach was performed by Dravniek. His study was gathering odor perception information of 144 monomolecular odor with 146 odor descriptors. The study resulted in an invention of odor perception space. Here, he performed sensory test to the trained panelists to give descriptors to the given monomolecular odors. The descriptors such as "sweet," "floral," and "minty" are given. Thus we can obtain 146-dimensional odor perceptual space to cover 144 monomolecular odors. This semantic labeling of monomolecular odorant with odor descriptors is widely used to identify the basis of odor space and odor primaries.

Similar approaches also performed by Keller and Vosshall (2016), in which data of 476 molecules, diverse both structurally and perceptually, were collected from 49 individuals. This provided information on 4884 physicochemical characteristics for each molecule as experienced by the participants. The dataset further explores the connection among the physical attributes of these molecules and their perceived qualities. An illustration of the several most common odorant stimuli along with their representative monomolecular odorants is presented in Fig. 8.3.

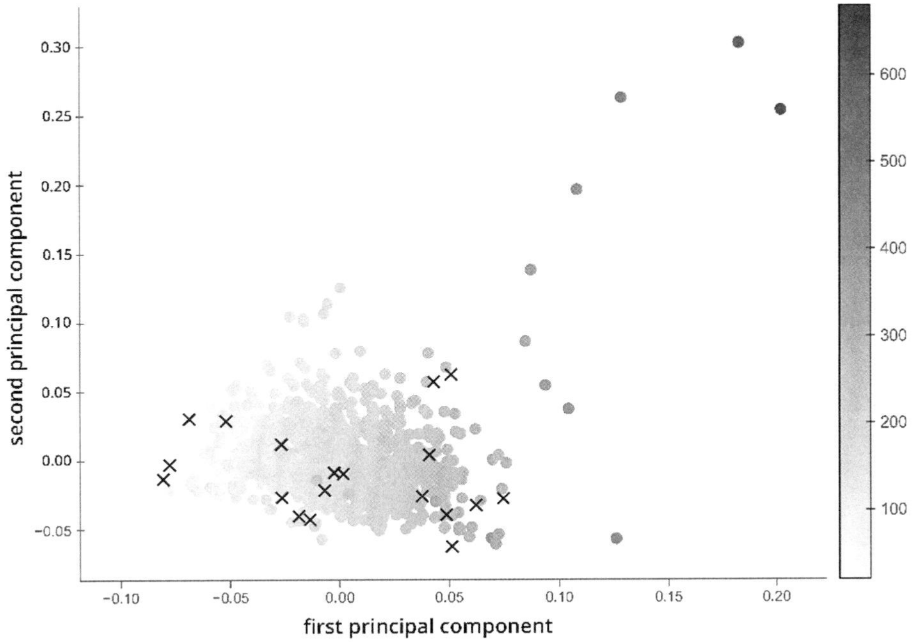

FIGURE 8.2 **Physicochemical odor space**. The graph shows the values for the first two principal components as obtained from PCA applied to the 1389 odor molecules. The color map represents odor molecules numbers included in the odor space. The highlighted "×" are samples used by Bierling et al. for further analysis. *PCA*, Principal component analysis. *Source: From Bierling, A. L., Croy, I., Hummel, T., Cuniberti, G., & Croy, A. (2021). Olfactory perception in relation to the physicochemical odor space.* Brain Sciences, 11*(5), 563. https://doi.org/10.3390/BRAINSCI11050563.*

Additionally, sensory testing uncovered intriguing links between odor descriptors. For instance, odors described as "fruit" were often also perceived as "sweet." This kind of correlation is also evident with other descriptors. Such correlation also occurred in the other descriptor, which is shown in Fig. 8.4. While some descriptors may be mutually exclusive or independent, others show strong associations with additional descriptors.

Further approach was performed by Keller et al., which was a creation of DREAM olfaction prediction model. The model incorporated 19 semantic descriptors (Keller et al., 2017). Interestingly, while a reduced number of descriptors can make predicting certain odors in odor space more challenging, it allows for the development of models capable of high-quality, personalized perceptual predictions in odor space. Additionally, these perceived descriptors effectively represent various olfactory perceptual qualities of a molecule with high precision in odor space. As a result, it made us easier to understand to predict the olfaction in the level of olfactory coding.

As highlighted in this section, research into the odor space using various methods has shown that the number of odorants requiring analysis is quite limited. Therefore it is beneficial to extend the range of odor samples as broadly as possible for odor analysis, as well as finding the dimensionality of odor space to uncover the true nature of different smells. Section 8.2.2 describes several researches that related to finding the dimensionality of odor space.

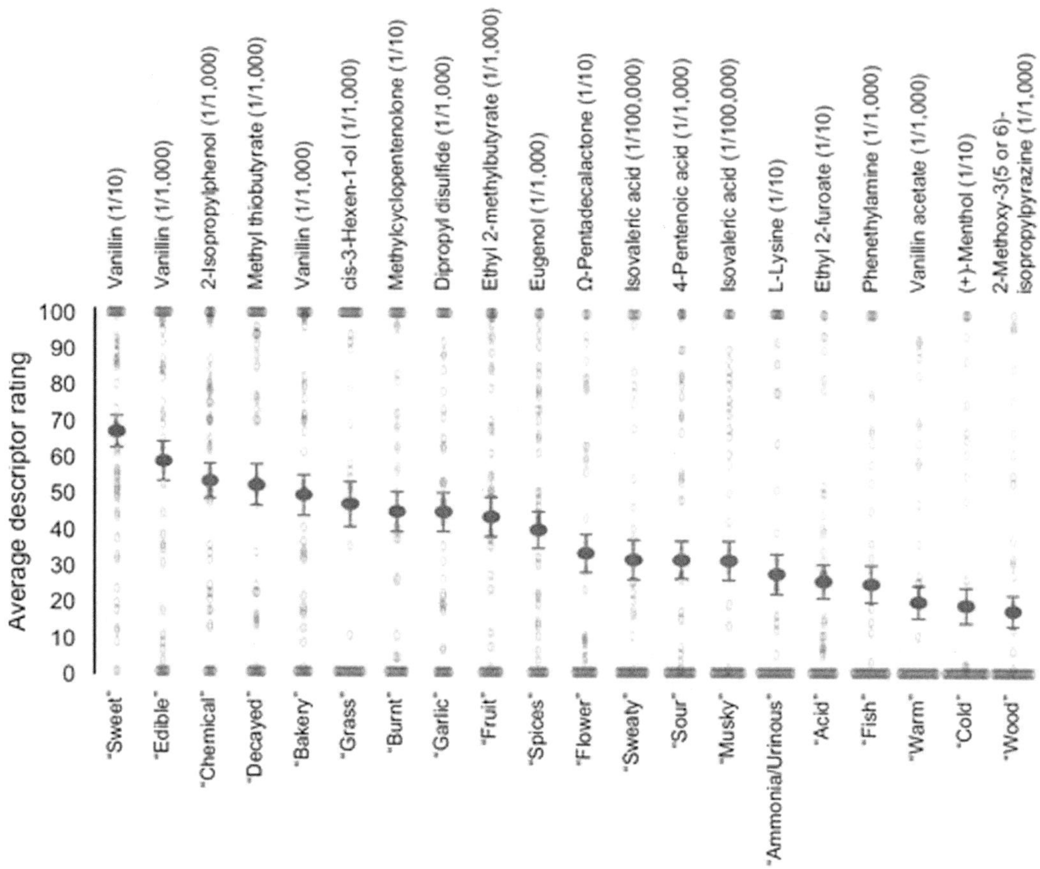

FIGURE 8.3 **Descriptor rating of the stimuli most representative of each of the 20 descriptors**. Individual ratings as well averages and standard errors (in red) are shown. *Source: From Keller, A., & Vosshall, L. B. (2016). Olfactory perception of chemically diverse molecules.* BMC Neuroscience, *17(1), 55. https://doi.org/10.1186/s12868-016-0287-2.*

8.2.2 Dimensionality of odor space

Since odor does not have "primary odors" that can fully span the odor space, several efforts to finding appropriate set to uncover odor components that can cover a wide range of odors were performed by researchers. Finding the dimensionality of odor space is deeply correlated to the possibility of finding a set of odor components to perform odor reproduction. Thus the explanation of several researches in this section might overlap with content in Chapter 11. However, this section focused on the researches that were performed to find the dimensionality of odor space, with or without feature to explore odor reproduction.

In an early study in 1960s, Amoore (1967, 1974) proposed a set of seven primary odors (camphoraceous, pungent, ethereal, floral, pepperminty, musky, and putrid) that can be used to represent the odor space. The proposed study gives us an early understanding of dimensionality of odor space, which is translated to odor space that can be spanned by seven-dimensional space (as the number of proposed seven primary odors). Although this initial study was not enough to cover a wide range of

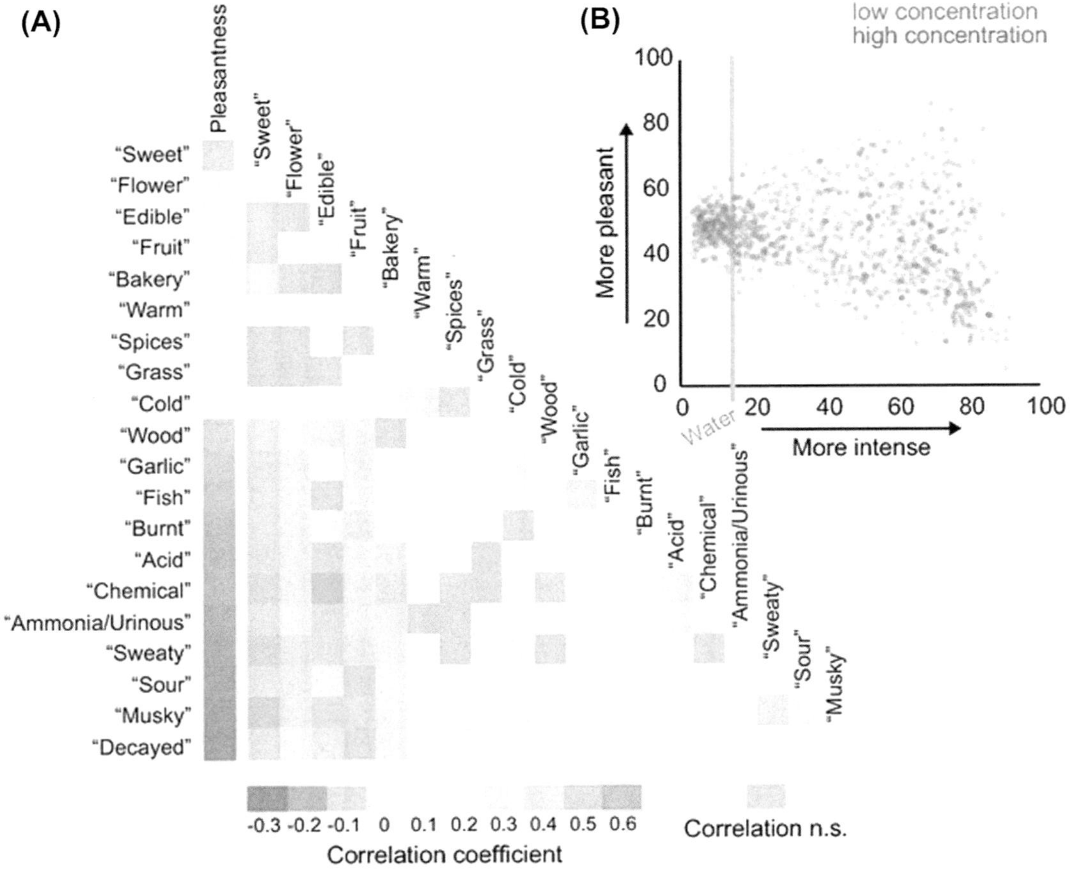

FIGURE 8.4 **Correlations between odor quality descriptors.** (A) Heat map of correlation between pleasantness ratings and the ratings of 20 descriptors. Correlations that are not statistically significant (after Bonferroni correction; $N = 41{,}289$, $P > 0.000238$) are indicated in gray. (B) Average intensity and pleasantness ratings for 1000 stimuli. *Source: From Keller, A., & Vosshall, L. B. (2016). Olfactory perception of chemically diverse molecules.* BMC Neuroscience, 17*(1)*, 55. *https://doi.org/10.1186/s12868-016-0287-2.*

odors, it paved the way for subsequent researchers to gain insights into odor space and the study of dimensionality of odor space as well as primary odor research.

Conversely, Castro et al. (2013) contributed a perspective on the categorical dimension of human odor descriptor space. Their early efforts in systematically characterizing odor space involved identifying a limited number of primary perceptual elements. These identified primaries were intended to cover the entire spectrum of olfactory experiences. Their analysis focused on the perceptual characteristics of 144 monomolecular odors, with each odor represented by a 146-dimensional vector known as an odor profile or odor descriptors. Each dimension in this vector correlates with the rated appropriateness of specific semantic labels or odor descriptors, such as "sweet," "floral," or "heavy." This odor profile data was initially compiled by Dravnieks (1992). Through their analysis, they explored potential foundational perceptual elements to structure the dataset.

The dataset was analyzed using nonnegative matrix factorization (NMF), a method for reducing dimensionality that imposes nonnegative constraints and extracts significant features from the applied dataset (Lee & Seung, 1999). Further details about NMF will be provided in Chapter 12. In NMF, the quantity of basis vectors is equivalent to the basic perceptual dimensions of the odor space. The researchers suggested that there should be 10 basis vectors as odor primaries were enough to cover 146 odor descriptors spaces. Fig. 8.5A and B illustrates the properties of these perceptual elements within the 10 basis vectors, with each vector displaying distinct standout descriptors.

A different approach was performed by Weiss et al. (2012). They explored perceptual convergence in sense of olfaction based on the number of components. This method draws inspiration from the sense of vision, where combining a set of independent wavelengths in various mixtures can yield a common color known as "white." The hypothesis is that a similar phenomenon occurs in the olfactory sense: With a certain number of odorant mixtures, a "whiteness" or analogous perception emerges. This phenomenon arises from the broad simultaneous activation of ORs, leading to a convergence point in the sense of smell. This concept is termed "olfactory white." Thus the study

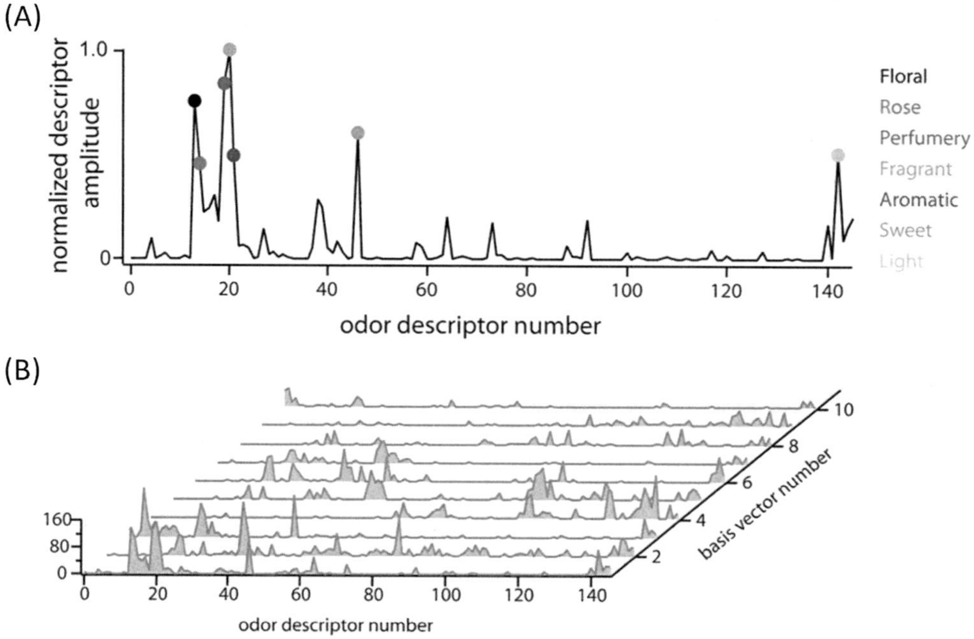

FIGURE 8.5 **Properties of the perceptual basis vectors**. (A) Plot of normalized odor descriptor amplitude versus odor descriptor number for the basis vector 1. Each point along the x-axis corresponds to a single odor descriptor, and the amplitude of each descriptor indicates the descriptor's relevance to the shown perceptual basis vector. The colored circles show the seven largest points within the basis vector, and descriptors corresponding to these points are listed to the right. (B) Waterfall plot of the 10 basis vectors constituting 10 basis vectors, used in subsequent analyses. Note that each vector contains many values close to or equal to zero. *Source: From Castro, J. B., Ramanathan, A., & Chennubhotla, C. S. (2013). Categorical dimensions of human odor descriptor space revealed by non-negative matrix factorization.* PLoS One, *8(9), e73289. Available from: https://dx.plos.org/10.1371/journal.pone.0073289, https://doi.org/10.1371/journal.pone.0073289.*

indirectly led to a possibility to reveal the limit of dimensionality of odor space.

They performed sensory tests to the participants who were trained with 146 odor descriptors. The four or five alternative forced-choice descriptor identification was done to 23 different, novel, but partially overlapping target odorant mixtures of 1, 4, 10, 20, 30, or 40 components. In this method, the components of the odorant mixtures must cover a wide range of odorant stimulus space as possible, and they should be of equal intensity. The sensory experiments in the research utilized 86 monomolecular odorants, evenly distributed across both perceptual and physicochemical stimulus spaces. Each component mixture was also designed to be as distinct from the others as possible. To avoid bias from the preconceived notion of the term "white" in olfactory contexts, the label "Laurax" was used instead. The illustration of their sensory test is shown in Fig. 8.6. It was revealed that the increasing the number of mixtures resulted in olfactory white, where convergence point was met. Thirty or more components with a good distribution in olfactory space will cause the olfactory white phenomena. Thus it can be implied that the redundancy of odor perceptual space increases much when more than 30 dimensional spaces are used.

On the other hand, Mamlouk et al. adopted a different method to examine the dimensionality of the olfactory perception space. They utilized a dataset comprising 851 stimuli, initially described by 278 descriptors (obtained by Aldrich chemical company) (Chee-Ruiter, 2000; Mamlouk & Martinetz, 2004), which they then narrowed down to 171 by eliminating descriptors unique to a single chemical. Their strategy involved employing dimensionality reduction techniques to determine a potentially smaller set of dimensions that could effectively represent the olfactory space.

First, PCA was utilized to determine the maximum limit of the odor space's dimensionality. The findings indicated that 68 dimensions constituted a reasonable upper limit for this space, as depicted in Fig. 8.7A. Additionally, MDS was employed on a 171×171 dissimilarity matrix of descriptors. MDS is a method of visualizing the level of similarity between different cases of datasets. The MDS analysis revealed that 32 dimensions were sufficient to represent the lower limit, as illustrated in Fig. 8.7B.

There are several research studies that aim to find an appropriate set of odor components to perform odor reproduction. Here, we can translate the analog to find an appropriate dimensionality of odor space to span the odor space. Since the odor samples that are used by their research are hypothetically spanned within odor space and the number of sets of odor components is the dimensionality of odor space.

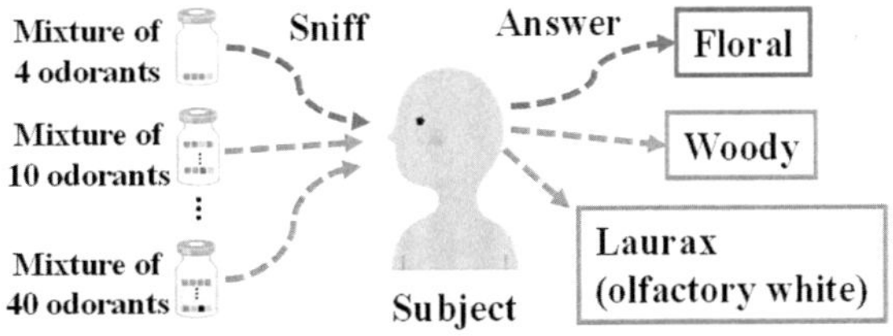

FIGURE 8.6 **Illustrated image of sensory test to confirm olfactory white performed by Weiss et al.** Here, the subject was exposed to odor samples (mixture of odorants) and predicting the descriptor of the sample. Mixture of 30 with good distribution in olfactory space tends to have answer of "Laurax" that means olfactory white. *Source: From Prasetyawan, D., & Nakamoto, T. (2023)Odor reproduction technology using a small set of odor components.* IEEJ Transactions on Electrical and Electronic Engineering. *Available from: https://onlinelibrary.wiley.com/doi/10.1002/tee.23915, https://doi.org/10.1002/TEE.23915.*

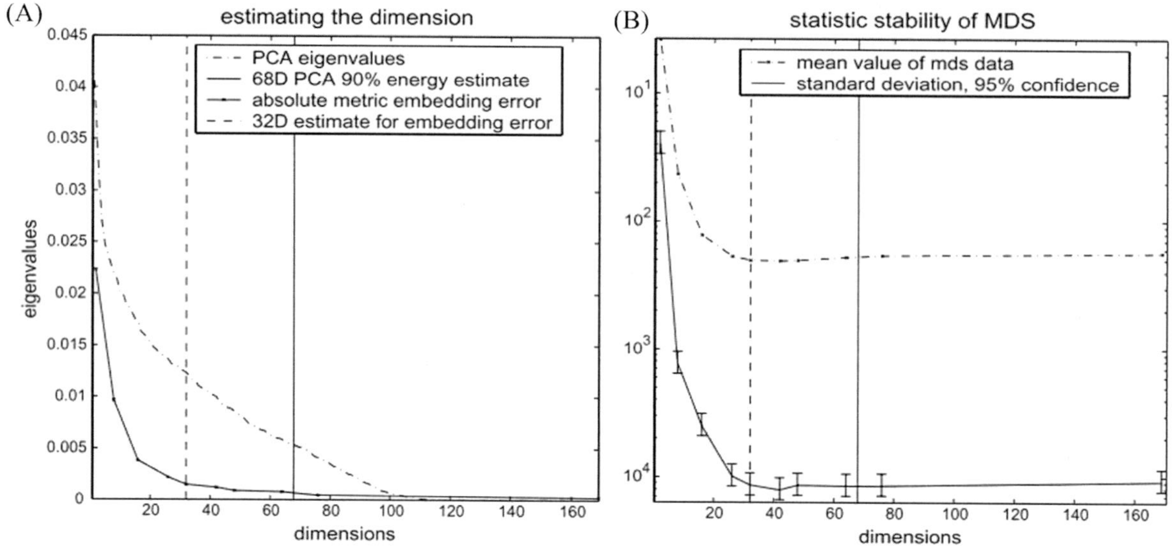

FIGURE 8.7 Result by Mamlouk et al. in finding dimensionality of odor space. (A) Dashdotted line shows the sorted eigenvalues for 170D, vertical line marks 90% energy cut-off (68D). The metric embedding error (solid curve) increases significantly for dimensions smaller than 32 (see dashed vertical line). The error was scaled to fit in the energy graph. (B) Standard deviation and mean values for simulation runs of MDS. Values become smaller the more the scaled distances correspond to the given dissimilarities. *MDS*, Multidimensional scaling. *Source: From Mamlouk, A. M., & Martinetz, T. (2004). On the dimensions of the olfactory perception space.* Neurocomputing, *58–60, 1019–1025. https://doi.org/10.1016/j.neucom.2004.01.161.*

An example of such approach was performed by Nicolaï et al. (2016). Initially, they suggested using 45 odor components to recreate various citrus scents. This number was effectively reduced to 12 key odor components for reconstructing citrus-based aromas [an example of such approach was performed by Nicolaï et al. (2016)]. Initially, they suggested using 45 odor components to recreate various citrus scents. This number was effectively reduced to 12 key odor components for reconstructing citrus-based aromas (Nicolaï et al., 2016). These 12 components were termed "odor primaries" for the samples they tested. From their research, it was evident that these 12 independent variables are capable of reconstructing five distinct citrus odors. Their research indicates that 12 independent variables are sufficient to reconstruct five distinct citrus alike odors.

On the other hand, the research group by Nakamoto et al. (2012) tried to reconstruct 158 essential oils as odor samples using mass spectrometry data. Although the main focus of their research was on reproducing odors, it provided valuable information about the complexity of the odor space. The study revealed that a combination of 30 odor components was sufficient to accurately replicate 158 essential oils. In contrast, a combination of just 12 odor components fell short. Consequently, this suggests that to cover the range of 158 different odors, at least 30 independent odor components are necessary.

Prasetyawan et al. improved the study originally conducted by the Nakamoto group by employing various cost functions in their numerical calculations to identify the optimal set of odor components to reconstruct odor profiles of 185 essential oils (Iseki et al., 2022; Prasetyawan & Nakamoto, 2019; Prasetyawan & Nakamoto, 2020). They discovered that an improved reconstruction of these 185 essential oils as odor samples could potentially be achieved using just 10 odor components. The further explanation regarding their method is explained in Chapters 11 and 12.

Despite different approaches for finding the dimensionality of odor space, most of the research studies were done by using a different set of odor samples that span the odor space. Eventually, it resulted in a distinct size of odor space dimensionality depending on the sample used. Nevertheless, we still need more studies to uncover the dimensionality of odor space that can be universally accepted.

TABLE 8.1 Odor space and its features.

Method	Feature of odor space
Molecular structure parameter	More than 1 trillion odors (Bushdid et al., 2014)
ORs data	10,000 odorants with 400 ORs
Sensory data	146 descriptors × 144 odors (Dravnieks, 1992), 4884 physicochemical structure (Keller & Vosshall, 2016), 278 descriptors × 851 chemicals (Chee-Ruiter, 2000; Mamlouk & Martinetz, 2004)
Sensor data (MS)	(158–185) × 201 dimension of MS data (Nakamoto et al., 2012; Prasetyawan & Nakamoto, 2020)

ORs, Olfactory receptors; *MS*, mass spectrometry.

8.3 Summary of concept of odor space

Currently, the precise number of odors distinguishable by an individual remains a mystery. Researchers have employed a variety of methods, including analyzing the structural forms of chemicals, exploring perception space, and conducting sensory tests. Yet, many potential odorants may not be included in these obtained datasets of samples. A summary of these findings is presented in Table 8.1. Additionally, factors such as a subject's social background, gender, and habits can introduce subjectivity into sensory test results, potentially resulted in bias outcomes. While increasing the number of test subjects could address this problem, it would also demand considerable time and resources.

Ongoing research is also focused on understanding the adequate dimensionality to cover the entire range of odor space, with several studies summarized in Table 8.2. These

TABLE 8.2 Possible dimensionality of odor space.

Research groups	Feature	Possibility of odor space dimensionality
Amoore (1967, 1974)	600 stereochemical structure of odorants	Around 7
Berglund et al. (1973)	420 pairs of 21 odorants	Around 21
Castro et al. (2013)	146 descriptors × 144 odors	Around 12
Weiss et al. (2012)	86 monomolecular odorants	Less than 30
Mamlouk and Martinetz (2004)	851 odorant stimuli × 171 descriptors	Between 32 to 68
Nicolaï et al. (2016)	45 citruses odorant compounds	Around 12
Nakamoto et al. (2012), Prasetyawan and Nakamoto (2020)	Mass spectrometry data of 158–185 odorants	Around 10–20

investigations explore various approaches, such as finding a correlation of the structural shapes of odorants with their perceived qualities, to predict the basis of these odors. Although the list of odors identified with specific perceptual descriptors and those included in the broader odor space remains limited, these efforts pave the way to determine the primary dimensions of odors. However, more research is essential to ascertain the true dimensionality of odors. Such studies may eventually identify a set of "primary odors" that are based on the number of dimensions in the odor space.

References

Amoore, J. E. (1967). Specific anosmia: A clue to the olfactory code. *Nature*, *214*(5093), 1095–1098. Available from https://doi.org/10.1038/2141095a0.

Amoore, J. E. (1974). Evidence for the chemical olfactory code in man. *Academy of Sciences*, *237*(1), 137–143. Available from https://doi.org/10.1111/j.1749-6632.1974.tb49850.x.

Auffarth, B. (2013). Understanding smell—The olfactory stimulus problem. *Neuroscience and Biobehavioral Reviews*, *37*, 1667–1679.

Berglund, B., Berglund, U. L. F., Engen, T., & Ekman, G. Ö. (1973). Multidimensional analysis of twenty-one odors. *Scandinavian Journal of Psychology*, *14*(2), 131–137. Available from https://doi.org/10.1111/J.1467-9450.1973.TB00104.X, https://pubmed.ncbi.nlm.nih.gov/4705857/.

Bierling, A. L., Croy, I., Hummel, T., Cuniberti, G., & Croy, A. (2021). Olfactory perception in relation to the physicochemical odor space. *Brain Sciences*, *11*(5), 563. Available from https://doi.org/10.3390/BRAINSCI11050563.

Brookes, J. C., Horsfield, A. P., & Stoneham, A. M. (2009). Odour character differences for enantiomers correlate with molecular flexibility. *Journal of the Royal Society Interface*, *6*(30), 75–86. Available from https://doi.org/10.1098/rsif.2008.0165.

Bushdid, C., Magnasco, M. O., Vosshall, L. B., & Keller, A. (2014). Humans can discriminate more than 1 trillion olfactory stimuli. *Science (New York, N.Y.)*, *343*(6177), 1370–1372. Available from https://doi.org/10.1126/science.1249168.

Castro, J. B., Ramanathan, A., & Chennubhotla, C. S. (2013). Categorical dimensions of human odor descriptor space revealed by non-negative matrix factorization. *PLoS ONE*, *8*(9), e73289. Available from https://doi.org/10.1371/journal.pone.0073289, https://dx.plos.org/10.1371/journal.pone.0073289.

Chee-Ruiter, C.W. J. (2000). *The biological sense of smell: Olfactory search behavior and a metabolic view for olfactory perception* (Dissertation Ph.D.). California Institute of Technology. Available from https://doi.org/10.7907/WDN1-5V55.

Cleland, T. A. (2010). Early transformations in odor representation. *Trends in Neurosciences*, *33*(3), 130–139. Available from https://doi.org/10.1016/J.TINS.2009.12.004.

Dravnieks, A. (1992). Atlas of odor character profiles. ASTM International. Available from http://doi.org/10.1520/DS61-EB.

Genva, M., Kemene, T., Kenne, D., Magali, L., Laurence, F., & Marie, L. (2019). Is it possible to predict the odor of a molecule on the basis of its structure? *International Journal of Molecular Sciences*, *20*(12). Available from https://doi.org/10.3390/ijms20123018.

Grabe, V., & Sachse, S. (2018). Fundamental principles of the olfactory code. *BioSystems*, *164*, 94–101.

Haddad, R., Khan, R., Takahashi, Y. K., Mori, K., Harel, D., & Sobel, N. (2008). A metric for odorant comparison. *Nature Methods*, *5*(5), 425–429. Available from https://doi.org/10.1038/NMETH.1197.

Iseki, M., Prasetyawan, D., Yokoshiki, Y., & Nakamoto, T. (2022). A study of odor reproduction using multi-component olfactory display. *IEEJ Transactions on Sensors and Micromachines*, *142*(4), 63–70. Available from https://doi.org/10.1541/IEEJSMAS.142.63.

Keller, A., Gerkin, R. C., Guan, Y., Dhurandhar, A., Turu, G., Szalai, B., Mainland, J. D., Ihara, Y., Yu, C. W., Wolfinger, R., Vens, C., Schietgat, L., Grave, K. D., Norel, R., Stolovitzky, G., Cecchi, G. A., Vosshall, L. B., Meyer, P., Bhondekar, A. P., et al. (2017). Predicting human olfactory perception from chemical features of odor molecules. *Science*, *355*(6327), 820–826. Available from https://doi.org/10.1126/science.aal2014.

Keller, A., & Vosshall, L. B. (2016). Olfactory perception of chemically diverse molecules. *BMC Neuroscience*, *17*(1), 55. Available from https://doi.org/10.1186/s12868-016-0287-2.

Lee, D. D., & Seung, H. S. (1999). Learning the parts of objects by non-negative matrix factorization. *Nature*, *401*(6755), 788–791. Available from https://doi.org/10.1038/44565, https://www.nature.com/articles/44565.

Malnic, B., Godfrey, P. A., & Buck, L. B. (2004). The human olfactory receptor gene family. *Proceedings of the National Academy of Sciences of the United States of America*, *101*(8), 2584–2589.

Mamlouk, A., Madany, H., & Martin, M.T. (2017). Perception space analysis: From color vision to odor perception. In *Proceedings of the International Joint Conference on Neural Networks* (pp. 689–696). Institute of Electrical and

Electronics Engineers Inc. Available from https://doi.org/10.1109/IJCNN.2017.7965919.

Mamlouk, A. M., & Martinetz, T. (2004). On the dimensions of the olfactory perception space. *Neurocomputing, 58-60*, 1019–1025. Available from https://doi.org/10.1016/j.neucom.2004.01.161.

Meister, M. (2015). On the dimensionality of odor space. *eLife, 4*, e07865. Available from https://doi.org/10.7554/eLife.07865, http://www.pubmedcentral.nih.gov/articlerender.fcgi?artid = PMC4491593.

Mori, K., & Yoshihara, Y. (1995). Molecular recognition and olfactory processing in the mammalian olfactory system. *Progress in Neurobiology, 45*(6), 585–619.

Nakamoto, T., Ohno, M., & Nihei, Y. (2012). Odor approximation using mass spectrometry. *IEEE Sensors Journal, 12*(11), 3225–3231. Available from https://doi.org/10.1109/JSEN.2012.2190506.

Nicolaï, B. M., Micholt, E., Scheerlinck, N., Vandendriessche, T., Hertog, M. L. A. T. M., Ferguson, I., & Lammertyn, J. (2016). Real time aroma reconstruction using odour primaries. *Sensors and Actuators, B: Chemical, 227*, 561–572. Available from https://doi.org/10.1016/j.snb.2015.12.074.

Prasetyawan, D., & Nakamoto, T. (2019). Comparison of NMF with Kullback–Leibler divergence and Itakura–Saito divergence for odor approximation. In *IEEE International Symposium on Olfaction and Electronic Nose (ISOEN)* (pp. 1–3). IEEE. Available from https://doi.org/10.1109/ISOEN.2019.8823186, https://ieeexplore.ieee.org/document/8823186/.

Prasetyawan, D., & Nakamoto, T. (2020). Sensory evaluation of odor approximation using NMF with Kullback–Leibler divergence and Itakura–Saito divergence in mass spectrum space. *Journal of The Electrochemical Society, 167*(16), 167520. Available from https://doi.org/10.1149/1945-7111/abd110.

Prasetyawan, D., & Nakamoto, T. (2023). Odor reproduction technology using a small set of odor components. *IEEJ Transactions on Electrical and Electronic Engineering*. Available from https://doi.org/10.1002/TEE.23915, https://onlinelibrary.wiley.com/doi/10.1002/tee.23915.

Sarafoleanu, C., Mella, C., Georgescu, M., & Perederco, C. (2009). The importance of the olfactory sense in the human behavior and evolution. *Journal of Medicine and Life, 2*(2), 196–198.

Schiffman, S. S. (1974). Physicochemical correlates of olfactory quality. *Science, 185*(4146), 112–117. Available from https://doi.org/10.1126/science.185.4146.112.

Secundo, L., Snitz, K., & Sobel, N. (2014). The perceptual logic of smell. *Current Opinion in Neurobiology, 25*, 107–115. Available from https://doi.org/10.1016/j.conb.2013.12.010.

Secundo, L., Snitz, K., Weissler, K., Pinchover, L., Shoenfeld, Y., Loewenthal, R., Agmon-Levin, N., Frumin, I., Bar-Zvi, D., Shushan, S., & Sobel, N. (2015). Individual olfactory perception reveals meaningful nonolfactory genetic information. *Proceedings of the National Academy of Sciences of the United States of America, 112*(28), 8750–8755. Available from https://doi.org/10.1073/pnas.1424826112.

Snitz, K., Perl, O., Honigstein, D., Secundo, L., Ravia, A., Yablonka, A., Endevelt-Shapira, Y., & Sobel, N. (2019). Smellspace: An odor-based social network as a platform for collecting olfactory perceptual data. *Chemical Senses, 44*(4), 267–278. Available from https://doi.org/10.1093/chemse/bjz014.

Snitz, K., Yablonka, A., Weiss, T., Frumin, I., Khan, R. M., & Sobel, N. (2013). Predicting odor perceptual similarity from odor structure. *PLoS Computational Biology, 9*(9). Available from https://doi.org/10.1371/JOURNAL.PCBI.1003184.

Weiss, T., Snitz, K., Yablonka, A., Khan, R. M., Gafsou, D., Schneidman, E., & Sobel, N. (2012). Perceptual convergence of multi-component mixtures in olfaction implies an olfactory white. *Proceedings of the National Academy of Sciences of the United States of America, 109*(49), 19959–19964. Available from https://doi.org/10.1073/pnas.1208110109.

CHAPTER

9

Odor impression prediction using multidimensional data analysis

Tanoy Debnath
Yokogawa Electric Corporation, Tokyo, Japan

9.1 Introduction

The sense of smell plays an important role in our daily lives, influencing our choices, emotions, and even memories. From the captivating aroma of freshly brewed coffee to the beautiful scent of a blooming flower, smells have the power to evoke powerful impressions and shape our experiences. Understanding and predicting these olfactory impressions has long been an interesting challenge in the fields of neuroscience, psychology, and sensory science. Odor impression prediction is the result of receiving chemical stimuli using the sense of smell (Nakamoto, 2016). Odor "impression" and odor "descriptor" are common terms as people express their feeling of smell or scent with a descriptive word (e.g., the odor or smell of an orange is sweet). We will use odor impression interchangeably with odor descriptor.

In recent years, advances in technology and data analysis techniques have opened new paths for investigating the complex nature of odors. An interesting area of research is the development of methods for predicting odor impressions using multidimensional data analysis (Nakamoto, 2019) using data from a variety of sensory modalities, such as chemical composition, physicochemical properties, and human perception. Researchers aim to study the complex relationship between odorant molecules and the subjective experiences they elicit.

This chapter presents a comprehensive investigation of odor impression prediction using multidimensional data analysis. We delve into the underlying principles of olfactory perception, highlighting the complex relationship between the subjective (human evaluation of odor) and objective (sensor data) information. We review existing methods employed in odor impression prediction, from traditional machine learning models to deep learning techniques.

In this chapter, initially, we discuss about the input and output data for the odor impression predictive model in Section 9.2. After that, we will review several multidimensional data analysis techniques in Section 9.3 (including data preprocessing steps in Sections 9.3.1 and 9.3.2). Finally, we will explore the deep learning methods to predict the odor impression prediction in Section 9.4.

Digital Technologies in Olfaction
DOI: https://doi.org/10.1016/B978-0-443-15721-9.00009-9

9.2 Input and output of the odor impression–predictive model

Odor is another form of a scent. For example, the scent of an orange is sour and sweet. Thus we express the verbal impression of scent (odor). As a supervised odor descriptor predictive model, it is necessary to have a knowledge of the input and output of the odor perception model. However, the database in this domain is not as huge as other machine learning fields such as computer vision, and natural language processing. Input of an odor impression–predictive model can be the molecular fingerprint, physicochemical properties of a flavor molecule, and different sensors data (such as quartz crystal microbalance (QCM) and metal oxide sensor (MOS) sensor). The information on various sensors in detail can be found in Chapter 3. Fig. 9.1A–C depicts different inputs for the odor impression–predictive model. The task of odor impression predictive models is to learn the relationship between those input features and target vectors to predict the odor descriptor (e.g., "fruity" and "bitter") of an unknown odorant molecule.

Two different types of target (output) data are used in odor impression–predictive models. One is binary odor descriptors (as shown in Fig. 9.1E) where only odor descriptors (e.g., "fruity" and "minty") are recorded for a flavor molecule. Good scents (The good scents company—Flavor, fragrance food, and cosmetics ingredients information, n.d.), Leffingwell PMP 2001 (Leffingwell & Associates, 2001), Sigma-Aldrich (n.d.) are several examples of binary odor descriptors datasets. Another type of target vector is continuous odor impression (as shown in Fig. 9.1D) where human participants provide a score from 0 to full scale. The Dravnieks dataset is the oldest dataset in this domain and has been used in several papers to understand subjective and objective relationships for odor perception. This dataset was the result of the ASTM Odor Profiling Project led by several US research institutes reported by Dravnieks (1985). They presented 144 monomolecular odors to the trained panelists to give the perception score on a continuous scale. A total of 146 descriptors were prepared to evaluate the odor properties of the sample. In the answer sheet given to the subject in the examination, a number line for evaluating the fit to the descriptor on a 6-point scale from 0 to 5 is given directly below this descriptor, and the examiner uses this to evaluate the nature of the odor. Here, 0 means that the descriptor does not apply at all, and 5 means that it applies very well. Creating a large and quality continuous sensory dataset is difficult because it takes a lot of time and experienced human noses.

9.3 Odor impression prediction using multidimensional data analysis

Here, we describe several traditional methods for predicting odor impressions using both continuous and binary sensory data. First, we describe the data-processing procedures before we feed the data into the odor-predictive model. Next, regression and classification models will be discussed. Here we describe only the methods that were used for the odor impression–predictive model.

9.3.1 Data preprocessing

Two types of nonparametric (no assumption about the statistical distributions of data) methods are available, both in supervised and unsupervised ways. Supervised models require target vectors for learning data, while unsupervised methods do not require any data labels. In recent times, unsupervised methods have become popular due to the lack of data labeling. Before we feed any data to the odor-predictive model, we need to process the data.

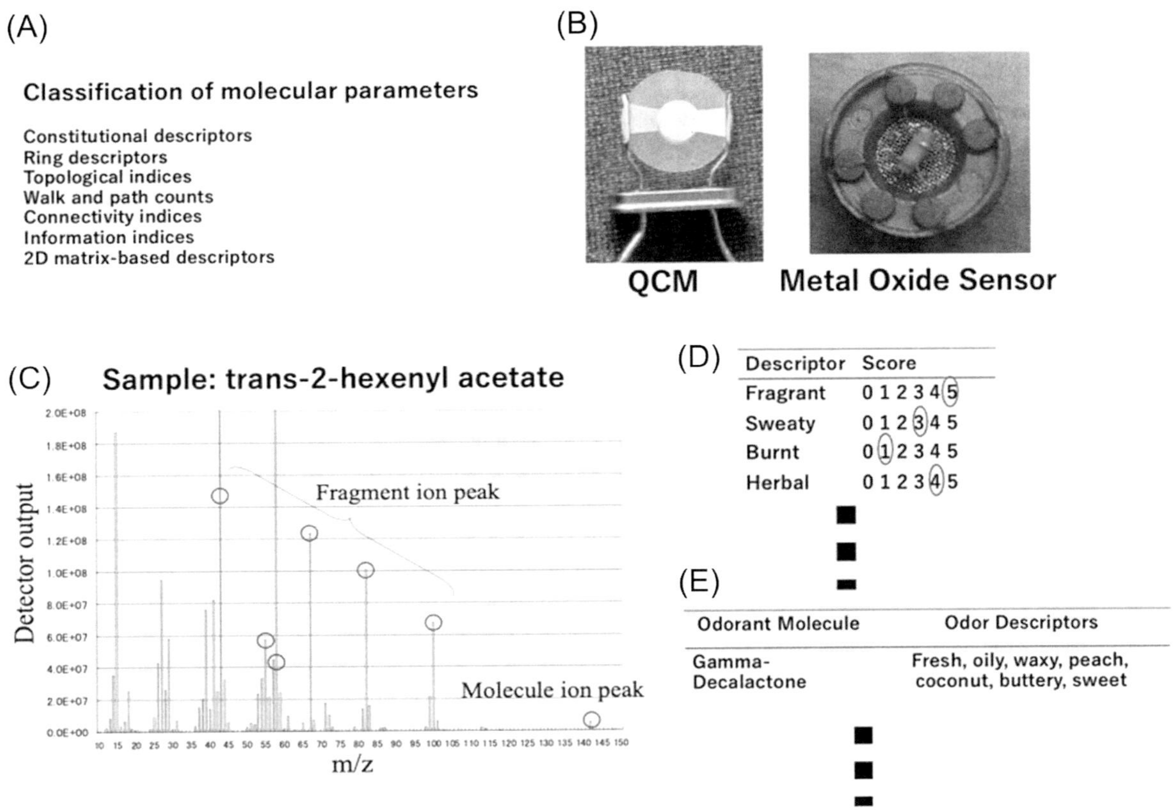

FIGURE 9.1 **Several types of inputs and outputs for the chemoinformatic model.** (A) Molecular structure parameters, (B) QCM and metal oxide sensors, (C) mass spectrum, (D) continuous value of odor perception, and (E) binary value of odor perception. *QCM*, Quartz crystal microbalance.

Both linear and nonlinear dimension reduction methods can be used. However, to learn the complex relationships between subjective (human evaluation of odor) and objective (sensor's data) information, nonlinear dimensionality reduction methods learn and extract better features from highly dimensional datasets that can be further used to make predictions of the odor impression (Nozaki & Nakamoto, 2016). There are also other methods that can be used as a dimensionality reduction method. However, we only describe the odor impression–based predictive model approach.

Principal component analysis (PCA) is the most widely used method of unsupervised linear dimensionality reduction. As will be introduced in the next section, PCA has been widely used in odor research. PCA finds a new axis (direction) with maximum variance in the high-dimensional data and projects it into a new subspace composed of lower dimensions than the original. A new axis determined to maximize this variance is called a principal component. The axis that explains the most variance is called the first principal component, followed by $(n-1)$ principal components. These principal components are chosen to be orthogonal to each other. PCA is affected by the scale of the original data, so if the features are measured at different scales, it is necessary to standardize the features before performing PCA. By compressing the high-dimensional data to 3 or fewer dimensions, it is possible to visualize the low-dimensional distribution of the samples. The detail of PCA is described in Chapter 7.

An autoencoder proposed by Hinton and Salakhutdinov (2006) is a neural network with nonlinear dimensionality reduction capability. This neural network has a small intermediate layer to reconstruct the high-dimensional input layer into the output layer. The intermediate layer can reduce the high-dimensional data of the input layer into the low-dimensional data, often called the bottleneck hidden layer. This method has been used in the image recognition domain because of its high reconstruction capability with a small number of hidden nodes. The autoencoder with the nonlinear activation function followed by the multilayer perception model appeared in Chapter 3 and will be discussed in a later section for the odor descriptor predictive model. *t*-Distributed stochastic neighbor embedding (*t*-SNE) is also a nonlinear dimensionality reduction (Van Der Maaten & Hinton, 2008) technique that can be used for visualizing high-dimensional data into a lower dimensional space to find important clusters or groups in the data. In the next section, we demonstrate the use of both these nonlinear dimensionality reduction methods as preprocessing steps for odor impression prediction.

9.3.2 Methods for odor impression–predictive model

9.3.2.1 Combination of principal component analysis and regression model using continuous sensory data

Linear prediction model, a method of supervised learning, is widely used for classification and regression. Regression analysis is a supervised machine learning technique that predicts a continuous target variable. Regression analysis learns the relationship between variables so that a continuous target variable (response variable) can be predicted from multiple explanatory variables (or predictor variables). In linear regression, which is often used in regression analysis, x is the explanatory variable, y is the objective variable, and the mean squared error is used to fit a regression line so that the error between the sample points can be minimized. The objective variable of new data is predicted using the straight line and the slope of the intercept learned from the data, but it cannot be modeled correctly when the relationship between the variables is represented by a nonlinear function.

Sobel et al.'s research team, consisting of researchers from Israel and the United States, predicted the odor pleasantness from the odorant structure. They used PCA and linear models to explain the relationship between the physicochemical parameters of odorants and odor properties (Khan et al., 2007). They used Dragon (Mauri et al., 2006), a molecular modeling software, to calculate 1513 physicochemical parameters (these are called molecular properties, including molecular weight, atom types, functional groups, topological and geometrical properties) of odorants and used a linear model to find out the relationship between these parameters and sensory test of Dravnieks dataset (Dravnieks, 1985). A total of 146 odor descriptors and 1513 physicochemical parameters were dimensionally compressed by PCA and the correlation between each principal component and large-grained descriptors such as "pleasantness" and "edibility" was investigated. As a result, it was reported that the first principal component of the reduced physicochemical parameters had a positive correlation with the odor pleasantness–unpleasantness (Fig. 9.2). So, just linear model was used to predict the first PC in the perceptual space, which correlates to pleasantness.

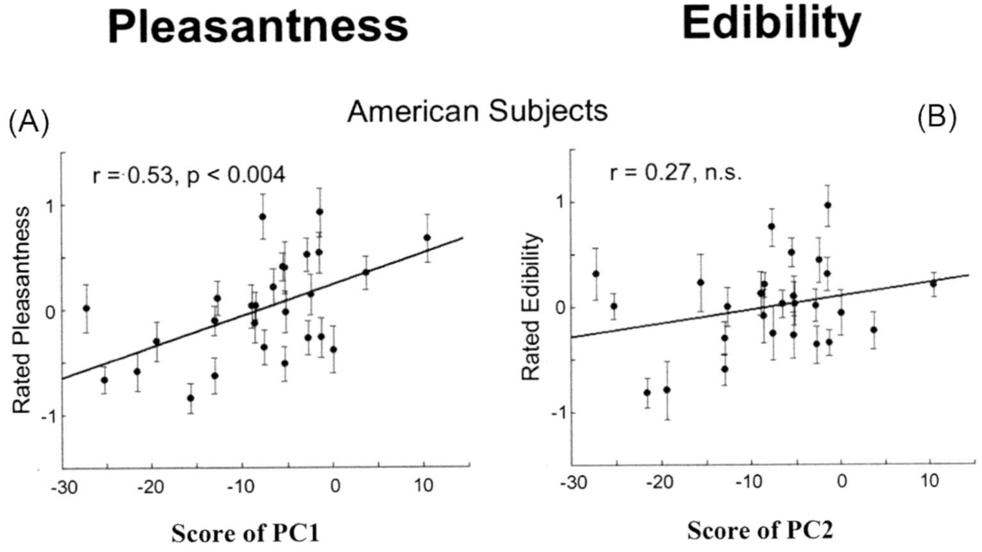

FIGURE 9.2 **Results of odor impression of prediction for unknown molecules.** Twenty-seven odorous molecules not commonly used in olfactory studies, and not previously tested by us, were presented to three cultural groups of naive subjects: Americans (23 subjects). Their predicted PC1 values were significantly better correlated with judgments of pleasantness (A) than judgments of edibility (B). *Reproduced with permission from: Khan, R. M., Luk, C.H., Flinker, A., Aggarwal, A., Lapid, H., Haddad, R., & Sobel, N. (2007). Predicting odor pleasantness from odorant structure: Pleasantness as a reflection of the physical world.* Journal of Neuroscience, *27(37), 10015–10023, https://doi.org/10.1523/JNEUROSCI.1158-07.2007. Copyright 2007.*

9.3.2.2 Odor prediction using binary sensory data

Shang et al. (2017) studied the method of predicting odor descriptors related to odor molecules using parameters of molecular structure. This is a binary value prediction, that is, classification, since data from the Sigma-Aldrich company were used. Sigma-Aldrich is an international company headquartered in St. Louis, USA, and is a chemical distributor created by the merger of Sigma Chemical Company and Aldrich Chemical Company in 1975. In this catalog, more than 1200 kinds of single-molecule odorants are listed, and the odor properties of each substance are expressed by the presence or absence of a descriptor. The physicochemical molecular parameters (MPs) of several flavors and fragrances obtained by means of gas chromatography coupled with mass spectrometry were used to train supervised machine-learned classifiers. The algorithms were provided with GC/MS-derived MPs of 1026 different odorants annotated with 160 different odor descriptors (where they took the 10 most frequent odor descriptors).

A prediction model with different feature extraction methods from MPs and machine learning algorithms were used in this study. Different types of feature extraction include all parameters, PCA, Boruta algorithm (BR), and machine learning algorithms include support vector machine (SVM), random forest (RF), extreme learning machine (ELM), and a combination of these was used to train the model. The best classifier was based on ELM, which is a special form of neural network with single or multiple layers of hidden nodes with the advantage over other network approaches that the parameters of hidden nodes do not need to be tuned. ELM provided the best identification (odor descriptors) accuracy (97.53%) followed by SVM (97.19%) and RF (92.79%). They suggested this research as an alternative way to help the panelists to perceive the odor impression. However, it is unclear whether the model reproduces the same level of highest accuracy when predicting descriptors when the number of odor descriptors will be increased.

The same research team used Pearson correlation coefficient maps (PCC maps) and the *t*-SNE method to study the relationship between the odor map (OM) and the MPs (MP) of odorants (Shang et al., 2018). The response pattern of the olfactory bulb (OM), which is triggered based on the characteristics of the molecule to be received, shows basic information on odor quality. They proposed a method to project both relations on a two-dimensional space and they used neural network to specify functional groups from OM and feature quantity of MPs. In this research, using the image (178×357 pixel) of the olfactory bulb for 178 types of odorants acquired from OdorMapDB, the relationship between the OM and MPs of odorants was mathematically obtained. PCC maps create a matrix with correlation coefficients calculated from the OM matrix and the MP matrix. The correlation coefficient matrix generated by PCC maps is analyzed by hierarchical clustering and the reaction of the olfactory bulb is divided into seven clusters according to the MP. They also used a method using *t*-SNE to represent odorous substances in a two-dimensional space by applying PCA to each of the OM dataset and the molecular dataset, using 80 principal components and 23 principal components, respectively. Fig. 9.3 is a mapping of odorants clustering in a two-dimensional space. Fig. 9.3A is a *t*-SNE map created from OM and Fig. 9.3B is a map created from molecular information. In both parts (A) and (B) of Fig. 9.3, a portion surrounded by an ellipse indicates 14 functional groups. In Fig. 9.3A, samples from small aliphatic ester ($C \leqq 8$), aliphatic or alicyclic hydrocarbon, aromatic, and carboxylic acid were clustered in multiple groups. Carbon-chain length is a vital factor for aliphatic esters (cluster 1) and alcohols (cluster 9), and branched chains play a role in distinguishing odorants in hydrocarbon alkanes (cluster 6) and carboxylic acids (cluster 12). Compared with cluster map generated by MPs, olfactory images contained more detailed information, such as the lengths of carbon and branched chains because some clusters (Fig. 9.3B), such as 1, 9, and 12, were combined into multiple groups in OM but were clustered into a single group in molecular map. A neural network model based on OM and MP features identified the odorant functional groups with an accuracy of 93.02% by the OM−PCA−ELM model.

9.3.2.3 Dream Olfaction prediction challenge

Dream olfactory database is another source of continuous sensory dataset. The Dream Olfaction prediction challenge was organized by Pablo Meyer, an IBM researcher, to build a predictive model based on data collected by Keller and Vosshall (2016). They collected data of 480 diverse molecules from 61 different individuals. A total of 20 out of 480 molecules were tested twice at different concentrations, so there were a total of 1000 stimuli. They used these 1000 stimuli so that subjects could perceive intensity, pleasantness, and familiarity. They also asked subjects to rate 20 semantic odor descriptors on a scale from 0 to 100. The 20 descriptors were "edible," "bakery," "sweet," "fruit," "fish," "garlic," "spices," "cold," "sour," "burnt," "acid," "warm," "musky," "sweaty," "ammonia/urinous," "decayed," "wood," "grass," "flower," and "chemical." The procedure of data collection of this research is depicted in Fig. 9.4. They studied the correlation between odor quality descriptors. Odor stimuli perception such as "fruit" was more likely to be perceived as

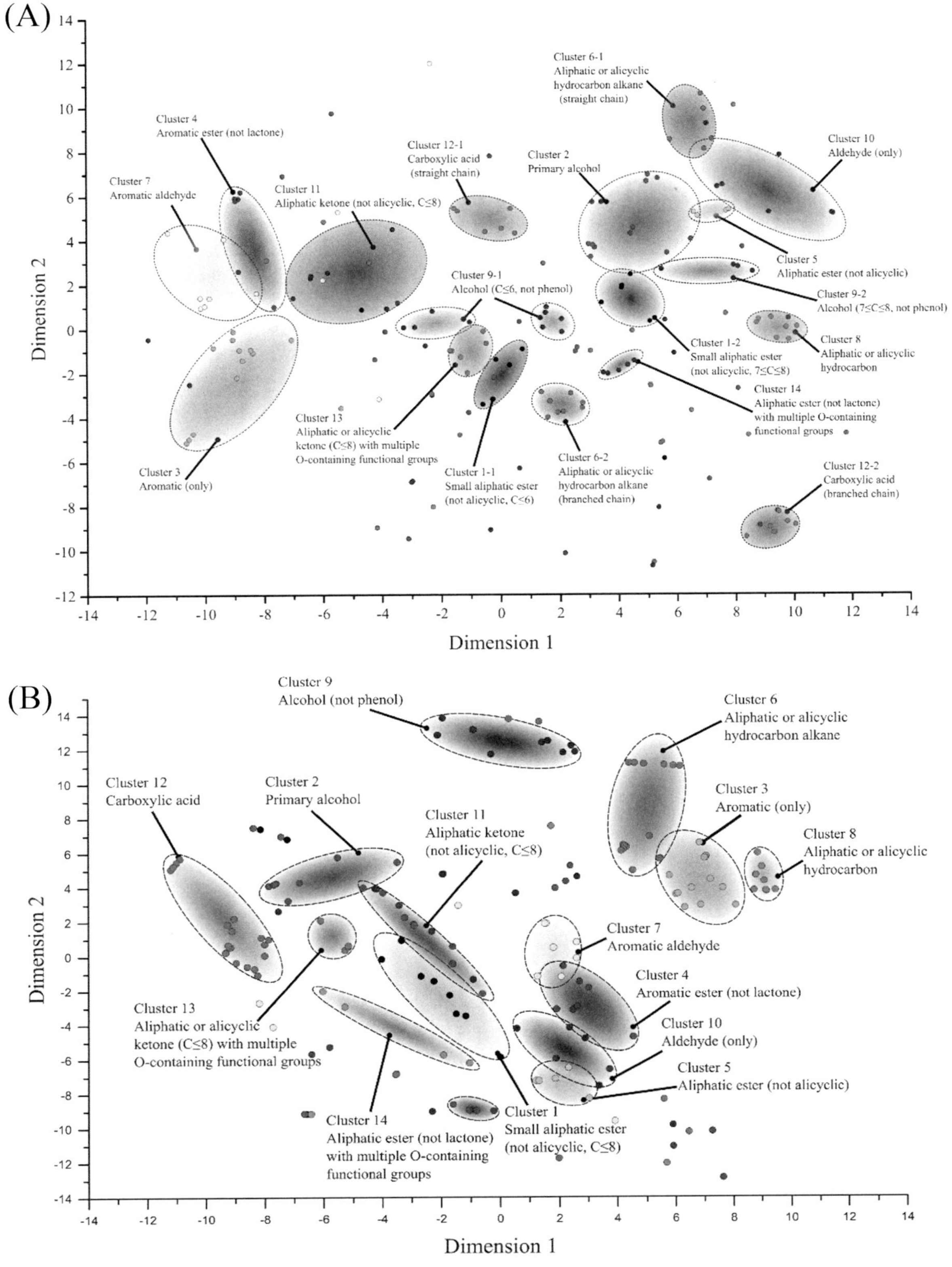
(A)
Cluster 6-1
Aliphatic or alicyclic hydrocarbon alkane (straight chain)
Cluster 4
Aromatic ester (not lactone)
Cluster 10
Aldehyde (only)
Cluster 12-1
Carboxylic acid (straight chain)
Cluster 2
Primary alcohol
Cluster 7
Aromatic aldehyde
Cluster 11
Aliphatic ketone (not alicyclic, C≤8)
Cluster 5
Aliphatic ester (not alicyclic)
Cluster 9-1
Alcohol (C≤6, not phenol)
Cluster 9-2
Alcohol (7≤C≤8, not phenol)
Cluster 8
Aliphatic or alicyclic hydrocarbon
Cluster 1-2
Small aliphatic ester (not alicyclic, 7≤C≤8)
Cluster 14
Aliphatic ester (not lactone) with multiple O-containing functional groups
Cluster 13
Aliphatic or alicyclic ketone (C≤8) with multiple O-containing functional groups
Cluster 6-2
Aliphatic or alicyclic hydrocarbon alkane (branched chain)
Cluster 3
Aromatic (only)
Cluster 1-1
Small aliphatic ester (not alicyclic, C≤6)
Cluster 12-2
Carboxylic acid (branched chain)
Dimension 2
Dimension 1
(B)
Cluster 9
Alcohol (not phenol)
Cluster 6
Aliphatic or alicyclic hydrocarbon alkane
Cluster 12
Carboxylic acid
Cluster 2
Primary alcohol
Cluster 3
Aromatic (only)
Cluster 11
Aliphatic ketone (not alicyclic, C≤8)
Cluster 8
Aliphatic or alicyclic hydrocarbon
Cluster 7
Aromatic aldehyde
Cluster 4
Aromatic ester (not lactone)
Cluster 13
Aliphatic or alicyclic ketone (C≤8) with multiple O-containing functional groups
Cluster 10
Aldehyde (only)
Cluster 5
Aliphatic ester (not alicyclic)
Cluster 14
Aliphatic ester (not lactone) with multiple O-containing functional groups
Cluster 1
Small aliphatic ester (not alicyclic, C≤8)
Dimension 2
Dimension 1

"sweet." They found a complex relationship between perceived intensity and pleasantness.

Following the previous study, the Dream Olfaction prediction challenge was called, and the result was published in Science Journal (Keller et al., 2017) demonstrating that it is possible to predict odor from molecular structures. In this work, their team explains their capability to build a model to predict the odor qualities of a certain molecule, just by using its structure. They found that regularized linear regression models gave the best predictions. They predicted 8 among 19 rated semantic descriptors (garlic, fish, sweet, fruit, burnt, spices, flower, and sour). These results represent a giant leap toward solving a major challenge of understanding olfaction.

9.4 Deep learning method

Deep learning, a subset of machine learning, has emerged as a powerful tool for odor prediction due to its ability to automatically learn complex patterns and representations from large datasets. This section explores the application of deep learning methods for odor impression prediction.

9.4.1 Deep learning framework for odor descriptor rating prediction

Dream dataset was used by Guo et al. (2021) for an odor descriptor rating prediction task using the deep learning method. They studied mapping from QCM and MOS responses to nine odor descriptors (pleasantness, sweet, fruit, fish, garlic, grass, burnt, musky, and decayed) using convolutional long short-term memory (LSTM) model for 44 odorants. Convolutional neural network (CNN) utilizes the convolutional layers for extracting the spatial features of signals. LSTM is a variation of recurrent neural network that has loops. This is suitable for processing series data, and it has the benefit of a memory unit that can reflect past inputs for a long time. ConvLSTM, or convolutional LSTM, is a neural network architecture that combines the spatial information processing capabilities of CNNs with the temporal modeling capabilities of LSTM networks. This architecture is particularly useful for tasks that involve both spatial and temporal dependencies.

The neural network proposed here to predict odor descriptors from the E-nose signal by regression to a pair or a single odor descriptor. There are two types of regression models for odor descriptors: a single model predicting a single descriptor (one-to-one model [ODRP_Sgl]) and a single model predicting all descriptors (one-to-all model [ODRP_All]). One-to-combination model can combine multiple odor descriptors that have high correlation and predict it by the same model. The E-nose used was the commercially available MOSES II (equipped with an eight-sensor quartz-microbalance module and an eight-sensor metal-oxide) with 16 sensors. A total of 44 odorants were all measured for 250 seconds at a sampling frequency of 1 Hz. This was repeated 5 or 6 times for each molecule, for a total of 228 samples. This means that each sample has 16×250-dimensional vector. The model was trained with approximately 70% of the samples as training data and 30% as validation data. The 16×250 E-nose signals were sliced into

◀ FIGURE 9.3 **Odorant clustering generated in an OM (A) and MP (B) spaces by using the *t*-SNE method**. Dimensions 1 and 2 indicate two values calculated by t-SNE. Each point indicates an odorant. ***MP*, Molecular parameters**; ***OM*, odor map;** *t-SNE*, *t*-Distributed stochastic neighbor embedding. *Reproduced with permission from*: *Shang, L., Liu, C., Tomiura, Y., & Hayashi, K. (2018). Odorant clustering based on molecular parameter-feature extraction and imaging analysis of olfactory bulb odor maps.* Sensors and Actuators, B: Chemical, 255, *508–518. https://doi.org/10.1016/j.snb.2017.08.024. Copyright © 2017, Elsevier*

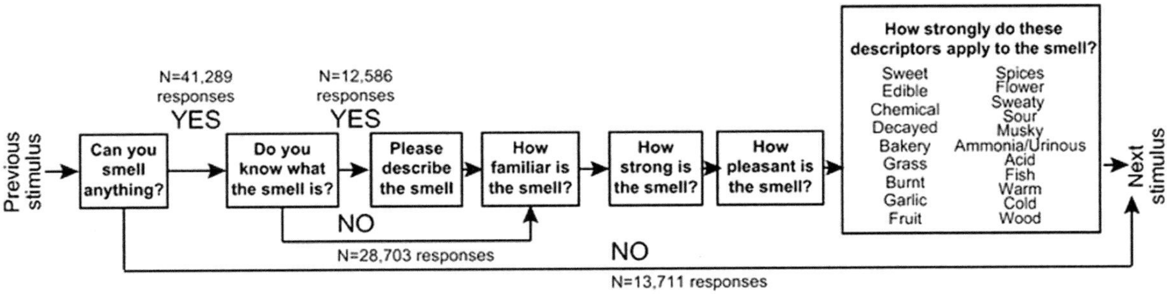

FIGURE 9.4 **Sequence of prompts for each stimulus.** *N* Denotes the total number of responses across all stimuli and all subjects. *Reproduced with permission from: Keller, A., & Vosshall, L.B. (2016). Olfactory perception of chemically diverse molecules.* BMC Neuroscience, 17*(1), https://doi.org/10.1186/s12868-016-0287-2. Copyright 2016.*

16×25 sizes as an input to the model. The ConvLSTM has a single hidden layer with 16 input channels, 14 hidden channels, a convolutional kernel size (16,4), and a stride size of 1. Adaptive moment estimation, an extension of the stochastic gradient descent optimization algorithm that computes adaptive learning rates for each parameter, was used to optimize the model, and the training was repeated 700 times. The correlation coefficients between true and predicted values were less than 0.8 except for the descriptor "garlic" even if the proposed methods were superior to ordinary CNN, LSTM, backpropagation, and PCA/RF. In summary, they demonstrated that we could reduce the number of parallel models constructing whole prediction model and improve the prediction accuracy by predicting combined odor descriptors that have high similarities. However, they combined only two descriptors with high correlation. If the sensors have more discrimination capability, higher prediction accuracy is expected.

9.4.2 Odor impression prediction using a neural network with two autoencoders

Nozaki and Nakamoto (2016) proposed a nonlinear deep neural network model for predicting odor impressions from mass spectra, which are physicochemical features of single-molecule odorants. Mass spectrum data and a deep learning technique (autoencoder followed by multilayer perceptron model) were used to predict the odor impression. Due to the availability of a large-scale mass spectrum dataset (a figure is shown in Fig. 9.1C) from the National Institute of Standards and Technology (NIST) (NIST Chemistry WebBook, 2017), it was used in this work. The mass spectrum is considered for its ability to ionize pure molecules or complex mixtures as m/z (mass-to-charge ratio) at multiple fragments ion peaks with molecular ion peaks. When a vaporized sample passes into the mass spectrometer, the sample molecules disassociate during the ionization process, which is called fragmentation. Peaks from 51 m/z to 261 m/z included in the mass spectrum were used in this study for predicting the odor impression because it is considered that the peaks at 50 m/z or less contain many odorless molecules and humans can only perceive odorous molecules in the air. It is considered that the peak at 300 m/z or more, which has low volatility under atmospheric pressure, hardly contributes to the impression of odor. The benefit of using mass spectrum dataset is that it can be used both for monomolecule and molecular mixtures because linear superposition is valid for mass spectrum, which means, without measuring

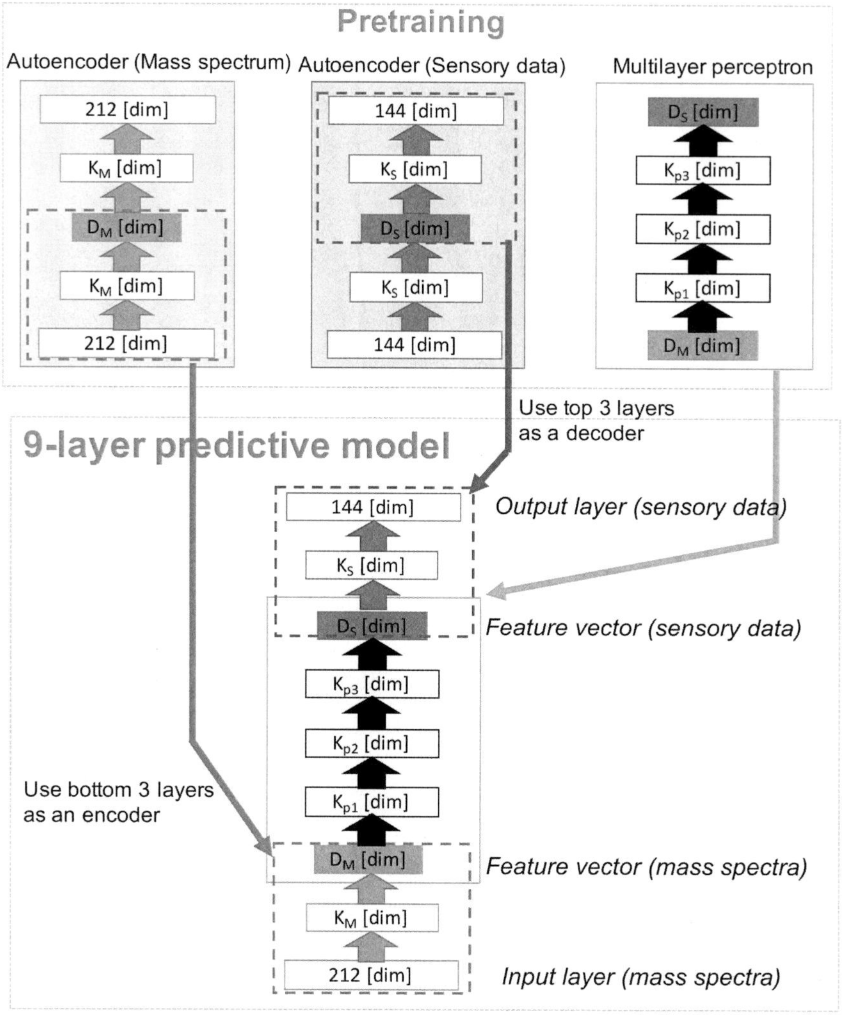

FIGURE 9.5 **Schematic diagram of the predictive model.** The character or number in each box gives the name of the layer or the number of units (dimensionality), respectively. *Reproduced with permission from: Nozaki, Y., & Nakamoto, T. (2016). Odor impression prediction from mass spectra.* PLoS ONE, 11*(6), https://doi.org/10.1371/journal.pone.0157030. Copyright 2016.*

the mass spectrum of a chemical mixture, we can obtain mass spectrum of two or more component mixture.

A 212-dimensional mass spectrum dataset was obtained according to the listed molecules from the Dravnieks dataset. This mass spectrum dataset and continuous sensory information from Dravnieks (144-dimensional data) were used as an input and the target vectors of the neural network model, respectively. Although mass spectrum is one of the structural information of the molecule of an odor, the relation between mass spectra and the impression of the smell has a complex nonlinear relationship. For this reason, the high-dimensional mass spectrum dataset was initially processed using a nonlinear dimensional reduction method. A schematic diagram is depicted in Fig. 9.5. This proposed method consists of two five-layer autoencoders, one for mass spectrum (input data) and the other for sensory

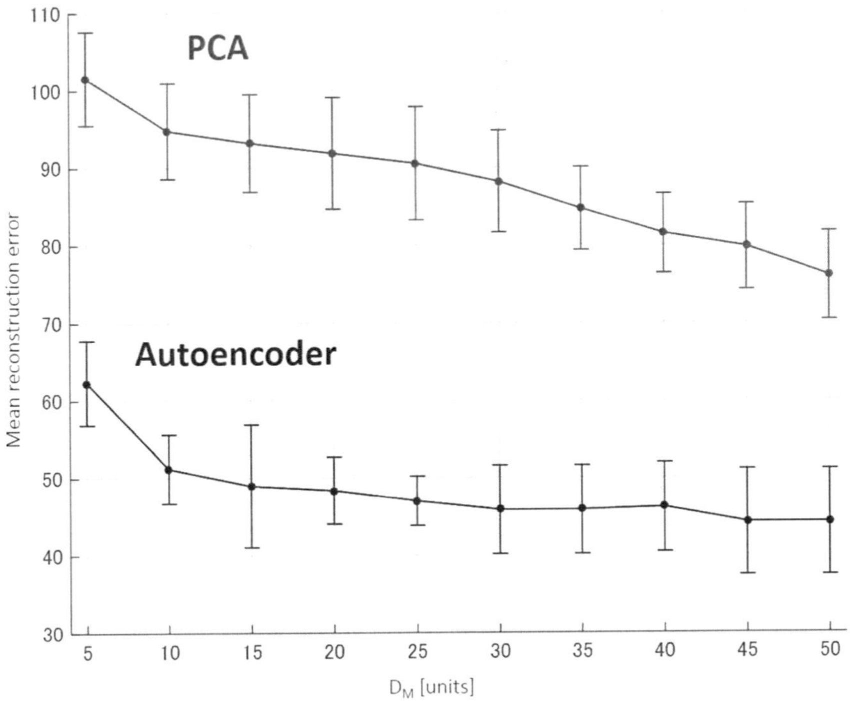

FIGURE 9.6 Mean reconstruction errors in cross-validation. Error for mass spectra with respect to the number of neurons in the hidden layer (in autoencoder) and error with respect to the number of principal components (in PCA). *PCA*, Principal component analysis. *Reproduced with permission from: Nozaki, Y., & Nakamoto, T. (2016). Odor impression prediction from mass spectra.* PLoS ONE, 11*(6), https://doi.org/10.1371/journal.pone.0157030. Copyright 2016.*

data (output data), to extract the feature vector from the bottleneck hidden layer. Here K_m is for the first hidden layer, D_m is for the bottleneck hidden layer, and K_{p1}, K_{p2}, and K_{p3} are the first, second, and third hidden neurons of the five-layer (multilayer) perceptron model.

Then, this feature vector of the mass spectrum was projected onto the feature vector of sensory data using a five-layer (multilayer) perceptron model. The optimal dimension of feature vectors for mass spectrum and sensory data were 45 and 30, respectively. The architecture of the five-layer autoencoders for mass spectrum and sensory data were 212−85−45−85−212 and 144−65−30−65−144, respectively. The structure of the multi-layer perceptron (MLP) between two autoencoders was 45−50−55−50−30. Thus a nine-layer neural model was used in total. L1-norm regularization, with the absolute value of the coefficient as a penalty term, was used with the objective function to avoid the overfitting of the learning model. A detailed explanation of the model training was reported in the article of Nozaki and Nakamoto (2019). The model was evaluated 60 times (sixfold cross-validation × 10 times) to get the average performance. They found that the reconstruction error using autoencoder was much smaller than that of PCA because of the nonlinear characteristics as illustrated in Fig. 9.6.

The odor impression prediction model proposed here was compared using a linear regression method, the partial least squares regression (PLS method) (Geladi & Kowalski, 1986), and PCC. While the correlation coefficient of the linear PLS method was $R \sim 0.61$ (B), the correlation coefficient of the proposed nonlinear model was $R \sim 0.76$ (A), suggesting the existence of a nonlinear relationship between

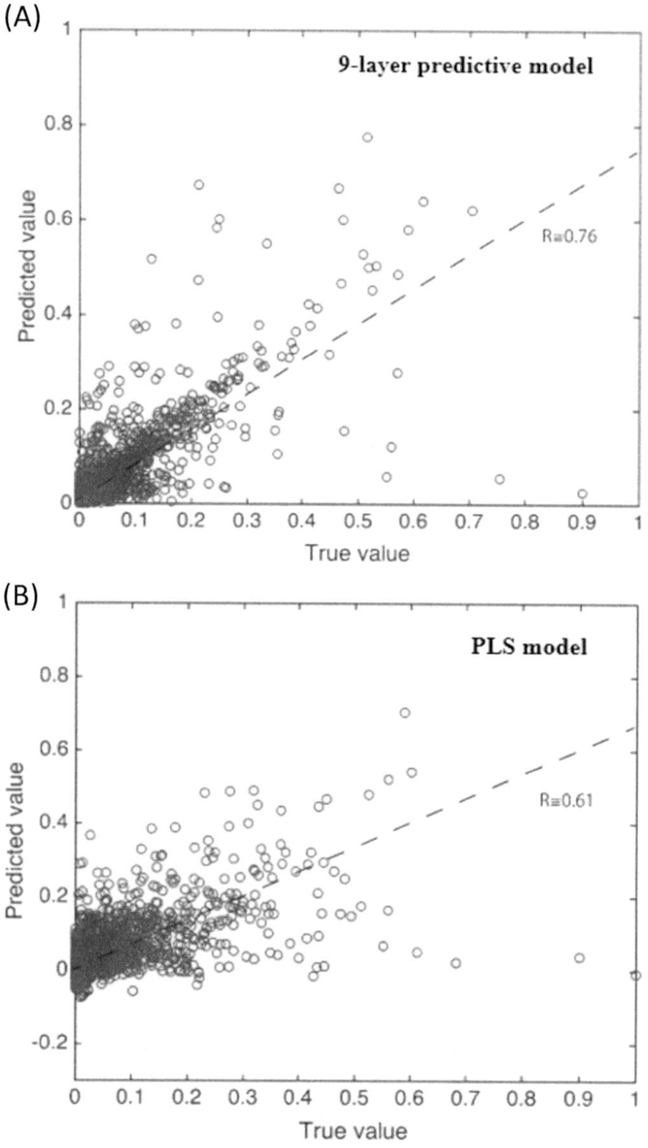

FIGURE 9.7 Comparison of the proposed model and the prediction model based on the PLS method. (A) Result for the nine-layer predictive model ($R \sim 0.76$) and (B) result for the PLS model ($R \sim 0.61$). Each horizontal axis represents the true value of the sensory test data, and the vertical axis represents the sensory test data predicted by the model. *PLS*, Partial least squares regression. *Reproduced with permission from: Nozaki, Y., & Nakamoto, T. (2016). Odor impression prediction from mass spectra.* PLoS ONE, 11*(6), https://doi.org/10.1371/journal.pone.0157030. Copyright 2016.*

mass spectrum and olfactory perception as illustrated in Fig. 9.7. This research showed that olfactory perception can be predicted by mass spectra. This result was further improved when more data were used for training the model.

9.4.3 Improvement of odor Impression predictive model using mass spectrum

Ito and Nakamoto (2020) improved the performance of the odor impression–predictive model using a large-scale mass spectrum database in addition to the dataset utilized by Nozaki and Nakamoto (2016). In addition to the 121 kinds of single molecules corresponding to the sensory data reported by Dravnieks, 2335 mass spectrums of flavor molecules were used from the Flavor Database. They collected the corresponding mass spectrum dataset from the NIST WebBook. This is to be noticed that the sensory dataset (target vector) was not increased. Large mass spectrum dataset was used to train an autoencoder for extracting the mass spectrum using Itakura–Saito divergence (IS divergence) as a cost function because the dynamic range of the mass spectrum is large. IS divergence was effective to extract tiny peak features (as reported in the article of Nozaki and Nakamoto (2017)) from the mass spectrum. The same nine-layer deep learning architecture (hidden neurons were different as the dataset was large) as discussed before was used to predict the odor impressions. The correlation coefficient of the odor impression–predictive model was improved to $R \sim 0.90$ from $R \sim 0.76$ (Nozaki & Nakamoto, 2016).

9.4.4 Graph neural network for odor impression prediction

A graph is a data structure type in which objects called nodes are connected by relationships called edges. Relationships between among data can be represented graphically in computer vision, molecular chemistry, molecular biology, pattern recognition, data mining, and others. The existing machine learning model fails to perform on graphs because of the arbitrary size of the graph and the complex topology. This data structure graph can be used as an input of the machine learning model without preprocessing. Graph neural network (GNN) can directly input various types of practical graphs (cyclic/acyclic graphs, directed/undirected graphs, etc.).

A research team from Google utilized the concept of GNN (Sanchez-Lengeling et al., 2019) for predicting the odor perception since the molecule's atoms and bonds can be considered nodes and edges, respectively. They used the whole structure of the molecular graph as input to directly predict the odor descriptors for each individual flavor molecule as shown in Fig. 9.8. They used an expert-labeled set of 5030 molecules from two different sources (Good Scent and Leffingwell PMP 2001 database). These two websites provide odor descriptors where more than one odor impression is recorded. They gathered a total of 138 odor descriptors for 5030 molecules. They used several handcrafted features as a baseline for making the comparison with their GNN-based deep neural model.

They said that GNN captured a meaningful relationship between chemical structure and odor, as demonstrated by transfer learning tasks using the odor embedding in the bottleneck representation of the network. Their proposed method showed that the mean area under the receiver-operating characteristic curve score across 138 odor descriptors is 0.894 compared to conventional featurization techniques (Mordred features, molecular fingerprints).

9.4.5 Odor descriptor prediction using imbalanced dataset

In the domain of odor descriptor prediction, the important issue to be solved is the imbalance problem, that is, it means the number of target odor descriptors (e.g., peach odor

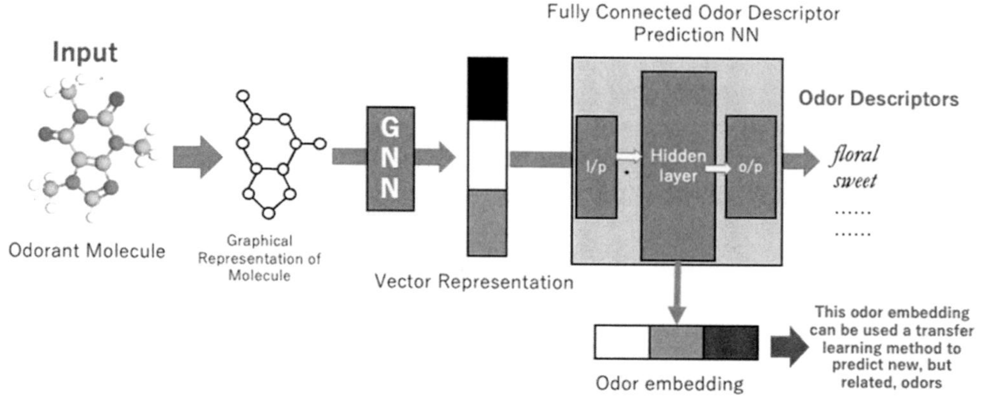

FIGURE 9.8 **The predictive model architecture of the** GNN**-based odor prediction**. Graphical representation of molecule is input to GNN. *GNN*, **Graph neural network**.

descriptor prediction) we want to predict is much smaller than the negative samples (non-peach samples). Creating artificial samples can duplicate samples from the negative class and this increases the likelihood of overfitting especially for high oversampling rates when class skew was severe. For this reason, we solved the imbalance problem for the mapping of the mass spectrum of odorant molecules into the binary odor perceptual space without generating the artificial samples. Although there are different algorithms for imbalanced binary data classification, we utilized the method of one class support vector machine (OCSVM) because it does not require equal proportions of positive–negative samples and neural network trained with weighted neural network to handle these large numbers of negative samples with small target odors. We predicted a total of 89 odor descriptors where each odor impression was highly imbalanced and had a different occurrence frequency (Debnath & Nakamoto, 2022). In Fig. 9.9, we depicted the 2-D PCA visualization of three odor descriptors fruity, Banana, and peach, and the prediction (using OCSVM) results had been visualized in the bottom row. We investigated experimental results to select the correct algorithm as a function of the number of positive (target odor) and negative (nontarget odor) samples. We empirically divided these odor descriptors into three categories, "large," "medium," and "small," based on the number of positive samples.

9.5 Summary

Odor impression prediction from the binary or continuous olfactory sensory information is a difficult task compared to the typical classification of several kinds of odors. The main difficulty in this domain is data scarcity. However, researchers developed the fundamental concept for predicting odor descriptors using machine learning models. In this chapter, we discussed several data-preprocessing steps for the olfactory dataset followed by the multidimensional data analysis methods to predict the odor impression. Moreover, we explored the deep learning methods for extracting features from its nonlinear high-dimensional sensor

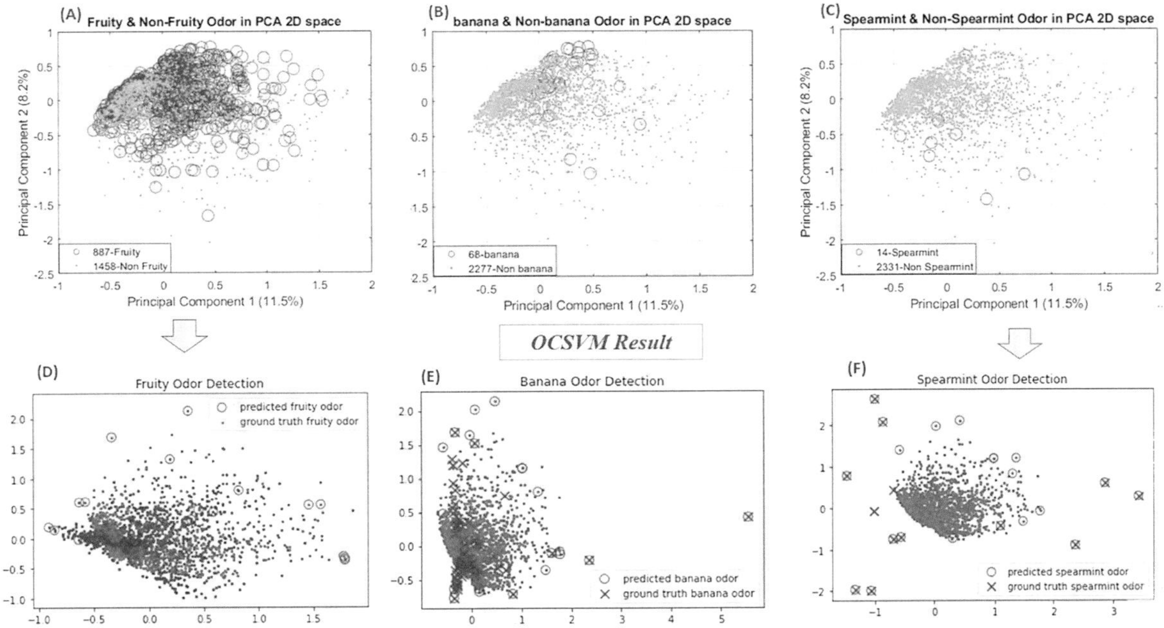

FIGURE 9.9 **PCA and OCSVM analysis for imbalanced dataset.** Top row: PCA in 2D space for each odor category. (A) Fruity, (B) banana, and (C) spearmint odor samples for large, middle, and small categories, respectively. Bottom row: Parts (D)–(F) depict the predicted versus ground truth odor detection for fruity, banana, and spearmint using OCSVM where green dots indicate the nontarget odor samples for each example. *PCA*, Principal component analysis; *OCSVM*, One class support vector machine. *Reproduced with permission from: Debnath, T., & Nakamoto, T. (2022). Predicting individual perceptual scent impression from imbalanced dataset using mass spectrum of odorant molecules.* Scientific Reports, 12*(1), https://doi.org/10.1038/s41598-022-07802-3.*

dataset. We found that the nonlinear architecture of the model can better predict the odor impression than the linear predictive model.

Reference

Debnath, T., & Nakamoto, T. (2022). Predicting individual perceptual scent impression from imbalanced dataset using mass spectrum of odorant molecules. *Scientific Reports., 12* (1). Available from https://doi.org/10.1038/s41598-022-07802-3, http://www.nature.com/srep/index.html.

A. Dravnieks. (1985). Atlas of odor character profiles. ASTM.

Geladi, P., & Kowalski, B. R. (1986). Partial least-squares regression: A tutorial. *Analytica Chimica Acta, 185*(C), 1–17. Available from https://doi.org/10.1016/0003-2670(86)80028-9.

Guo, J., Cheng, Y., Luo, D., Wong, K. Y., Hung, K., & Li, X. (2021). ODRP: A deep learning framework for odor descriptor rating prediction using electronic nose. *IEEE Sensors Journal, 21*(13), 15012–15021. Available from https://doi.org/10.1109/JSEN.2021.3074173, http://ieeexplore.ieee.org/xpl/RecentIssue.jsp?punumber = 7361.

Hinton, G. E., & Salakhutdinov, R. R. (2006). Reducing the dimensionality of data with neural networks. *Science (New York, N.Y.), 313*(5786), 504–507. Available from https://doi.org/10.1126/science.1127647.

Ito, K., & Nakamoto, T. (2020). Improvement of odor impression predictive model using machine learning. *Proceedings of IEEE Sensors, 2020*. Available from https://doi.org/10.1109/SENSORS47125.2020.9278592, http://www.ieee.org/sensors.

Keller, A., Gerkin, R. C., Guan, Y., Dhurandhar, A., Turu, G., Szalai, B., Mainland, J. D., Ihara, Y., Yu, C. W., Wolfinger, R., Vens, C., Schietgat, L., De Grave, K., Norel, R., Stolovitzky, G., Cecchi, G. A., Vosshall, L. B., Meyer, P., Bhondekar, A. P., ... Zupan, B. (2017). Predicting human olfactory perception from chemical features of odor molecules. *Science (New York, N.Y.), 355*(6327), 820–826. Available from https://doi.org/10.1126/science.aal2014, http://science.sciencemag.org/content/sci/355/6327/820.full.pdf.

Keller, A., & Vosshall, L. B. (2016). Olfactory perception of chemically diverse molecules. *BMC Neuroscience, 17*(1). Available from https://doi.org/10.1186/s12868-016-0287-2, http://www.biomedcentral.com/bmcneurosci/.

Khan, R. M., Luk, C. H., Flinker, A., Aggarwal, A., Lapid, H., Haddad, R., & Sobel, N. (2007). Predicting odor pleasantness from odorant structure: Pleasantness as a reflection of the physical world. *Journal of Neuroscience., 27*(37), 10015–10023. Available from https://doi.org/10.1523/JNEUROSCI.1158-07.2007, http://www.jneurosci.org/cgi/reprint/27/37/10015.

Leffingwell & Associates. (2001). *Leffingwell PMP database*. Leffingwell. http://www.leffingwell.com/bacispmp.htm.

Van Der Maaten, L., & Hinton, G. (2008). Visualizing data using *t*-SNE. *Journal of Machine Learning Research., 9*, 2579–2625. Available from http://jmlr.csail.mit.edu/papers/volume9/vandermaaten08a/vandermaaten08a.pdf.

Mauri, A., Consonni, V., Pavan, M., & Todeschini, R. (2006). DRAGON software: An easy approach to molecular descriptor calculations. *Match, 56*(2), 237–248.

Nakamoto, T. (2016). *Essentials of machine olfaction and taste* (Vol. 1). Wiley. Available from 10.1002/9781118768495.

Nakamoto, T. (2019). Odor sensing system with multidimensional data analysis. *Japanese Journal of Applied Physics, 58*(SB), SB0804. Available from https://doi.org/10.7567/1347-4065/ab0740.

NIST Chemistry WebBook. (2017). *8 10. Unpublished content*. http://webbook.nist.gov/chemistry/.

Nozaki, Y., & Nakamoto, T. (2016). Odor impression prediction from mass spectra. *PLoS ONE, 11*(6). Available from https://doi.org/10.1371/journal.pone.0157030, http://journals.plos.org/plosone/article?id = 10.1371/journal.pone.0157030.

Nozaki, Y., & Nakamoto, T. (2017). Itakura–Saito distance based autoencoder for dimensionality reduction of mass spectra. *Chemometrics and Intelligent Laboratory Systems, 167*, 63–68. Available from https://doi.org/10.1016/j.chemolab.2017.05.002.

Nozaki, Y., & Nakamoto, T. (2019). An olfactory sensor array for predicting chemical odor characteristics from mass spectra with deep learning. *Methods in Molecular Biology, 2027*. Available from https://doi.org/10.1007/978-1-4939-9616-2_3, http://www.springer.com/series/7651.

Sanchez-Lengeling, B., Wei, J. N., Lee, B. K., Gerkin, R. C., Aspuru-Guzik, A., & Wiltschko, A. B. (2019). Machine learning for scent: Learning generalizable perceptual representations of small molecules. *arXiv*. Available from https://doi.org/10.48550/arxiv.1910.10685, https://arxiv.org.

Shang, L., Liu, C., Tomiura, Y., & Hayashi, K. (2017). Machine-learning-based olfactometer: Prediction of odor perception from physicochemical features of odorant molecules. *Analytical Chemistry, 89*(22), 11999–12005. Available from https://doi.org/10.1021/acs.analchem.7b02389, http://pubs.acs.org/journal/ancham.

Shang, L., Liu, C., Tomiura, Y., & Hayashi, K. (2018). Odorant clustering based on molecular parameter-feature extraction and imaging analysis of olfactory bulb odor maps. *Sensors and Actuators, B: Chemical, 255*, 508–518. Available from https://doi.org/10.1016/j.snb.2017.08.024.

Sigma-Aldrich, (n.d.). *Flavors and fragrances*. https://www.scribd.com/doc/121621778/Flavors-and-fragrances-catalog.

The good scents company—Flavor, fragrance food, and cosmetics ingredients information. (n.d.). https://www.thegoodscentscompany.com/.

CHAPTER

10

Odor impression prediction using natural language processing

Tanoy Debnath
Yokogawa Electric Corporation, Tokyo, Japan

10.1 Introduction

Odor impression prediction is the task of predicting the perceived odor quality (e.g., fruity and sweet). This problem is challenging due to the complexity and subjectivity of odor perception. Each individual may have different associations and experiences with certain odor descriptors, making it difficult to generalize across a population (Arshamian et al., 2022). To tackle this problem, researchers have developed methods for making a sematic relationship among odor descriptors.

Rule-based approaches rely on predefined rules and are empirical to map odor descriptors to understand the association among odor impressions. These rules are often developed on the basis of expert knowledge or empirical data. For example, the odor wheel (Suffet et al., 1988) is a widely used rule-based approach for describing and categorizing odors. Although rule-based approaches are simple and interpretable, they may not be able to capture the complexity and variability of odor perception, especially when dealing with novel odor descriptors.

Machine learning–based approaches, on the other hand, learn the mapping between subjective (human evaluation of odor impressions) and objective information of data. "Spicy" is a popular odor descriptor term that can be altered with "hot" and "pepper" because people express their perception of smell in different ways. How can we project those similar odor descriptors on odor descriptor space so that those descriptors appear together? It will help to understand the verbal descriptors of a flavor molecule irrespective of one's cultural background. Thus researchers utilized the natural language processing (NLP) (LeCun et al., 2015) technique for clustering the similar kinds of smell descriptors to make an association among smell impressions. One approach is to use pretrained word embedding, such as word2vec (Mikolov, Chen, et al., 2013; Mikolov, Sutskever, et al., 2013) or fastText (Bojanowski et al., 2017) to capture the semantic relationships between words based on their cooccurrence in odor descriptors dataset. This is considered a downstreaming task as we applied the already pretrained word vectors to our own dataset (odor descriptors).

Digital Technologies in Olfaction
DOI: https://doi.org/10.1016/B978-0-443-15721-9.00010-5

We compared the performance of word similarity matrix (based on NLP methods) with the correlation coefficient similarity matrix. However, there is a limitation of making odor descriptor cluster using binary sensory information compared to continuous sensory data because similar descriptors, such as "rose" and "floral," are exclusively used, resulting in vanishing of correlation between descriptors.

In this chapter, we begin with Section 10.2 by introducing the concept of odor impression predictive model, including (1) difference between binary and continuous sensory data (Section 10.2.1), (2) problem of binary odor descriptors (Section 10.2.2), (3) odor descriptors clustering method (Section 10.2.3), and (4) odor descriptor predictive model (Section 10.2.4). In Sections 10.3 and 10.4 we provide two case studies for odor descriptor cluster prediction (Debnath & Nakamoto, 2020; Nozaki & Nakamoto, 2018). Specifically, in Section 10.4, we introduce the method to obtain scent impression scores with continuous values from the score with binary value (Debnath & Nakamoto, 2020).

10.2 Concept of odor impression predictive model

Odor descriptor is a verbal impression that describes the perceived characteristics of an odor. Odor descriptor clustering is a technique used to classify and organize different odor descriptors based on their perceived similarity. For example, "fruity," "floral," "woody," "spicy," and "earthy" are all odor descriptors. The process of odor descriptor clustering involves collecting a large number of odor descriptors and then grouping them into clusters based on their perceived similarity. Mass spectrum (it is the physicochemical properties that represent the molecular properties of a chemical) dataset was used as an input for the model to predict the presence or absence of an odor descriptor group from the mass spectrum of a flavor molecule. The input units of the neural network correspond to the m/z (mass-to-charge ratio) values of the mass spectrum and the output units correspond to the odor descriptors.

10.2.1 Difference between binary and continuous sensory data

Two types of sensory datasets are available in the olfactory perception domain. One is the continuous sensory data where "continuous" means panelists give a sensory marking from 0 to full scale (Dravnieks, 1985; Keller & Vosshall, 2016). Another one is binary value, which means a set of several odor descriptors (Leffingwell & Associates, n.d.; Sigma-Aldrich, n.d.; The Good Scents Company, n.d.) are used to express each flavor molecule. We illustrated the difference between binary and continuous sensory information in Fig. 10.1.

Leffingwell, Good Scents, and the catalog published from Sigma-Aldrich are the popular

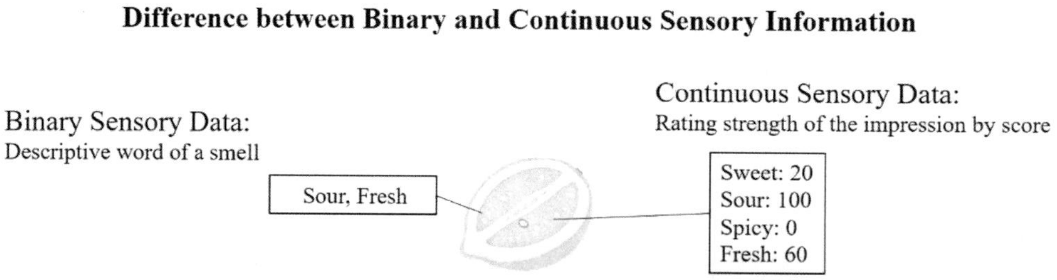

FIGURE 10.1 Difference between binary and continuous sensory information. Binary information consists of a set of odor descriptors, whereas continuous value is a score given by a panelist.

binary odor descriptor databases where odorants are listed and the odor properties of each substance are expressed by the presence or absence of a descriptor, whereas Dravnieks' sensory datasets are given on a scale from 0 to 5. Essential oils dataset, which is considered a chemical mixture, was collected in-house that will be described in Section 10.4 (case study 2).

10.2.2 Problem of binary odor descriptors

When we have continuous sensory data, we can directly apply a hierarchical clustering method using a dendrogram on the matrix of correlation coefficient based on continuous values. But when the database indicates a set of odor descriptors, only binary data is available and the correlation between the similar descriptors disappears.

For example, "Balsam" and "Balsamic" were very similar impressions, and the correlation coefficient between these in the Sigma-Aldrich catalog data was less than 0.01. Correlation coefficients between each of these three odor descriptors, "Rose," "Violet," and "Lavender," indicate relatively similar smell impressions, calculated for both the catalog data and Dravnieks' sensory test data, as shown in Table 10.1(a) and (b), respectively.

This result indicates that the catalog data is compiled in such a way that only descriptors that are more applicable are selected in a mutually exclusive manner, even if they have very similar impressions. From this result, we can consider the possibility that the neural network may have difficulty in learning because it gives an output 0 even for odorants that have some fit to the "Rose" descriptor.

10.2.3 Odor descriptor clustering method

A hierarchical clustering method was applied to the dataset to investigate the similarities among odor descriptors. A dendrogram helps to determine the number of clustering that is most appropriate for an application. Two types of clustering methods are compared using hierarchical clustering. Dendrogram clustering based on the unweighted average distance method was performed on both the correlation coefficient matrix and the cosine similarity matrix (Szekely & Rizzo, 2005). Cosine similarity measures the cosine of the angle between two vectors (in this

TABLE 10.1 Correlation coefficients among three-odor descriptors: (a) binary value data (Sigma-Aldrich), (b) continuous-value data (Dravnieks).

	Rose	Violet	Lavender
(a)			
Rose	1	0.02	0.08
Violet	0.02	1	0.00
Lavender	0.08	0.00	1
(b)			
Rose	1	0.95	0.79
Violet	0.95	1	0.84
Lavender	0.79	0.84	1

Reproduced with permission from: Nozaki, Y., & Nakamoto, T. (2018). Predictive modeling for odor character of a chemical using machine learning combined with natural language processing. PLoS ONE, 13(6). https://doi.org/10.1371/journal.pone.0198475. Copyright 2018 Nozaki, Nakamoto.

case, word vectors) that will be projected into a multidimensional space:

1. Similarity matrix was obtained by calculating the correlation coefficient between all possible vector pairs.
2. Cosine similarity matrix to show similarity of words calculated by the natural language modeling method. Word2vec and fastText will be used for making similarity of words of odor descriptor group.

Word similarity matrix was generated using a word embedding method called word2vec. It is a language modeling approach proposed by Bojanowski et al. (2017). The model uses a three-layer neural network (Fig. 10.2A) called a skip-gram model trained with a large text corpus and used to obtain vector representations of words. Input layer is the word to be predicted, and the output layer is the context words. The hidden layer represents the embedding of the input word learned during training. Fig. 10.2B shows word vectors of country names and city names mapped into a two-dimensional principal component analysis space.

The words included in the training corpus (in NLP, a corpus is a large set of machine-readable texts) are projected into the word vector space as the corresponding vectors. Similarity between words was calculated as the cosine distance between word vectors by word2vec, and clustering was performed using a tree diagram based on distance. The procedure for constructing the similarity matrix is given in the following:

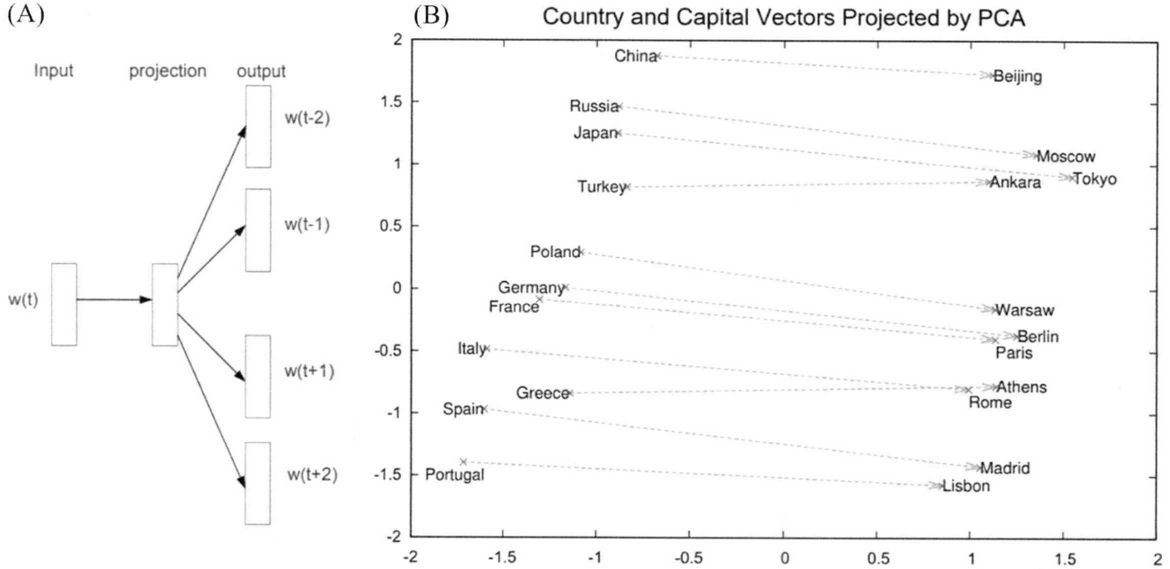

FIGURE 10.2 Model used in NLP and vector representation of word. (A) Skip-gram model used in word2vec from Mikolov, Chen, et al. (2013) and Mikolov, Sutskever, et al. (2013). This model predicts surrounding words $w(t-2)$ to $w(t+2)$ from a word (input) $w(t)$ given to the model. (B) Vector representation of words mapped into two dimensions by PCA where countries and cities are located close to each other. *NLP*, Natural language processing; *PCA*, principal component analysis. *Reproduced with permission from: Mikolov, T., Chen, K., Corrado, G., & Dean, J. (2013).* Efficient estimation of word representations in vector space. *First International Conference on Learning Representations, ICLR 2013—Workshop Track Proceedings. ICLR, United States. https://dblp.org/db/conf/iclr/iclr2013w.html; Mikolov, T., Sutskever, I., Chen, K., Corrado, G., & Dean, J. (2013).* Distributed representations of words and phrases and their compositionality. *Proceedings of the 26th International Conference on Neural Information Processing Systems (Vol. 2).*

1. The full text of the English Wikipedia corpus (The Wikipedia Corpus, 2017) was used as a training corpus.
2. After that, the cosine similarity of odor descriptors was calculated using a trained wiki model.
3. Finally, hierarchical clustering was carried out based on the cosine similarity matrix.

The hyperparameters of the skip-gram model are shown in Table 10.2, where window size determines the maximum distance between words within a sentence. Same procedures have been followed in both case studies.

10.2.4 Odor descriptor cluster predictive model

To evaluate the clustering performance of abovementioned methods, we built a predictive model using odor descriptor clusters as target vectors. To generate targets for the predictive model, an output vector indicating the presence or absence of the odor descriptor group is required. Descriptors for which each example matches the original catalog data are grouped into clusters. If any one of the plurality of descriptors is included in a particular cluster, the corresponding sample belongs to that cluster as depicted in Fig. 10.3.

The predictive model in both case studies is a six-layer neural network that predicts the presence or absence of odor descriptor group from the mass spectrum as depicted in Fig. 10.4. The input units of the neural network correspond to the m/z values of the mass spectrum, and the output units correspond to the descriptors of the odor impression. The benefit of using mass spectrum dataset is that it can be used both for monomolecule and molecular mixtures because linear superposition is valid for mass spectrum, which means, without measuring the mass spectrum of a chemical mixture, we can obtain mass spectrum of two or more component mixtures. Mass spectrum dataset was normalized by dividing by the maximum value in the dataset to have a value between 0 and 1. As mass spectra are high dimensional, an autoencoder was used first for dimensionality reduction (Hinton & Salakhutdinov, 2006). The resulting mass spectrum feature vector space and cluster vector space are mapped with two hidden layers. In both case studies in this chapter, we will follow the same strategy.

The performances of the predictive models were evaluated in terms of accuracy. Let the input be $\mathbf{x}$, the weights and biases of the predictive model be $\mathbf{w}$ and $\mathbf{b}$, the model's output vector be $f(\mathbf{x}; \mathbf{w}, \mathbf{b})$, and the desired output be $\mathbf{y}$. $\mathbf{w}$ and $\mathbf{b}$ include all weights and biases in all layers, respectively. Then the accuracy of the predictive model is calculated as the rate of concordance between the model's output and desired output.

The main hyperparameters of the model proposed here include the number of neurons in each layer (K_m for first hidden layer and D_m for

TABLE 10.2 Hyper-parameter used in the skip-gram model (Nozaki & Nakamoto, 2018).

Hyperparameter	Value
No. of units in Hidden Llayer	400
Window Size	10
Minimum Count	5

Reproduced with permission from: Nozaki, Y., & Nakamoto, T. (2018). Predictive modeling for odor character of a chemical using machine learning combined with natural language processing. PLoS ONE, 13(6). https://doi.org/10.1371/journal.pone.0198475.

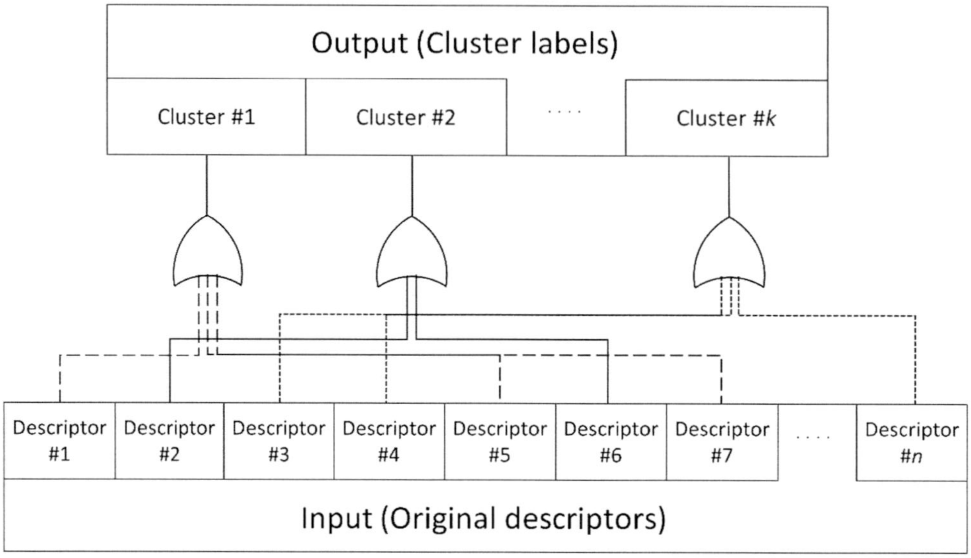

FIGURE 10.3 Odor character clustering. When any of the descriptors in a cluster has a value of 1, the value of the corresponding cluster that includes the sample is set to 1, for example, if descriptor 5 has a value of 1, and cluster 1 is set to 1. *Reproduced with permission from: Nozaki, Y., & Nakamoto, T. (2018). Predictive modeling for odor character of a chemical using machine learning combined with natural language processing.* PLoS ONE, *13(6). https://doi.org/10.1371/journal.pone.0198475.*

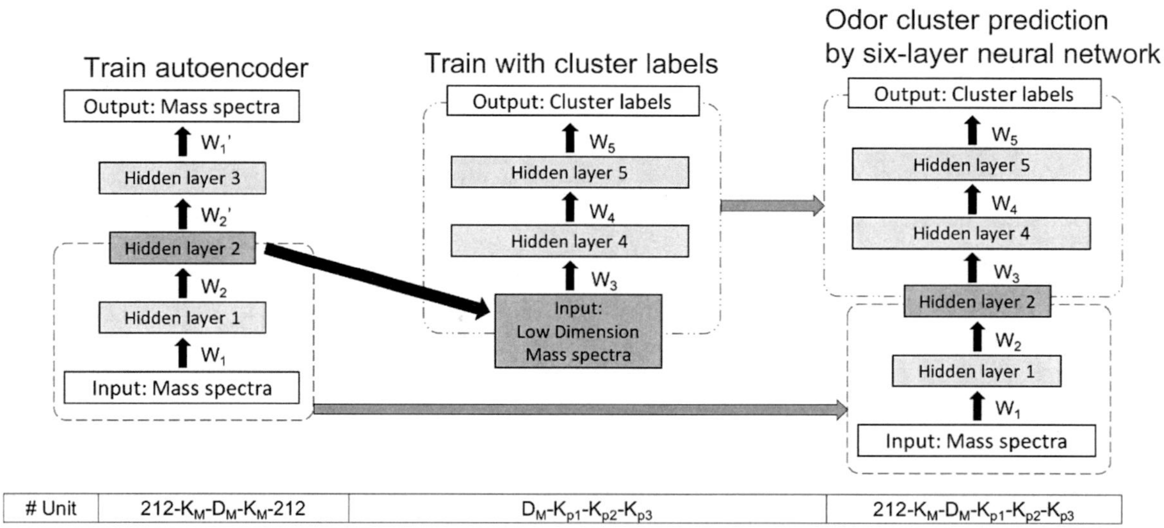

FIGURE 10.4 Odor cluster predictive model. Extracted feature of mass spectrum is classified into odor cluster. (Nozaki & Nakamoto, 2018). *Reproduced with permission from: Nozaki, Y., & Nakamoto, T. (2018). Predictive modeling for odor character of a chemical using machine learning combined with natural language processing.* PLoS ONE, *13(6). https://doi.org/10.1371/journal.pone.0198475*

bottleneck hidden layer), including the number of clusters and the learning coefficient when updating weight vectors. In Eq. (10.1), E_n (**w**) is the error at the nth neuron in the output layer, N is the number of the neurons in the output layer, and **w** is all the weights in the predictive model. Here, λ is the regularization constant, and p is the norm of the regularization term:

$$E(w) = \frac{1}{N}\sum_{n}^{N} E_n(\mathbf{w}) + \lambda W^p \tag{10.1}$$

The other hyperparameters of the predictive model, namely, the number of clusters, the number of neurons in each layer, the learning rate η, the momentum constant α were optimized in a series of simulations and are summarized in the main paper (Debnath & Nakamoto, 2020; Nozaki & Nakamoto, 2018). Since the predictive model includes two neural networks (the autoencoder and the mapping neural network), hyperparameters for the autoencoder are displayed with a subscript of m, whereas the hyperparameters for the mapping neural network are displayed with subscript p. The number of output clusters K_{p3} is determined on the basis of experimental results.

10.3 Case study 1: odor impression prediction using binary sensory data

Odor impression prediction using binary sensory data can be achieved using machine learning algorithms. Binary sensory data refers to data that is represented in binary form (0 or 1) indicating the presence or absence of a particular sensory attribute. In this case study, we will describe a method for predicting the odor descriptor cluster using binary sensory data where the mass spectrum was used as an input. NLP was used for clustering the odor descriptors followed by the neural network to evaluate the clustering performance by predicting the presence or absence of the odor descriptor cluster for a given flavor molecule. This is the first study using the combination of machine learning and NLP for odor impression prediction (Nozaki & Nakamoto, 2018).

10.3.1 Dataset

There were total 999 samples collected from the Sigma-Aldrich catalog dataset. A total of 138-odor descriptors out of 150 were chosen from the catalog since the other 12 descriptors appear three times or less in the catalog. We also collected the corresponding mass spectrum dataset from the Chemistry Webbook provided by National Institute of Standards and Technology (NIST Chemistry WebBook, n.d.) where mass spectrum was acquired with electron ionization of 70 [eV]. Thus we obtained a matrix of 138×999 sensory data and a matrix of 251×999 mass spectrum dataset because intensities between 51 and 262 m/z were extracted from the original dataset of mass spectrum. Here, sensory dataset will be the output of the model, and mass spectrum dataset was used as an input. We will use a model that connects these two datasets.

10.3.2 Methodology

Using the Sigma-Aldrich catalog data, we consider a prediction model that classifies the odor impression represented by a binary value of 0 or 1 from the mass spectrum. Consider a neural network that classifies and predicts whether or not a molecule belongs to 138 types of descriptors when the mass spectrum of a molecule is given. For example, similar descriptors, such as "rose," "violet," and "lavender," are all grouped into the same cluster. This cluster may represent applicability to "flowers." This work can be divided into two parts where odor descriptors were clustered followed by the neural network model that will predict a particular odor cluster. Thus the target is to predict the presence or absence of an odor descriptor cluster for a given flavor molecule.

10.3.2.1 Clustering of odor descriptors

To know how the similarity between odor descriptors created by word2vec was different from the similarity matrix created from the correlation coefficient between descriptors in the catalog dataset, multidimensional scaling (MDS) (MDS can determine pairwise distances) (Borg & Groenen, 2003) was utilized to visualize the positional relationship of the descriptors in two-dimensional space. Parts (A) and (B) of Fig. 10.5 show the results of applying metric MDS to two matrices: the dissimilarity created by subtracting the correlation coefficient between descriptors from a unit distance and the distance created from the cosine distance between word vectors of the descriptors obtained by word2vec.

It was confirmed that the calculated similarity between descriptors was greatly different between the two methods. As shown in Fig. 10.5B, in the two-dimensional image obtained from word2vec, descriptors such as "Rose," "Jasmine," "Lily," "Iris," and "Violet," and descriptors such as "Milk," "Cheese," and "butter" are placed closer to the figure, although those are scattered in Fig. 10.5A. Thus we can expect that groups will be formed on the basis of the impression of "flower" and "dairy product" in the two-dimensional scatter image obtained from word2vec. To get a better understanding, we provided the odor descriptors for each cluster index for six-odor descriptor groups and also gave a common name depending on the contained descriptors of that particular index of cluster in Table 10.3.

10.3.3 Predictive model and results

To compare the generalization performance of the models depicted in Fig. 10.4, fivefold cross-validation (it is a method for evaluating machine learning (ML) models by training several ML models on subsets of the available input data and evaluating them on the remaining subset of the data) was performed on 999 types of samples. Since a remainder is generated when the number of samples is divided by 5, the number of samples included in each divided set was [200, 200, 200, 200, and 199]. Mean squared error was used as an error function, and training was performed using stochastic gradient descent (Cun et al., 1990). Fig. 10.6 shows the accuracy of the predictive models (word2vec- and correlation-based) as a function of K_{p3}. The number of clusters K_{p3} in the output layer is plotted on the horizontal axis, and the value of true positive (target vector 1 and model output is also 1, in this case, it is the target odor descriptor group) or true negative (model outputs 0 when the target output is 0) is plotted on the vertical axis. Based on the results of this experiment, the number of clusters K_{p3}, which is a hyperparameter, was set to 6 to ensure the classification accuracy. The number of odor clusters K_{p3} has a trade-off with prediction ability. When the number of clusters was 6, the result was that the true positive prediction was around 80% and the true negative prediction was around 90%. But when the number of odor descriptor group increases, the true position rate was decreased because of the number of training instances during the model training. Although there is a trade-off between the number of odor descriptor group and the prediction performance, this study successfully predicts the odor descriptor cluster. We expect that dedicated text corpus of olfactory dataset will improve the model generalization capability.

Moreover, there is a limitation of making odor descriptor group using continuous sensory information compared to binary sensory data because only word similarity can be calculated using binary data. In the following case study, we will describe the procedure of conversion of binary data to continuous sensory information followed by the prediction model.

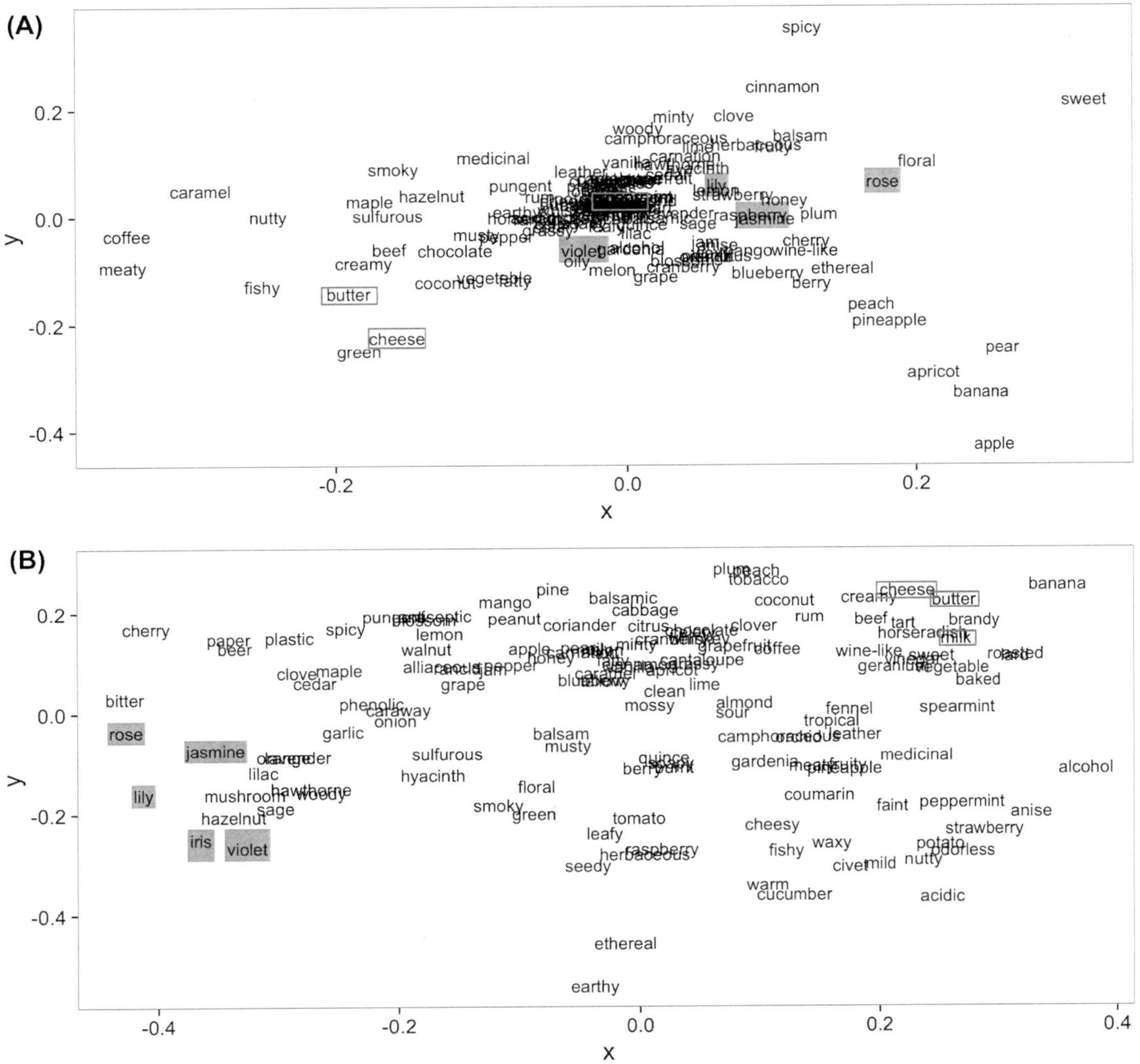

FIGURE 10.5 MDS diagrams based on correlation and cosine distance followed by NLP. MDS diagrams based on (A) one minus the sample correlation between points, (B) cosine distance based on word2vec modeling. Descriptors considered belonging to the same group, for example, "rose," "jasmine," "lily," "iris," and "violet" are marked in blue. "milk," "cheese," and "butter" are enclosed in red. *NLP*, Natural language processing. *Reproduced with permission from: Nozaki, Y., & Nakamoto, T. (2018). Predictive modeling for odor character of a chemical using machine learning combined with natural language processing.* PLoS ONE, 13*(6). https://doi.org/10.1371/journal.pone.0198475.*

TABLE 10.3 Descriptors included in odor clusters and a common name for each cluster index.

Cluster index	Odor descriptors clustered	Common name given
1	Hazelnut and sage	–
2	Apricot, jam, musty, cheesy, meaty, berry, quince, tomato, burnt, earthy, ethereal, civet, soapy, raspberry, and seedy	Vegetables and Grassy
3	Minty, creamy, tart, orchid, leather, and pineapple	–
4	Peppermint, strawberry, anise, medicinal alcohol, faint, floral, fishy, waxy, herbaceous, leafy, cucumber, acidic, potato, nutty, odorless, mild, warm, fennel, and fruity	Fruity and grassy
5	Caramel, pear, blueberry, tallow, grape, sulfurous, coumarin, gardenia, sour, spearmint, almond, grassy, cinnamon, vanilla, cantaloupe, chocolate, vegetable, horseradish, coffee, geranium, cranberry, celery, fatty, cabbage, tobacco, apple, peanut, alliaceous, pungent, honey, blossom, spicy, lemon, pepper, carnation, oily, melon, mossy, peach, plum, clover, mango, balsamic, pine, sweet, lime, clean camphoraceous, tropical, citrus, coriander, vinegar, milk, butter, cheese, brandy, lard, banana, baked, roasted, grapefruit, wine-like, coconut, beef, rum, and whiskey	Fruity, spicy, and animals
6	Balsam, hyacinth, caraway, rose, mushroom, woody, iris, violet, jasmine, lily, orange, garlic, lavender, lilac, beer, antiseptic, rancid, clove, bitter, plastic, cherry, paper, green, smoky, cedar, maple, phenolic, walnut, hawthorne, onion	Floral and grassy

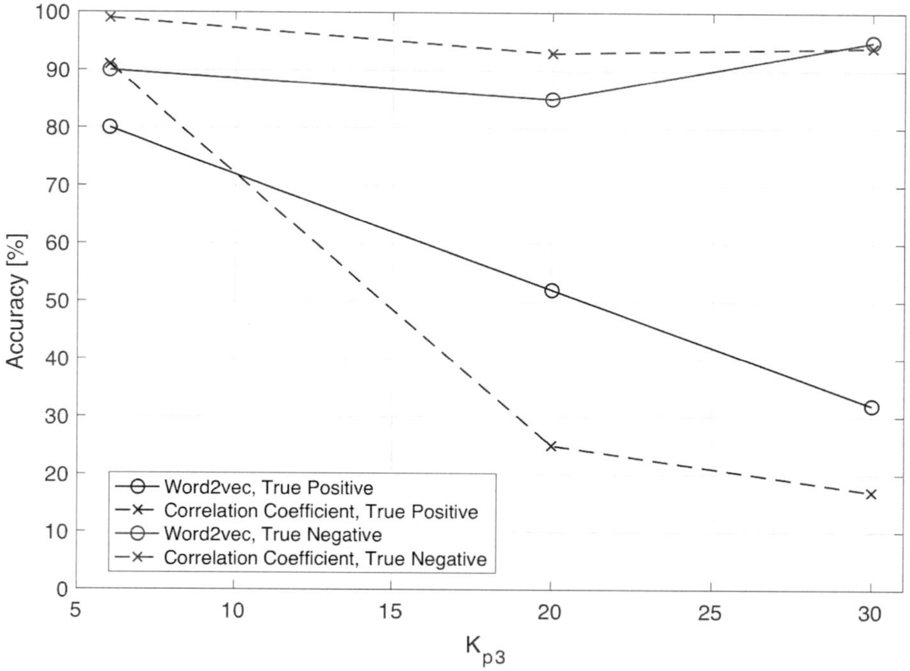

FIGURE 10.6 Accuracy of the predictive model. Accuracy comparison between NLP-based model and correlation-based one. *NLP*, Natural language processing. *Reproduced with permission from: Nozaki, Y., & Nakamoto, T. (2018). Predictive modeling for odor character of a chemical using machine learning combined with natural language processing.* PLoS ONE, 13*(6). https://doi.org/10.1371/journal.pone.0198475.*

10.4 Case study 2: odor impression prediction with continuous values from the score with binary sensory data

In above case study 1, we described a method to calculate the correlation coefficient among odor descriptors only using binary data, and this method was compared with NLP method word2vec to predict odor characters of chemicals. Since the limitation creates for binary representation data (described in Section 10.2.2), we proposed a method to convert those binary values to continuous sensory information.

This work is an extended version of case study 1 where we proposed the method to obtain scent impression scores with continuous values from the score with binary value. Here, the method using continuous value-based correlation coefficient is compared with NLP method fastText (an extension of word2vec) (Bojanowski et al., 2017). Although the correlation coefficient based on binary value was also used in case study 1, it was different from the case study 2, which is based on continuous value. We replaced the correlation coefficient with continuous value-based correlation coefficient in our case study 2 to clarify this point. We depicted two proposed approaches to obtain odor descriptor group in Fig. 10.7.

Moreover, previous work did not describe any procedure to overcome the imbalanced ratio of positive (minority samples with specific odor descriptor group) and negative samples (majority samples with nonspecific odor descriptor group) that influenced the model predictive performance. Synthetic minority oversampling technique (SMOTE) (Chawla et al., 2002) was used to balance the positive and negative samples for both correlation similarity and word similarity-based odor descriptor group prediction. It is a statistical method for increasing the number of training instances in the dataset for balancing the positive and negative samples.

10.4.1 Dataset

Mass spectrum dataset of essential oils collected by us using the gas chromatography–mass spectrometry [Agilent technologies, 7890 GC system, and 5977 MSD using electron ionization (70 eV)]. As this work is concentrated only on mass spectrum, the column without

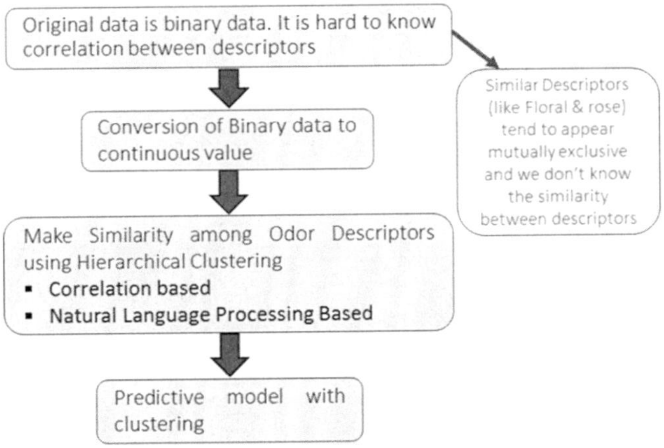

FIGURE 10.7 Two proposed approaches to obtain odor descriptor groups. Correlation-based and NLP-based methods. *NLP*, Natural language processing. *Reproduced with permission from: Debnath, T., & Nakamoto, T. (2020). Predicting human odor perception represented by continuous values from mass spectra of essential oils resembling chemical mixtures.* PLoS ONE, *15(6). https://doi.org/10.1371/journal.pone.0234688.*

coating was used in this experiment. A total of 96 mass spectra of essential oils can be divided into seven groups: "Herbal," "Citrus," "Floral," "Exotic," "Resin," "Spicy," and "Woody" according to "The Directory of Essential Oil" by Sellar (1992). About 50–250 m/z intensities were extracted from the original data and used them for these experiments because ethanol (MW: 46) was used as solvent, and intensities at m/z below 50 mainly originate from odorless molecules such as oxygen, nitrogen, and intensities at high m/z originate from molecules with low volatility. The data matrix is expressed by rows of 96 samples and columns of 201 intensities.

10.4.2 Method to create continuous values from binary sensory data for odor descriptor

Even if the original dataset contains only binary values, we aim to have a dataset with continuous values. After preparing odor descriptors for each essential oil, a table was then created for 96 essential oils assigned to one of the 7 essential oil groups and examining how often a descriptor appeared in one of each essential oil group. Thus we obtained a seven-dimensional vector for each odor descriptor. We got total 38-odor descriptors and made seven-dimensional vectors for 38-odor descriptors. Procedure of making seven-dimensional vectors for 38-odor descriptors has been shown in Fig. 10.8.

10.4.3 Odor descriptor clustering

10.4.3.1 Odor descriptor clustering using correlation similarity matrix

One essential oil can be described with several descriptors, and it is necessary to consider the similarities among those 38-odor descriptors of 7 essential oil groups. For making the clustering of similar odor descriptors by using correlation coefficient matrix, several odor descriptor groups are possible. Total five-odor descriptor groups were considered in this study. Fig. 10.9 shows a hierarchical structure appearing in the dendrogram obtained from the correlation coefficient matrix. The horizontal axis of the dendrogram represents the dissimilarity between descriptors. Therefore the number of clusters to be separated and descriptors contained in each cluster are determined at a cutoff point.

10.4.3.2 Odor descriptor clustering using word similarity matrix

FastText model was used to assess accurately the semantic similarity among odor descriptors. Pretrained word vectors trained on English Wikipedia using fastText were used. A total of 300-dimensional word vectors were obtained using the skip-gram model described in Bojanowski et al. (2017) with parameters mentioned in Table 10.2. English pretrained word vector model was downloaded (4.20 GB), and there are 2000,000 tokens (words) available. After that, the cosine similarity of 38-odor descriptors was calculated using pretrained English wiki fastText model. Finally, hierarchical clustering was carried out based on cosine similarity matrix. Descriptors, such as oily, fatty, earthy, pungent, peppery, and spicy, clustered closely with each other in Fig. 10.9B than continuous value-based correlation coefficient matrix in Fig. 10.9A as those descriptors are divided into several clusters.

10.4.4 Predictive model and results

By following Fig. 10.3 in Section 10.2.4, we prepared the target vector for our predictive model. For example, when "orange," "citrus," and "lemon" belong to certain cluster, a sample having at least one of these descriptors in the original descriptor table of essential oils was regarded as the odor character of that cluster.

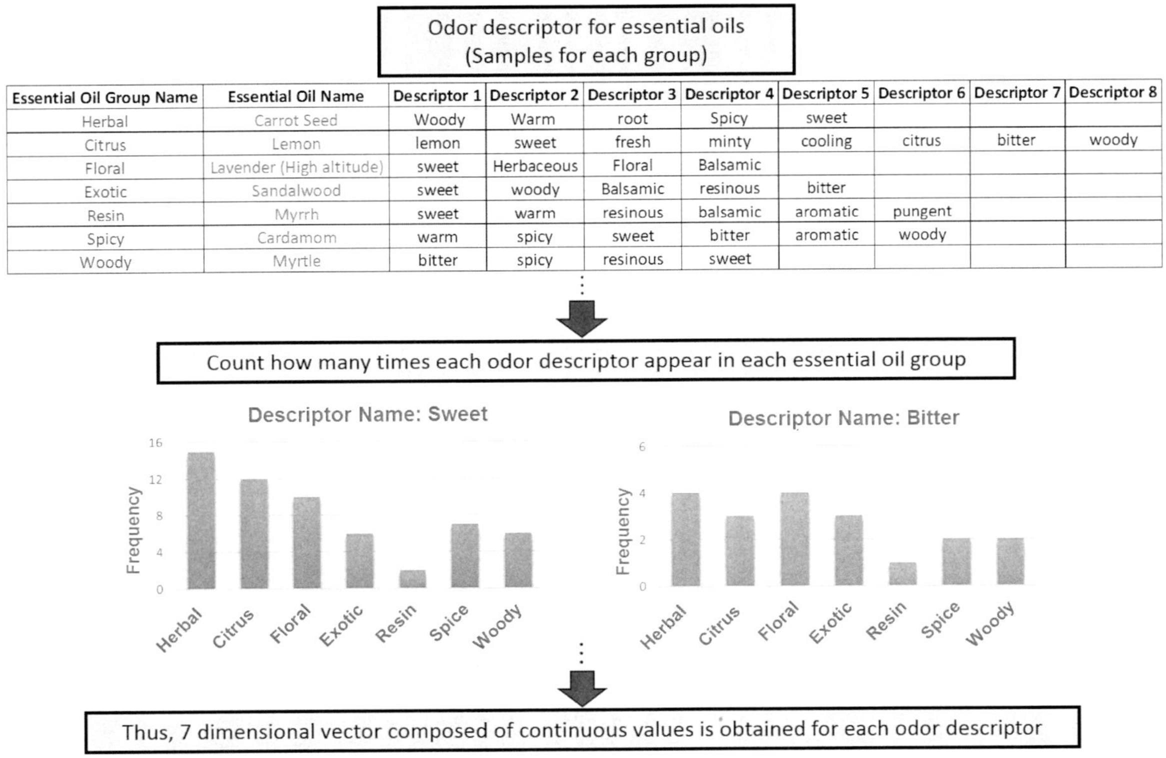

Essential Oil Group Name	Essential Oil Name	Descriptor 1	Descriptor 2	Descriptor 3	Descriptor 4	Descriptor 5	Descriptor 6	Descriptor 7	Descriptor 8
Herbal	Carrot Seed	Woody	Warm	root	Spicy	sweet			
Citrus	Lemon	lemon	sweet	fresh	minty	cooling	citrus	bitter	woody
Floral	Lavender (High altitude)	sweet	Herbaceous	Floral	Balsamic				
Exotic	Sandalwood	sweet	woody	Balsamic	resinous	bitter			
Resin	Myrrh	sweet	warm	resinous	balsamic	aromatic	pungent		
Spicy	Cardamom	warm	spicy	sweet	bitter	aromatic	woody		
Woody	Myrtle	bitter	spicy	resinous	sweet				

FIGURE 10.8 Method to create vector made up of continuous values. Procedure to make 7-dimensional continuous vector from 38 binary odor descriptors. (Debnath & Nakamoto, 2020). *Reproduced with permission from: Debnath, T., & Nakamoto, T. (2020). Predicting human odor perception represented by continuous values from mass spectra of essential oils resembling chemical mixtures.* PLoS ONE, 15*(6). https://doi.org/10.1371/journal.pone.0234688.*

The dimension of the feature vector was set to 20 through an optimization process, where the residual between the input and output was evaluated by sixfold cross-validation. After the original mass spectra of 201 dimensions are converted into feature vectors of 20 dimensions, each feature vector is fed to the four-layer multilayer perceptron that gives odor descriptor group as the desired output. The mass spectra data was randomly separated into 6 subsets, 5 of which (80 samples) were used for the model, and other 16 samples were used for the evaluation of the model. The cross entropy (it takes the output probabilities and measures the distance from the actual values) was used as the error function of the neural networks, and a regularization term was added to the error function to avoid overfitting of the model. A sigmoid function was used as the activation function in the hidden layer and the output layer of both the autoencoder and the mapping neural network. Training was carried out by error backpropagation using scaled conjugate gradient (Møller, 1993). After reducing the dimension of the original mass spectra from 201 to 20, then the neural network was trained to output odor descriptor group from a feature vector. The final output of the neural network is converted to 0 or 1 by a soft-max function (Goodfellow et al., 2016) (it is a function placed at the end of the neural network model to convert the output into classification probabilities).

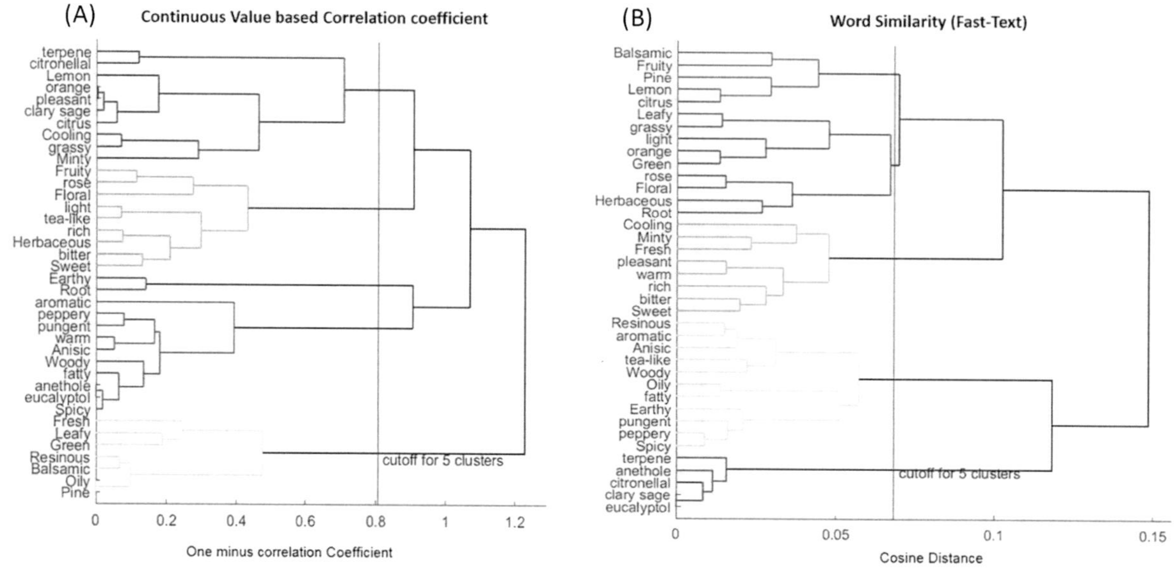

FIGURE 10.9 Odor Descriptors clustering. Dendrograms (unweighted average distance method) based on (A) one minus the sample correlation between two-odor descriptors, (B) cosine distance based on fastText modeling. The vertical line in both figures indicates the cut off point for five clusters. (Debnath & Nakamoto, 2020). *Reproduced with permission from: Debnath, T., & Nakamoto, T. (2020).Predicting human odor perception represented by continuous values from mass spectra of essential oils resembling chemical mixtures.* PLoS ONE, 15*(6). https://doi.org/10.1371/journal.pone.0234688.*

The number of odor descriptor group K_{p3} must be determined carefully, since the distribution of samples strongly affects the performance of the predictive model. "True positive" indicates the rate at which the predicted output is 1 when the desired output is 1. The predictive model tends to output "0" because most of the desired outputs were zeros (imbalanced dataset with respect to positive and negative targets), resulting in lower accuracy of true positives. This was improved using SMOTE technique for balancing the imbalanced dataset (Debnath & Nakamoto, 2020).

We reported the results of the true positive rate before and after SMOTE in Table 10.4. When we used only 58 original training datasets, we got only 33.44% and 63.68% true positive rates for the continuous value-based correlation coefficient and word similarity-based clustering model, respectively, for five-odor perception groups. Prediction performance was improved to 58.82% and 71.14% for continuous value-based correlation coefficient model and word similarity-based clustering, respectively, after performing SMOTE on the training dataset. This work showed the feasibility of using machine learning combined with NLP and SMOTE to predict odor perception groups from chemical mixtures (essential oils) and binary sensory data.

In addition to the above works, an IBM research group (Gutiérrez et al., 2018) showed that natural language semantic representations can be applied to a small number of olfactory perceptual descriptors to accurately infer the perceptual ratings of monomolecular odorants for a large and potentially arbitrary set of descriptors.

TABLE 10.4 Comparison between the true positive results before and after SMOTE (Debnath & Nakamoto, 2020).

	Correlation coefficient model (%)	Word similarity model (%)
Original data	33.44	63.68
SMOTE	58.82	71.14

SMOTE, Synthetic minority oversampling technique.
Reproduced with permission from: Debnath, T., & Nakamoto, T. (2020).Predicting human odor perception represented by continuous values from mass spectra of essential oils resembling chemical mixtures. PLoS ONE, 15(6). https://doi.org/10.1371/journal.pone.0234688.

10.5 Summary

NLP is the subfield of machine learning domain that enables computers to understand the human language. In this chapter, we described the procedure for odor descriptor clustering using NLP. Initially, we demonstrated the difference between the binary and continuous sensory information followed by the limitation of the binary odor descriptors. Most importantly, we showed the procedure for the conversion to continuous sensory information from binary odor descriptors. Here, we used a six-layer neural network that was used in both case studies for predicting the odor descriptor group. In summary, this chapter shows the fundamental methods to predict the odor impression using linguistic approach. We hope that more research will be done using advanced NLP methods such as Bidirectional Encoder Representations from Transformers (Devlin et al., 2019).

References

Arshamian, A., Gerkin, R. C., Kruspe, N., Wnuk, E., Floyd, S., O'Meara, C., Garrido Rodriguez, G., Lundström, J. N., Mainland, J. D., & Majid, A. (2022). The perception of odor pleasantness is shared across cultures. *Current Biology.*, *32*(9), 2061–2066. Available from https://doi.org/10.1016/j.cub.2022.02.062, e3, http://www.elsevier.com/journals/current-biology/0960-9822.

Bojanowski, P., Grave, E., Joulin, A., & Mikolov, T. (2017). Enriching word vectors with subword information. *Transactions of the Association for Computational Linguistics*, *5*, 135–146. Available from https://doi.org/10.1162/tacl_a_00051.

Borg, I., & Groenen, P. (2003). Modern multidimensional scaling: Theory and applications. *Journal of Educational Measurement.*, *40*(3), 277–280. Available from https://doi.org/10.1111/j.1745-3984.2003.tb01108.x.

Chawla, N. V., Bowyer, K. W., Hall, L. O., & Kegelmeyer, W. P. (2002). SMOTE: Synthetic minority over-sampling technique. *Journal of Artificial Intelligence Research*, *16*, 321–357. Available from https://doi.org/10.1613/jair.953.

Cun, Y. L., Boser, B., Denker, J. S., Howard, R. E., Habbard, W., Jackel, L. D., & Henderson, D. (1990). Handwritten digit recognition with a back-propagation network. *Advances in Neural Information Processing Systems*, *2*.

Debnath, T., & Nakamoto, T. (2020). Predicting human odor perception represented by continuous values from mass spectra of essential oils resembling chemical mixtures. *PLoS ONE*, *15*(6). Available from https://doi.org/10.1371/journal.pone.0234688.

Devlin, J., Chang, M.W., Lee, K., & Toutanova, K. (2019). *BERT: Pre-training of deep bidirectional transformers for language understanding*. NAACL HLT 2019—2019 Conference of the North American Chapter of the Association for Computational Linguistics: Human Language Technologies—Proceedings of the Conference. (Vol. 1, pp. 4171–4186). Association for Computational Linguistics (ACL) undefined.

Dravnieks, A. (1985). Atlas of odor character profiles.

Leffingwell & Associates. (n.d.). *Flavor-base 10*. http://www.leffingwell.com/flavbase.htm.

Goodfellow, I., Bengio, Y., & Courville, A. (2016). *Deep learning*. MIT Press.

Gutiérrez, E. D., Dhurandhar, A., Keller, A., Meyer, P., & Cecchi, G. A. (2018). Predicting natural language descriptions of mono-molecular odorants. *Nature*

Communications, 9(1). Available from https://doi.org/10.1038/s41467-018-07439-9, http://www.nature.com/ncomms/index.html.

Hinton, G. E., & Salakhutdinov, R. R. (2006). Reducing the dimensionality of data with neural networks. *Science (New York, N.Y.), 313*(5786), 504–507. Available from https://doi.org/10.1126/science.1127647.

Keller, A., & Vosshall, L. B. (2016). Olfactory perception of chemically diverse molecules. *BMC Neuroscience, 17*(1). Available from https://doi.org/10.1186/s12868-016-0287-2, http://www.biomedcentral.com/bmcneurosci/.

LeCun, Y., Bengio, Y., & Hinton, G. (2015). Deep learning. *Nature, 521*(7553), 436–444. Available from https://doi.org/10.1038/nature14539.

Mikolov, T., Chen, K., Corrado, G., & Dean, J. (2013). *Efficient estimation of word representations in vector space*. First International Conference on Learning Representations, ICLR 2013—Workshop Track Proceedings. ICLR, United States. https://dblp.org/db/conf/iclr/iclr2013w.html.

Mikolov, T., Sutskever, I., Chen, K., Corrado, G., & Dean, J. (2013). *Distributed representations of words and phrases and their compositionality*. Proceedings of the 26th International Conference on Neural Information Processing Systems (Vol. 2).

Møller, M. F. (1993). A scaled conjugate gradient algorithm for fast supervised learning. *Neural Networks., 6*(4), 525–533. Available from https://doi.org/10.1016/s0893-6080(05)80056-5.

NIST Chemistry WebBook. (n.d.). *NIST standard reference database number 69*. http://webbook.nist.gov/chemistry/.

Nozaki, Y., & Nakamoto, T. (2018). Predictive modeling for odor character of a chemical using machine learning combined with natural language processing. *PLoS ONE, 13*(6). Available from https://doi.org/10.1371/journal.pone.0198475.

Sellar, W. (1992). The directory of essential oils.

Sigma-Aldrich. Flavors and fragrances. n.d. https://www.scribd.com/doc/121621778/Flavors-and-fragrances-catalog.

Suffet, I. H., Brady, B. M., Bartels, J. H. M., Burlingame, G., Mallevialle, J., & Yohe, T. (1988). Development of the flavor profile analysis method into a standard method for sensory analysis of water. *Water Science and Technology, 20*(8–9), 1–9. Available from https://doi.org/10.2166/wst.1988.0217.

Szekely, G. J., & Rizzo, M. L. (2005). Hierarchical clustering via joint between-within distances: Extending Ward's minimum variance method. *Journal of Classification., 22*(2), 151–183. Available from https://doi.org/10.1007/s00357-005-0012-9.

The good scents company—flavor, fragrance, food, and cosmetics ingredients information. (n.d.). https://www.thegoodscentscompany.com/.

The Wikipedia Corpus. (2017). https://corpus.byu.edu/wiki/.

CHAPTER

11

Odor reproduction using odor components

Dani Prasetyawan

Institute of Integrated Research, Institute of Science Tokyo, Yokohama, Japan

11.1 Introduction

An odor component composed of chemical compounds plays a significant part in how we perceive a smell or fragrance. Usually, these components are small molecules with a high propensity for evaporation, enabling them to become airborne and stimulate our sense of smell through our olfactory receptors. An odor component could be a single molecule or a blend of various molecules, and each odor component contributes to a unique scent profile. The molecular structure of an odorant molecule, including its size, shape, and the functional groups it contains, significantly influences its odor characteristics. These factors dictate how odorant molecule interacts with our olfactory receptors, thereby shaping our perception of the scent (Grabe & Sachse, 2018; Nakamoto, 2012; Nakamoto & Murakami, 2009; Tu et al., 2002).

In practice, it is common to create new smells by changing the blending ratio and the number of odor components included during odor approximation. However, approximating the same odor as we sense is an uncommon practice. Approximating odor to the target odor is called odor reproduction (Liu et al., 2021; Nakamoto, 2016a). An illustration of how odor reproduction is performed is shown in Fig. 11.1. The similarity between the target and approximated odor is essential in odor reproduction. Therefore an approximated odor in odor reproduction should be similar to the target odor. The procedure of odor reproduction is similar to visual displays, which display pictures or videos with the same color as it is recorded.

There is a distinguished feature in the visual sense in reproducing different colors compared to the sense of smell. The visual sense has a set of "primary colors" that allows the reproduction of a wide range of colors. On the other hand, in sense of olfaction, no "primary odors" are yet found (Mamlouk et al., 2017). The absence of "primary odor" makes odor reproduction challenging. Moreover, in the visual sense, the "primary colors" are separable between the wavelengths. In the sense of olfaction, the study of smell requires chemistry, biology, and sensory science, which adds complexity to finding "primary odors" (Genva et al., 2019; Snitz et al., 2013).

It is important to note that it is difficult to achieve an exact match of odor. The human

Digital Technologies in Olfaction
DOI: https://doi.org/10.1016/B978-0-443-15721-9.00011-7

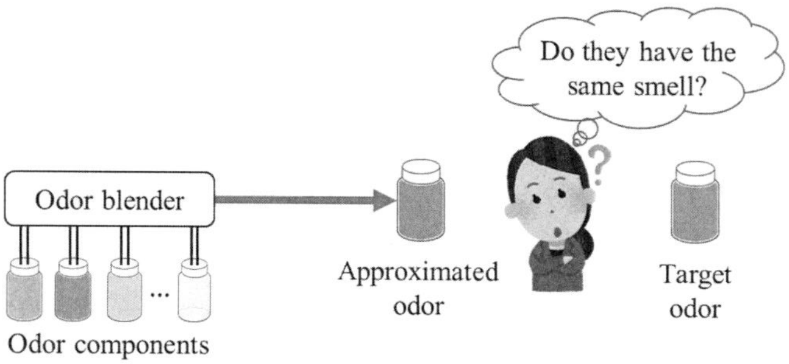

FIGURE 11.1 Illustration of blending odor component to approximate odor for odor reproduction. The similarity between target odor and approximated odor is important.

olfactory system is highly sensitive, and small variations in the composition or concentration of odor components can lead to significant differences in perception. Moreover, some natural odors may contain hundreds of individual components, making it challenging to recreate them accurately. Despite the challenges in odor reproduction, the research and development of odor reproduction are being studied.

This section briefly explains several achievements by various groups in finding a set of odor components to perform odor reproduction. Different sets of odor components were developed and proposed by various groups. Some sets of odor components performed odor reproduction with the number of approximated samples less than the number of odor components. On the other hand, there is a set of odor components that can reproduce a greater number of samples than the number of odor components itself.

Researchers performed different approaches to find an appropriate set of odor components. For example, some researchers used sensor responses followed by numerical analysis to find an appropriate set of odor components and utilized sensor response again to evaluate the reproduced odor. On the other hand, several groups of researchers performed human sensory tests to validate odor reproduction performance. Nevertheless, the aim to reproduce odor as same as it is sensed using odor components is the same.

Furthermore, research related to the use of odor components is also ongoing (Liu et al., 2021; Nakamoto, 2016a). A device that can deliver certain odors is called an olfactory display by using odor components stored within the device. A detailed explanation of olfactory displays is explained separately in fundamental of olfactory display in Chapter 14 and multicomponent olfactory display in Chapter 15. In this chapter, the explanation of olfactory display for odor approximation will be briefly explained in Section 11.3.

There are a lot of practical uses of odor reproduction once we can find an appropriate set of odor components to perform odor reproduction. Using odor components stored within olfactory displays can enhance user experiences in various fields, such as gaming, virtual reality, movies, marketing, and education, by integrating the sense of smell into these experiences (Kato & Nakamoto, 2019; McMillan, 2020). This makes the experience more immersive and realistic. For example, we could smell the ocean while playing a beach-themed video game or smell freshly baked bread in a virtual bakery.

11.2 Finding an appropriate set of odor components for odor reproduction

As previously stated, color reproduction is relatively straightforward and easily performed due to the existence of primary colors. However, odor reproduction is different from color ones, as no definitive primary odors exist. Therefore reproducing a particular scent as perceived remains a significant challenge (Prasetyawan & Nakamoto, 2023). For instance, we wish to share a specific fragrance from our environment with a distant relative; recreating that exact scent elsewhere is difficult because there are no primary odors to aid in the process. Hence, finding an appropriate set of odor components to perform odor reproduction becomes essential.

Determining the exact number of odor components that can act as "primary odors" is challenging. In the case of visual sense, the number of primary colors can easily be determined by the basic wavelength that the receptors in our eyes can sense. However, suppose we choose the all types of olfactory receptors as the primary odors (Malnic et al., 2004; Secundo et al., 2015). Since the number of olfactory receptors is approximately 400, it would be impractical. Hence, many researchers tried to find fewer candidates for a set of odor components for practical applications. Different approaches by researchers lead to the finding of several "representative" sets of odor components for odor reproduction. Most of the study related to odor components and odor reproduction explained in this section is from numerical analysis followed by the reproduction of the actual sample for validation.

11.2.1 Early study on odor reproduction

An early study of finding a set of odor components for odor reproduction was performed by Amoore (1967) during the 1960s. The author suggested that seven distinct odors (camphoraceous, pungent, ethereal, floral, pepperminty, musky, and putrid) might serve as standards for odor qualities, potentially hinting at the existence of primary odors. Amoore's proposal was based on the stereochemical structure of the odor molecule, as depicted in Fig. 11.2, implying that the shape and size of an odorant should align with its receptors.

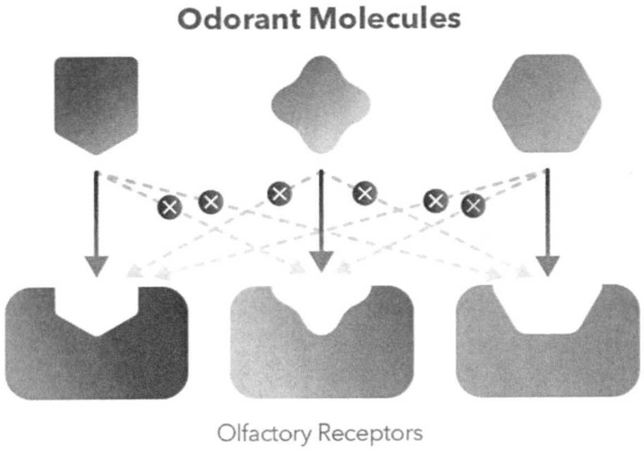

FIGURE 11.2 **Illustration of stereochemical structure of odorant and its relation with odorant receptors.** The shape and size of an odorant should align with its receptors for docking.

However, further study revealed that these primary odors were not enough to represent all possible odor characteristics due to smell interference caused by of specific anosmia. This suggests the potential existence of a larger set of primary odors. Furthermore, the physiological responses and their perception might be different in each individual human (Koulakov, 2011; Trimmer et al., 2019). Although this early research could not cover an extensive spectrum of odors, it paved the way to the subsequent studies, offering valuable insights into the investigation of odor space and primary odors.

11.2.2 Further studies in odor reproduction analysis

Other research regarding finding a set of odor components for odor reproduction was performed by Nicolaï et al. (2016), which tried to perform odor reproduction of citrus-related smells in real-time using odor components. Their research utilized odor samples analyzed by gas chromatography/mass spectrometry (GC/MS). Impressively, they managed to narrow down 45 odor compounds typically found in citrus scents to just 12 odor components for an approximation of five citrus varieties (lime, lemon, orange, grapefruit, and mandarin). The theoretical classification accuracy of their approximated odors was 80%, with an actual classification accuracy rate of 73%. They utilized partial least squares-discriminant analysis for the classification of odor compounds in this study. Despite the approximation requiring more odor components than those present in the target scent, their level of accuracy was deemed acceptable.

An interesting approach was also proposed by Xu et al. They discussed the concept of "quasi" primary odors as an alternative approach in the absence of definitive primary odors (Xu et al., 2021). The proposed method involves selecting quasiprimary odors from within the collected odor dataset (Nakamoto & Murakami, 2009) and using machine learning for endpoint detection (Nakamoto, 2012). Endpoint detection method is a method that examines the data points that are located in the edge of the dimensional space featured in the study. These quasiprimary odors are chosen based on their endpoint scores within the collected dataset. The odor dataset utilized in their study consisted of 12 different odors (fish sauce, cooking sake, pure water, mentsuyu, soy sauce, grain vinegar, lemon juice, ketchup, mayonnaise, grilled meat sauce, oyster sauce, and Worcestershire sauce). Despite the limitation of being confined to the dataset, any other odor in the dataset can be recreated using a combination of three quasiprimary odors (fish sauce, cooking sake, and pure water) in certain ratios. To visually illustrate this concept, the authors use the top three quasiprimary odors, assigning them to the vertices of a chromaticity triangle with red, green, and blue. Other odors within the dataset are then projected onto this triangle, giving them unique colors. The 12 odor samples, projected in three quasiprimary odor color map by Xu et al. is shown in Fig. 11.3. While the number of quasiprimary odors is not restricted to three and can be arbitrarily set, this approach allows the identification of "extreme" odors within a given dataset. These "extreme" odors are the representative of the "quasiprimary odors." Consequently, any other odor in the dataset can be reproduced, or a new arbitrary odor can be synthesized by combining these quasiprimary odors in specific ratios.

On the other hand, the Nakamoto group did another research regarding finding a set of odor components. The early study by Yamanaka

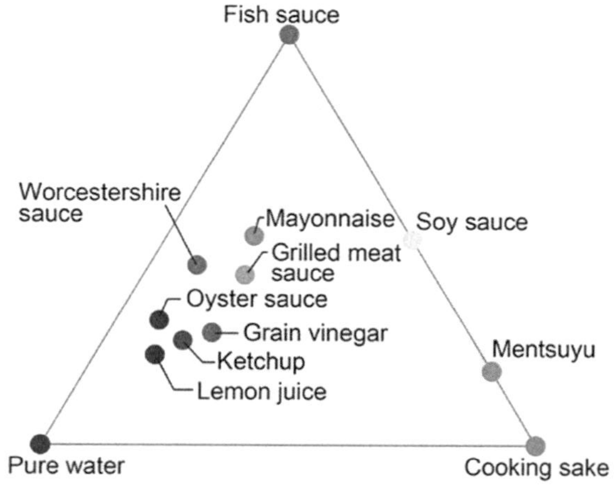

FIGURE 11.3 Color map of projection of pure water and 11 seasonings. Fish sauce, cooking sake, and pure water were selected as quasiprimary odors and represented by red, green, and blue colors. Other odor samples can be reproduced using the three quasiprimary odors and represented in different colors, showing the blending ratio of the three quasiprimary odors. Here, the position of each sample within the map represents the composition of each quasiprimary odors needed to be mixed to reproduce the sample. *Source: From Xu, H., Kitai, K., Minami, K., Nakatsu, M., Yoshikawa, G., Tsuda, K., Shiba, K., & Tamura, R. (2021). Determination of quasi-primary odors by endpoint detection.* Scientific Reports, *11(1). https://doi.org/10.1038/s41598-021-91210-6.*

et al. from the Nakamoto group proposed an odor recorder system aimed at sensing the target odor and reproducing the smell. Detailed explanations of how the odor recorder worked will be explained in Section 11.3. Yamanaka, Matsumoto, et al. (2003) and Yamanaka, Wyszynski et al. (2003) performed odor reproduction of orange and apple-like smell with an odor recorder. They also successfully reduced the number of odor components for odor reproduction. For example, typically used 14 odor components to reproduce orange-like smells were successfully reduced to three odor components. On the other hand, only five odor components were needed to reproduce apple smells. However, odor components for different samples are not always common.

11.2.3 Mass spectrometry–based odor reproduction analysis

The early study of odor reproduction by the Nakamoto group needed a greater number of odor components than the number of reproduced odors. Further study by the same group further refined the odor approximation technique. They increased the number of odor samples for odor reproduction using 158 essential oils (Nakamoto et al., 2012; Nakamoto & Nihei, 2013). Each essential oil had its own smell characteristic originating from plants. Hence, odor reproduction in this study is not restricted to certain groups of smells. They obtained MS data of 158 essential oils to perform odor reproduction analysis. They utilized GC/MS to obtain the MS of essential oils. Despite they did not use GC part for the study, the column in the GC cannot be eliminated because of the need to maintain large pressure difference between inside and outside of the Ms. Their idea for odor reproduction was utilizing mass spectrum. If we can approximate an odor with the same mass spectrum as the target odor, we can reproduce the same smell. The example of MS of odorant samples for odor reproduction is illustrated in Fig. 11.4.

They successfully approximated the mass spectra of 158 essential oils by numerical simulation. They also approximated real samples utilizing a set of 30 and 12 odor components. They also performed classification test on the 172 subjects using three samples (organic orange, peppermint, and black pepper) to validate the performance of odor reproduction. Additional sensory tests (triangle test) also performed using the blended essential oils by them to evaluate the odor approximation accuracy. As a result, the target odors were satisfactorily reproduced using 30 odor components,

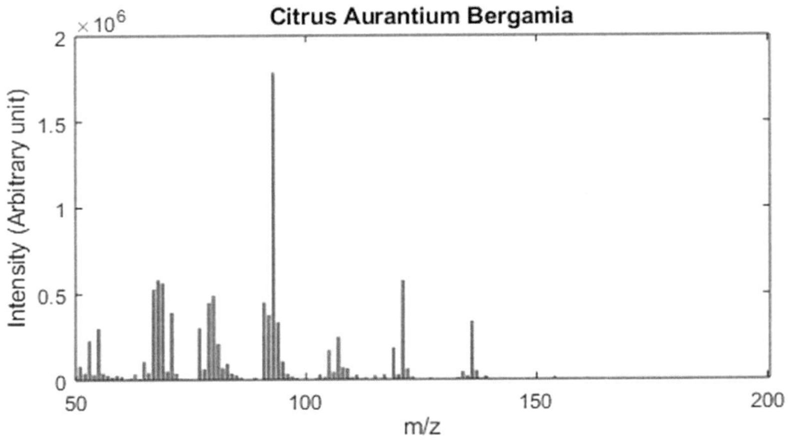

FIGURE 11.4 Example of mass spectrometry of odorant sample. Each odorant has its own pattern of mass spectrometry. *Source: Obtained from Prasetyawan D.*

whereas a set of 12 odor components were not enough for odor reproduction.

The latest odor reproduction analysis by Prasetyawan et al. from the group of Nakamoto utilized an increased number of odorant samples, which is 185 essential oils (Prasetyawan & Nakamoto, 2019; Prasetyawan & Takamichi, 2020). Here, they incorporated tried to reduce the number of odor components by using different cost functions during numerical simulations of odor reproduction analysis. A more detailed explanation of numerical simulations will be provided in Chapter 12. Here, they found two sets of 20 and 10 odor components were sufficient for odor reproduction during the numerical simulation in the MS data. Examples of comparison of approximated and target odors in odor reproduction using MS are shown in Figs. 11.5 and 11.6. The comparison shows that the high similarity of MS patterns between approximated and target odor is achieved in odor reproduction. The average correlation coefficient between target and approximated odor for 185 essential oils samples was 0.971 showing high similarity between target and approximated odor. The study was also followed by sensory evaluation with the human as subjects for the validation of odor reproduction performance. The sensory test of odor approximation of 20 odor components was performed by Iseki et al. by utilizing a multicomponent olfactory display that will be explained in Section 11.3.

The sensory test of the duo–trio test was conducted by Prasetyawan et al. from the Nakamoto group using odors reproduced from 10 odor components. The duo–trio test is a sensory test in which the subject is required to

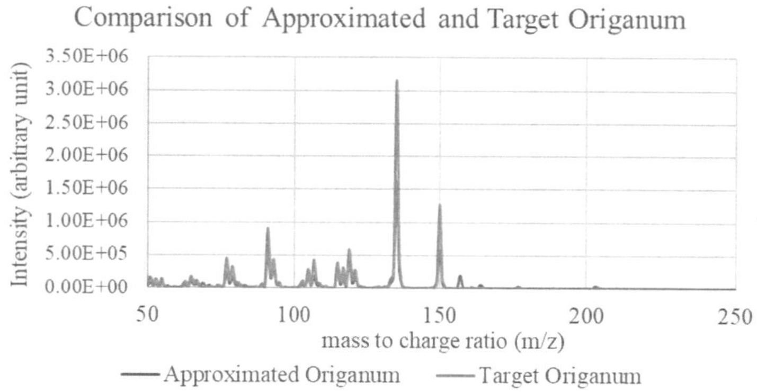

FIGURE 11.5 Mass spectrometry comparison of approximated and target odor samples. Odorant samples of *Origanum*, which is approximated using a set of 20 odor components, are shown. *Source: Obtained from Prasetyawan D.*

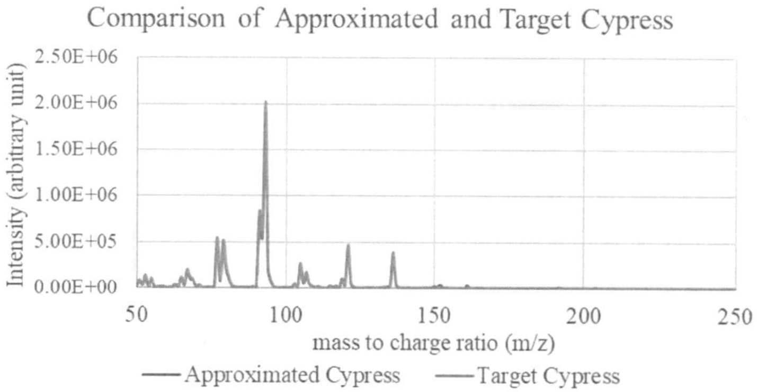

FIGURE 11.6 Mass spectrometry comparison of approximated and target odor samples. Odorant samples of Cypress, which is approximated using a set of 20 odor components, are shown. *Source: Obtained from Prasetyawan D.*

choose whether the given two samples are closer to the target odor. The sensory test result by Prasetyawan et al. is detailed in Table 11.1. The test involved six samples representing different essential oils groups. They used the z score to evaluate the results of the sensory test. Here, z score representing how close the tested sample with the target sample. For the evaluation of their study, the z score of more than "$+2$" means the approximated odor based on Itakura–Sato divergence is significantly closer to the target odor, whereas the z score of less than "-2" means that approximated odor by Kullback–Leibler (KL) divergence is closer to the target odor. Their study showed that Itakura–Saito divergence during numerical calculations (NMF-IS) improved the reconstruction of 185 essential oils using the set of 10 odor components compared to KL divergence (NMF-KL), as reflected in the total z score of 2.417.

Conclusions from the studies from the Nakamoto group, including their methodology for identifying a set of odor components for odor reproduction, are shown in Fig. 11.7.

TABLE 11.1 Sensory test result (duo–trio sensory test) from Prasetyawan et al. for evaluation of a set of 10 odor components for odor reproduction.

	Number of subjects	Number of subjects choose NMF-IS	Number of subjects choose NMF-KL	
Orange	24	19	5	2.654
Mint	24	20	4	3.062
Origanum	24	12	1	−0.204
Ylang-ylang	24	18	6	2.245
Clove	24	11	13	−0.612
Cypress	24	7	17	−2.245
Total	144	87	57	2.417

Source: From Prasetyawan, D., & Takamichi, N. (2020). Sensory evaluation of odor approximation using NMF with Kullback–Leibler divergence and Itakura–Saito divergence in mass spectrum space. Journal of the Electrochemical Society, 167(16). https://doi.org/10.1149/1945-7111/abd110.

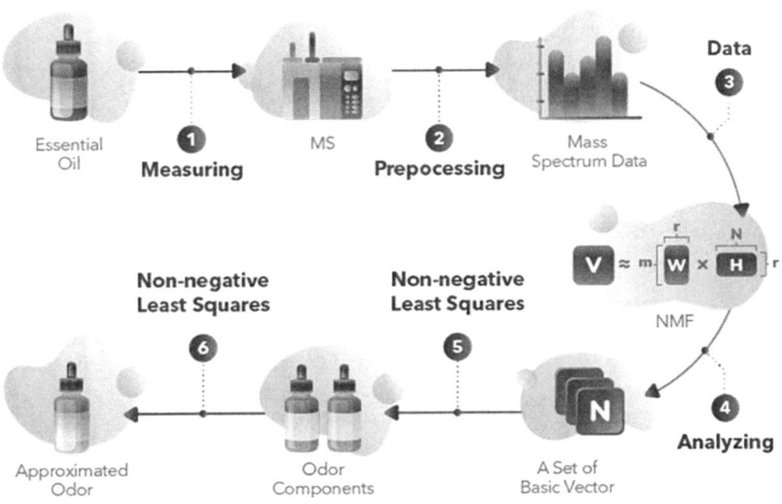

FIGURE 11.7 Procedure to find odor components for odor approximation in odor reproduction by group of Nakamoto et al. The number of odor components and approximated odor can easily be adjusted. *Source: From Prasetyawan, D., & Takamichi, N. (2020). Sensory evaluation of odor approximation using NMF with Kullback–Leibler divergence and Itakura–Saito divergence in mass spectrum space.* Journal of the Electrochemical Society, 167*(16). https://doi.org/10.1149/1945-7111/abd110.*

First, the mass spectra of odor samples are obtained and preprocessed. Then, a large-scale mass spectrum database can be obtained and processed with an appropriate method (such as NMF) to find a set of basis vectors. This set of basis vectors is used as the base of odor components for odor reproduction. However, to make actual odor components additional method such as nonnegative least squares method are needed to calculate the recipe to approximate odor components. Once a set of actual odor components is made, additional nonnegative least squares method is performed to obtain recipe of actual odor approximation using actual odor components. Their approach allows a different quantity of odor sample and a number of odor components used for odor reproduction.

11.2.4 Other odor reproduction analysis and conclusion

A different approach was proposed by Castro et al. Here, they utilized sensory test data from Dravnieks to search for the odor primaries. As a result, their simulated model could obtain 10 odor primaries that can reproduce 144 different odors. Illustration of 144 odors reproduced by 10 odor primaries by Castro et al. is shown in Fig. 11.8. The color difference represents the intensity of each odor primary in contribution to reproduce 144 odors. However, unlike several methods proposed previously in this section in which numerical simulation was followed by actual samples for validation, Castro et al. did not perform the sensory test to validate the numerical simulation result. Nevertheless, like other research related to odor reproduction, Castro et al.'s findings were essential to give us insight into the primary odor for odor reproduction.

A brief conclusion in the relation of odor components and odor reproduction by different group mentioned above is listed in Table 11.2. Even though the exact number of "primary odors" is not yet determined, their research gave us the insight of the possibility of reproducing different odors using certain set of odor components.

11.3 Odor reproduction in olfactory display and odor recorder

The concept of an "odor recorder" is a device that can capture, store, and reproduce smells, much like a video recorder does for visual and auditory information. It is essential for an odor recorder to reproduce the target smell as its senses. The concept of an odor recorder is shown in Fig. 11.9. The flow starts with the target odor captured by the sensing system in the odor recorder. Information related to the target odor sensed by the sensing system is translated into computational information and stored in the computer (Gong et al., 2022; Nakamoto, 2016a; Yamanaka, Matsumoto, et al., 2003). Then, the computer calculates the appropriate blending recipe for odor components to reproduce the target odor. During the odor reproduction process, the

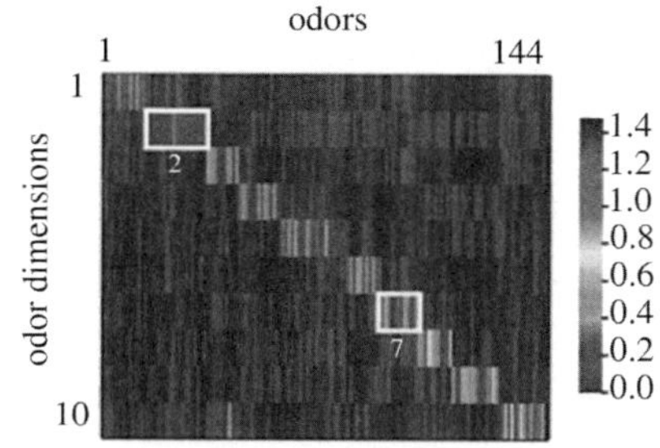

FIGURE 11.8 Illustration of 10 odor primaries in reproducing 144 odors. Each block represents each odor, and color difference representing the intensity of each odor primary in contribution to reproduce 144 odors. *Source: From Castro, J. B., Ramanathan, A., Chennubhotla, C. S., & Schaefer, A. (2013). Categorical dimensions of human odor descriptor space revealed by non-negative matrix factorization.* PLoS ONE, 8*(9). https://doi.org/10.1371/journal.pone.0073289.*

TABLE 11.2 Relation of odor components and odor reproduction by different groups.

Research groups	Number of samples	Number of odor components needed
Amoore (1967)	600 stereochemical structure of odorants	Around 7
Nicolaï et al. (2016)	45 citruses odorant compound	Around 12
Xu et al.	12 odor samples	Around 3
Nakamoto group (Prasetyawan & Nakamoto, 2023)	158–185 mass spectrum data	Around 10 to 20
Castro et al. (2013)	146 descriptors × 144 odors	Around 10

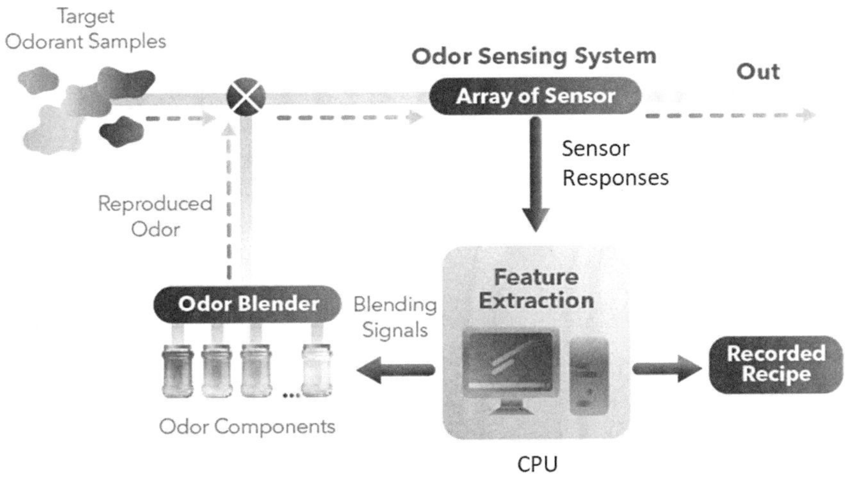

FIGURE 11.9 The concept of odor recorder by the Nakamoto group. Similarity of reproduced odor is sensed again to confirm the similarity to the target odor.

reproduced odor is sensed again to confirm the similarity to the target odor. Once the similarity is confirmed, the reproduced odor is released to the user that requested the reproduction of the odor. It is easy to say that an odor recorder is a combination of an odor sensing system and an olfactory display.

On the other hand, as mentioned before, an olfactory display is a device to generate and control scents or odors in user's environment using odor components stored within it. The number of odorants that the olfactory display can generate is at least the same as the number of odor components stored within. Although the number of odor components that can be stored within the olfactory display is limited, the number of odors that can be generated by the olfactory display can be greatly increased once we add the capability of blending odor components to the olfactory display. It is possible to approximate different odors once we have the appropriate odor components and their blending ratio.

Olfactory displays can be divided into two types: a wearable olfactory display and a desktop olfactory display. Wearable olfactory displays have advantages in their size that they are small enough and easy to be used anywhere.

However, wearable olfactory displays also have disadvantages, such as the limitation in the number of odor components that can be stored. Due to the small size of wearable olfactory displays, most of them can only store a few odor components, and the amount of odor that can be generated is also limited (Liu et al., 2021; Nakamoto, 2016b). On the other hand, the desktop olfactory display has advantages in the number of odor components that can be stored and generally can generate more odor than wearable olfactory display. However, desktop olfactory displays are not portable and need more setup for preparations. More explanation of olfactory display is provided in Chapter 14. This section will briefly explain the olfactory display that can be used for odor reproduction.

Several examples of odor recorders and olfactory displays have already developed. One example of an odor recorder was proposed by Yamanaka et al. from the Nakamoto group at the Tokyo Institute of Technology, Japan. They developed an odor recorder that could store the information and reproduce the smell of oranges and apples. Quartz crystal microbalance sensors were utilized as the odor sensing system in the odor recorder by Yamanaka, Matsumoto, et al. (2003) and Yamanaka, Wyszynski et al. (2003). The odor recorder can reproduce different types of orange- and apple-like smells using three and five odor components, respectively. They also performed a triangle test sensory test to validate the odor reproduction result. A success rate of 33.3% indicates that the odor is identical to the target one in the triangle test. The triangle test resulted in a high similarity between reproduced and target odors, with the best success rate of 38%.

On the other hand, Li et al. (2017) also proposed an odor display device that can record and reproduce five different target odors (banana, lemon, apple, tomato, and cantaloupe). They utilized 16 components to reproduce the target odors. In addition, they performed odor reproduction in different concentrations of 100, 500, and 1000 ppm. Fifty volunteers participated in the sensory test to validate the odor reproduction performance, and the odor display device can achieve the similarity between target and reproduced odor more than 90%.

Even though several odor recorders had been proposed, incorporating an odor sensing system makes it bulkier than an olfactory display. Thus more prototypes of olfactory displays have been developed than odor recorders. However, most olfactory displays only emit and reproduce the same number of odors as the number of odor components stored within them to the user. Only a handful of olfactory displays can blend odor components to approximate many types of odors.

An example of an olfactory display that can hold many odor components is "aroma-card," a card-based olfactory display developed by Kim et al. (2009). It used an "aroma chip," a card-like transparent acrylic board containing odors. A set of 15 "aroma chips" called "aroma-card" olfactory display. It can generate up to 15 different odors based on stored "aroma-chips." Illustration of "aroma-card" olfactory display is shown in Fig. 11.10. The user can select an aroma by heating a specific chip.

Examples of olfactory displays mentioned before do not have the capability to mix or blend different odor components in certain ratios to approximate different odors. An example of an olfactory display that can

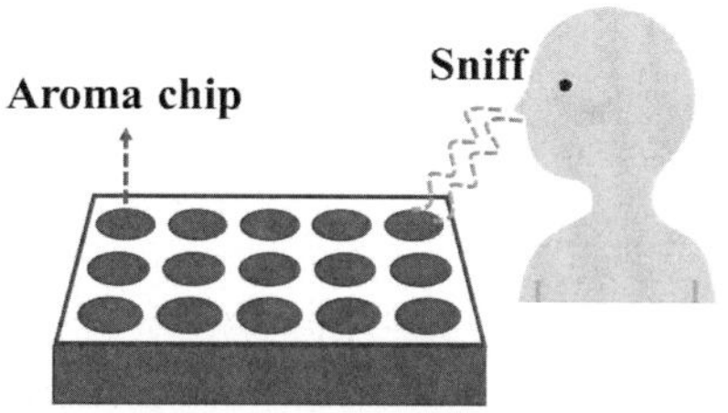

FIGURE 11.10 Illustration of the aroma-card olfactory display. The odorant is stored within aroma chips on the aroma-card olfactory display and presented to the user.

FIGURE 11.11 Multicomponent olfactory display developed by Iseki et al. A total of 20 components are stored within the olfactory display. Here, the approximated odor by 20 odor components is presented to the user. *Source: From Demo of odor reproduction using 20-component olfactory display n. (2023). April, 12. https://doi.org/10.2312/egve20221309 (Original work published in 2023).*

approximate different odors by blending odor components was proposed by Iseki et al. (2022). Iseki et al. proposed a multicomponent olfactory display that stored 20 different odor components to approximate different essential oil samples. Microdispensers utilized in the olfactory display allowed the possibility of blending odor components in different ratios. Each microdispenser can eject certain amounts of liquid independently, thus realizing different blending ratios to approximate different odors. The odorant liquid is ejected onto the surface acoustic wave (SAW) device, and then the SAW device atomizes the liquid. The atomized liquid is then directed to the user for odor presentation. Iseki et al. performed a sensory test to evaluate the odor approximation. The olfactory display developed by Iseki et al. is shown in Fig. 11.11. Their findings demonstrated that there was no notable distinction between the approximated and original samples, as evaluated by the sensory test. Its detail is described in Chapter 15.

The olfactory display proposed by Iseki et al. itself does not have the capability to calculate the blending ratio of odor components for odor approximation. The information on blending ratio is called blending recipe. Thus the blending recipe of odor components was studied a priori, and then, the information on the blending recipe was given to the olfactory display. As a result, their olfactory display had the capability to approximate 185 different essential oils. However, the range of odors that can be approximated can be increased as long as the olfactory system is given the new information of odor components blending recipe. Thus it is possible to reproduce a wider range of odors.

Moreover, the proposed olfactory display was also exhibited at the International Conference on Artificial Reality and Telexistence and Eurographics Symposium on Virtual Environments (ICAT-EGVE 2022), drawing conference participants' attention.

11.4 Conclusion

Understanding olfaction sense is still far from mature. Despite the limitation, many researchers attempted to understand the principle of olfactory sense and uncover the "primary odors" behind it. Numerical simulation and sensory evaluation of odor components for odor reproduction gave us insight into how a set of "primary odors" would be. Various researchers are working on developing devices for odor reproduction. These advancements are accelerating the progress of research in odor reproduction, particularly concerning sensory tests of reproduced odors. Nevertheless, several candidates for odor components have already been proposed, and the number of odors that can be reproduced is also increasing. Thus we might see those sets of odor components widely available in the future.

References

Amoore, J. E. (1967). Specific anosmia: A clue to the olfactory code. *Nature*, *214*(5093), 1095–1098. Available from https://doi.org/10.1038/2141095a0.

Castro, J. B., Ramanathan, A., Chennubhotla, C. S., & Schaefer, A. (2013). Categorical dimensions of human odor descriptor space revealed by non-negative matrix factorization. *PLoS ONE*, *8*(9). Available from https://doi.org/10.1371/journal.pone.0073289.

Genva, M., Kemene, T. K., Deleu, M., Lins, L., & Fauconnier, M. L. (2019). Is it possible to predict the odor of a molecule on the basis of its structure? *International Journal of Molecular Sciences*, *20*(12). Available from https://doi.org/10.3390/ijms20123018, https://www.mdpi.com/1422-0067/20/12/3018/pdf.

Gong, N. Aleixandre, M., & Nakamoto, T. (2022). *Odor recorder based on an array of QCM sensors using frequency shifts and resistance changes of multiple harmonics*. Proceedings of IEEE Sensors. http://www.ieee.org/sensors 10.1109/SENSORS52175.2022.9967066.

Grabe, V., & Sachse, S. (2018). Fundamental principles of the olfactory code. *BioSystems*, *164*, 94–101. Available from https://doi.org/10.1016/j.biosystems.2017.10.010.

Iseki, M., Prasetyawan, D., Yokoshiki, Y., & Nakamoto, T. (2022). A study of odor reproduction using multi-component olfactory display. *IEEJ Transactions on Sensors and Micromachines*, *142*(4), 63–70. Available from https://doi.org/10.1541/ieejsmas.142.63.

Kato, S., & Nakamoto, T. (2019). *Wearable olfactory display with less residual odor*. ISOEN 2019—18th International Symposium on Olfaction and Electronic Nose, Proceedings http://ieeexplore.ieee.org/xpl/mostRecentIssue.jsp?punumber = 8809667. 10.1109/ISOEN.2019.8823231.

Kim, D. W., Nishimoto, K. Kunifuji, S., Cho, Y. H., Kawakami, Y., & Ando, H. (2009). *Development of aroma-card based soundless olfactory display*. 16th IEEE International Conference on Electronics, Circuits and Systems (pp. 703–706), ICECS 2009. 10.1109/ICECS.2009.5410784.

Koulakov, A. A. (2011). In search of the structure of human olfactory space. *Frontiers in Systems Neuroscience*, *5* (2011). Available from https://doi.org/10.3389/fnsys.2011.00065.

Li, H., Sun, Y., & Luo, D. (2017). *A method of olfactory display: Odor characterization and reproduction*. ISOEN 2017—ISOCS/IEEE International Symposium on Olfaction and Electronic Nose, Proceedings. 10.1109/ISOEN.2017.7968893.

Liu, Q., Luo, D., Wen, T., Mo, Z., Li, J., & Li, Q. (2021). The odor characterizations and interactive olfactory display: A survey. Proceedings—5th IEEE international conference on smart internet of things (pp. 337–341), Smart IoT 2021. http://ieeexplore.ieee.org/xpl/mostRecentIssue.jsp?punumber = 9556111 10.1109/SmartIoT52359.2021.00061.

Malnic, B., Godfrey, P. A., & Buck, L. B. (2004). The human olfactory receptor gene family. *Proceedings of the National Academy of Sciences of the United States of America*, *101*(8), 2584–2589. Available from https://doi.org/10.1073/pnas.0307882100.

Mamlouk, A. M., Haker, M., & Martinetz, T. (2017). *Perception space analysis: From color vision to odor perception*. Proceedings of the international joint conference on neural networks 2017 (pp. 689–696). 10.1109/IJCNN.2017.7965919.

McMillan, C. (2020). *Aura: Maton: A wearable olfactory display for immersive scentscapes*. TEI 2020—Proceedings of the 14th international conference on tangible, embedded, and embodied interaction (pp. 677–682). http://dl.acm.org/citation.cfm?id = 3374920 10.1145/3374920.3375290.

Nakamoto, T. (2012). Technology frontier of augmented reality (6); Olfactory display and its application. *Kyokai Joho Imeji Zasshi/Journal of the Institute of Image Information and Television Engineers*, *66*(6), 478–483. Available from https://doi.org/10.3169/itej.66.478.

Nakamoto, T. (2016a). *Olfactory display and odor recorder. Essentials of machine olfaction and taste* (pp. 247–314). Wiley. Available from http://onlinelibrary.wiley.com/book/10.1002/9781118768495, 10.1002/9781118768495.ch7.

Nakamoto, T. (2016b). *Essentials of machine olfaction and taste*. Wiley. Available from 10.1002/9781118768495.

Nakamoto, T., & Murakami, K. (2009). *Selection method of odor components for olfactory display using mass spectrum database*. Proceedings—IEEE Virtual Reality (pp. 159–162). 10.1109/VR.2009.4811016.

Nakamoto, T., & Nihei, Y. (2013). Improvement of odor approximation using mass spectrometry. *IEEE Sensors Journal*, *13*(11), 4305–4311. Available from https://doi.org/10.1109/JSEN.2013.2267728.

Nakamoto, T., Ohno, M., & Nihei, Y. (2012). Odor approximation using mass spectrometry. *IEEE Sensors Journal*, *12*(11), 3225–3231. Available from https://doi.org/10.1109/jsen.2012.2190506.

Nicolaï, B. M., Micholt, E., Scheerlinck, N., Vandendriessche, T., Hertog, M. L. A. T. M., Ferguson, I., & Lammertyn, J. (2016). Real time aroma reconstruction using odour primaries. *Sensors and Actuators, B: Chemical, 227*, 561–572. Available from https://doi.org/10.1016/j.snb.2015.12.074.

Prasetyawan, D., & Nakamoto, T. (2019). *Comparison of NMF with Kullback–Leibler divergence and Itakura–Saito divergence for odor approximation*. ISOEN 2019—18th International Symposium on Olfaction and Electronic Nose, Proceedings http://ieeexplore.ieee.org/xpl/mostRecentIssue.jsp?punumber = 8809667 10.1109/ISOEN.2019.8823186.

Prasetyawan, D., & Nakamoto, T. (2023). Odor reproduction technology using a small set of odor components. *IEEJ Transactions on Electrical and Electronic Engineering*. Available from https://doi.org/10.1002/tee.23915.

Prasetyawan, D., & Takamichi, N. (2020). Sensory evaluation of odor approximation using NMF with Kullback–Leibler divergence and Itakura–Saito divergence in mass spectrum space. *Journal of the Electrochemical Society, 167*(16). Available from https://doi.org/10.1149/1945-7111/abd110.

Secundo, L., Snitz, K., Weissler, K., Pinchover, L., Shoenfeld, Y., Loewenthal, R., Agmon-Levin, N., Frumin, I., Bar-Zvi, D., Shushan, S., & Sobel, N. (2015). Individual olfactory perception reveals meaningful non-olfactory genetic information. *Proceedings of the National Academy of Sciences, 112*(28), 8750–8755. Available from https://doi.org/10.1073/pnas.1424826112.

Snitz, K., Yablonka, A., Weiss, T., Frumin, I., Khan, R. M., Sobel, N., & Diedrichsen, J. (2013). Predicting odor perceptual similarity from odor structure. *PLoS Computational Biology, 9*(9). Available from https://doi.org/10.1371/journal.pcbi.1003184.

Trimmer, C., Keller, A., Murphy, N. R., Snyder, L. L., Willer, J. R., Nagai, M. H., Katsanis, N., Vosshall, L. B., Matsunami, H., & Mainland, J. D. (2019). Genetic variation across the human olfactory receptor repertoire alters odor perception. *Proceedings of the National Academy of Sciences of the United States of America, 116* (19), 9475–9480. Available from https://doi.org/10.1073/pnas.1804106115, https://www.pnas.org/content/pnas/116/19/9475.full.pdf.

Tu, N. T. M., Onishi, Y., Choi, H. S., Kondo, Y., Bassore, S. M., Ukeda, H., & Sawamura, M. (2002). Characteristic odor components of *Citrus sphaerocarpa* Tanaka (Kabosu) cold-pressed peel oil. *Journal of Agricultural and Food Chemistry, 50*(10), 2908–2913. Available from https://doi.org/10.1021/jf011578a.

Xu, H., Kitai, K., Minami, K., Nakatsu, M., Yoshikawa, G., Tsuda, K., Shiba, K., & Tamura, R. (2021). Determination of quasi-primary odors by endpoint detection. *Scientific Reports, 11*(1). Available from https://doi.org/10.1038/s41598-021-91210-6.

Yamanaka, T., Matsumoto, R., & Nakamoto, T. (2003). Study of recording apple flavor using odor recorder with five components. *Sensors and Actuators B: Chemical, 89*(1-2), 112–119. Available from https://doi.org/10.1016/s0925-4005(02)00451-3.

Yamanaka, T., Wyszynski, B., & Nakamoto, T. (2003). *Study of odor recorder for recording recipe of orange flavor*. TRANSDUCERS 2003—12th international conference on solid-state sensors, actuators and microsystems, digest of technical papers 2 (pp. 1140–1143) 10.1109/SENSOR.2003.1216971.

CHAPTER

12

Odor reproduction based on several divergences

Dani Prasetyawan

Institute of Integrated Research, Institute of Science Tokyo, Yokohama, Japan

In Chapter 11, odor reproduction using odor components and its realization has been explained. In this chapter, we go deeper into the explanation of several numerical equations utilized for odor reproduction analysis. Odor reproduction is regarded as successful if the reproduced odor matches or is similar to the target odor that we wanted to reproduce. Moreover, for practical use, the number of odor components should be as few as possible, while it can reproduce a wide range of odors. Hence, most of the numerical analysis that is being used in the odor reproduction analysis utilizes dimensional reduction techniques where the number of reduced dimensions represents the number of odor components. Even though the set of odor components found by utilizing those techniques still has limitations in the accuracy of odor reproduction, different divergence and methods have been proposed to overcome the limitations.

12.1 Introduction

Odor reproduction in the olfaction sense is analogous to color reproduction in the sense of vision. However, as mentioned in Chapter 11, while primary colors exist, primary odors are yet to be found (Nakamoto, 2016; Schiffman & Pearce, 2002). For example, the sense of vision has a well-known set of three primary colors (red, green, and blue). Let us denote those three primary colors as dimensional spaces that can span the whole color space. Hence, all the colors in the color space can be reproduced by changing the combination of the mixing ratio of those three primary colors. We can also think of that in the olfaction sense. However, human has roughly 400 olfactory receptors. If we interpret that number as the number of primary odors, it is impractical to realize it (Nakamoto et al., 2012; Secundoa et al., 2015). Hence, most of the odor reproduction analyses that many researchers have done utilized dimensional reduction techniques to uncover the number of odor components that are appropriate for odor reproduction (Castro et al., 2013; Nicolaï et al., 2016; Prasetyawan & Nakamoto, 2020).

A lot of research has been done to uncover the exact dimensionality of odor space for a better understanding of odor space. However, most of the research was performed by numerical analysis and stopped after finding the

Digital Technologies in Olfaction
DOI: https://doi.org/10.1016/B978-0-443-15721-9.00012-9

dimensionality of odor space (Berglund et al., 1973; Grabe & Sachse, 2018; Koulakov, 2011; Mamlouk & Martinetz, 2004; Mamlouk et al., 2017; Meister, 2015). Only a handful of researchers perform numerical analysis followed by odor reproduction. This chapter will explain several numerical analyses that have been developed for odor reproduction. Numerical analysis with dimensional reduction properties such as nonnegative matrix factorization (NMF), principal component analysis (PCA), partial least squares discriminant analysis, and nonnegative least squares method are commonly used for odorant analysis (Delgado-Rodríguez et al., 2012; Nakamoto et al., 2012; Nicolaï et al., 2016; Prasetyawan & Nakamoto, 2019). However, this chapter focused on NMF, utilized in different studies for odor reproduction analysis in numerical analysis and then actual samples.

An illustration of dimensional reduction is shown in Fig. 12.1A–C. Here, the same dataset appeared in three different ways, differing in the number of dimensions to represent, where Fig. 12.1A represents the dataset in the three-dimensional space, (B) two-dimensional space, and (C) one-dimensional space. Dimensional reduction techniques enable us to reduce the number of dimensions, which makes the data storage smaller in exchange for a possible loss of information. Suppose the loss of the information is redundant data. In that case, the dimensional reduction technique can optimize the use of the dataset using reduced dimensions while retaining important information related to the original dataset.

As mentioned before, the similarity between target odor and reproduced odor is essential in odor reproduction. Hence, during the numerical analysis of odor reproduction, target data (target odor) and reconstructed data (reproduced odor) should be similar. Thus the accuracy of data reconstruction is significant. A simple example of numerical analysis with data reconstruction property is matrix factorization. When nonnegative constraint is added to the matrix factorization, we call it NMF. Here, an original matrix **W** with the size of $m \times n$ is decomposed into two smaller matrices of **X** and **Y**. Then, the two smaller matrices should be able to reconstruct the original matrix similarly (Févotte et al., 2009; Lee & Seung, 1999). The equation of matrix factorization is shown in Eq. (12.1) and illustrated in Fig. 12.2. The size of decomposed matrices can be determined by a number of basis vectors **r**. The matrix factorization can enable us to adjust the size of factorized matrices dynamically in exchange for reconstruction accuracy.

Nonnegativity constraints of NMF only allow additive combinations and do not allow

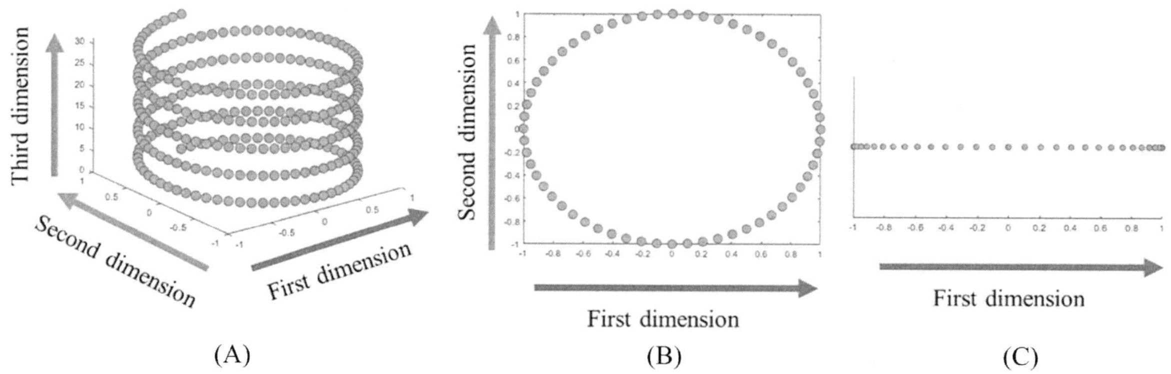

(A) (B) (C)

FIGURE 12.1 Example of dimensional reduction from (A) three-dimensional data to (B) two-dimensional data and (C) one-dimensional data. The dimensional reduction technique can reduce the dimension while retaining some information from the original data.

subtraction combinations. NMF allows a combination of basis vectors to represent reconstructed data. Hence, it makes NMF suitable for data analysis when a dataset has nonnegative properties and does not allow a subtraction procedure to reconstruct it. Thus it is very suitable to perform odor reproduction analysis using mass spectrometry performed by the Nakamoto group (Harada & Nakamoto, 2014; Nakamoto et al., 2012; Prasetyawan & Nakamoto, 2019).

Interestingly, the nonnegativity constraints of NMF are compatible with the intuitive notion of combining parts to reconstruct original data. It was demonstrated by Lee et al. using the facial image as a dataset to be analyzed by NMF. NMF can learn a parts-based representation of the given dataset, such as the nose, mouth, and eyes as part of a facial image (Lee & Seung, 1999). NMF is apart from PCA, which PCA could not intuitively learn the holistic, part-based, and representation learning.

Furthermore, numerical analysis optimization can also be performed by changing the cost function to minimize equations during the training procedure. Utilizing optimal divergence for cost function can lead to a higher accuracy of reproduced odor during odor reproduction analysis. More about optimization using different cost functions will be explained in Section 12.3.

$$\mathbf{W} \approx \mathbf{XY} \tag{12.1}$$

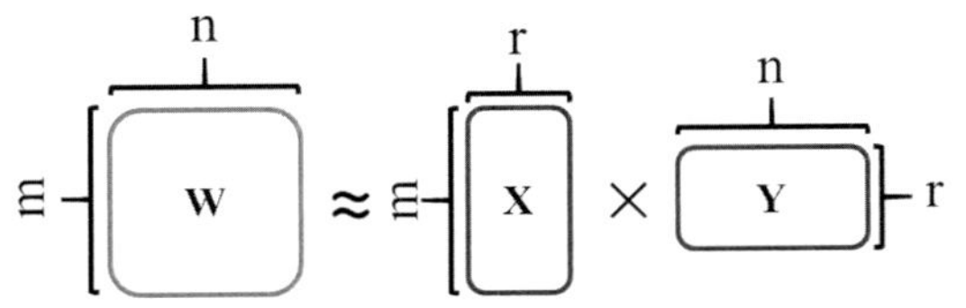

FIGURE 12.2 Illustration of nonnegative matrix factorization (NMF). The original matrix *V* is decomposed into two smaller **X** and **Y** matrices. The size of decomposed matrices can be controlled by changing the number of basis vectors **r**.

12.2 Methods for odor reproduction analysis

Several methods can be used for odor reproduction analysis. Numerical analysis, such as NMF, is commonly used for odor reproduction analysis. Castro et al. and the Nakamoto group utilized NMF to uncover the basis vectors that can be used as odor components for odor reproduction. Both groups utilized different datasets for the NMF, where the number of basis vectors became the set of odor components for their respective studies.

Castro et al. utilized Dravnieks data and tried to extract the representation of the human odor descriptor space using NMF. They systematically characterize odor space by focusing on identifying small numbers of perceptual primaries. Here, one of the factorized matrices represents "odor primaries," their terms of odor components, which should be able to span the full range of olfactory experiences as possible. Their NMF method uncovered 10 odor primaries were enough to reproduce 144 different odors (Castro et al., 2013).

On the other hand, the Nakamoto group utilized mass spectrometry data of essential oils and tried to extract an appropriate set of odor components for odor reproduction using NMF. Their method utilized NMF to decompose matrix **W** of essential oils data and extract one of decomposed matrix **Y** as the candidate for the set of odor components (Nakamoto et al., 2012). The denotation of NMF here is based on Fig. 12.2. Here, the size of *m* is equal to the number of samples used in the study, and *n* is the dimensionality of the data. Note that the odor component obtained by the Nakamoto group is not defined by choosing one of the essential oils as a candidate or representative of the odor component but by approximating the mass spectrum pattern extracted by NMF. Moreover, obtaining odor components that match the mass spectrometry pattern obtained by NMF is difficult. Hence, the nonnegative least squares method was needed for the adjustment in approximating odor components in the actual sample.

After finding candidate odor components from NMF, the nonnegative least squares method was utilized to calculate the mixing ratio to approximate odor components in the sample. The odor component was approximated by blending essential oils based on a mixing ratio obtained by the nonnegative least squares method (Nakamoto et al., 2012). The nonnegative least squares method utilized by the Nakamoto group is shown in Eq. (12.2). Here, the target of odor components to be approximated is defined by **Y**, which is $n \times r$ matrix (as in Eq. 12.1), where **a** is $r \times m$, which is a matrix representing the mixing ratio of essential oils **W** to approximate odor components **Y**. Here, the actually approximated odor components based on Eq. (12.2) is defined as $\mathbf{Y}'$.

Moreover, an additional nonnegative least squares method was utilized again to calculate the mixing ratio of approximated odor components to reproduce the target odor. The second nonnegative least squares method is needed to adjust the mixing recipe to compensate for the deviation that occurs during the approximation of odor components $\mathbf{Y}'$. The second nonnegative least squares method is defined in Eq. (12.3). Here, the target of odor to be reproduced is defined by **W**, an $m \times n$ matrix of essential oils (as in Eq. 12.1), where $\mathbf{X}'$ is an $m \times r$ matrix representing the mixing ratio of approximated odor components $\mathbf{Y}'$ to reproduce the target odor. The actually reproduced target odor based on Eq. (12.3) is called $\mathbf{W}'$:

$$\mathbf{Y}^{\mathrm{T}} = \mathbf{a}\mathbf{V}^{\mathrm{T}} \tag{12.2}$$

$$\mathbf{W} \approx \mathbf{W}' = \mathbf{X}'\mathbf{Y}' \tag{12.3}$$

The early study of the group of Nakamoto by Nihei et al. used 158 essential oils data to the NMF, in which they extracted a set of 12 odor components and 30 odor components. They made the actual odor components and reproduced the target odor using the method previously mentioned. They found that 30 odor components extracted by NMF could perform appropriate odor reproduction. Moreover, the study was improved by Prasetyawan et al. from the same group utilizing 185 essential oils (Prasetyawan & Nakamoto, 2019, 2020). The improvement of odor reproduction analysis using NMF was to change the cost function for minimizing equations during the training of NMF and the nonnegative least squares method. They utilized different divergences for the optimization of NMF and nonnegative least squares method, which detailed explanation is provided in Section 12.3.

12.3 Different divergences for odor reproduction analysis

In this section, several divergences utilized for NMF are explained. Especially divergence that is being utilized in NMF for odor reproduction analysis.

The most common divergence utilized for the cost function of NMF is Euclidean distance (EUC) and Kullback–Leibler divergence (KL). The early adaptation of NMF by Lee and Seung introduced gradient descent algorithms for minimizing equations using EUC and KL divergence (Févotte et al., 2009; Hien & Gillis, 2021; Lee & Seung, 1999). The gradient descent update rules can be transformed into multiplicative rules by selecting an appropriate step size. These rules ensure that the cost function decreases over iterations. The simplicity of these update rules has contributed significantly to the widespread adoption of NMF.

EUC is a commonly used distance measure between two points in space, derived from the Pythagorean theorem, and it can be used in spaces of any dimensionality. A two-dimensional space EUC between two points is defined in Eq. (12.4). In simple terms, it is the straight-line distance between two points. Its linear property makes it easier to converge during the training of NMF.

Another common divergence for NMF is the KL divergence. KL-divergence, also referred to as *I*-divergence, is defined in Eq. (12.5) in a one-dimensional case. Both KL-divergence and EUC are positive and take the zero value only if $x = y$. The logarithmic function in KL-divergence makes it more sensitive to small changes than the EUC as a cost function (Févotte et al., 2009; Kameoka et al., 2017).

On the other hand, Itakura–Saito divergence (IS) is also being used as a cost function for minimizing equations in NMF. IS-divergence is derived from maximum likelihood estimation (Févotte et al., 2009; Ito & Nakamoto, 2020; Prasetyawan & Nakamoto, 2019). It is widely used in speech spectra–based data. An interesting property of IS-divergence is that it is a particular scale-invariant, which treats small-intensity of peaks as same as high-intensity peaks. The use of IS-divergence is relevant when a dataset with a large dynamic range is used in the analysis. Unlike the other two divergences mentioned previously, which tend to have linear properties that treat high-intensity peaks significantly more than small peaks intensity, IS-divergence tends to be more sensitive when small peak intensities are present in the dataset. The equation of IS-divergence in a one-dimensional case is defined in Eq. (12.6).

A more straightforward understanding of the difference between each cost function mentioned above is shown in Fig. 12.3. Here, Fig. 12.3 represents Euclidean, KL, and IS costs for $x = 1$, where each cost function shows the different characteristics in treating different costs of y. Here, EUC is convex on zero to infinite $(0, \infty)$, in which KL-divergence tends to do so. It means that Euclidean and KL-divergence tend to have linear propositions. However, the IS-divergence differs from the other two divergences, which tend to be convex on ($[0, 2\ x\ ]$ and concave on $[2\ x, \infty]$). It means that the IS-divergence has a different proposition before and after x. It tends to be reactive to the change in value below x but lenient to the change of value after x. Interestingly, KL-divergence tends to be in the middle between Euclidean and IS-divergence at almost all costs.

Furthermore, there are other well-known divergences, such as α- and β-divergence. Both divergences measure how close two probability distributions diverge from each other, where 0 means that two distributions are identical. Here, α-divergence is a divergence that

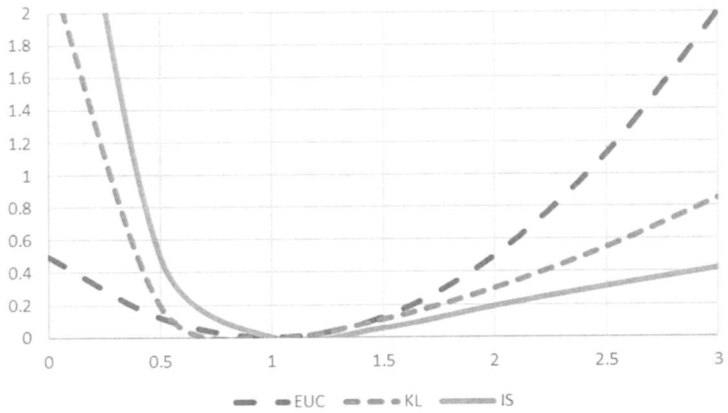

FIGURE 12.3 Reconstructed image from Fevotte et al., showing Euclidean, KL, and IS costs $d\ (x\,|\,y)$ as a function of y and for $x = 1$. The image shows the difference in characteristics of each divergence in treating given costs. *KL*, Kullback–Leibler divergence; *IS*, Itakura–Saito divergence. *Source: From Févotte, C., Bertin, N., & Durrieu, J. L. (2009). In Neural computation (Vol. 21, Issue 3, pp. 793–830). MIT Press. https://doi.org/10.1162/neco.2008.04-08-771.*

tends to focus on the overall characteristic of distribution, whereas β-divergence is more focused on pinpoint changes in the distribution. Interestingly, α-divergence includes KL-divergence (where $\alpha = 1$) in a particular case. On the other hand, β-divergence includes well-known divergences like KL-divergence (where $\beta = 1$) and EUC (where $\beta = 2$) as particular cases:

$$d_{\mathrm{EUC}}(x|y) = \frac{1}{2}(x-y)^2 \tag{12.4}$$

$$d_{\mathrm{KL}}(x|,y) = x\log\frac{x}{y} - x + y \tag{12.5}$$

$$d_{\mathrm{IS}}(x|,y) = \frac{x}{y} - \log\frac{x}{y} - 1 \tag{12.6}$$

There is a case when the divergences are used in a multidimensional case instead of a one-dimensional case, as explained before. Let us expect that the number of dimensions to be included in the equation is n; the equations of EUC, KL-divergence, and IS-divergence are defined as follows:

$$d_{\mathrm{EUC}}(x|y) = \sum_{i=1}^{n}\frac{1}{2}(x_i-y_i)^2 \tag{12.7}$$

$$d_{\mathrm{KL}}(x|,y) = \sum_{i=1}^{n} x_i\log\frac{x_i}{y_i} - x_i + y_i \tag{12.8}$$

$$d_{\mathrm{IS}}(x|,y) = \sum_{i=1}^{n}\frac{x_i}{y_i} - \log\frac{x_i}{y_i} - 1 \tag{12.9}$$

Odor reproduction analysis from the Nakamoto group utilizing NMF successfully reproduced the target odors with different sets of odor components. They achieved the optimization of odor reproduction by utilizing different divergence as cost functions mentioned above during NMF training. Nihei et al. utilized a dataset of 158 essential oils mass spectrum data for odor reproduction analysis using NMF in their early study. They utilized NMF with EUC and KL-divergence as cost functions to extract a set of odor components in odor reproduction analysis. Then, they calculated the Euclidean-based nonnegative least squares method to obtain a blending ratio to approximate actual odor components. They calculate nonnegative least squares again to calculate the blending ratio of actual odor components to approximate target odor for odor reproduction. Here, they obtained a set of 12 and 30 odor components by utilizing the mentioned method. The fidelity of approximated odor components was tested by approximating three odors (organic orange, peppermint, and black pepper) using both sets of odor components. High accuracy in odor approximation was obtained when utilizing 30 odor components.

Further odor reproduction analysis was performed by Harada et al. and Prasetyawan et al. from the Nakamoto group. They utilized IS-divergence as a cost function for minimizing equations during the training procedure of NMF (Harada & Nakamoto, 2014; Prasetyawan & Nakamoto, 2019). Harada et al. utilized insect olfactory receptor neuron (ORN) response data to be analyzed by NMF. They found that the 10 extracted basis vectors by NMF were found to be located according to the spatial distribution of 110 odor chemicals. The method successfully reduced the residuals of the reconstruction error of NMF, and the response pattern of 24 ORNs can be approximated using the extracted basis vectors.

On the other hand, Prasetyawan et al. utilized 185 essential oils mass spectra with the measured range of 50–250 m/z region data for odor reproduction analysis (185 samples $\times$ 201-dimensional data). They compared the performance of IS- and KL-divergence as the cost function of NMF in odor reproduction analysis (Prasetyawan & Nakamoto, 2019, 2020). The comparison of NMF with both divergences in reconstructing original data is shown in Fig. 12.4. IS-divergence as residuals was calculated for the evaluation of reconstruction error. The figure shows that the approximation by NMF with IS-divergence was better than that by widely used NMF with KL-divergence, especially when the number of odor components was smaller. Thus NMF with IS-divergence enables Prasetyawan et al. to optimize odor reproduction while using a fewer number of odor components than a previous study by Nihei et al. from the same group. Eventually, they chose the 10 odor components for odor approximation using actual samples since the reconstruction error difference between NMF with IS- and KL-divergence is significant when the number of basis vectors is less than 10. Their result was obtained by Prasetyawan et al., as is shown in Fig. 12.4.

Furthermore, Prasetyawan et al. also utilized IS-divergence as the cost function of the nonnegative least squares method for optimal recipe extraction to approximate odor components and recipe to approximate the target odor. IS-divergence was utilized as a cost function for minimizing equations during the nonnegative least squares method training procedure. Here, IS-divergence as the optimization of nonnegative least squares can enhance the approximation accuracy compared to ordinary nonnegative least squares, which is shown by lower residual as shown in Fig. 12.5. IS-divergence as residuals was calculated between the original and approximated odor.

A different comparison was also performed by Prasetyawan et al. Here, NMF with KL-divergence utilizes the ordinary nonnegative least squares method compared with NMF with IS-divergence utilizing IS-divergence-based nonnegative least squares. The comparison is shown in Fig. 12.6. As expected, NMF with IS-divergence utilizing IS-divergence-based nonnegative least squares has lower residual form than its counterpart. Since NMF and nonnegative least squares methods utilize similar cost functions, the optimization for the odor reproduction analysis with

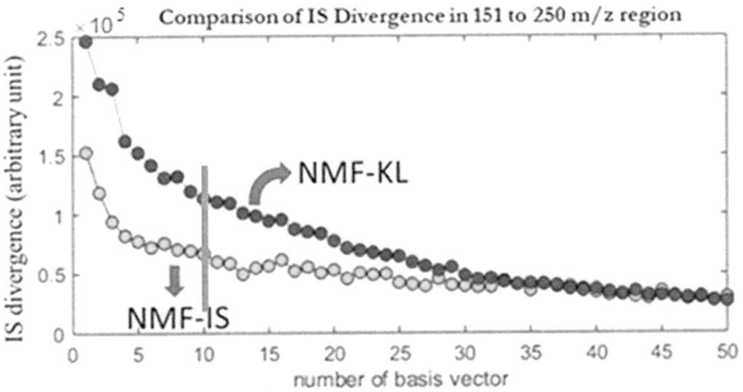

FIGURE 12.4 Reconstruction error of nonnegative matrix factorization (NMF) with Itakura–Saito (IS) and Kullback–Leibler (KL)-divergence using different numbers of basis vectors. Here, a set of 10 odor components chosen by Prasetyawan et al. for the evaluation of fewer odor components for odor approximation is shown in the green line. *Source: From Prasetyawan, D., & Nakamoto, T. (2023). IEEJ Transactions on Electrical and Electronic Engineering. https://doi.org/10.1002/TEE.23915.*

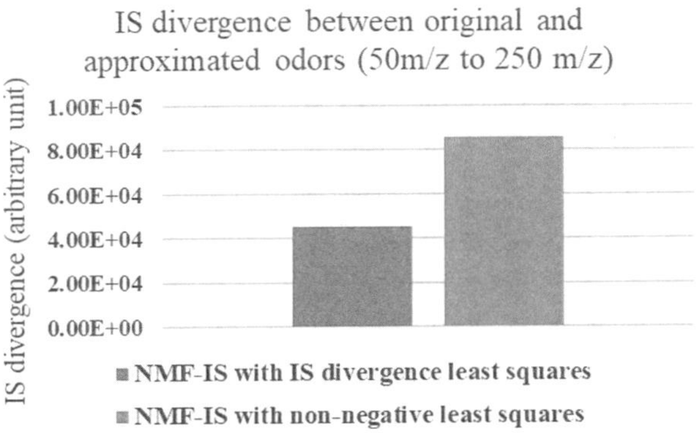

FIGURE 12.5 Comparison of nonnegative matrix factorization: Itakura–Saito divergence (NMF-IS) in reconstructing original data with different nonnegative least squares methods in 50–250 m/z region. The results of the lowest residuals are shown (number of basis vectors = 10). *Source: From Dani, P. (2021). http://t2r2.star.titech.ac.jp/cgi-bin/publicationinfo.cgi?lv = en&q_publication_content_number = CTT100863231.*

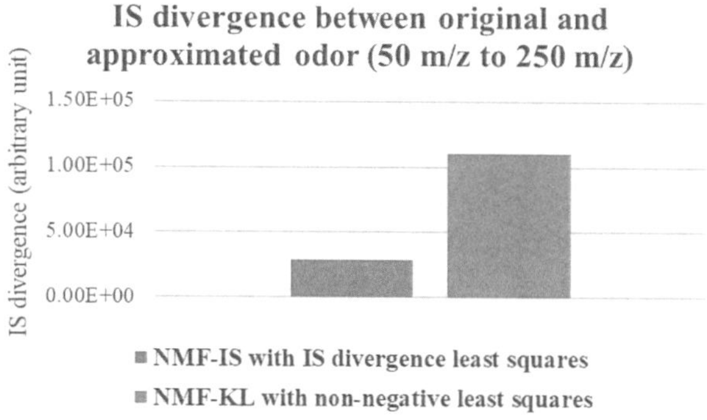

FIGURE 12.6 Comparison of nonnegative matrix factorization: Itakura–Saito divergence (NMF-IS) and NMF-Kullback–Leibler divergence (KL) in reconstructing original data in 50–250 m/z region utilizing different nonnegative least squares method. The results of the lowest residuals are shown (number of basis vectors = 10). *Source: From Dani, P. (2021). http://t2r2.star.titech.ac.jp/cgi-bin/publicationinfo.cgi?lv = en&q_publication_content_number = CTT100863231.*

respective divergence is possible (Prasetyawan & Nakamoto, 2020; Prasetyawan, 2021).

Prasetyawan et al. extracted 10 basis vectors that have a significant difference in reconstruction error between NMF with KL-divergence and IS divergence and approximate actual odor components. They also approximated six target odors to compare the performance of both divergences in actual samples. The comparison result is that NMF with IS-divergence resulted in higher accuracy in odor reproduction, especially in the fewer number of basis vectors. Further explanation related to the actual sample and human sensory test result is mentioned in Section 11.2.

Moreover, it is known that a mass spectrum peak in the high m/z region (151–250 m/z) typically originate from a molecule with large molecular weight, which typically has a low human olfactory threshold. It is considered to have a significant contribution to smell quality despite the small peak value. Thus IS-divergence, which tends to be more sensitive to small intensity changes compared to KL-divergence, is suitable for this task.

12.4 Conclusion

Odor reproduction analysis can be performed in different numerical analyses. Despite the promising results in numerical analysis, approximating actual samples derived from numerical results is challenging. A method such as NMF was utilized by several studies to extract odor components as a representative of "primary odors" of a particular set of odorants. Optimization of NMF for odor reproduction analysis can be performed utilizing different divergences, as shown by Nihei et al. and Prasetyawan et al. Interestingly, choosing the appropriate divergence might increase the accuracy of odor reproduction and reduce the number of odor components. Despite the challenge of realizing the actual samples, the utilization of appropriate divergence helped in the realization of odor reproduction not only in numerical equations but also in actual samples.

References

Berglund, B., Berglund, U., Engen, T., & Ekman, G. (1973). Multidimensional analysis of twenty-one odors. *Scandinavian Journal of Psychology, 14*(2), 131–137. Available from https://doi.org/10.1111/J.1467-9450.1973.TB00104.X, https://pubmed.ncbi.nlm.nih.gov/4705857/.

Castro, J. B., Ramanathan, A., & Chennubhotla, C. S. (2013). Categorical dimensions of human odor descriptor space revealed by non-negative matrix factorization. *PLoS ONE, 8*(9), e73289. Available from https://doi.org/10.1371/journal.pone.0073289, -e73289, https://dx.plos.org/10.1371/journal.pone.0073289.

Delgado-Rodríguez, M., Ruiz-Montoya, M., Giraldez, I., López, R., Madejón, E., & Díaz, M. J. (2012). Use of electronic nose and GC–MS in detection and monitoring some VOC. *Atmospheric Environment., 51*, 278–285. Available from https://doi.org/10.1016/j.atmosenv.2012.01.006.

Févotte, C., Bertin, N., & Durrieu, J. L. (2009). Nonnegative matrix factorization with the Itakura–Saito divergence: With application to music analysis. *Neural Computation, 21* (3)), 793–830. Available from http://direct.mit.edu/neco/article-pdf/21/3/793/820289/neco.2008.04-08-771.pdf.

Grabe, V., & Sachse, S. (2018). Fundamental principles of the olfactory code. *BioSystems, 164*, 94–101.

Harada, Y., & Nakamoto, T. (2014). Analysis of an insect's olfactory receptor neuron response by NMF method for odor approximation. *Sensors and Materials, 26*(3)), 181–187.

Hien, L. T. K., & Gillis, N. (2021). Algorithms for nonnegative matrix factorization with the Kullback–Leibler divergence. *Journal of Scientific Computing, 87*(3), 93. Available from https://doi.org/10.1007/s10915-021-01504-0, –93, https://link.springer.com/10.1007/s10915-021-01504-0.

Ito, K., & Nakamoto, T. (2020). Improvement of odor impression predictive model using machine learning IEEE. *IEEE Sensors autoencoder deep learning Itakura-odor smell*. IEEE (pp. 1–4).

Kameoka, H., Kagami, H., & Yukawa, M. (2017). *Complex NMF with the generalized Kullback–Leibler divergence IEEE*. ICASSP, IEEE International Conference on Acoustics, Speech and Signal Processing—Proceedings (pp. 56–60). http://ieeexplore.ieee.org/document/7952117/ 10.1109/ICASSP.2017.7952117.

Koulakov, A. A. (2011). In search of the structure of human olfactory space. *Frontiers in Systems Neuroscience, 5* (2011), 65. Available from https://doi.org/10.3389/fnsys.2011.00065, –65, http://journal.frontiersin.org/article/10.3389/fnsys.2011.00065/abstract.

Lee, D. D., & Seung, H. S. (1999). Learning the parts of objects by non-negative matrix factorization. *Nature, 401* (6755), 788–791. Available from https://doi.org/10.1038/44565, https://www.nature.com/articles/44565.

Mamlouk, A. M., Haker, M., & Martinetz, T. (2017). *Perception space analysis: From color vision to odor perception*. Proceedings of the International Joint Conference on neural networks (pp. 689–696). Available from https://doi.org/10.1109/IJCNN.2017.7965919.

Mamlouk, A. M., & Martinetz, T. (2004). On the dimensions of the olfactory perception space. *Neurocomputing., 58-60*, 1019–1025. Available from https://doi.org/10.1016/j.neucom.2004.01.161.

Meister, M. (2015). On the dimensionality of odor space. *eLife., 4*, e07865. Available from https://doi.org/10.7554/eLife.07865, -e07865, http://www.pubmedcentral.nih.gov/articlerender.fcgi?artid = PMC4491593.

Nakamoto, T. (2016). *Essentials of machine olfaction and taste essentials of machine olfaction and taste* (pp. 1–322). John Wiley & Sons Singapore Pte Ltd. Available from http://doi.wiley.com/10.1002/9781118768495, 10.1002/9781118768495.

Nakamoto, T., Ohno, M., & Nihei, Y. (2012). Odor approximation using mass spectrometry. *IEEE Sensors Journal, 12*(11), 3225–3231. Available from https://doi.org/10.1109/JSEN.2012.2190506.

Nicolaï, B. M., Micholt, E., Scheerlinck, N., Vandendriessche, T., Hertog, M. L. A. T. M., Ferguson, I., & Lammertyn, J. (2016). Real time aroma reconstruction using odour primaries. *Sensors and Actuators, B: Chemical, 227*, 561–572. Available from https://doi.org/10.1016/j.snb.2015.12.074.

Prasetyawan, D. (2021). *Exploration of odor components for odor reproduction based on mass spectrometry*. [Unpublished content exploration of odor components for odor reproduction based on mass spectrometry]. Tokyo Institute of Technology. http://t2r2.star.titech.ac.jp/cgi-bin/publicationinfo.cgi?lv = en&q_publication_content_number = CTT100863231.

Prasetyawan, D., & Nakamoto, T. (2019). *Comparison of NMF with Kullback–Leibler divergence and Itakura–Saito divergence for odor approximation IEEE*. IEEE International Symposium on olfaction and electronic nose (ISOEN) (pp. 1–3). https://ieeexplore.ieee.org/document/8823186/ 10.1109/ISOEN.2019.8823186.

Prasetyawan, D., & Nakamoto, T. (2020). Sensory evaluation of odor approximation using NMF with Kullback–Leibler divergence and Itakura–Saito divergence in mass spectrum space. *Journal of the Electrochemical Society, 167*(16), 167520. Available from https://doi.org/10.1149/1945-7111/abd110, –167520, https://iopscience.iop.org/article/10.1149/1945-7111/abd110/meta.

Schiffman, S. S., & Pearce, T. C. (2002). Handbook of machine olfaction introduction to olfaction: Perception, anatomy, physiology, and molecular biology. In T. Nakamoto (Ed.), *Handbook of machine olfaction*. Wiley. Available from https://doi.org/10.1002/3527601597.

Secundoa, L., Snitz, K., Weissler, K., Pinchover, L., Shoenfeld, Y., Loewenthal, R., Agmon-Levin, N., Frumin, I., Bar-Zvi, D., Shushan, S., & Sobel, N. (2015). Individual olfactory perception reveals meaningful non-olfactory genetic information. *Proceedings of the National Academy of Sciences of the United States of America, 112* (28), 8750–8755. Available from https://doi.org/10.1073/pnas.1424826112.

CHAPTER

13

Analysis for interference removal in odor reproduction

Dani Prasetyawan

Institute of Integrated Research, Institute of Science Tokyo, Yokohama, Japan

13.1 Introduction

In odor reproduction, a set of odor components should reproduce as many odorants as possible. Hence the increasing number of odorant samples to be analyzed is significant during odor reproduction analysis (Nakamoto et al., 2012; Prasetyawan & Takamichi, 2020). There are a lot of odorant samples that are available in our life. Perfumes, detergent, cologne, food, and many other things utilized odor-related samples to make them more interesting and compelling to the customers (Rodrigues et al., 2021; Teixeira et al., 2013; vom Ende et al., 2017). Determining an odor component that can reproduce such vast odorant samples as much as possible would be a feasible application of odor reproduction in our life. However, commercially available odorant samples might include more odorless compounds besides odorous compounds. Those additional compounds, mostly consisting of preservatives called fixatives, are added to the products to prolong the shelf-life. Such additional compounds, especially odorless ones, can cause interference during odorant analysis.

Even though odorant samples are widely used in everyday life, they can pose challenges during odorant analysis, particularly when using tools such as electronic noses to measure them (Prasetyawan & Nakamoto, 2021; Scott et al., 2006). For example, if the odorant sample is measured when an ambient smell is lingering around, the human nose can easily ignore ambient smells and recognize the target samples. However, the electronic nose cannot easily separate the ambient smells without a method that can separate ambient smells and target odors (Meng, 2010; Prasetyawan et al., 2023; Tian et al., 2016).

Another example is sensing an odorant sample with fixatives using machine olfaction. While the human nose cannot pick up odorless fixatives, sensors in the machine olfaction can pick up odorless fixatives during the measurement (Prasetyawan & Nakamoto, 2021; vom Ende et al., 2017). Hence it can lead to discrepancies between the data obtained from machine-based olfaction and human sensory evaluations. Thus comparing the two sets of results can cause more complications in the odorant analysis process. Therefore eliminating interference such as fixatives before odorant analysis becomes essential.

Digital Technologies in Olfaction
DOI: https://doi.org/10.1016/B978-0-443-15721-9.00013-0

Different researchers have performed and studied some methods of interference or noise suppression analysis during feature extraction of odor-related samples (Liang et al., 2018; Scott et al., 2006). Most of the studies on interference suppression in olfaction utilized the source separation technique. First, the source separation technique is utilized to separate components from given data that interferences are included within. Next, the separated data is identified and separated into wanted and unwanted data. Then we can extract the wanted data to be utilized for odorant analysis and omit the unwanted ones. An illustration of the source separation procedure is shown in Fig. 13.1. Here, three different sources of mixture signals are obtained, and a source separation technique is performed on the signals. Then three different signals are obtained, in which those three signals are characteristically different. At last, we can identify the separate signals that belong to interference or data that we actually need. Note that during a source separation technique, the number of source signals must equal or exceed the number of sought signals to get an optimal source separation technique.

A commonly used source separation technique for interference suppression is independent component analysis (ICA). ICA is commonly used in analyzing data obtained from electrocardiograms (ECGs), where it helps in analyzing and interpreting the data more effectively (Mi, 2014; Tanskanen et al., 2012). It can be used to perform noise reduction and artifact removal, in which ECG signals often contain noise and artifacts from various sources such as muscle contractions, electrode movement, or external electromagnetic fields. ICA can separate these unwanted components from the actual heart signal, enhancing the clarity and accuracy of the ECG reading.

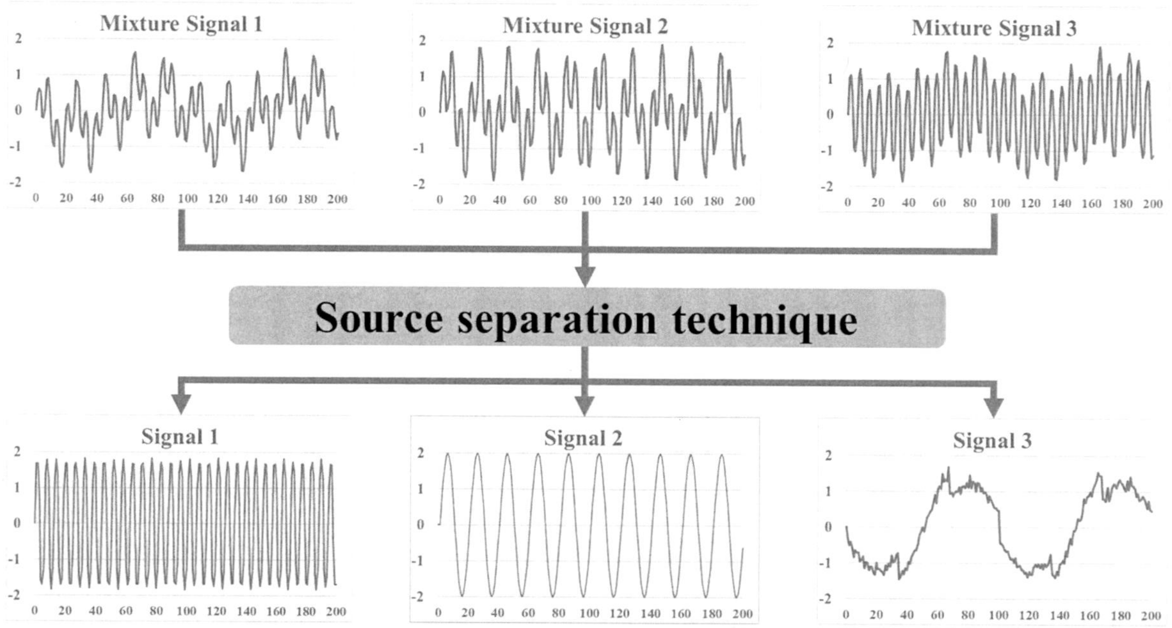

FIGURE 13.1 An example of a source separation technique. Three mixture signals as source signals are processed using the source separation technique, and then three different signals are obtained.

Moreover, ICA is a blind source separation technique, which is an unsupervised technique that uncovers independent components (ICs), which are hidden within the observed data (Hyvärinen et al., 2009; Stone, 2002). The formula of the ICA is shown in Eq. (13.1). Here, let us denote a linear mixture of m-dimensional observed data of $\mathbf{S} = (\mathbf{s}_1, \mathbf{s}_1, \ldots \mathbf{s}_k)$, $\mathbf{S} \in R^{k \times m}$, m-dimensional ICs of $\mathbf{V} = (\mathbf{v}_1, \mathbf{v}_2, \ldots \mathbf{v}_n)$, $\mathbf{V} \in R^{n \times m}$, and $\mathbf{a}(\mathbf{a} \in R^{n \times k})$ is a demixing matrix for $\mathbf{S}$ to obtain $\mathbf{V}$, where k is the number of observed data and n is the number of ICs sought by ICA:

$$\mathbf{V} = \mathbf{aS} \tag{13.1}$$

ICA has superiority in separating the observed data as source data without a priori information of the observed data and any supervision or prior stimulus. Here, ICA can extract the number of ICs at the most, the same number as the observed data. Interestingly, we can use these independencies of ICs extracted by ICA to isolate the actual data, which may have been blended with possible noises and interferences in the observed data. While it is possible to enhance the amount of observed data, excessive data could lead to a prolonged extraction process and might be sensitive to data distribution.

Unlike other commonly used methods, such as principal component analysis (PCA), which is explained in Chapter 7, ICA strives to identify ICs that are not correlated with each other (Aguilera et al., 2012). Conversely, the principal components (PCs) derived via PCA tend to assess the variance within the observed data, in which each PC is orthogonal to each other (Scott et al., 2006). One significant distinction between ICA and PCA is that ICA utilizes the non-Gaussian structure of the data. This feature is critical for unveiling the ICs inherent in the observed data. Here, Fig. 13.2 illustrates the difference of PCs obtained by PCA and ICs obtained by ICA. This chapter will focus on explaining the utilization of ICA in interference suppression during odorant analysis, specifically in odor reproduction analysis.

13.2 Overview of interference and its removal using independent component analysis in odorant analysis

In an odorant analysis, a set of data related to the odorant samples are needed to be obtained beforehand. The odorant dataset can be obtained using sensory tests by panelists or by measuring odorant samples using sensors

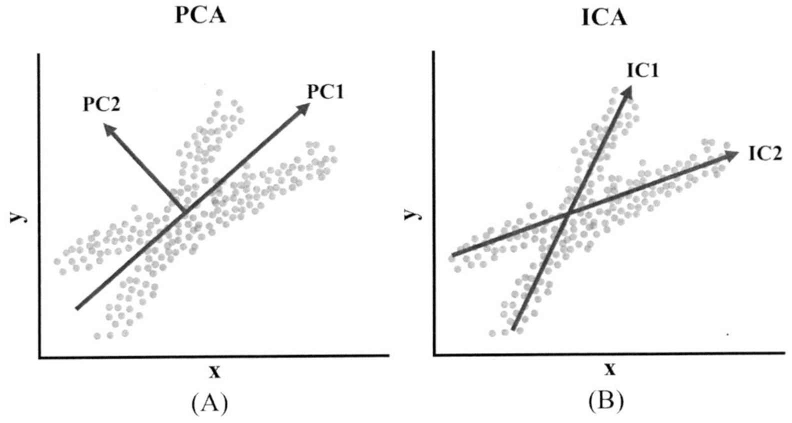

FIGURE 13.2 Illustration of PCA and ICA in treating a distribution of data. Here, (A) orthogonality of each PC obtained from PCA and (B) independency of each IC obtained from ICA are shown. *IC*, Independent component; *ICA*, independent component analysis; *PCA*, principal component analysis.

(e.g., chemical sensors) and analytical instruments. However, some unwanted samples or disturbances might be included during the measurement of samples by analytical instruments, which cause interference with the data obtained by the measurement. That interference can cause problems during the odorant analysis because hindering us from analyzing the meaningful parts of the data.

The unwanted data that cause interference in the measurement of odorant samples can come in many forms, such as sensor drift, sensor deviation, background interference, and odorless compounds included in the odorant sample. In most cases, those interferences that occur in gas and odor-related variables are difficult to separate using conventional filtering or wavelet transformations, as these interferences are thoroughly mixed with the target samples intended for detection in both frequency and time domains (Tian et al., 2016). Hence addressing the challenge of suppressing interference in an electronic nose (e-nose) becomes a significant factor in odorant analysis.

An example of interference commonly occurring during measurement is the background interference caused by surrounding environmental conditions. In a practical application of an electronic nose, the isolation of the electronic nose from the surrounding environment is critical. The ideal condition of the electronic nose can be maintained by controlling parameters such as temperature, humidity, and surrounding gas (Di Natale et al., 2002; Liang et al., 2018; Tian et al., 2016). However, several parameters cannot be contained in practical applications due to the isolation limitation, which makes direct interference almost negligible. Interference caused by temperature or humidity can be compensated once the precise model of the electronic nose as parameter of temperature and humidity is found.

On the other hand, interference caused by surrounding gases during measurement is rather challenging to compensate for. This is due to changing conditions whenever the measurement is performed in different locations of applications. Suppose we measured food a sample in a controlled room, where no odors are lingering inside the room. The result would be different from that when we measured food samples directly in a kitchen where other smells are lingering around when ideal isolation of the measured sample cannot be performed. The overall sensor signal might be influenced by a combination of different gases and odors that contribute to obscure the information relevant to the sample.

Many research studies in the suppression of interference of background interference were performed by utilizing methods based on correlation, ICA, and machine learning (Liu et al., 2023; Tian et al., 2016). Software interference suppression is affordable and easier to perform than changing the hardware of analytical tools. An example of interference suppression using a machine learning–based method was performed by Zhang et al. (2013). Here, data obtained by the sensor array is analyzed using machine learning–based methods such as support vector machine (SVM) and backpropagation (BP) networks. First, determine the interference that needs to be suppressed. Then SVM and BP networks are utilized to categorize the sensor response, which is replaced by the most nearby signal if it was determined as an interference.

Different researchers studied and performed interference suppression using ICA in odorant analysis (Di Natale et al., 2002; Liang et al., 2018; Meng, 2010; Prasetyawan & Nakamoto, 2021, 2022; Prasetyawan et al., 2023; Tian et al., 2016; Wang et al., 2020). ICA has been proven to be capable of decomposing matrix data into ICs.

Here, independence implies that the information carried by one component cannot be deduced from the rest. In statistical terms, this means the joint probability of independent quantities is the product of each individual's probability. During the interference suppression in odorant using ICA, data obtained in measurement is decomposed by ICA into several ICs. Then the correlation coefficient is calculated between ICs obtained by ICA and references to identify each IC. Here, the references can be interferences that need to be suppressed or meaningful signals we need.

For example, let us think that the references we have are interferences that we want to suppress. First, we identify interference-related ICs by calculating the correlation-based algorithm between each IC and reference data. The remaining ICs that were not identified as interference-related IC became the meaningful signal we sought. Since ICs are independent to each other, the remaining ICs that are related to the meaningful signals should be uncorrelated to the interference. Hence the suppression of interference is done by this method.

On the contrary, if the reference is meaningful signal that we seek, we can obtain the meaningful signal from extracted ICs by identifying the ICs using the correlation between each IC and reference. The remaining ICs that are uncorrelated to the meaningful signals are regarded as interference that we can discard. Here, since the ICs correlated to the meaningful signals are independent of the remaining ICs that are regarded as interference, the ICs correlated to the meaningful signals can be said to be data free from interference.

The abovementioned methods are illustrated in Fig. 13.3A and B for easier understanding. Fig. 13.3A shows when interference data is being used as reference data, and Fig. 13.3B shows when meaningful signals are being used as reference data. Here, suppose we have m number of reference data, and ICA is being used to extract n number of ICs (where $n > m$). The number of ICs extracted by ICA must be at least $m+1$ to cover the number of reference data. Researchers, such as Di Natale et al. (2002), Liang et al. (2018, 2022), Prasetyawan and Nakamoto (2021, 2022), Prasetyawan et al. (2023), Tian et al. (2016), used interferences as a reference and successfully extracted meaningful signals that are without disturbance of interference using ICA.

Several studies using ICA for interference suppression were performed to extract meaningful signals separated from sensor signals disturbed by background interference and sensor drift (Di Natale et al., 2002; Tian et al., 2016; Zhang et al., 2013). Background interference occurs when the disturbance by the surrounding environment that is not correlated to the meaningful signals that we sought is included during measurement. For example, Di Natale et al. (2002) suppress the interference caused by disturbance from the surrounding environment and sensor drift in the sensors in the electronic nose. Initially, they accumulated information on background noise and sensor drift. Then the measurement of intended samples is performed (even if background interference and sensor drift are present) and obtained the measurement data. Then ICA is applied to the obtained measurement data. Since we already know the interference a priori as reference data, it is possible to separate the meaningful signals by identifying the interference first and the remaining ICs uncorrelated to the reference as meaningful signals.

Another example is performed by Prasetyawan et al. Here, Prasetyawan and Nakamoto (2021) performed an interference removal in odorant samples using ICA in mass spectrum space. They utilized 185 pure essential oils and 5 fixatives as interference. They also obtained mass spectrometry of each essential oil and fixative for the study. Mass spectrum of essential oils and five fixatives

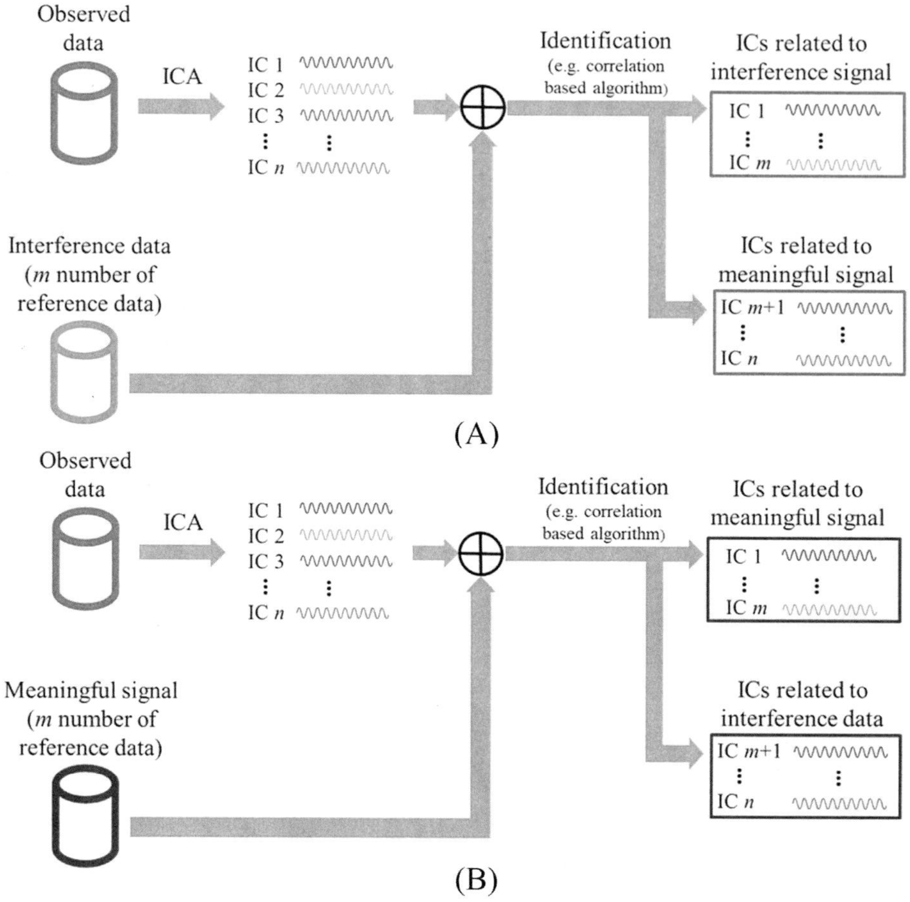

FIGURE 13.3 Illustration of ICA for interference suppression. (A) When interference data is being used as reference data and (B) when meaningful signal is being used as reference data. *ICA*, Independent component analysis.

is shown in Fig. 13.4. The concept of the study is shown in Fig. 13.5.

First, they tried to imitate widely available odorant samples by creating an artificial mixture of odorant samples. Each mixture data contains an essential oil and five fixatives in random ratio. Here, ICA was applied to the 100 mixture data to obtain the ICs. The number of ICs to be sought was six, which corresponds to five fixatives and a pure sample. Then the identification of ICs correlated to the fixatives was obtained first using Pearson's correlation coefficient. After identifying five fixatives, one remaining IC was considered a pure odor sample without the influence of fixatives. An example of the identification result is shown in Table 13.1. Here, the highest correlation coefficient in each column indicates the IC corresponding to the odor sample or fixative in the row. They repeated the same procedure for 185 essential oil samples. For easier understanding, Fig. 13.6 shows the illustration of their procedure of removal of interference using ICA in odorant samples. Here, they successfully

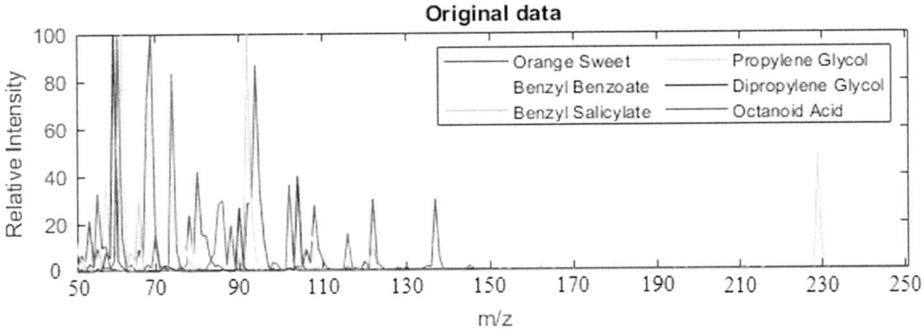

FIGURE 13.4 Mass spectrum of essential oil (*orange* sweet) and fixatives. Note the difference of essential oil and fixatives mass spectra. *Source: From Prasetyawan, D., & Nakamoto, T. (2021).* Suppression of interference of fixative with odorant samples in mass spectrum using ICA. *IEEE Access (pp. 1), https://doi.org/10.1109/ACCESS.2021.3115760.*

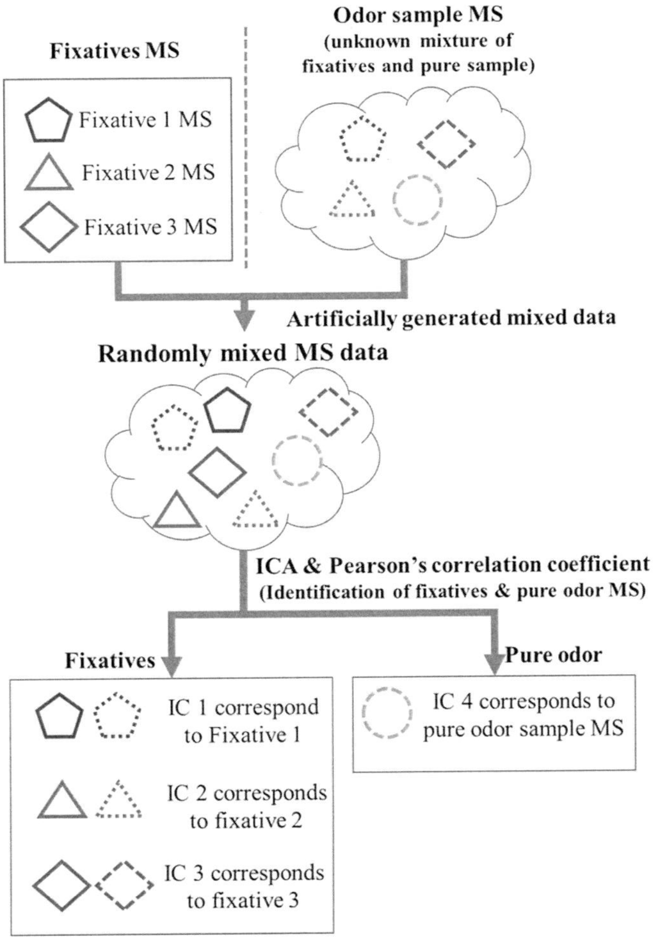

FIGURE 13.5 Concept of removal of interference using ICA in odorant samples. Although separation of interferences can be performed using ICA, additional method is needed to identify the interferences. *ICA*, Independent component analysis. *Source: From Prasetyawan, D., & Nakamoto, T. (2021).* Suppression of interference of fixative with odorant samples in mass spectrum using ICA. *IEEE Access (pp. 1), https://doi.org/10.1109/ACCESS.2021.3115760.*

TABLE 13.1 Correlation coefficient between independent components and original data (orange sweet and five fixatives).

	IC1	IC2	IC3	IC4	IC5	IC6
Orange sweet	0.19	0.98	0	0.01	0	0.07
Benzyl benzoate	0.51	0.01	0	0.01	0.01	0.86
Benzyl salicylate	0.99	0.03	0	0.01	0.01	0.06
Propylene glycol	0	0	0.1	0.99	0.05	0
Dipropylene glycol	0	0.01	0.99	0.02	0.03	0
Octanoic acid	0.01	0.01	0.02	0.1	0.99	0.01

Source: From Prasetyawan, D., & Nakamoto, T. (2021). Suppression of interference of fixative with odorant samples in mass spectrum using ICA. IEEE Access (pp. 1), https://doi.org/10.1109/ACCESS.2021.3115760.

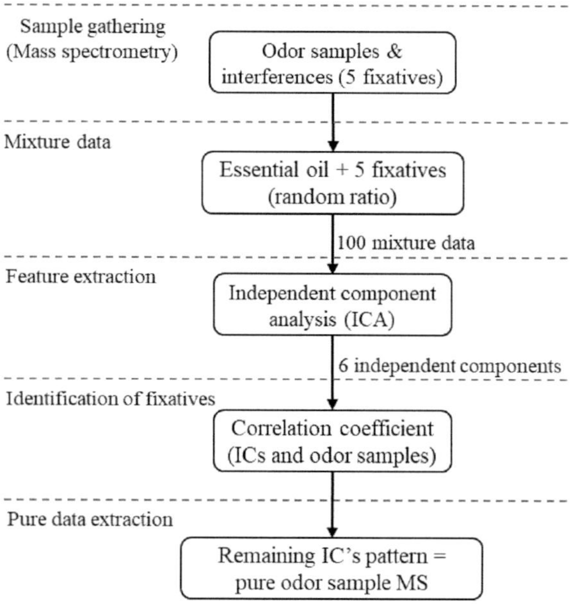

FIGURE 13.6 Procedure of removal of interference using ICA. Remaining IC that is not identified as interference is mass spectrum of pure odor sample. *IC*, Independent component; *ICA*, independent component analysis. *Source: From Prasetyawan, D., & Nakamoto, T. (2021).* Suppression of interference of fixative with odorant samples in mass spectrum using ICA. *IEEE Access (pp. 1), https://doi.org/10.1109/ACCESS.2021.3115760.*

separated the pure odor samples without the influence of fixatives with correlation coefficient of 0.949 (average of 185 samples) (Prasetyawan & Nakamoto, 2021).

The advantage of interference suppression using ICA in machine olfaction is the robustness of ICA in extracting meaningful signal using different methods of sensor measurements. The method enhanced the sample identification in machine olfaction because of the absence of influence caused by interference. This was shown by Di Natale et al. (2002), which can increase the classification rate from 69% before the removal of disturbance by ICA to 89% after the removal of disturbance of ICA. On the other hand, the disadvantage of ICA in interference suppression is when the cross-sensitivity between interference and signals that we seek occurs during measurement. It can cause meaningful signal that might overlap with interference in specific properties to be regarded as interference and to be suppressed.

13.3 Interference removal in odor reproduction analysis using independent component analysis

Previously, the overview of ICA in interference suppression in machine olfaction is briefly explained. This section explains the odor reproduction analysis that utilizes an interference-suppressed odorant sample. As mentioned before, there is a problem when we want to increase the number of odorant samples being used in odorant reproduction. Odorant samples widely available daily are not typically pure odorous compounds in their ingredients (Teixeira et al., 2013; vom Ende et al., 2017). Hence performing interference removal is significant before performing odor reproduction analysis.

In the previous section, interference suppression using ICA performed by Prasetyawan et al. is explained. Their study went further by performing interference suppression using ICA with physical mixture samples, followed by odor reproduction analysis (Prasetyawan et al., 2023). This physical mixture is referred to as an imitation of actual odorant samples where odorless fixatives are mixed within. They created mixture samples using 180 essential oils and 5 fixatives. Here, they successfully extracted essential oil samples without the influence of interference using ICA with a correlation coefficient between the pure sample and the extracted sample of 0.874.

The following odorant reproduction analysis was performed by calculating nonnegative matrix factorization with Itakura–Saito divergence in numerical space, followed by the nonnegative least squares method to obtain a physical recipe for the odor reproduction analysis (Prasetyawan & Takamichi, 2020). They performed odor reproduction using 20 odor components obtained from both original essential oils and extracted essential oils by ICA.

The odor reproduction analysis was evaluated by comparing the target and approximated odors. Here, the performance of odor reproduction by a set of 20 odor components was obtained from both original essential oils and extracted essential oils by ICA. Seven different reference samples (clove bud, elemi, orange sweet, sandalwood, chamomile maroc, *cajeput*, and *mentha pipetha mitchan)* were approximated, and a duo–trio sensory test was performed for the evaluation. The aim of the duo–trio sensory test was to evaluate whether the smell difference existed between approximated odor from original essential oils and extracted essential oils (Meilgaard et al., 2016). Illustration of duo–trio sensory test is shown in Fig. 13.7.

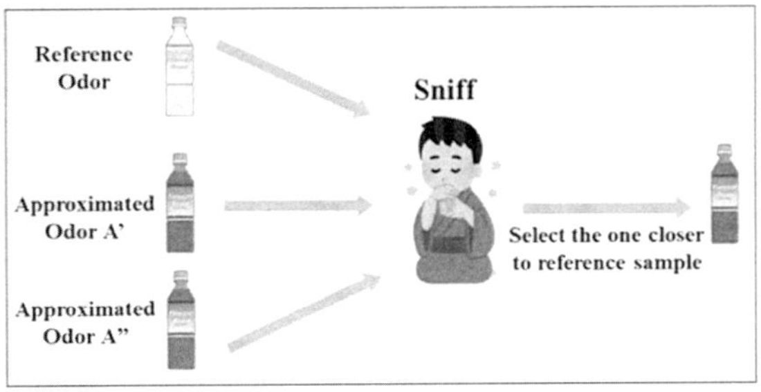

FIGURE 13.7 Illustration of duo–trio test. In the duo–-trio sensory test, a subject was first exposed to the reference odors. Then the subject needed to sniff two approximated odors. Later, the subject needed to choose one of the approximated odors closer to the reference odor. *Source: From Prasetyawan, D., Kashiwagi, Y., & Nakamoto T. (2023).* Sensory evaluation of interference suppression in odor reproduction using mass spectrometry. *IEEE Access. https://doi.org/10.1109/ACCESS.2023.3253891.*

An example of a comparison of the target and approximated odor from their study is shown in Fig. 13.8A and B. Here, we can notice that both sets of odor components can approximate target odor with similar performance (Prasetyawan et al., 2023).

Moreover, their duo–trio sensory result is shown in Table 13.2. *z*-Score was calculated to evaluate if there was any significant difference in tested samples. The *z*-score is a statistical measurement that describes a relationship between samples. Here, the *z*-score ($P < .05$) must be below 1.717 to prove that there is no significant difference between tested samples in the duo–-trio test. Since the *z*-score was below 1.717, it was proved that there is no significant difference between tested samples in the duo–trio test. Their method proved that the odor reproduction analysis could be done without performance deterioration from the case of original pure samples even after the suppression of interference (Prasetyawan et al., 2023).

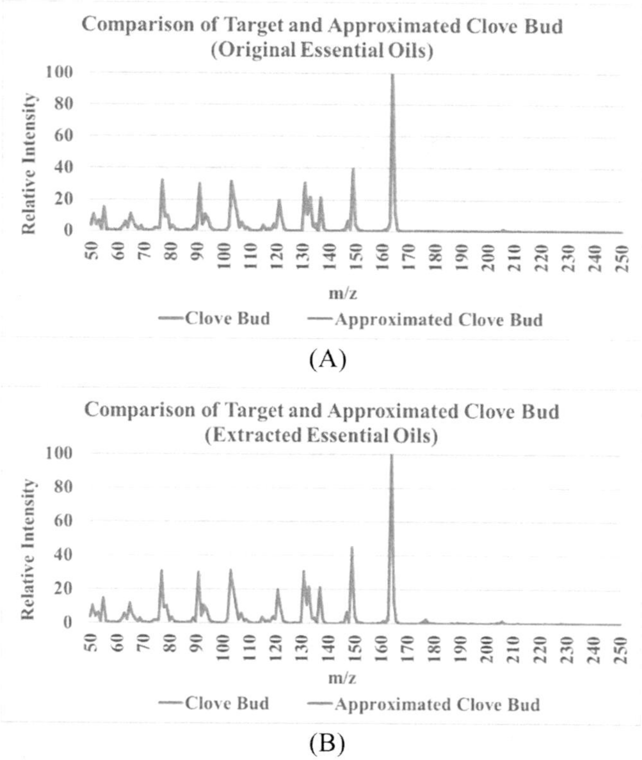

FIGURE 13.8 Comparison of clove bud reference odor and approximated odor. The odor approximation using odor components based on (A) original essential oils and (B) extracted essential oils. *Source: From Prasetyawan, D., Kashiwagi, Y., & Nakamoto T. (2023).* Sensory evaluation of interference suppression in odor reproduction using mass spectrometry. *IEEE Access. https://doi.org/10.1109/ACCESS.2023.3253891.*

TABLE 13.2 Duo–trio sensory test result by Prasetyawan D, et al.

	Subject response		
Reference odor (target essential oil)	**Approximated by original essential oil**	**Approximated by extracted essential oil**	***z*-Score**
Clove bud	14	8	1.279
Elemi	8	14	1.279
Orange sweet	12	10	0.426
Sandalwood	9	13	0.853
Chamomile (Maroc)	10	12	0.426
Cajeput	7	15	1.706
Mentha piperita mitcham	9	13	0.853
Total test number	69	85	1.289

Source: From Prasetyawan, D., Kashiwagi, Y., & Nakamoto T. (2023). Sensory evaluation of interference suppression in odor reproduction using mass spectrometry. IEEE Access. https://doi.org/10.1109/ACCESS.2023.3253891.

13.4 Conclusion

Several studies related to interference suppression in machine olfaction are presented in this chapter. Studies that mainly utilize ICA for interference suppression are introduced. The advantage of ICA in performing source separation is that it is very useful in interference suppression. As long as the reference data to be suppressed or kept is present, interference suppression can be performed successfully. It is a beneficial method in suppressing interference in numerical space without additional filter to the machine olfaction, which can cause complexity in the measurement system. Moreover, in part of odor reproduction analysis, it opens the way to explore a much wider range of odorant samples to be analyzed without being hindered by the influence of interferences.

References

Aguilera, T., Lozano, J., Paredes, J. A., Álvarez, F. J., & Suárez, J. I. (2012). Electronic nose based on independent component analysis combined with partial least squares and artificial neural networks for wine prediction. *Sensors*, *12*(6), 8055–8072. Available from https://doi.org/10.3390/S120608055.

Di Natale, C., Martinelli, E., & D'Amico, A. (2002). Counteraction of environmental disturbances of electronic nose data by independent component analysis. *Sensors and Actuators B: Chemical*, *82*(2-3), 158–165. Available from https://doi.org/10.1016/S0925-4005(01)01001-2.

Hyvärinen, A., Hurri, J., & Hoyer, P. O. (2009). *Independent component analysis* (pp. 151–175). Springer. Available from http://doi.org/10.1007/978-1-84882-491-1_7.

Liang, Z., Tian, F., Yang, S. X., Zhang, C., Sun, H., & Liu, T. (2018). Study on interference suppression algorithms for electronic noses: A review. *Sensors*, *18*(4), 1179. Available from https://doi.org/10.3390/S18041179.

Liang, Z., Xue, Q., Tian, F., Xu, C., Wang, C., Yang, L., & Guo, T. (2022). A sparse reconstruction domain transfer method for interference suppression in artificial olfactory system. *IEEE Sensors Journal*, *22*(7), 6717–6730. Available from https://doi.org/10.1109/JSEN.2022.3154647.

Liu, T., Guo, L., Wang, M., Su, C., Wang, D., Dong, H., Chen, J., & Wu, W. (2023). Review on algorithm design in electronic noses: Challenges, status, and trends. *Intelligent Computing*, 2. Available from https://doi.org/10.34133/ICOMPUTING.0012.

Meilgaard, M. C., Civille, G. V., & Carr, B. T. (2016). *Overall difference tests: Does a sensory difference exist between samples? Sensory evaluation techniques* (5th, pp. 79–122). CRC Press.

Meng, X. (2010). *ICA algorithm based on intelligent electronic nose in the mixed gas of feature extraction.* 2010 6th International Conference on Wireless Communications, Networking and Mobile Computing, WiCOM 2010. 9781424437092. Available from https://doi.org/10.1109/WICOM.2010.5600613.

Mi, J. X. (2014). A novel algorithm for independent component analysis with reference and methods for its applications. *PLoS ONE*, *9*(5), 93984. Available from https://doi.org/10.1371/journal.pone.0093984.

Nakamoto, T., Ohno, M., & Nihei, Y. (2012). Odor approximation using mass spectrometry. *IEEE Sensors Journal*, *12*(11), 3225–3231. Available from https://doi.org/10.1109/JSEN.2012.2190506.

Prasetyawan D., Y. Kashiwagi, T. Nakamoto. (2023). Sensory evaluation of interference suppression in odor reproduction using mass spectrometry. IEEE Access, doi:10.1109/ACCESS.2023.3253891.

Prasetyawan D., T. Nakamoto, (2021). Suppression of interference of fixative with odorant samples in mass spectrum using ICA. IEEE Access, 1, Available from https://doi.org/10.1109/ACCESS.2021.3115760.

Prasetyawan, D., & Nakamoto, T. (2022). *Separation of pure perfume mass spectrum from interferences using ICA.* International Symposium on Olfaction and Electronic Nose, ISOEN2022—Proceedings. 9781665458603. Available from https://doi.org/10.1109/ISOEN54820.2022.9789671.

Prasetyawan, D., & Takamichi, N. (2020). Sensory evaluation of odor approximation using NMF with Kullback–Leibler divergence and Itakura-Saito divergence in mass spectrum space. *Journal of the Electrochemical Society*, *167*(16), 167520. Available from https://doi.org/10.1149/1945-7111/abd110.

Rodrigues, A. E., Nogueira, I., & Faria, R. P. V. (2021). Perfume and flavor engineering: A chemical engineering perspective. *Molecules (Basel, Switzerland)*, *26*(11), 3095. Available from https://doi.org/10.3390/MOLECULES26113095.

Scott, S. M., James, D., & Ali, Z. (2006). Data analysis for electronic nose systems. *Microchimica Acta*, *156*(3-4), 183–207. Available from https://doi.org/10.1007/S00604-006-0623-9.

Stone, J. V. (2002). Independent component analysis: An introduction. *Trends in Cognitive Sciences*, *6*(2)), 59–64. Available from https://doi.org/10.1016/S1364-6613(00)01813-1.

Tanskanen, J. M. A., Viik, J. J., Tanskanen, J. M. A., & Viik, J. J. (2012). Independent component analysis in ECG signal processing. *Advances in Electrocardiograms—Methods and Analysis*. Available from https://doi.org/10.5772/

22719, 978-953-307-923-3. Available from, http://undefined/chapters/27022.

Teixeira, M. A., Rodríguez, O., Gomes, P., Mata, V., & Rodrigues, A. E. (2013). Design of perfumes. *Perfume Engineering*, 15–60. Available from https://doi.org/10.1016/B978-0-08-099399-7.00002-X.

Tian, F., Zhang, J., Yang, S. X., Zhao, Z., Liang, Z., Liu, Y., & Wang, D. (2016). Suppression of strong background interference on e-nose sensors in an open country environment. *Sensors (Switzerland)*, *16*(2). Available from https://doi.org/10.3390/S16020233.

vom Ende, M., Sturm, W., & Peters, K. (2017). *Perfumes Ullmann's Encyclopedia of industrial chemistry* (pp. 1–7). Wiley-VCH Verlag GmbH & Co. KGaA. Available from http://doi.org/10.1002/14356007.a19_171.pub2.

Wang, H., Zhao, Z., Wang, Z., Xu, G., & Wang, L. (2020). Independent component analysis-based baseline drift interference suppression of portable spectrometer for optical electronic nose of Internet of Things. *IEEE Transactions on Industrial Informatics*, *16*(4), 2698–2706. Available from https://doi.org/10.1109/TII.2019.2939645.

Zhang, L., Tian, F., Dang, L., Li, G., Peng, X., Yin, X., & Liu, S. (2013). A novel background interferences elimination method in electronic nose using pattern recognition. *Sensors and Actuators A: Physical*, *201*, 254–263. Available from https://doi.org/10.1016/J.SNA.2013.07.032.

SECTION 4

Olfactory displays

CHAPTER

14

Fundamental techniques of olfactory displays

Takamichi Nakamoto

Institute of Integrated Research, Institute of Science Tokyo, Yokohama, Japan

14.1 Background

An olfactory display is a device to present a scent to a human. Frequently asked question is as follows:

Is this a television to generate a smell?

Does a smell come from a computer screen?

Although a display can provide some sensation for a human, it provides visual sensation in most of cases. However, there is a haptic display to give a haptic sensation to a human. In the case of olfaction, we use an olfactory display to provide olfactory sensation for a human.

History of olfactory display is relatively new compared with other sensations. Heilig (1962) proposed Sensorama, including the smell presentation. He tried to simulate experience when he rode on a motorcycle to go to many places around a city. The simulator included olfactory experience such as food scents. Later, Toyota Motor Corporation presented smells at their showroom. Shiseido, perfume company, made Kaori on demand to emit scents. Kaori is a Japanese word meaning a scent. The equipment was aimed to generate scents at the remote site through the web.

Other smell presentation equipment called olfactometer has been developed. It is intended to give an olfactory stimulus to a subject accurately (Livneh & Paz, 2010; Johnson & Sobel, 2007). Since it is expensive and bulky, only big companies and research institutes own the olfactometers. It is preferable to have cheap and easy-to-use apparatus. Thus an olfactory display has been studied by researchers in virtual reality.

The olfactory display consists of odor generation part and odor delivery one. The methods to generate and deliver odors to a human are described in this chapter. Then, the actual developed devices of olfactory displays are explained. Moreover, the applications of the olfactory display are shown although the digital contents with scents are described in Chapter 17.

14.2 Odor generation method

14.2.1 Aroma diffuser

An aroma diffuser is the gadget to generate a smell (Du et al., 2022). It is widely spread on the

Digital Technologies in Olfaction
DOI: https://doi.org/10.1016/B978-0-443-15721-9.00014-2

market, and many products are commercially available. People can enjoy a scent diffused in a room or car cabin using the aroma diffuser. Especially, the aroma diffuser is used to spread a scent of an essential oil since the essential oil has various effect, such as relaxation, better sleep, pain relief, and improved mood (Lee et al., 2017). One of examples of the aroma diffusers is shown in Fig. 14.1. It is similar to a humidifier but its appearance is more fashionable.

FIGURE 14.1 **Photo of aroma diffuser**. Aroma diffuser is a gadget to generate a smell.

There are several types of aroma diffusers. Ultrasonic method is to atomize water with scent. Water is necessary for this method. Reed aroma diffuser uses a stick which sucks up essential oil, followed by its evaporation. Aroma pot type heats a glass bottle with essential oil. It is also called aroma light diffuser since it can be also used as light. Moreover, fan can be used to spread the aroma.

Although the aroma diffuser is simple and easy-to-use, they cannot control the intensity of aroma precisely. Moreover, it does not have blending function described in Chapter 15.

14.2.2 Diffusion and permeation methods

There is vapor generation equipment called permeator. That equipment enables both diffusion and permeation methods, which allow us to generate a vapor with relatively low concentration precisely (Barratt, 1981; Nakamoto, 2003). The diffusion method is illustrated in Fig. 14.2. A glass bottle with thin tube is placed in the enclosure where the air flows very fast. The vapor evaporated from the liquid surface diffuses through thin tube little by little. Since the flow rate in the enclosure is very high, the vapor can be highly diluted with the air with

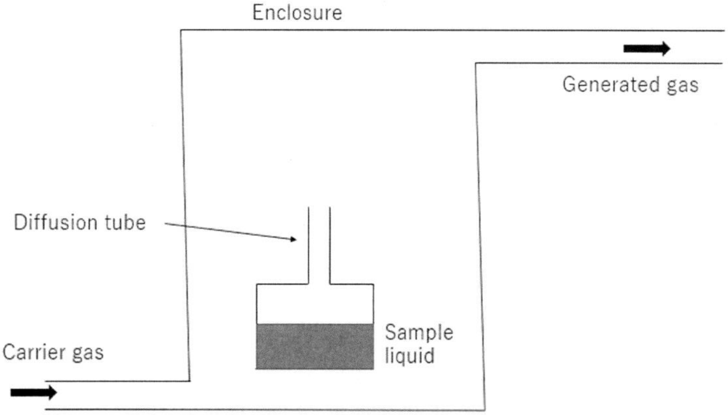

FIGURE 14.2 **Principle of diffusion method to generate vapor**. This method is suitable for generating a vapor with low concentration.

high accuracy. The temperature of the enclosure should be kept constant. A user can adjust the vapor concentration with the air flow rate and the enclosure temperature. Moreover, the inner diameter of the diffusion tube and its length influence the vapor concentration. Typically, ppm-range concentration can be realized using the diffusion method. The drawback of this method is that long time is required for the stabilization of the concentration. Moreover, the concentration cannot be easily adjusted, and the concentration range cannot be freely selected.

Another method is to use permeation tube. It is difficult to generate the vapor of highly volatile compound with high stability and accuracy using diffusion method. It is very difficult to keep the concentration of highly volatile compound, that is, methyl mercaptan, hydrogen sulfide, acetaldehyde, and so on, constant. Those compounds are often deleterious. The liquid compound is enclosed in a tube, called permeation tube, made of fluorine-based resin. Its structure is shown in Fig. 14.3. That compound moves from high concentration to low concentration across its wall. Its permeation rate across the tube wall is constant if the temperature is kept the same. The vapor can be diluted with the air in the same way as the diffusion method. Thus the permeator can be used both for diffusion and permeation methods.

The calibration method is to measure the weight change of diffusion or permeation tube when the air flows for long time although the manufacturer prepares for the equation to calculate the vapor concentration. The equipment to realize diffusion method as well as permeation method is commercially available.

Those methods are suitable for generating vapor with low concentration stably. However, it takes long time to change the vapor concentration and a user cannot freely select the concentration.

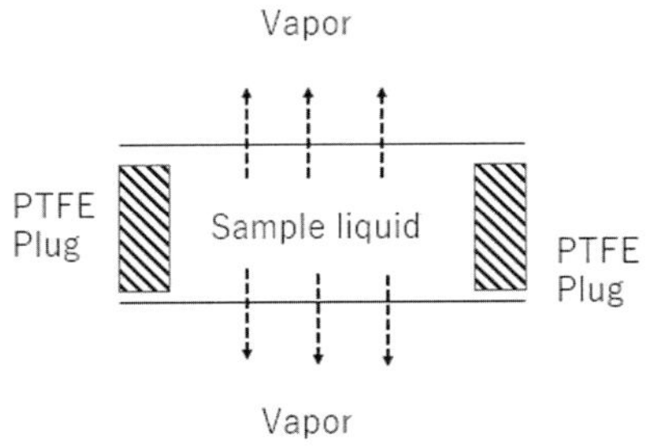

FIGURE 14.3 Permeation tube used for vapor generation. Permeation tube is often used to generate a highly volatile compound.

14.2.3 Headspace and bubbling method

It is easy to generate the vapor using headspace and bubbling method. Fig. 14.4 shows a headspace method. The headspace is the space just above the liquid in a bottle. The carrier gas such as dry air or nitrogen flows into the bottle through the tube and pushes the vapor evaporated from the liquid surface toward the outlet. This method is often used to supply the vapor to sensors.

The headspace vapor is not typically saturated and its concentration depends upon the temperature and flow rate. Moreover, it is necessary to keep the distance between the liquid surface and the tips of the tubes constant when we keep the vapor concentration accurately. The vapor concentration varies according to its distance.

The headspace method is an easy method and has been used for sensor measurements

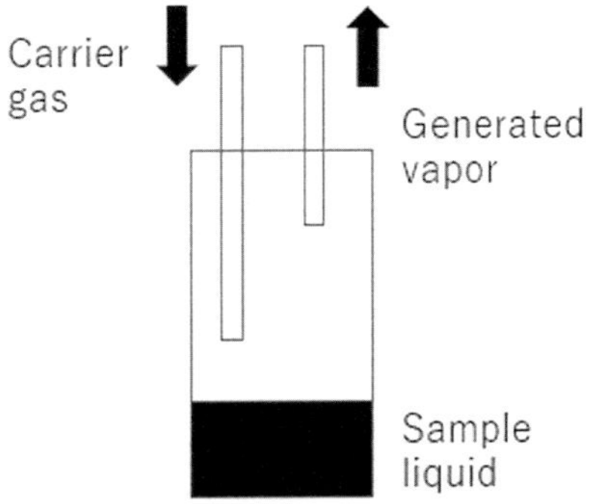

FIGURE 14.4 Headspace method to generate vapor. The headspace is the space just above the liquid in a bottle.

(Gardner et al., 1994; Nakamoto et al., 1993). However, the vapor concentration is not known and it varies during the vapor generation.

When the inlet tube is inserted into the liquid, we can make a bubbler as is illustrated in Fig. 14.5. A carrier gas is passed though the liquid in the bottle and the generated vapor is carried to the outlet. Even if the bubbling method is used, the headspace over the liquid is not always saturated. Glass particles are sometimes put in the liquid to increase the area of contact between the liquid and the carrier gas. Another problem is that tiny liquid particles, not vaporized ones, are sometimes generated due to heavy bubbling at a fast flow rate.

14.2.4 Sampling bag method

The sampling bag is used when we would like to generate the vapor at the specified concentration. The vapor is generated after a liquid sample is injected to the sampling bag by a syringe, followed by its evaporation in the bag. The vapor in the bag is then sucked out using an air pump as is illustrated in Fig. 14.6. The vapor concentration in the sampling bag is determined by the injected liquid volume and the air volume in the sampling bag. Mass flow controller (MFC) is sometimes used to keep the flow rate constant (Nakamoto et al., 1994). In the case of sensor measurement, a sensor chamber is located between MFC and the air pump to avoid the influence of vapor adsorption at the internal wall of the air pump.

The material of the sampling bag should be carefully selected to avoid the permeation of water and other molecules through its wall. The generation of the vapor from the internal wall of a certain plastic bag can also occur. Adsorption of the sampled vapor inside the bag cannot be ignored in case of low vapor concentration. A fluorine-containing resin bag is the best one owing to low permeability of vapor and low adsorption characteristic although it is expensive. Instead of sampling bag, a glass vessel is sometimes used. However, careful handling is required since it can be easily broken. A canister is also used since its handling is easier than the glass vessel.

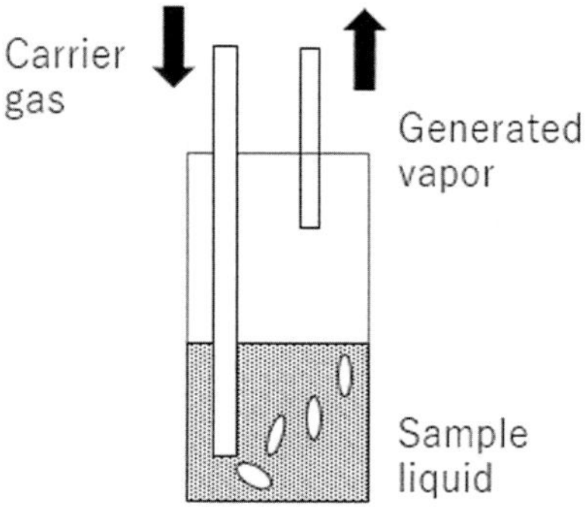

FIGURE 14.5 Bubbling method to generate vapor. A carrier gas gets through the liquid sample and the generated vapor is carried to the outlet.

14.2.5 Heating method

Since most of scents are generated from the liquid, it can be evaporated using a heater. The simple method is illustrated in Fig. 14.7. If it is necessary to take constant volume of the liquid precisely, a sample loop is often used (Somboon et al., 2007). Micro dispensers based on solenoid valve or inkjet device can be used to supply the liquid to the heater.

A mesh heater can be used if it is necessary to pass the vapor though the heater (Nakamoto

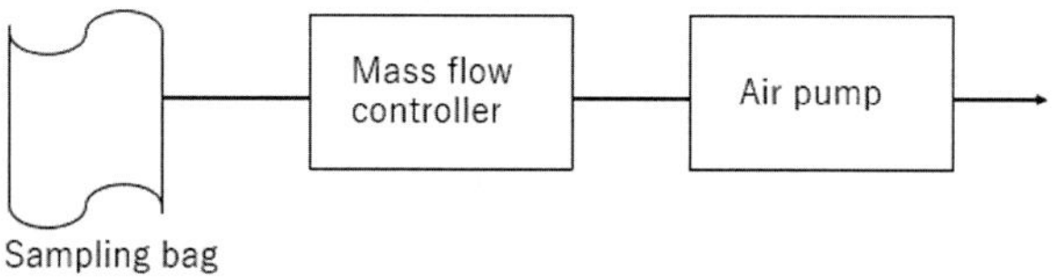

FIGURE 14.6 Sampling bag method to generate vapor. The vapor in the sampling bag is sucked out using an air pump.

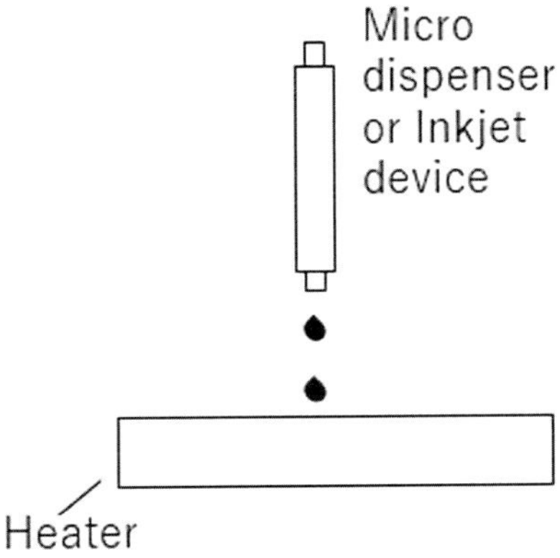

FIGURE 14.7 **Heating method to generate vapor.** Vapor is generated by heating a liquid droplet.

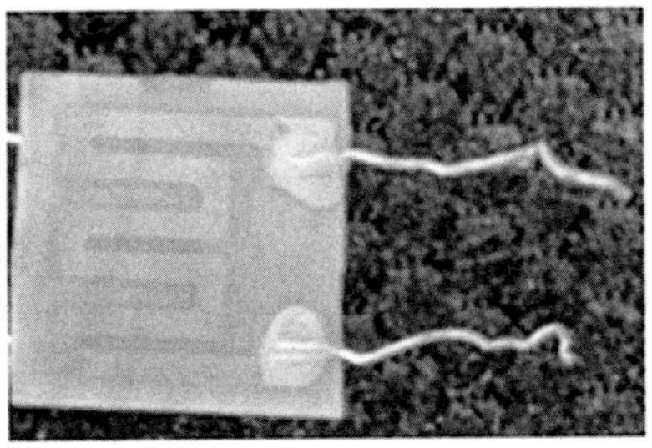

FIGURE 14.8 **Heater electrode formed at the back side of substrate.** Meandering electrode is used to increase its resistance as well as to form uniform temperature distribution over the substrate.

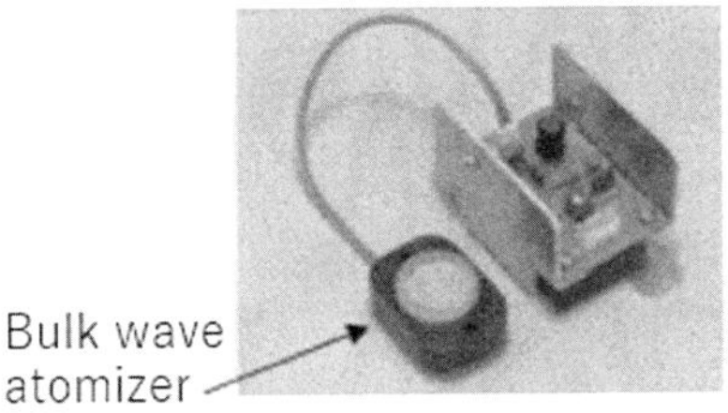

FIGURE 14.9 **Untrasonic atomization device.** Bulk wave is used for atomization.

et al., 2004). A variety of mesh sizes are available from mm range down to a few tens of micro meter range. Since mesh material is stainless in most of cases, its resistance is low. Thus high current is necessary enough to obtain the power to evaporate the liquid sample. The vapor generation using the mesh heater in detail is described in Chapter 15.5.

Since the boiling point of the volatile compound ranges up to 300°C, it is preferable to achieve that temperature. However, the temperature between 150°C and 200°C is often used because it is easily realized. A few-mm-size ceramic heater is often used. Micro electro mechanical system (MEMS) is sometimes used. A meandering electrode is formed backside of the substrate as heater. Its photo is shown in Fig. 14.8. A heating element suspended by beams is sometimes used to reduce the heat capacity. Several olfactory-display examples using MEMS heaters are shown in Section 14.4.

14.2.6 Atomization method

Atomization is used to generate the mist. The most popular atomization method is to spray the liquid. Other method is to take up the liquid by chemisorption. In the case of an olfactory display, an ultrasonic method is often used.

There are two types of ultrasonic atomization. The mist is generated instantaneously and no heat is required in the ultrasonic atomization. One is to use a bulk wave device and the other is to use a surface acoustic wave (SAW) device (Kurosawa et al., 1995). In the case of bulk wave device, the mist is generated when the bulk acoustic wave is illuminated from the bottom to the liquid surface. Its device is shown in Fig. 14.9. The atomization devices are commercially available. The application of the atomization is nebulizer for medical use, the coating of a sensing film onto the gas sensor device such as quartz crystal microbalance (Muñoz-Aguirre et al., 2005), and so on.

A SAW atomizer is the other type of the ultrasonic atomization (Kurosawa et al., 1995; Chono et al., 2004; Yabe et al., 2014). Although SAW device is explained as a sensor in Chapter 3, SAW atomizer works as an actuator for liquid droplet. SAW from inter digital transducer propagates along the piezoelectric substrate, for example, lithium niobate. When a liquid droplet is placed onto the substrate along the SAW propagation path, it causes liquid motions, such as vibration, flow, soaring, and atomization. The vibration occurs in case of small power, whereas atomization occurs in case of large SAW power. The application of SAW atomizer to olfactory display is shown in Section 15.6.

Example of SAW atomization is shown in Fig. 14.10. When water soaked into a cotton swab is put on a surface of a piezoelectric substrate, its mist is instantaneously generated by SAW energy. The atomization method can be used without tubing for odor delivery. Therefore it is possible to achieve the quick odor delivery.

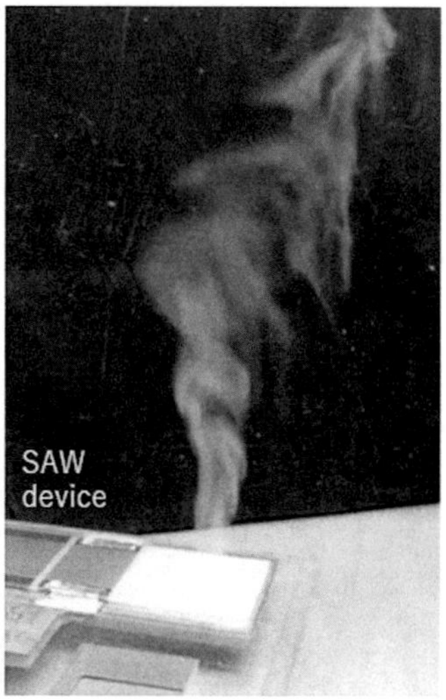

FIGURE 14.10 **Photo of atomization behavior caused by a SAW device**. The mist is instantaneously generated. *SAW*, Surface acoustic wave.

14.3 Odor delivery method

The odor delivery system, which delivers the odor to human nose, is also important for realizing an olfactory display. This section describes several odor delivery methods.

14.3.1 Tube method

The most popular method to deliver odor to human nose is to utilize a tube. A tube is used to deliver the odor from the odor generation part to the human nose as is illustrated in Fig. 14.11. The tube is placed along the headset so that its tip can be kept close to the nose. The olfactory display using the solenoid valves was connected to the tube in this figure. The vapor

FIGURE 14.11 **Odor delivery system to utilize tube**. The tip of the tube is placed close to her nose.

generation methods, such as the diffusion method, permeation method, head space, bubbler, and sampling bag methods, can be used when the odor is delivered via tube.

Although we can easily deliver the odor to the human nose via tube, there is a problem when the low-volatile odor is generated. There is no problem when the vapor with an appropriate volatility is generated. However, the tube causes the significant delay for the low-volatile compound.

The behavior of low-volatile compound is illustrated in Fig. 14.12. The behavior similar to gas chromatography occurs even if there is no coating at the internal wall of the tube. The detail is explained in the reference (Nakamoto, 2016). The vapor close to the wall interacts with the wall material and the delay occurs at the tube. The delay occurs even if we use Teflon or stainless tube. If the vapor pulse is applied, its peak becomes dull and spreads broadly as is shown in Fig. 14.12. This behavior can be avoided when the tube temperature is kept high, for example, 200°C. However, it is very difficult to keep the temperature high along the all the path. It is dangerous if the heated tube is located close to the human nose. Moreover, the vapor path often includes solenoid valves, which cannot work at the high temperature.

The behavior similar to gas chromatography occurs only for low-volatile compound. We can deliver the odor via tube if the sample volatility is appropriate. However, the odor with high odor intensity often includes low-volatile compounds. For example, a perfume typically includes many low-volatile compounds. Thus it is necessary to avoid tubing if many samples include low-volatile compounds.

14.3.2 Wind method

In some cases, we can utilize the wind to deliver the odor to the human nose. A fan is often used for this purpose. The chromatographic behavior does not occur when we use the wind. Thus it can be combined with the atomization method. When we combine SAW

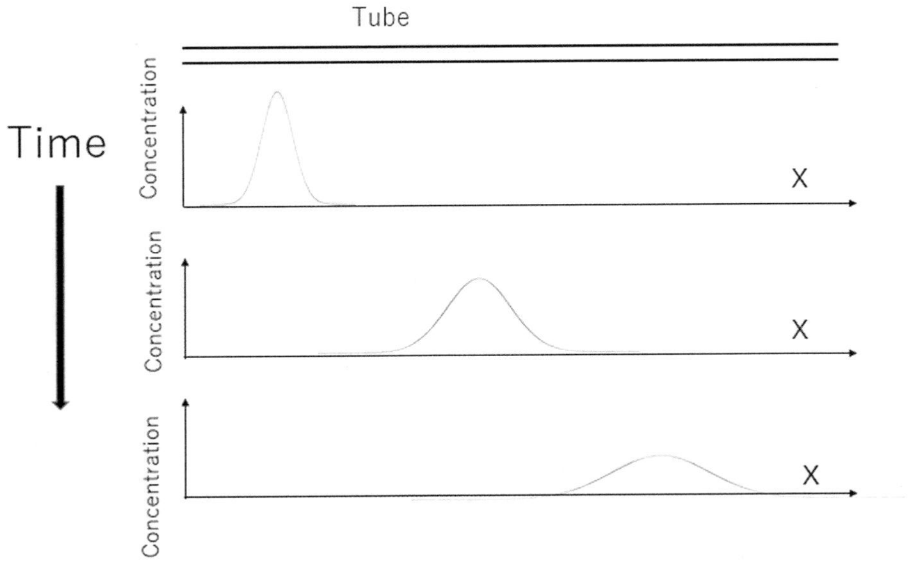

FIGURE 14.12 **Odor delivery via tube for low-volatile compound.** Peak concentration becomes smaller and its profile becomes dull as time increases.

atomizer with the fan, the fan can be placed at the rear of SAW atomizer or the front of SAW atomizer as is illustrated in Fig. 14.13. When the fan is placed at the rear of SAW atomizer, the odor is not attached to the fan. The fan is always clean. However, it is sometimes difficult to use the blending capability. The mixing is insufficient when the location of each odor source (atomization point) is a little different. When the fan is placed between the odor source and human nose as is illustrated in Fig. 14.13B, the odor might attach the fan wings. Thus the fan should be periodically cleaned. However, the mixing capability is sufficient when the fan at the font location is adopted.

The other point to be considered when we use the fan is the wind speed. Several olfactory displays use high wind speed. We can feel the wind together with the scents at that time. The smell persistence is somewhat better when we use high-speed wind. However, we should think about the effect of the wind as well as smell effect. Especially, the sensory test result is sometimes influenced by the wind speed. Thus the wind speed kept at the appropriate level is recommended.

14.3.3 Air cannon

The air cannon is sometimes used to deliver the odor to the human nose (Yanagida, 2012). It fires air vortices at a user away from the device. Its structure is illustrated in Fig. 14.14. The cannon is armed by pulling the diaphragm out and then is fired by releasing the diaphragm. The diaphragm quickly pushes the air out of the open end, creating a vortex ring.

The air cannon enables an odor to be delivered to only a single person without spreading it in space. People around the target person do not perceive odors, whereas the target person sniffs it. Therefore one Professor says that it can be used to wake up a sleeping student at a class!

The problem of the air cannon method is to feel the wind together with the odor. The wind might let the user feel the impression different from that of the only odor. To avoid the wind influence, it is possible to use two air cannons to collide two vortex rings in front of a user so that the wind from the air cannon can be reduced.

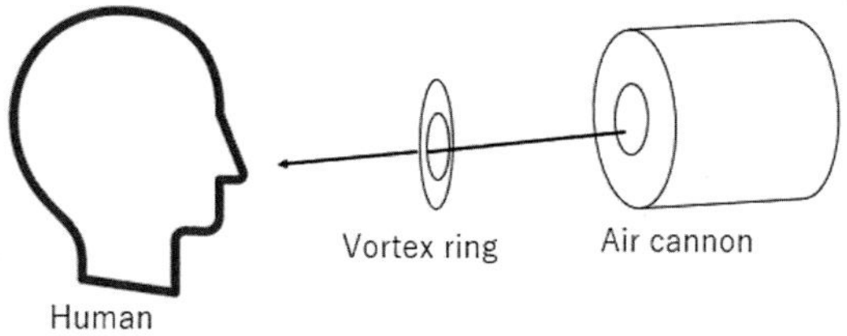

FIGURE 14.14 Air cannon used for odor delivery. Air cannon enables an odor to be quickly delivered to a target person.

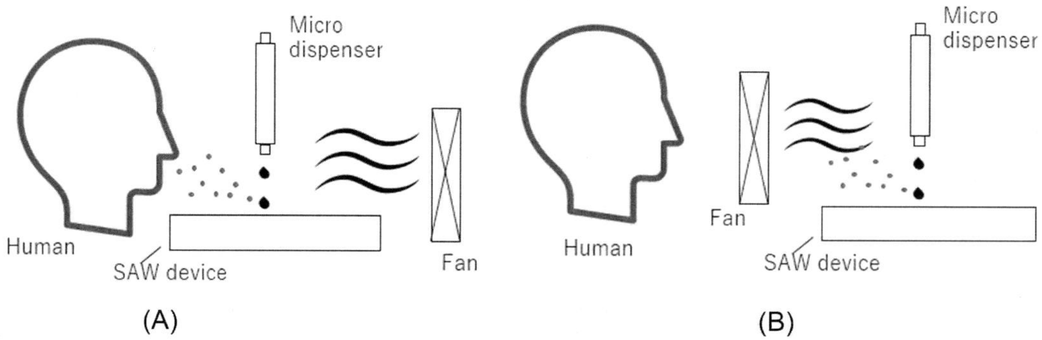

FIGURE 14.13 Location of fan. (A) Front of odor generation part and (B) rear of odor generation part.

The details of air-cannon-based devices are described in the Section 14.4.6.

14.4 Olfactory display devices

A variety of olfactory display devices have been so far proposed since Kaye (2004) made a simple perfume bottle with solenoids. The olfactory display devices except a multicomponent olfactory display are explained here. The multicomponent olfactory display is described in Chapter 15.

14.4.1 Micro electro mechanical system devices

Kim et al. (2011) proposed a concept of a two-dimensional array of aroma release devices. Its element device consists of cone-shaped microchamber made of polydimethylsiloxane. The odorant liquid is stored in the chamber as is illustrated in Fig. 14.15.

There is a matrix made up of these devices with its rows and columns composed of heater lines below the micro chambers. The micro chamber has a small hole on the top of the chamber and it can release an odor due to the increase in the pressure inside the chamber caused by heating. The pressure increase caused by two heater lines can release the odor, whereas its increase caused by a single heater line is not enough. Therefore only the micro chamber at the cross point of the activated row and column lines can release the odor. In this case, the micro chamber located at $X=4$, $Y=3$ releases the aroma. Although Fig. 14.15 shows the concept of a large array of odor-releasing devices, they only fabricated the system with two micro chambers.

Another example of MEMS olfactory display consists of a heater, a reservoir, and a mechanical valve as is shown in Fig. 14.16A (Ishizuka et al., 2018). The perfume in the reservoir is heated, followed by the evaporation. Although the valve is initially closed based on electrostatic force, it is open when the perfume is vaporized as is illustrated in Fig. 14.16B. The amount and the intensity of the released perfume are controlled by the heat time and ON time of the valve. The device size was 3.5 cm × 1.5 cm and the valve size was 1 cm × 1 cm at this stage. The photo of the valve is shown in Fig. 14.16C. They confirmed the vapor release using a metal oxide sensor.

Liu et al. (2023) developed soft and miniaturized olfactory display. They made a tiny odor generator and implemented nine odor generators into the soft face mask for VR application. The odor generator is shown in Fig. 14.17A. Odorous paraffin wax was put into the heating chamber. Thermal actuator based on polyimide supported gold traces with a thermistor was used for melting and solidification of odorous paraffin. A cantilever structured mechanical actuator based on the electromagnetic induction

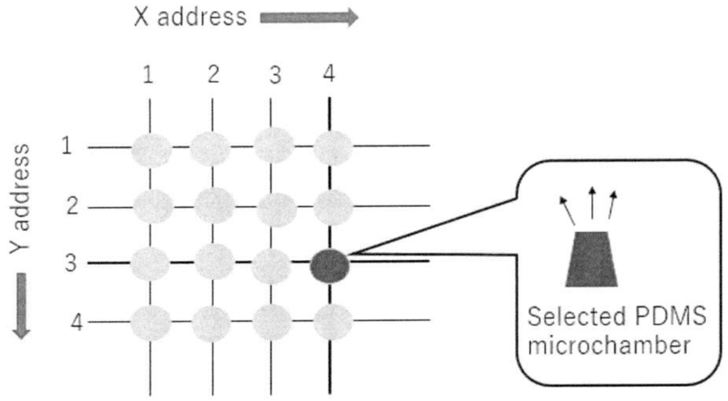

FIGURE 14.15 ***X–Y*** **addressable odor releasing MEMS system**. The micro chamber located at $X=4$ and $Y=3$ releases the odor. *MEMS*, Micro electro mechanical system.

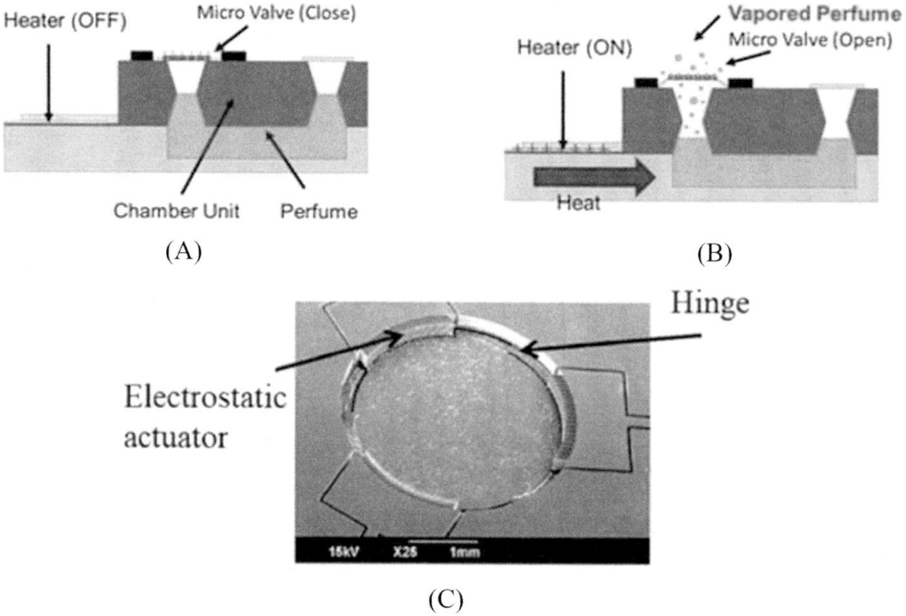

FIGURE 14.16 **Heater-based MEMS olfactory display**. (A) Heater and valve are closed, (B) heater is ON and the valve is open, and (C) photo of valve part. *MEMS*, Microelectromechanical system. *Source: From Ishizuka, H., Kashiwagi, K., Terao, K., Takao, H., & Shimokawa, F. (2018).* Development of a heat-driven-type MEMS olfactory display. *In Proceedings MDPI (pp. 839). MDPI.*

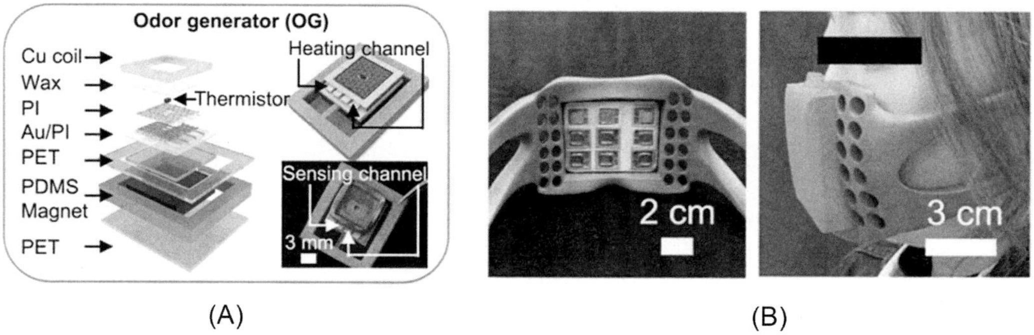

FIGURE 14.17 **Soft and miniaturized olfactory display** (A) Odor generator and (B) optical images of face mask based olfactory display equipped with nine odor generators. *Source: From Liu, Y., Yiu, C. K., Zhao, Z., Park, W., Shi, R., Huang, X., Zeng, Y., Wang, K., Wong, T. H., Jia, S., Zhou, J., Gao, Z., Zhao, L., Yao, K., Li, J., Sha, C., Gao, Y., Zhao, G., Huang, Y., … Yu, X. (2023). Soft, miniaturized, wireless olfactory interface for virtual reality.* Nature Communications. *14(1). https://doi.org/10.1038/s41467-023-37678-4.*

between a copper coil and permanent magnet was used to turn on and off the odor generator. A soft silicone-based frame acting as the shell for the odor generating unit was used to mount whole odor generator onto the face mask fitted into the curved human skin as is shown in Fig. 14.17B. The circuit to drive the odor generators was made on a flexible printed circuit board, including a microcontroller unit, a Bluetooth module, analog circuits, and a battery. They tried to apply that olfactory interface to VR/AR-based online teaching. MEMS technology was used to fabricate the thermal actuator made of polyimide and gold traces.

14.4.2 Heater-based devices

Since the heater-based MEMS devices were explained in the previous section, the heater-based olfactory devices without MEMS are described here.

Kim et al. (2009) developed an aroma-card based olfactory display. They used a chemical container made of temperature responsive hydrogel. It has reversible phase transition between sol and gel, which can be controlled by its temperature. The Peltier module embedded at the bottom of the container was used to determine the temperature. They claimed that the device works soundlessly although the devices based on solenoid valves are noisy.

Covington et al. developed a simple aroma dispenser based on capillary action and a thermal heating element as is illustrated in Fig. 14.18. Two glass capillaries were used and longer one with nichrome coil was immersed in a fragrance (Covington et al., 2018). They put eight dispensers into the system and release the specified smell by heating it even if solenoid valves were not used.

They made another type of simple thermal-based olfactory display (Tiele et al., 2020). They made the aroma capsule made of wax and fragrance. Its capsule was cooled or heated using a Peltier device. They could use up to twelve aromas.

Abid et al. developed heat based delivery apparatus. They developed the rotatable scent cartridge with solid scent materials. Laser illuminated one of the solid scent materials and then generated heat, causing the evaporation of the scent (Abid et al., 2015b).

Dobbelstein et al. (2017) made a wearable olfactory display (iScent) for mobile notification. It can be worn as a pendant around neck contains up to eight scents. The glass fiber was used as adsorbing layer, which delivered the liquid to the heating coil. Although precise control of aroma is not possible, it can be used to notify the information by changing the delivered scent.

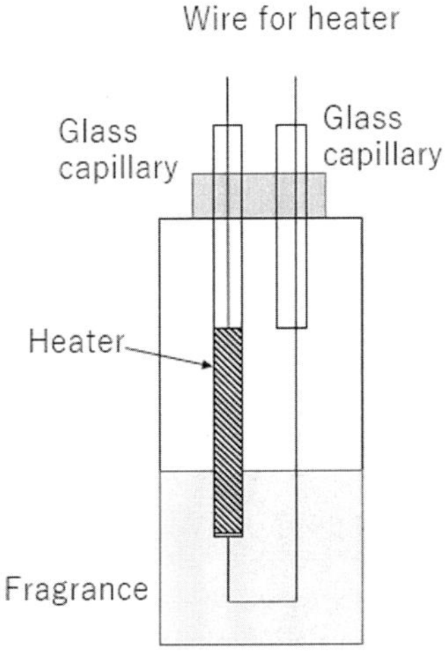

FIGURE 14.18 Capillary arrangement for aroma dispenser. Thermal heating element is used to generate the aroma.

14.4.3 Atomization-based devices

Several researchers have utilized the atomization to generate scents. There are two types of ultrasonic atomization techniques. One is an atomizer using ultrasonic bulk wave and the other utilizes SAW as is described in Section 14.2.6. Since the device utilizing SAW atomization is described in Chapter 15 (multicomponent olfactory display), the device using bulk wave is mainly described here.

Abid et al. (2015a) studied scent dispensing module using microporous piezoelectric film. Its diameter was 13.8 mm and its operation frequency was 113 kHz. The center of this film was a metal mesh with microporous piezoelectric film. The film was directly above the cotton core and the piezoelectric film atomized the scented liquid absorbed by the cotton core by capillary action. Javerliat et al. (2022) studied low-cost open source olfactory display for VR headset using the ultrasonic atomizer.

Amores & Maes (2017) developed a necklace with smell generation device called Essence. Its device is illustrated in Fig. 14.19. A cover on the top of the bottle holds a piezoelectric film just above the cotton stick fiber, soaking up the perfume from the container. They extended it for monitoring physiological information during the olfactory stimulation (Amores et al., 2018; Amores et al., 2018). Moreover, they used that device to release a scent based on physiological state during sleep (Amores et al., 2019).

14.4.4 Inkjet-based devices

An inkjet device can be used for dispensing liquid. Since very tiny droplet can be generated using the inkjet device, the dispensing amount of the liquid can be precisely controlled. The inkjet-based device for multicomponent olfactory display is described in Chapter 15. Several olfactory displays based on inkjet devices are explained here.

FIGURE 14.19 **Necklace type scent generator.** Atomization-based device is placed above the container.

Hayes & Taylor (2005) developed the vapor generator using the inkjet devices. Its principle is shown in Fig. 14.20. The inkjet device ejects drops and they hit a heater. It evaporates the liquid droplets and the vapor flows toward the outlet. The heated wall avoids the condensation of the liquid.

Sugimoto et al. studied inkjet olfactory display (Sugimoto et al., 2010). They intended to achieve instantaneous switch of the scents to avoid smell persistence. They claimed that their olfactory display quickly presented and erased the scent. They neither used the heater nor atomizer and the liquid droplet evaporated naturally due to small picoliter order size of the droplet. They applied a head of an inkjet printer to the olfactory display. Moreover, they also studied synchronization of the scent with the breath cycle (Ohtsu et al., 2009). Yamada et al. (2006) also studied the inkjet olfactory display using the direct-injection method.

14.4.5 Air flow based olfactory display

Several olfactory displays just use the air flow. Herrera developed a simple low-cost olfactory display. They used jar containing liquid odorant and a small fan (Herrera & Mcmahan, 2014).

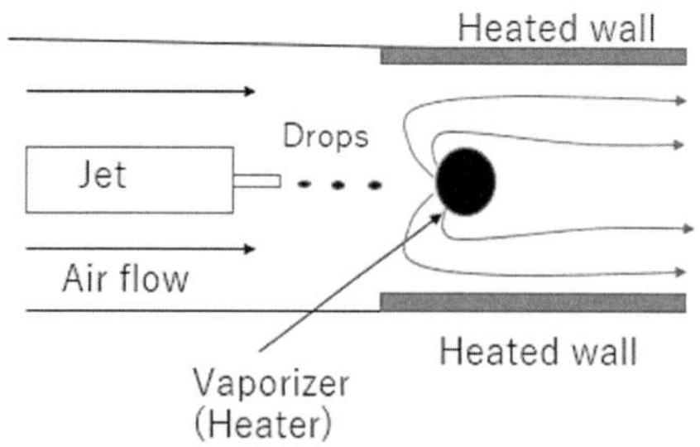

FIGURE 14.20 **Inkjet-based device with heater.** Heater evaporates drops from inkjet device.

Niedenthal et al. (2023) made a graspable olfactory display. A user wearing HMD (head-mount display) grasps the olfactory display, which is suitable for interaction with odorous objects in virtual space. Its photo is shown in Fig. 14.21. They used stepless valves to control scent magnitude and blending. Moreover, the fan was used to deliver the scent to the user.

Matsukura et al. (2013) studied the olfactory display called smelling screen. A user perceives the location of the odor source in the screen. The basic concept is shown in Fig. 14.22. When the wind from the right collides the wind from the left, the wind comes to the front direction in that figure. Thus the user perceives the location of the odor source as if it were located at the collision point. When the speed of the wind from the left becomes larger than that from the right, the collision point moves to the right. Thus the user perceives that the odor source location moves toward the right side. When the speed of the wind from the right becomes larger than that from the left, the user perceives the odor source location moves toward the left side. Therefore the odor source location the user perceives can be specified by controlling the wind speed ratio. It can be extended to two-dimensional space when four fans on the four corners of the screen are used.

14.4.6 Air canon-based devices

Yanagida et al. (2004) developed the olfactory display utilizing the air cannon (Nakaizumi et al., 2006). The odor is delivered to only a specified person, whereas people around him/her do not perceive it. However, there is a problem to perceive wind as well as odor. The wind might let the user feel differently from that of only odor. Therefore Yanagida et al. used two air cannons to collide two vortex rings just in front of the user's nose to reduce the influence of the wind from the air canon as is illustrated in Fig. 14.23. Moreover, they made the nose tracking system for a user to change his/her nose position at random.

They recently proposed a cluster digital air canon for delivering the scent to the specified person (Yanagida & Noma, 2020). It consists of an aggregate of small-size holes and the velocity control of vortex rings and their rapid shots can be achieved by controlling the air pressure and the timing of solenoid valves. Moreover, the equipment size became smaller compared with the

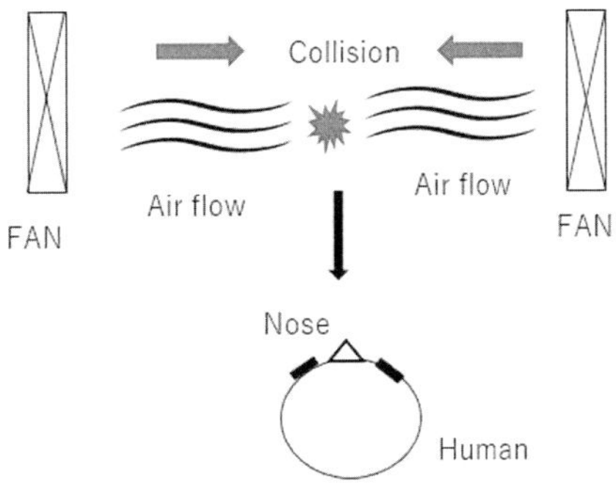

FIGURE 14.22 Principle of Smelling screen (top view). A user perceives the odor-source location as if it were located at the collision point.

FIGURE 14.21 Graspable olfactory display with HMD. Smell from olfactory display becomes stronger if the device approaches the face. *HMD*, Head-mount display. *Source: From Niedenthal, S., Fredborg, W., Lundén, P., Ehrndal, M., Olofsson, J. K. (2023). A graspable olfactory display for virtual reality.* International Journal of Human-Computer Studies. *169. https://doi.org/10.1016/j.ijhcs.2022.102928.*

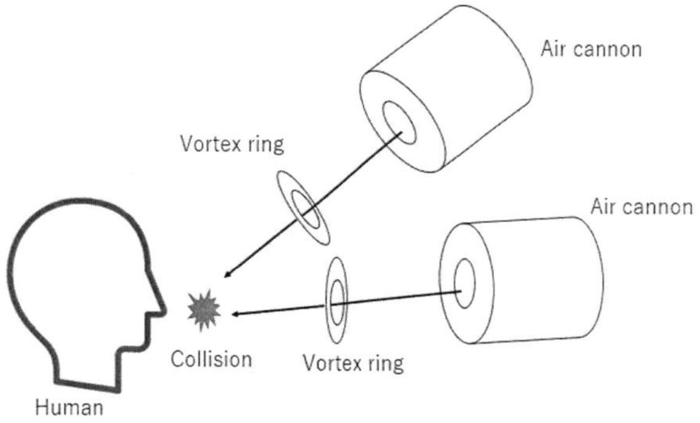

FIGURE 14.23 Method to reduce wind influence for air canon-based olfactory display. Two vortex rings collide with each other just in front of the user's nose.

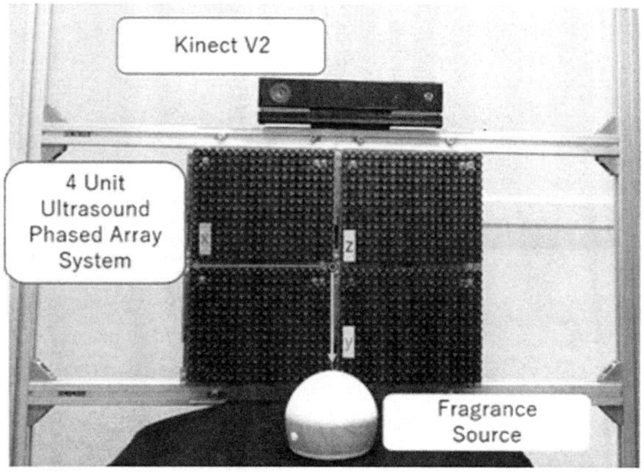

FIGURE 14.24 Ultrasound phased array to deliver scent to desired direction. Acoustic streaming effect causes mass flow along sound direction path. *Source: From Hasegawa, K., Qiu, L., Shinoda, H. (2018). Midair ultrasound fragrance rendering.* IEEE Transactions on Visualization and Computer Graphics. *24(4), 1477–1485. https://doi.org/10.1109/TVCG.2018.2794118.*

conventional air canon based system. Hu et al. (2021) developed the similar system using dual speakers.

Similar to the air cannon, the ultrasound-driven narrow air flow can be generated based upon acoustic streaming effect (Hasegawa et al., 2018). The acoustic streaming is nonlinear phenomenon to cause mass flow along the sound propagation path. Ultrasound phase array (40 kHz) is used to generate the air flow with its electronically steerable direction as is shown in Fig. 14.24. The phase difference among ultrasound transmitters controls the wave-front direction. Its sound direction can be controlled without physical movement of those transducers. It consists of 4 units, each of which contains 249 ultrasound transducers. The power consumption per unit was at most 50 W. Kinect V2 in the figure tracked the user's face in real time. The air flow generated by the ultra sound carries the scent from the fragrance source to the user's face as is shown in Fig. 14.25.

Other type of olfactory display is a wearable device for metaverse. Lukasiewicz et al. (2024) proposed a simple open-source kit for it.

14.5 Applications of olfactory displays

Although digital contents with scents are explained in Chapter 17, several applications of olfactory displays are explained here.

14.5.1 Entertainment/advertisement

The movie with scents is very popular one (see Chapter 17). The famous example is "Charlie and the Chocolate Factory." Chocolate scent lingers together with movie to increase the reality. In this application, the scent is released to a big theater. Since it is necessary to fill the large area with the scent within a short time, it is difficult to switch several smells in the theater. A large scent diffuser is normally used. Other application of the olfactory displays to entertainment is mainly for the events such as wedding ceremony and fashion show.

Scent diffusers are also used for sales promotion. When curry ingredients were sold at supermarket together with its smell from scent diffuser, its sales increased. They are sometimes used at hotel or resort facility, restaurant,

FIGURE 14.25 **Scented water vapor guided to the user's nose by the ultrasound driven air flow.** Scent direction can be controlled by adjusting phase difference among ultrasonic transmitters. *Source: From Hasegawa, K., Qiu, L., Shinoda, H. (2018). Midair ultrasound fragrance rendering.* IEEE Transactions on Visualization and Computer Graphics. *24(4), 1477–1485. https://doi.org/10.1109/TVCG.2018.2794118.*

shops, fitness club, game center, show room, seminar room, facilities for the elderly, and so on. Several companies have corporate scents. They are used to make their brand memorable.

In addition to scent diffusers, several gadgets to emit smells have appeared. Kim and Ando (2010) developed aroma shooter to release scents (Dmitrenko et al., 2016). They made the air flow with the scent from porous material such as calcium silicate soaked in odorant liquid. The gadget allows an odorant to be carried by directional air flow. Moreover, it is possible to switch several scents.

The early version of the gadget to emit scents for VR headset has appeared on the market (Smolin et al., 2020). The gadget is set beneath HMD and a user can perceive the scent in VR environment. Several scent cartridges can be put into it and the user can select the smell. Olfactory VR is explained in Chapter 16.

14.5.2 Medical field

Okada et al. developed inkjet-based olfactory display as is described in Section 14.4.4. Later they applied it to the medical field (Fukasawa & Okada, 2013; Matsuura et al., 2015). A decline in olfaction can be caused by disorders such as Alzheimer's dementia and Parkinson's disease (Baba et al., 2012). It can be detected at the early stage of the disease. Thus it is important to test olfactory capability periodically. T&T olfactometer, conventional olfactory test in Japan, is not often used because of long measurement time.

In that test, five types of chemicals at 7–8 concentration levels are used. A patient sniffs a smell of a paper with each sample. Since the experiment is conducted one by one, it takes much time to measure his/her detection threshold.

They applied their olfactory display using pulse ejection technique to measure the degree of olfactory loss of a patient. It is possible to dilute each sample quickly and accurately using the olfactory display. They claimed that there was good correlation between their measurement result and that by T&T olfactometer.

In 2023, SONY Corporation started selling the olfactory display for medical target (n.d.). They use the linear actuator technology and the characteristic of smell persistence seems good. It has forty channels that can realize five types of smells and eight concentration levels. However, it does not have a function of blending. It can be also used instead of conventional T&T olfactometer although it is different from the technology in Fukasawa and Okada (2013).

Since demand for medical use is now increasing because of our aging society, the application of olfactory display to the medical field is becoming important.

14.6 Summary

We have learned the fundamental technologies to generate smells, actual smell presentation devices, and their applications. The number of reports to propose the devices for odor presentation is recently increasing. However, most of them are intended to develop simple and low-cost devices. Currently, entertainment, advertisement, and medical fields are mainly considered application fields of olfactory displays. In addition to those, the olfactory display can be applied to an automobile industry (Wintersberger et al., 2019). The device and method to blend multiple ingredients are not shown in this chapter but is explained in Chapter 15.

References

Abid, S. H., Li, Z., Li, R., & Waleed, J. (2015a). A novel olfactory displays' scent dispersing module. *International Journal of Hybrid Information Technology, 8*(1), 435–442. Available from https://doi.org/10.14257/ijhit.2015.8.1.39.

Abid, S. H., Li, Z., Li, R., & Waleed, J. (2015b). Heat based odors delivery apparatus for interactive game playing. *International Journal of Multimedia and Ubiquitous Engineering, 10*(5), 331–342. Available from https://doi.org/10.14257/ijmue.2015.10.5.31.

Amores, J., Dotan, M., & Maes, P. (2019). An exploration of form factors for sleep-olfactory interfaces. In *Proceedings of the annual international conference of the IEEE engineering in medicine and biology society* (pp. 1456–1460). Institute of Electrical and Electronics Engineers Inc. EMBS. Available from https://doi.org/10.1109/EMBC.2019.8856805.

Amores, J., Hernandez, J., Dementyev, A., Wang, X., & Maes, P. (2018). BioEssence: A wearable olfactory display that monitors cardio-respiratory information to support mental wellbeing. In *Proceedings of the annual international conference of the IEEE engineering in medicine and biology society* (pp. 5131–5134). Institute of Electrical and Electronics Engineers Inc. EMBS. Available from https://doi.org/10.1109/EMBC.2018.8513221.

Amores, J., & Maes, P. (2017). Essence: Olfactory interfaces for unconscious influence of mood and cognitive performance. In *Conference on human factors in computing systems—Proceedings* (pp. 28–34). Association for Computing Machinery. Available from https://doi.org/10.1145/3025453.3026004.

Amores, J., Richer, R., Zhao, N., Maes, P., & Eskofier, B.M. (2018). Promoting relaxation using virtual reality, olfactory interfaces and wearable EEG. In *IEEE 15th international conference on wearable and implantable body sensor networks* (pp. 98–101). Institute of Electrical and Electronics Engineers Inc. BSN 2018. Available from https://doi.org/10.1109/BSN.2018.8329668.

Baba, T., Kikuchi, A., Hirayama, K., Nishio, Y., Hosokai, Y., Kanno, S., Hasegawa, T., Sugeno, N., Konno, M., Suzuki, K., Takahashi, S., Fukuda, H., Aoki, M., Itoyama, Y., Mori, E., & Takeda, A. (2012). Severe olfactory dysfunction is a prodromal symptom of dementia associated with Parkinson's disease: A 3 year longitudinal study. *Brain, 135*(1), 161–169. Available from https://doi.org/10.1093/brain/awr321.

Barratt, R. S. (1981). The preparation of standard gas mixtures a review. *The Analyst, 106*(1265), 817–849. Available from https://doi.org/10.1039/AN9810600817.

Chono, K., Shimizu, N., Matsui, Y., Kondoh, J., & Shiokawa, S. (2004). Development of novel atomization system based on SAW streaming. *Japanese Journal of Applied Physics, 43*(5S). Available from https://doi.org/10.1143/jjap.43.2987.

Covington, J. A., Agbroko, S. O., & Tiele, A. (2018). Development of a portable, multichannel olfactory display transducer. *IEEE Sensors Journal, 18*(12), 4969–4974. Available from https://doi.org/10.1109/JSEN.2018.2832284.

Dmitrenko, D., Vi, C.T., & Obrist, M. (2016). A comparison of scent-delivery devices and their meaningful use for in-car olfactory interaction. In *Automotiveui 2016—8th International conference on automotive user interfaces and interactive vehicular applications, proceedings* (pp. 23–26). Association for Computing Machinery, Inc. Available from https://doi.org/10.1145/3003715.3005464.

Du, B., Schwartz-Narbonne, H., Tandoc, M., Heffernan, E. M., Mack, M. L., & Siegel, J. A. (2022). The impact of emissions from an essential oil diffuser on cognitive performance. *Indoor Air, 32*(1). Available from https://doi.org/10.1111/ina.12919.

Dobbelstein, D., Herrdum, S.,& Rukzio, E. (2017). InScent: A wearable olfactory display as an amplification for mobile notifications. In *Proceedings—International symposium on wearable computers, ISWC* (pp. 130–137). Association for Computing Machinery Germany. Available from https://doi.org/10.1145/3123021.3123035.

Fukasawa, A., & Okada, K. (2013). Olfactory measurement method at health checkup with olfactory display using pulse ejection. *International Journal of Informatics Society, 5*, 13–19. Available from http://www.infsoc.org/journal/vol05/IJIS_05_1_013-019.pdf.

Gardner, J. W., Pearce, T. C., Friel, S., Bartlett, P. N., & Blair, N. (1994). A multisensor system for beer flavour monitoring using an array of conducting polymers and predictive classifiers. *Sensors and Actuators B: Chemical, 18*(1–3), 240–243. Available from https://doi.org/10.1016/0925-4005(94)87089-6.

Hasegawa, K., Qiu, L., & Shinoda, H. (2018). Midair ultrasound fragrance rendering. *IEEE Transactions on Visualization and Computer Graphics, 24*(4), 1477–1485. Available from https://doi.org/10.1109/TVCG.2018.2794118.

Hayes, D. J., & Taylor, D. W. (2005). Real-time calibration and testing of chemical sensor enabled by precision microdispensing technology. In Proceedings, SPIE Defense and Security Symposium. SPIE. https://doi.org/10.1117/12.604215.

Heilig,M. L. (1962). *Sensorama simulator* (U.S. Patent).

Herrera, N. S., & Mcmahan, R. P. (2014). *Development of a simple and low-cost olfactory display for immersive media experiencesin Proceedings of the 2nd ACM international workshop on immersive media experiences* (pp. 1–6). ACM. Available from https://doi.org/10.1145/2660579.2660584.

Hu, Y. Y., Jan, Y. F., Tseng, K. W., Tsai, Y. S., Sung, H. M., Lin, J. Y., & Hung, Y. P. (2021). ABio: Active bi-olfactory display using subwoofers for virtual reality. In *MM 2021–Proceedings of the 29th ACM international conference on multimedia* (pp. 2065–2073). Association for Computing Machinery, Inc. Available from https://doi.org/10.1145/3474085.3475678.

Ishizuka, H., Kashiwagi, K., Terao, K., Takao, H., & Shimokawa, F. (2018). Development of a heat-driven-type MEMS olfactory display. In *Proceedings MDPI* (pp. 839). MDPI.

Johnson, B. N., & Sobel, N. (2007). Methods for building an olfactometer with known concentration outcomes. *Journal of Neuroscience Methods, 160*(2), 231–245. Available from https://doi.org/10.1016/j.jneumeth.2006.09.008.

Kaye, J. J. (2004). Making scents. *Interactions, 11*(1), 48–61. Available from https://doi.org/10.1145/962342.964333.

Javerliat, C., Elst, P.P., Saive, A.L., Lavoué, G., & Baert, P. (2022). An affordable open-source and autonomous olfactory display for VR headsets. In *Proceedings of the ACM symposium on virtual reality software and technology* (pp. 1–8). Association for Computing Machinery. VRST. Available from https://doi.org/10.1145/3562939.3565617.

Kim, D.W., & Ando, H. (2010). Development of directional olfactory display. In *Proceedings—VRCAI 2010, ACM SIGGRAPH conference on virtual-reality continuum and its application to industry, Japan* (pp. 143–144). Available from https://doi.org/10.1145/1900179.1900208.

Kim, D. W., Nishimoto, K., Kunifuji, S., Cho, Y. H., Kawakami, Y., & Ando, H. (2009). Development of aroma-card based soundless olfactory display. In *16th IEEE international conference on electronics, circuits and systems* (pp. 703–706). ICECS, Japan. Available from https://doi.org/10.1109/ICECS.2009.5410784.

Kim, H., Park, J., Noh, K., Gardner, C. J., Kong, S. D., Kim, J., & Jin, S. (2011). An X – Y addressable matrix odor-releasing system using an on–off switchable device. *Angewandte Chemie, 123*(30), 6903–6907. Available from https://doi.org/10.1002/ange.201102759.

Kurosawa, M., Watanabe, T., Futami, A., & Higuchi, T. (1995). Surface acoustic wave atomizer. *Sensors and Actuators A: Physical, 50*(1–2), 69–74. Available from https://doi.org/10.1016/0924-4247(96)80086-0.

Lee, M.-k, Lim, S., Song, J.-A., Kim, M.-E., & Hur, M.-H. (2017). The effects of aromatherapy essential oil inhalation on stress, sleep quality and immunity in healthy adults: Randomized controlled trial. *European Journal of Integrative Medicine, 12*, 79–86. Available from https://doi.org/10.1016/j.eujim.2017.04.009.

Liu, Y., Yiu, C. K., Zhao, Z., Park, W., Shi, R., Huang, X., Zeng, Y., Wang, K., Wong, T. H., Jia, S., Zhou, J., Gao, Z., Zhao, L., Yao, K., Li, J., Sha, C., Gao, Y., Zhao, G., Huang, Y., & Yu, X. (2023). Soft, miniaturized, wireless olfactory interface for virtual reality. *Nature Communications, 14*(1). Available from https://doi.org/10.1038/s41467-023-37678-4.

Livneh, U., & Paz, R. (2010). An implicit measure of olfactory performance for non-human primates reveals aversive and pleasant odor conditioning. *Journal of Neuroscience Methods, 192*(1), 90–95. Available from https://doi.org/10.1016/j.jneumeth.2010.07.027.

Lukasiewicz, M. S., Rossoni, M., Spadoni, E., Dozio, N., Carulli, M., Ferrise, F., & Bordegoni, M. (2024). An open-source olfactory display to add the sense of smell to the metaverse. *Journal of Computing and Information Science in Engineering, 24*(2). Available from https://doi.org/10.1115/1.4062889.

Matsukura, H., Yoneda, T., & Ishida, H. (2013). Smelling screen: Development and evaluation of an olfactory display system for presenting a virtual odor source. *IEEE Transactions on Visualization and Computer Graphics, 19* (4), 606–615. Available from https://doi.org/10.1109/TVCG.2013.40.

Matsuura, E., Homma, S., Horiguchi, S., Matsumoto, S., & Okada, K. I. (2015). Development of olfactory display using pulse ejection for medical applications. In *17th International conference on e-health networking, application and services*, (pp. 534–539). Japan. HealthCom: Institute of Electrical and Electronics Engineers Inc. Available from https://doi.org/10.1109/HealthCom.2015.7454560.

Muñoz-Aguirre, S., Nakamoto, T., & Moriizumi, T. (2005). Study of deposition of gas sensing films on quartz crystal microbalance using an ultrasonic atomizer. *Sensors and Actuators B: Chemical, 105*(2), 144–149. Available from https://doi.org/10.1016/j.snb.2004.05.063.

Nakamoto, T., Fukuda, A., & Moriizumi, T. (1993). Perfume and flavour identification by odour-sensing system using quartz-resonator sensor array and neural-network pattern recognition. *Sensors and Actuators B: Chemical, 10*(2), 85–90. Available from https://doi.org/10.1016/0925-4005(93)80030-f.

Nakaizumi, F., Yanagida, Y., Noma, H., & Hosaka, K. (2006). SpotScents: A novel method of natural scent delivery using multiple scent projectors. In *Proceedings—IEEE Virtual Reality* (pp. 207–218). IEEE Computer Society, Japan. Available from https://doi.org/10.1109/VR.2006.122.

Nakamoto, Takamichi, Takigawa, Hidehiko, & Yamanaka, Takao (2004). Fundamental study of odor recorder using inkjet devices for low-volatile scents. *Institute of Electronics Information and Communcation Engineers, E87-C*(12), 2081–2086.

Nakamoto, T., Ustumi, S., Yamashita, N., Moriizumi, T., & Sonoda, Y. (1994). Active gas/odor sensing system using automatically controlled gas blender and numerical optimization technique. *Sensors and Actuators B: Chemical, 20*(2–3), 131–137. Available from https://doi.org/10.1016/0925-4005(93)01193-8.

Nakamoto, T. (2003). Odor Handling and Delivery system System. In *Handbook of machine olfaction: Electronic nose technology*, (pp. 55–78). John Wiley & Sons.

Nakamoto, T. (2016). *Essentials of machine olfaction and taste*. Wiley. Available from http://doi.org/10.1002/9781118768495.

Niedenthal, S., Fredborg, W., Lundén, P., Ehrndal, M., & Olofsson, J. K. (2023). A graspable olfactory display for virtual reality. *International Journal of Human-Computer Studies, 169*. Available from https://doi.org/10.1016/j.ijhcs.2022.102928.

Ohtsu, K., Sato, J., Bannai, Y., & Okada, K. I. (2009). Scent presentation technique of pulse ejection synchronized with breathing. In *Proceedings—2009 9th annual international symposium on applications and the internet* (pp. 125–128). SAINT Japan. Available from https://doi.org/10.1109/SAINT.2009.28.

Somboon, P., Wyszynski, B., & Nakamoto, T. (2007). Realization of recording a wide range of odor by utilizing both of transient and steady-state sensor responses in recording process. *Sensors and Actuators B: Chemical, 124*(2), 557–563. Available from https://doi.org/10.1016/j.snb.2007.01.030.

Smolin, A., Malyshev, K., Spiridonova, A., Mironov, A., Shchekoldin, A., & Vorobyev, P. (2020). Physiological aspects of the user's immersion in the virtual reality environment on the example of creating a hardware-software complex for smell transmission.Proceedings of the 30th International Conference on Computer Graphics and Machine Vision (GraphiCon 2020). Part 2 https://doi.org/10.51130/graphicon-2020-2-3-3

Sugimoto, S., Noguchi, D., Bannnai, Y., & Okada, K. (2010). Ink jet olfactory display enabling instantaneous switches of scents. In *MM'10—Proceedings of the ACM multimedia 2010 international conference* (pp. 301–310). ACM International. Available from https://doi.org/10.1145/1873951.1873994.

Tiele, A., Menon, S., & Covington, J. A. (2020). Development of a thermal-based olfactory display for aroma sensory training. *IEEE Sensors Journal, 20*(2), 631–636. Available from https://doi.org/10.1109/jsen.2019.2945953.

Yabe, A., Hamate, Y., Hara, M., Oguchi, H., Nagasawa, S., & Kuwano, H. (2014). A self-converging atomized mist spray device using surface acoustic wave. *Microfluidics and Nanofluidics, 17*(4), 701–710. Available from https://doi.org/10.1007/s10404-014-1358-2.

Wintersberger, P., Frison, A. K., Dmitrenko, D., Maggioni, E., Riener, A., Schartmüller, C., & Obrist, M. (2019). S(c)entinel—Monitoring automated vehicles with olfactory reliability displays. In *International conference on intelligent user interfaces, proceedings IUI* (pp. 538–546). Association for Computing Machinery, Germany. Available from https://doi.org/10.1145/3301275.3302332.

Yamada, T., Yokoyama, S., Tanikawa, T., Hirota, K., & Hirose, M. (2006). Wearable olfactory display: Using odor in outdoor environment. In *Proceedings—IEEE virtual reality* IEEE Computer Society, Japan. Available from https://doi.org/10.1109/VR.2006.147.

Yanagida, Y. (2012). A survey of olfactory displays: Making and delivering scents. In *Proceedings of IEEE sensors*. IEEE, Japan. Available from https://doi.org/10.1109/ICSENS.2012.6411380.

Yanagida, Y., Kawato, S., Noma, H., Tomono, A., Tetsutani N. (2004). Projection-based olfactory display with nose tracking. In *Proceedings—Virtual reality annual international symposium* (pp. 43–50). IEEE, Japan. Available from https://doi.org/10.1109/VR.2004.1310054.

Yanagida, Y., & Noma, H. (2020). Clustered digital air cannon: A novel device for projection-based olfactory display. In *Proceedings of the international display workshops*. Available from https://doi.org/10.36463/idw.2020.0695.

CHAPTER

15

Multicomponent olfactory displays

Takamichi Nakamoto

Institute of Integrated Research, Institute of Science Tokyo, Yokohama, Japan

15.1 Introduction

It is one of important functions to blend multiple ingredients in an olfactory display since a variety of scents can be realized using the blending function. Moreover, odor reproduction techniques in Chapter 11 can easily be realized using the multicomponent olfactory display (Iseki et al., 2022). This chapter shows several techniques to blend multiple ingredients.

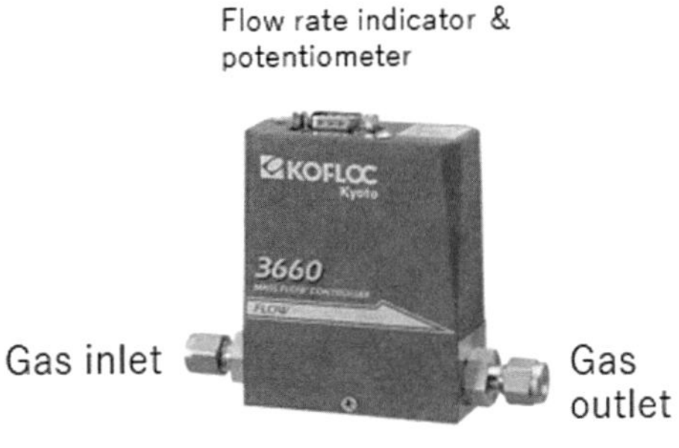

FIGURE 15.1 Photo of mass flow controller. Mass flow controller can regulate its flow rate regardless of the pressure at the outlet.

15.2 Mass flow controller

A mass flow controller is a device to control flow rate electronically as is shown in Fig. 15.1. Once the voltage corresponding to the specified flow rate is given, a solenoid control valve adjusts the flow rate according to the flow sensor readout in the mass flow controller. It can regulate flow rate regardless of pressure change at its outlet.

When we use several mass flow controllers, we can adjust the flow rate of each ingredient to obtain the specified mixing ratio (Grate et al., 1987; Johnson & Sobel, 2007; Nakamoto et al., 1994). Its concept is drawn in Fig. 15.2.

For example, just let us think about the case of binary mixture. The flow rate of one mass flow controller is 20 mL/minute and the other one is 80 mL/minute, and the mixture composition is 2:8. It can be extended to multiple ingredients. The mass flow controller uses mass flow, that is, mass of flow per unit time. It is not influenced by temperature and pressure. Mass is typically converted to the volume at standard condition (1 atom, 25°C). Sometimes 0°C is used as standard temperature. The unit of mass flow rate is "sccm," which stands for standard cubic centimeter per minute.

On the other hand, a typical flow meter measures that the volumetric flow is a volume of medium moving per unit time. Its photo is shown in Fig. 15.3. Since the volume expands

Digital Technologies in Olfaction
DOI: https://doi.org/10.1016/B978-0-443-15721-9.00015-4

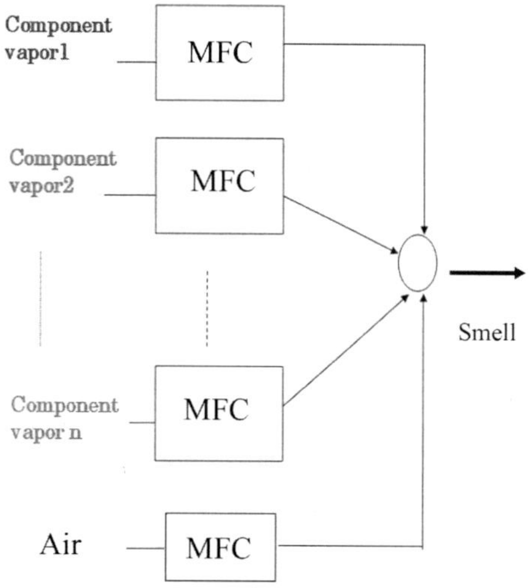

FIGURE 15.2 **Blending using multiple mass flow controllers.** *MFC*, Mass flow controller.

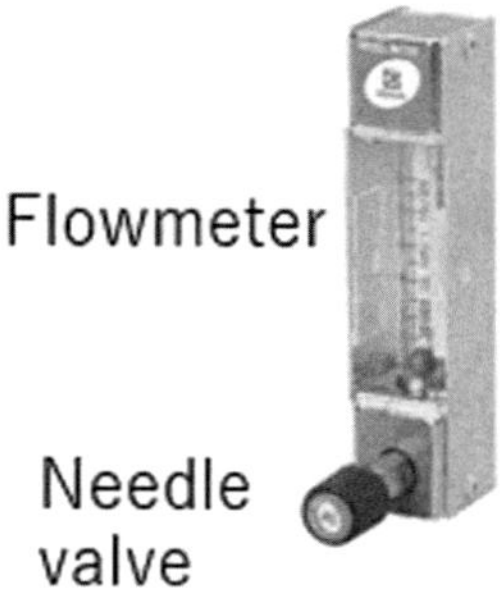

FIGURE 15.3 **Photo of flow meter with valve.** The volumetric flow can be manually adjusted using a needle valve.

or shrinks according to temperature and pressure, gas pressure and temperature should be taken into account when the volumetric flow is used. The volumetric flow can be manually adjusted using a volumetric flow meter with needle valve. The volumetric flow meter with valve is cheaper than mass flow controller.

However, the volumetric flow meter with valve has a problem. When that meter is connected in the same way as that in Fig. 15.2, the flow rate change at one channel influences the flow rate of the other channel because the pressure at the flow meter outlet also changes. We cannot set the flow rate of each channel independently. Since it takes much time to adjust the flow rate of each channel to desired value, the method to use volumetric flow meter is not realistic.

When we use mass flow controllers, on the other hand, there is no interference between channels and each mass flow controller can be used as open loop system even if closed loop system is embedded into each mass flow controller. We can specify the recipe, and it is not necessary to adjust the flow rate repeatedly. Thus the mass flow controllers are used instead of volumetric flow meters.

Note that the range of the recipe is limited when the flow rate of the whole blended odor is kept constant. In the case of binary mixture, three mass flow controllers are used. One of them is dedicated to the air channel. In this case, the range of the flow rate combination, that is, the recipe, is inside the triangle as is shown in Fig. 15.4. When we have n channels, the flow rate of each channel should be satisfied in the following equation:

$$F_{\text{total}} = F_1 + F_2 + \cdots F_n + F_{\text{air}} \tag{15.1}$$

and thus

$$F_{\text{total}} \geq F_1 + F_2 + \cdots F_n \tag{15.2}$$

where F_{total} is the rate of flow to the user, F_i is the flow rate of the ith channel, and F_{air} is the

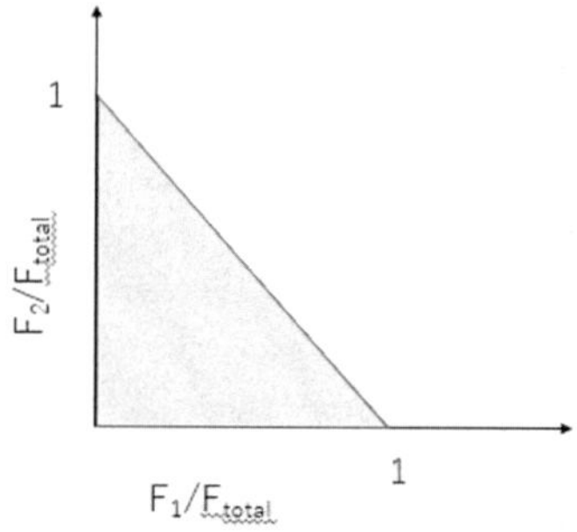

FIGURE 15.4 **Range of flow rate combination under the constant rate of flow to a user for binary mixture.** "F_i"/"F_{total}" means the recipe of ingredient i.

flow rate of the air channel. Therefore we cannot use the full range of flow rate combination at the constraint of the constant total flow rate.

In some mass flow controllers, it is difficult to realize zero [sccm] since the solenoid control valve in the mass flow controller cannot be closed completely. In this case, another solenoid valve is externally attached so that it can be shut down completely.

Photo of total system using mass flow controllers is shown in Fig. 15.5. The five mass flow controllers are used in this system. Although the accuracy of the mass flow controller is high, it is very expensive and bulky. Microelectromechanical system technique can be applied to realize a small mass flow controller. Although it was reported in the paper (Shoji & Esashi, 1994), we do not have commercially available one. If the small and cheap mass flow controller robust against physical damage appears, it is quite useful.

15.3 High-speed switching of solenoid valve

Solenoid valves are often used to control the flow path. In this scenario, they can be used to switch several vapors, and the only binary representation of the vapor is available. However, the concentration in the form of continuous-value representation is possible when the solenoid valve is switched very rapidly. In this section, two methods, such as delta-sigma modulation and pulse width modulation technique, are explained.

15.3.1 Solenoid valve

A solenoid valve is a valve driven by electromagnetic force. One of examples is shown in Fig. 15.6. It works as a fluidic switch in the same manner as that in an electronic circuit. Two types of solenoid valves, such as two-way and three-way ones, are available. The three-way solenoid valve is often used for the olfactory display.

A variety of solenoid valves are commercially available. Although a small solenoid valve is used for the olfactory display, its length is still a few centimeters. Since its drive voltage and current are not compatible with digital circuit, the drive circuit is required to drive the solenoid valve. Its typical voltage is between 10 and 30 V, and its typical current is a few tens of milliamperes. A typical drive circuit is shown in Fig. 15.7. Integrated circuit (IC), including several drive circuits, is commercially available.

FIGURE 15.5 **Multicomponent olfactory display using an array of mass flow controllers**. Five mass flow controllers were used in the photo.

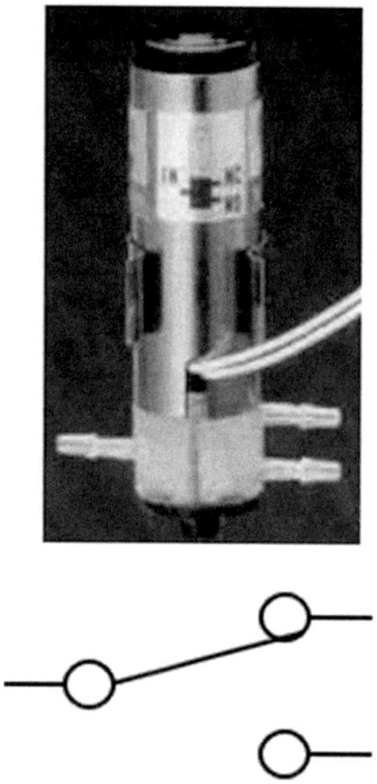

FIGURE 15.6 **Photo of three-way solenoid valve.** Three-way solenoid valve works as a fluidic switch.

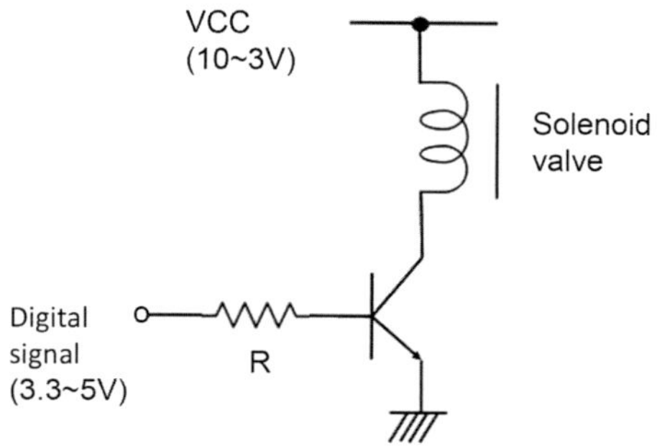

FIGURE 15.7 **Drive circuit of solenoid valve.** Current at the solenoid valve is typically a few tens of milliampere.

Since the solenoid valve has a mechanical movable portion, its switching speed is not so fast compared with electronic circuit. It is typically a few tens of milliseconds, and the fastest one is a millisecond. However, it is enough for the olfactory display since a human cannot detect very fast switching.

Then the principle to realize continuous-value representation is explained. In electronic circuit, there is a filter that processes an analog signal but is not continuous in operation. It is called a switched-capacitor filter, and its basic operation is shown in Fig. 15.8A. A capacitor C is repeatedly switched between input and output terminals. When the switching frequency is much higher than that of input signal, the switched-capacitor element works as a resistive element R_{eg} as is shown in Fig. 15.8B. This example suggests that it is possible to deal with analog signal even if the circuit is controlled only by digital signal, that is, binary value representation.

15.3.2 Delta-sigma modulation

The delta-sigma modulation is a method to convert an analog signal to 1-bit data stream (Norsworthy et al., 1996). It is known as a highly accurate analog-to-digital converter method. Since its accuracy is very high, it was adopted as the format of super audio compact disk.

First, delta modulation is explained before delta-sigma modulation. The analog signal is sampled much faster than the frequency of the analog signal, and only the sign of its derivative is recorded as is shown in Fig. 15.9. "H" means that signal goes up, and "L" means that the signal goes down. If the sampling is done much faster than the change of the analog

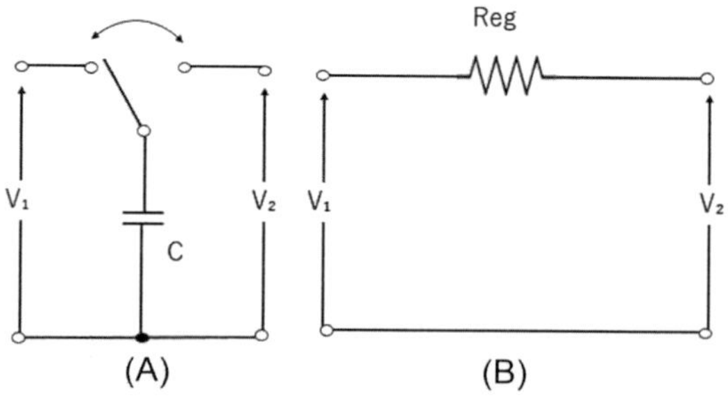

FIGURE 15.8 **Principle of switched-capacitor filter.** (A) Switched capacitor and (B) resistive element.

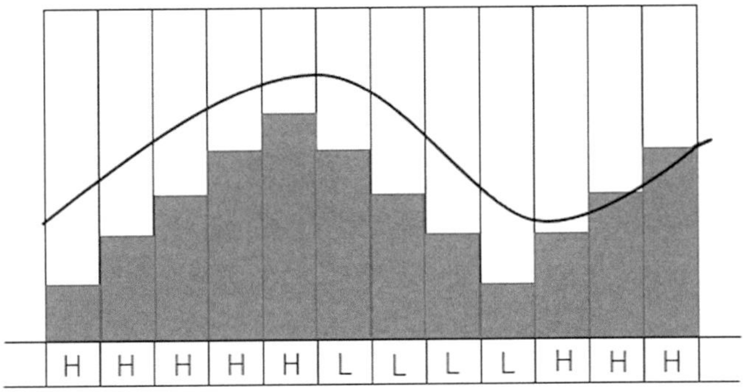

FIGURE 15.9 Principle of delta modulation. Method to covert an analog signal to 1-bit data stream.

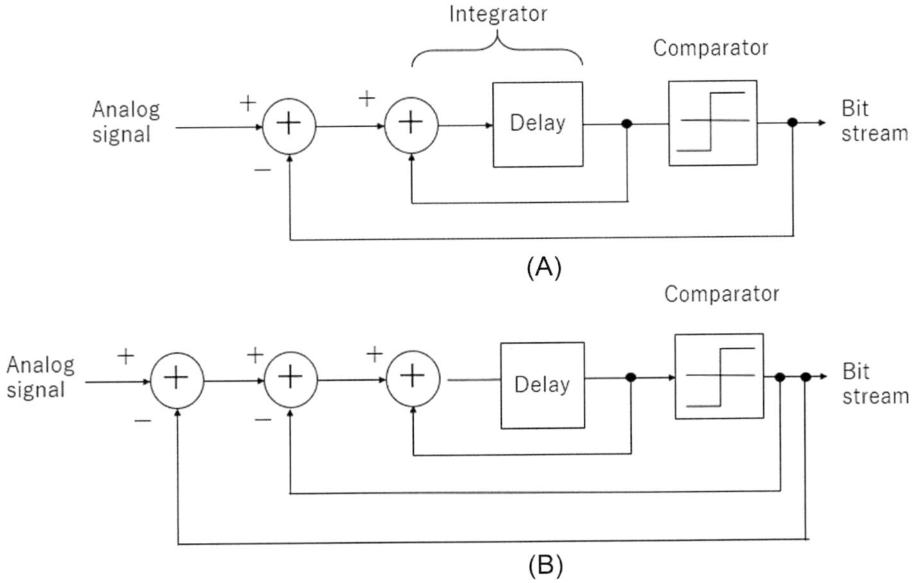

FIGURE 15.10 Principle of delta-sigma modulation. (A) first-order modulator and (B) second-order modulator.

signal, we can reconstruct the analog signal by summing the series of the data just made up of "H" and "L" as is illustrated in Fig. 15.9. If L level is assigned to "−1" and H level is assigned to "1," it is possible to represent the analog signal in the binary form.

However, this method can only restore alternating current (AC) component and direct current (DC) component cannot be restored. Thus we use delta-sigma modulation that add the integrator to delta modulator. The first-order delta-sigma modulation is illustrated in Fig. 15.10A. It consists of a subtractor, comparator, and integrator. The analog signal can be restored by simply passing the bit stream to LPF (Low Pass Filter) to remove the switching noise. Fig. 15.10B shows the second-order delta-sigma modulator. The second-order one was used in Yamanaka et al. (2002) because of its lower noise level.

The bit stream generated by delta-sigma modulation can control the solenoid valve as is illustrated in Fig. 15.11. The arbitrary (continuous-value) gas concentration can be realized by

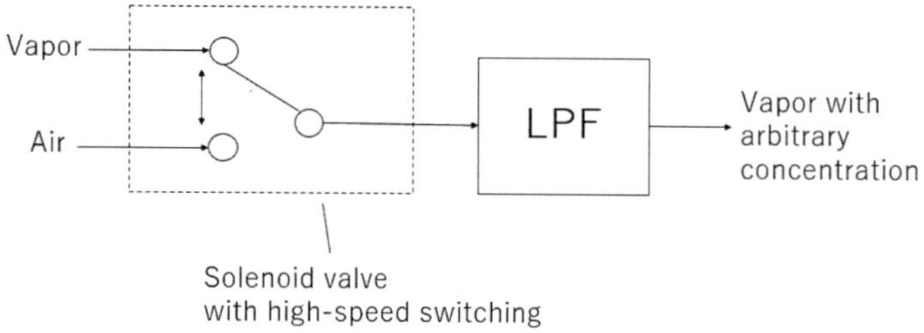

FIGURE 15.11 Principle of gas generation using high-speed switching of solenoid valve. LPF is fluidic. *LPF*, Low-pass filter.

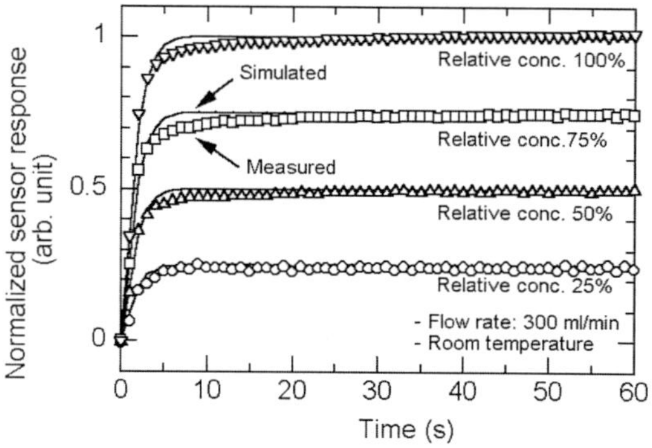

FIGURE 15.12 Comparison of experimental result with simulation one in case of step concentration change of 2-hexanone when second-order delta–sigma modulator was used. Buffer tank was not used. Flow rate: 300 mL/min, switching frequency: 20 Hz. *Source: From Yamanaka, T., Matsumoto, R., & Nakamoto, T. (2002). Study of odor blender using solenoid valves controlled by delta–sigma modulation method for odor recorder.* Sensors and Actuators B: Chemical, *87(3), 457–463, https://doi.org/10.1016/s0925-4005(02)00300-3.*

just switching the solenoid valve based on delta-sigma modulation (Yamanaka et al., 2002). LPF in Fig. 15.11 is fluidic. A simple empty bottle works as fluidic LPF, and even small dead space volume is enough to realize it.

The experimental and simulation results of delta-sigma modulation are shown in Fig. 15.12. The simple fluidic model was used in the simulation (Yamanaka et al., 2002). The relative concentration in the figure means the concentration relative to its full scale. Quartz Crystal Microbalance (QCM) sensor (20 MHz, AT-CUT, coating: UCON 90,000) was used to measure the gas concentration. The solenoid valve with delta-sigma modulation achieves the specified relative concentration without switching noise. In addition, the fluidic model explains the system behavior.

The multicomponent system is shown in Fig. 15.13. This is an eight-component system, and the bypath was used to keep the head space concentration in the vial constant. The empty vial was also used to keep the flow rates of main path and bypath constant. Each channel works independently without interference.

15.3.3 Pulse width modulation

Although the system in Fig. 15.13 works well, the number of solenoid valves and vials is twice as large as the number of channels. Moreover, the maximum concentration of each channel decreases as the number of channels increases as is illustrated in Fig. 15.14A. Each channel in Fig. 15.14A realizes the specified concentration, followed by mixing all the channels. Thus the maximum concentration of each channel becomes smaller when the number of channels becomes larger.

To solve those issues above, the blending method based on pulse width modulation (PWM) was investigated. Its principle is illustrated in Fig. 15.14B (Nakamoto & Minh, 2007). Only one odor component is allowed to go to the output at a certain time, and the time division multiplexing technique is adopted to blend multiple odor components. The time duration of each channel is proportional to the specified ratio. The maximum concentration of each channel is independent of number of

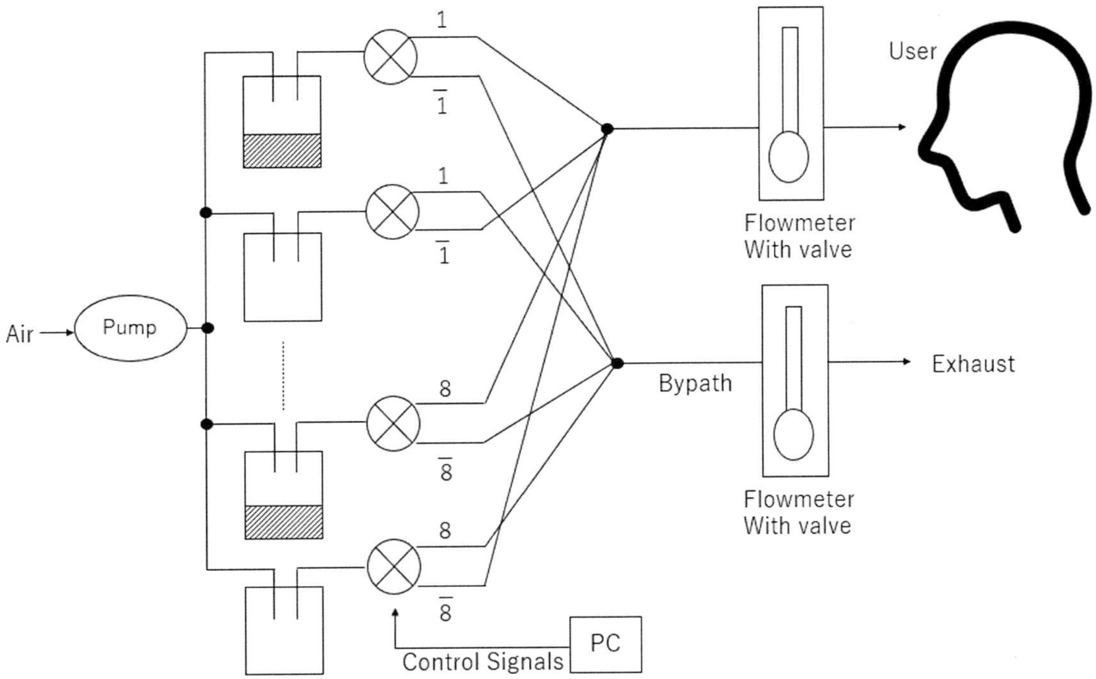

FIGURE 15.13 Multicomponent olfactory display using solenoid valve with delta-sigma modulation. Bypath is used to keep the head space concentration in the vial constant.

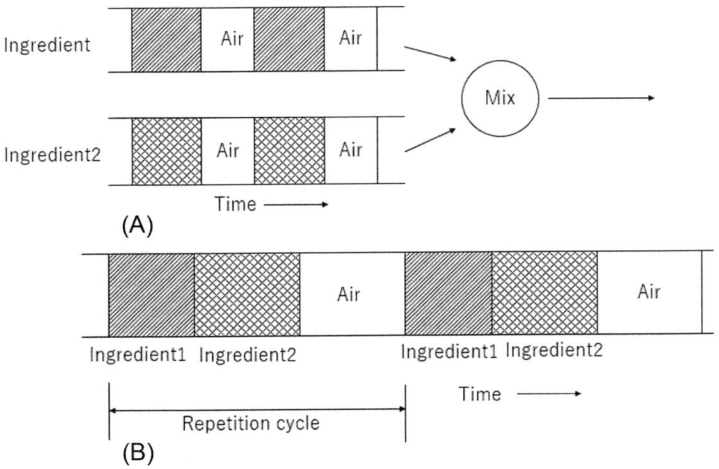

FIGURE 15.14 Blending method using solenoid valves. (A) delta-sigma modulator and (B) PWM-like method. *PWM*, Pulse width modulation.

channels. Therefore many channels can be implemented into the system.

A total of 32-channel olfactory display was realized on the basis of this technique as is illustrated in Fig. 15.15. The number of solenoid valves was the same as the number of the channels, and the number of empty vial for the air was only one. Therefore those numbers

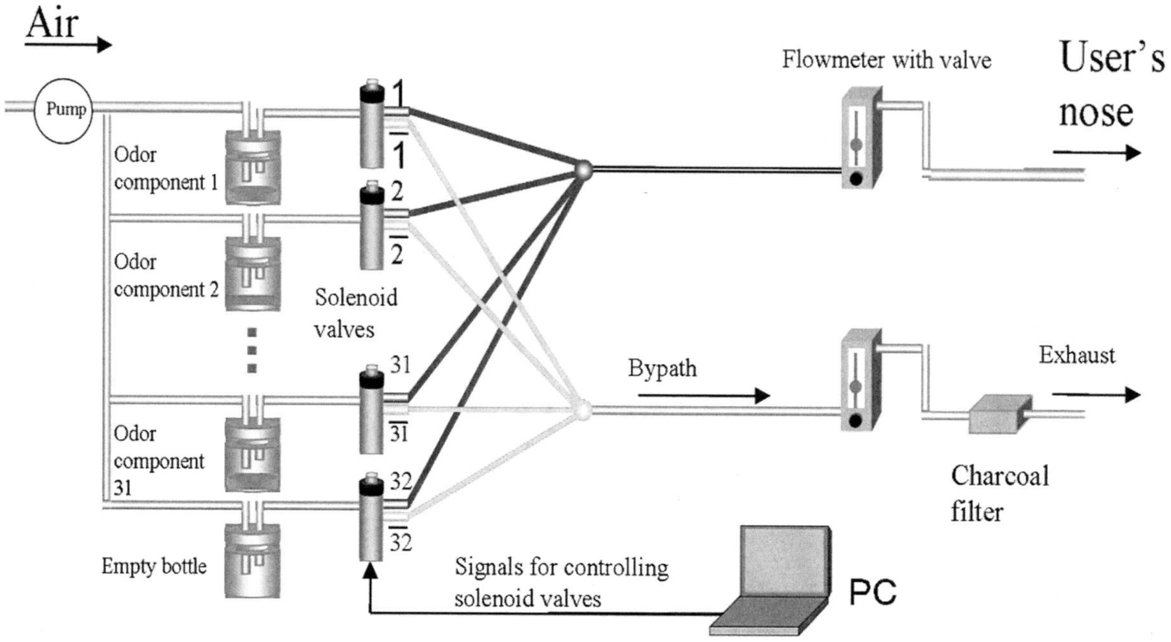

FIGURE 15.15 A total of **32-channel olfactory display** using solenoid valves 32-channel olfactory display using solenoid valves.

were reduced to almost a half of the system in Fig. 15.13. PWM-like method is superior to delta-sigma modulation method when the number of channels becomes larger.

Since 32-channel olfactory display was still large and the number of channels were too many in most of applications, 13-channel olfactory display was developed. The photo of 13-component olfactory display based on PWM-like method is shown in Fig. 15.16. Its principle is the same as 32-channel olfactory display in Fig. 15.15. The external air pump is necessary. This olfactory display has been used to develop many digital olfactory contents in Chapter 17.

15.4 Autosampler

When a perfumer creates a new scent, even 32 ingredients are sometimes not enough. For that purpose, autosampler normally used in gas chromatography/mass spectrometry (GC/MS) can be utilized. In GC/MS, many liquid samples are automatically injected in sequence using the autosampler (Somboon et al., 2007). It is indispensable for the actual GC/MS analysis.

The autosampler is illustrated in Fig. 15.17. The syringe in the autosampler precisely takes the specified amount of liquid sample and then transfers it to other vial. Since many vials with liquid samples are placed in a turn table, it is possible for the syringe to be located on the top of the vial by rotating the turn table. Then the syringe needle penetrates the septum on the vial, takes the liquid, and moves to the next vial. When the procedure is repeatedly performed, the autosampler can blend many ingredients. The autosampler can typically blend up to a few hundreds of ingredients. Although the autosampler is normally designed to be used in GC/MS system, some stand-alone autosamplers are available. Since its command system is open

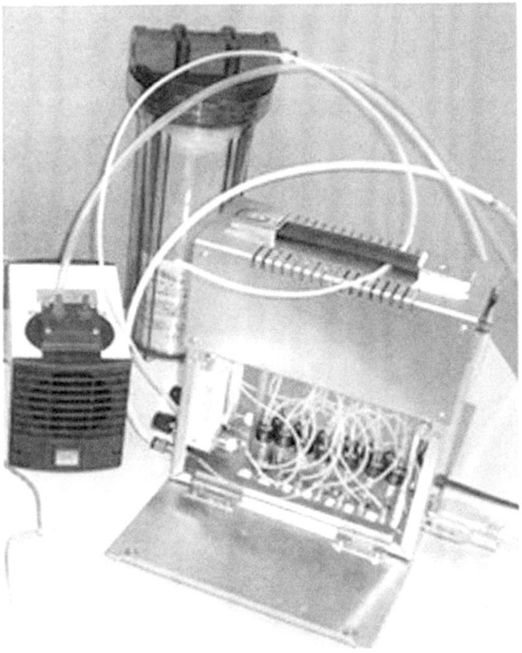

FIGURE 15.16 **Photo of 13-component olfactory display.** It is commercially available from Ono Denki Inc.

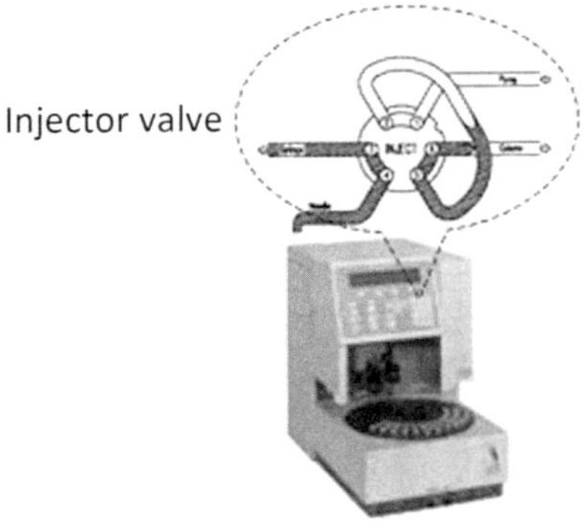

FIGURE 15.17 **Autosampler to blend many ingredients.** Sample loop is used to take liquid volume precisely.

to the public, we can build the blending system using the autosampler.

For the smell presentation, a tiny amount of the liquid with the volume of a few microliters is precisely taken by using a sample loop. It is injected onto the microheater, followed by the evaporation. The sequence of the operation is controlled by sending a series of commands from PC to the autosampler.

The advantage of the autosampler is to blend many ingredients. It is typically impossible for other method to blend more than 100 ingredients. Disadvantage is, on the other hand, to spend too long time to blend ingredients since only one syringe moves mechanically from vial to vial. It is impossible to blend them in real time. Previously, the autosampler was utilized to realize an odor recorder using QCM sensors (Somboon et al., 2007).

15.5 Inkjet devices

Although inkjet device is originally used in a printer, it is applicable to printed circuit board, microoptics, biosensors, drug delivery, bonding and packaging, fuel cell, etc. It can also be applied to olfactory display, which has appeared in Chapter 14. Although the inkjet device in Chapter 14 is used to generate a single component, the multicomponent olfactory display can be realized using inkjet devices (Nakamoto et al., 2004). Since its droplet size is down to the picoliter range, its blending accuracy is very high.

The inkjet device is illustrated in Fig. 15.18. The piezoelectric material is surrounded with glass tube. When the pulse voltage is applied between inner and outer electrodes of piezoelectric film, the deformation occurs, followed by the liquid drop ejection at the orifice.

Although the ejected volume from the inkjet device is very small, it takes still much time for a droplet to evaporate completely. Thus a heater is often used to accelerate the evaporation as is shown in Fig. 15.19. Mesh heater is used here since the vapor can pass through the heater. An appropriate back pressure should be applied to the inkjet device. The multicomponent

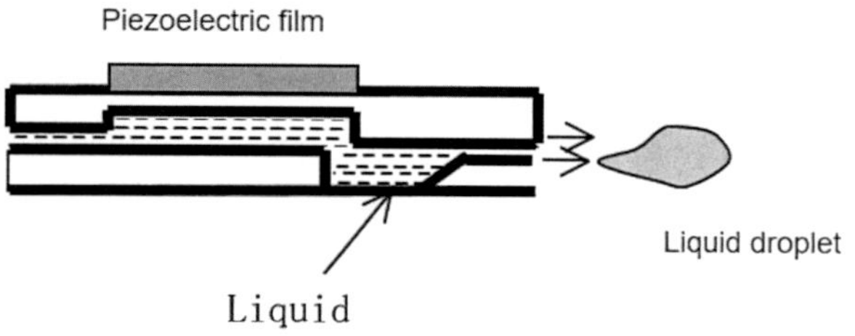

FIGURE 15.18 **Inkjet device**. Liquid droplet ejection occurs at the orifice.

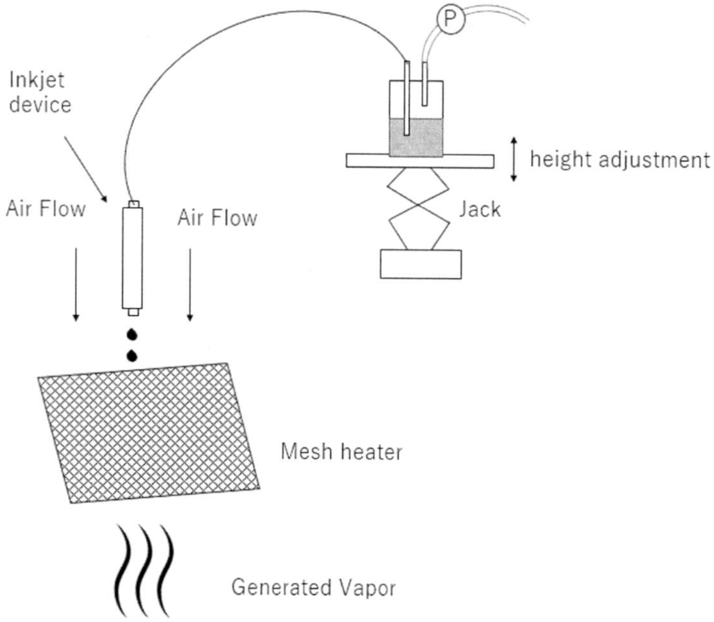

FIGURE 15.19 **Odor generator using inkjet devices**. Vapor can pass through a mesh heater.

olfactory display is realized when the multiple inkjet devices are used.

Although the droplet volume is smaller than that of other devices, we often encounter the instability problem of the inkjet device. Since the inkjet device does not have valve or shutter, its backpressure should be precisely controlled. If the back pressure is too small, the air bubble enters it and the droplet cannot be ejected as is shown in Fig. 15.20A. If the back pressure is too large, the liquid is accumulated at the tip as is shown in Fig. 15.20C. The back pressure should be kept in the situation as is shown in Fig. 15.20B. Since the back pressure should be independently adjusted for each device, simple and precise method to control the back pressure is required for the multicomponent olfactory display. One of the methods to adjust the back pressure is to adjust the height of the liquid position in the same way as intravenous drip in a hospital. Although the back pressure can be precisely controlled by this method, manual operation is required. Thus an appropriate pump is preferable to control the back pressure of the inkjet device (Hachiyama & Nakamoto, 2024).

Next, the inkjet device typically needs high voltage pulse (60–120 V) to drive it. Its pulse

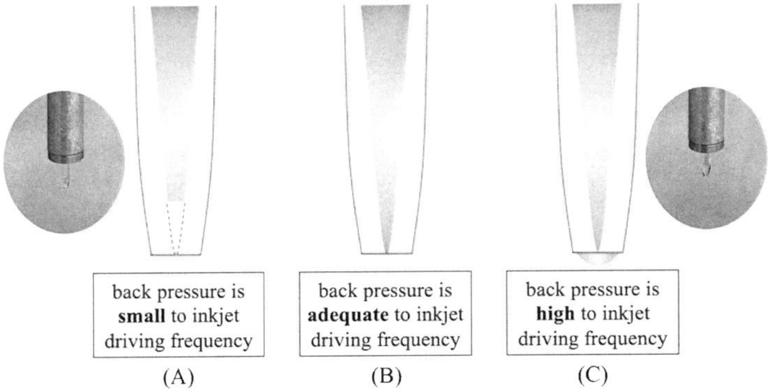

FIGURE 15.20 **State of tip of inkjet device**. (A) Too small back pressure, (B) appropriate back pressure, and (C) too high back pressure.

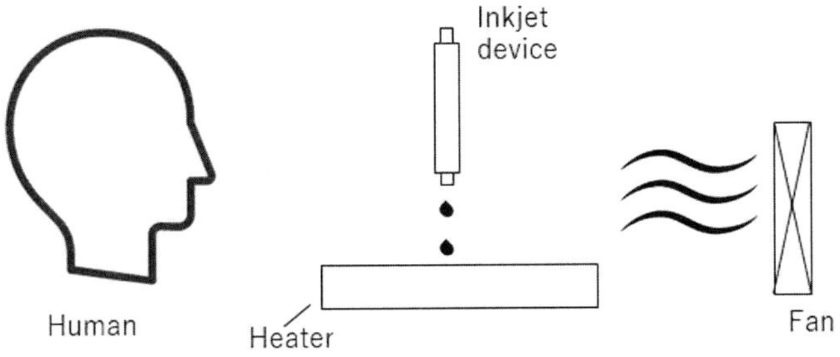

FIGURE 15.21 **Generated vapor at heater carried by air flow**. Heater is placed close to the inkjet device.

width is a few tens of microseconds. Power MOS FET can be used for this purpose. Since the gate capacitance of power MOS FET is large, the transistor to drive its gate capacitance is necessary.

Furthermore, the droplet is easily influenced by the air flow since its size is very small. Thus the heater for the evaporation should be close to the inkjet device when the configuration illustrated in Fig. 15.21 is used.

15.6 Surface acoustic wave atomizer

15.6.1 Feature and principle

Instead of heater, an ultrasonic device is sometimes used. It does not require heat and forcibly atomizes the liquid droplet. Thus the mixture composition after the atomization in the gas phase is the same as that composition of the liquid before the atomization and the specified mixture composition is guaranteed. On the other hand, it is not easy to obtain the specified composition from the olfactory display in Section 15.3 since the head space concentration in the vial is typically not saturated. Especially, the difference between them is dominant when low-volatile compound is used, although low-volatile compounds cannot be ignored because of their high odor intensity. Moreover, tubes are not used after vaporization in the atomizer method. Thus the remaining smell can be suppressed since the low-volatile compounds tend to stick to the internal wall of the tube.

There are two types of atomizers such as bulk and SAW (surface acoustic wave) atomizers. Although several types of bulk transducer are sometimes immersed in water (Muñoz-Aguirre et al., 2005), a SAW atomizer works in the gas phase (Kurosawa et al., 1995; Shiokawa et al., 1989; Yabe et al., 2014).

The principle of a SAW atomizer is based on SAW streaming as is shown in Fig. 15.22. SAW propagates along piezoelectric substrate. When it encounters the liquid droplet, SAW power leaks into the liquid droplet, followed by the conversion of longitudinal wave. Then the longitudinal wave excites a capillary wave at the surface of the droplet. As amplitude of the capillary wave increases, tiny droplets are generated from the surface of the droplet. This phenomenon is called atomization.

SAW can instantaneously atomize low-volatile odor compounds, which is difficult for a conventional olfactory display using air flow to evaporate. Thus this phenomenon is applicable to atomize odorants and is appropriate for multiple component olfactory display owing to forcible atomization of the liquid.

When we use a SAW atomizer, it is necessary to select a piezoelectric material, cut, and the orientation of SAW propagation. High efficiency of the atomization is obtained when the electromechanical coupling coefficient (Campbell & Jones, 1968) is high. Thus 128 degrees rotated *Y*-cut *X*-propagation $LiNbO_3$ is preferred because of its high electromechanical coupling coefficient even if its temperature characteristic is not always good.

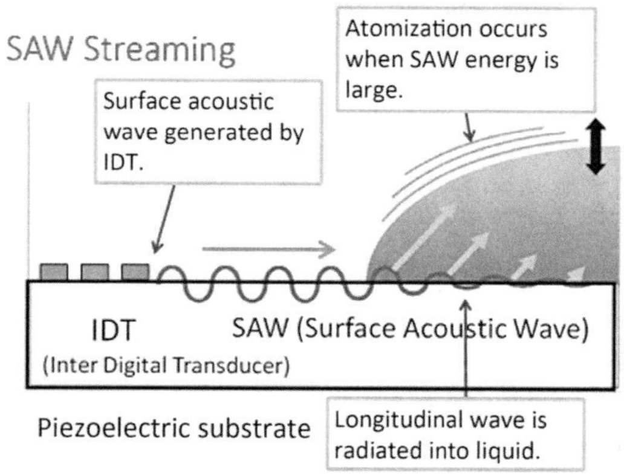

FIGURE 15.22 Principle of SAW atomizer. SAW power atomizes a liquid droplet. *SAW*, Surface acoustic wave.

SAW propagates to both directions, that is, front and its opposite. Thus the reflectors are placed back side of the excitation inter digital transducer (IDT) as is shown in Fig. 15.23. Thus SAW power toward the front direction becomes strong. In some cases, absorber is placed at the edge of the piezoelectric substrate to suppress triple transit echo. The SAW power increases as the number of electrode pairs increases. There is the trade-off between the SAW power and the device size.

The selection of SAW frequency can be done as follows. Currently, SAW frequency in an radio frequency (RF) filter becomes as high as GHz range. When SAW device is applied to SAW sensor (Chapter 3), high frequency is preferred because of its high sensitivity. However, a SAW atomizer is one type of actuator. Lower frequency provides higher efficiency of the atomization. Thus low frequency down to 10 MHz is preferred.

Moreover, hydrophobic film coated on the piezoelectric substrate is effective to increase the efficiency (Li et al., 2018; Nakamoto, Ollila, et al., 2018). Amorphous Teflon coating was used to enhance it. The supplemental material, that is, silane-coupling agent, is necessary to prevent Teflon coating from its atomization.

15.6.2 Micro pump

When we use a SAW atomizer, the method to supply the liquid onto the piezoelectric

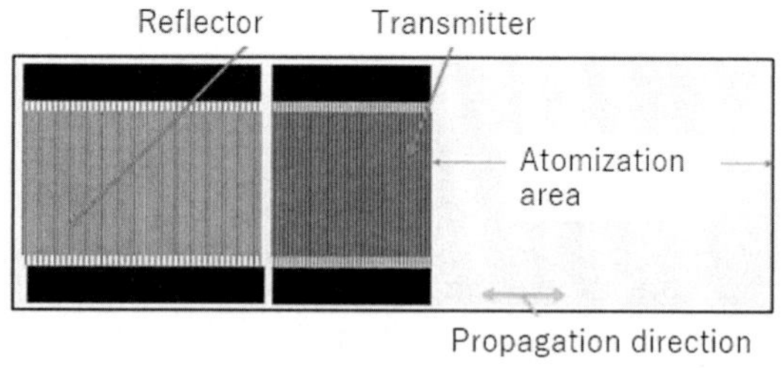

FIGURE 15.23 IDT with reflector (top view) Reflector is necessary since SAW propagates to both front and backside direction from transmitter. *SAW*, Surface acoustic wave.

substrate is necessary. Fig. 15.24A and B shows the configuration of SAW device and tube connected to the micropump.

The droplet is transferred from the tube to the piezoelectric substrate when the droplet is in contact with the substrate. Therefore the distance between the substrate and the tip of the tube should be smaller so that small droplet can be obtained. Considering this point, the configuration in Fig. 15.24B is preferable to obtain small droplet. Moreover, the tube with small inner diameter is better to obtain the small droplet. Stainless tube is typically used because of its rigid shape.

The most typical type of micropump is a diaphragm pump. The small diaphragm pump is commercially available. One of the examples of using miniaturized diaphragm pump is shown in the reference (Hashimoto & Nakamoto, 2016). Pairs of stainless tube and miniaturized pump were used to construct multicomponent olfactory display. However, the robustness of the micropump against the air bubble is insufficient, and the tube is often clogged since the drive power of the small diaphragm pump is weak.

Another type of micropump is electroosmotic (EO) pump (Ariyakul & Nakamoto, 2013). The motion of liquid is induced by applied voltage across porous material. Its drive power is strong, and the robustness against the air bubble is better than diaphragm pump, whereas its flow rate is slow.

However, micropump method has one drawback. Tiny stainless tube is used to suppress the liquid droplet volume. The liquid droplet gradually grows up until it touches the substrate surface as is illustrated in Fig. 15.25. Then it is transferred to the substrate. Since it is difficult to precisely keep the distance between the tip of the thin stainless tube and the substrate surface constant, the amount of liquid droplet changes according to the situation.

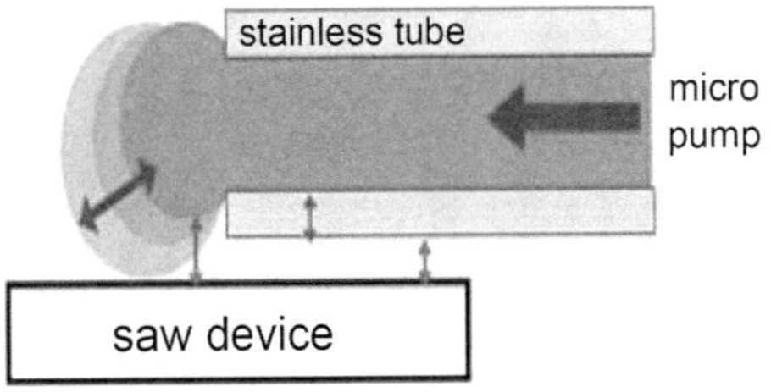

FIGURE 15.25 **Problem of micropump method**. It is difficult to precisely keep the distance between tip of thin stainless tube and substrate constant. *Source: From Nakamoto, T., Ito, S., Kato, S., & Qi, G. P. (2018). Multicomponent olfactory display using solenoid valves and SAW atomizer and its blending-capability evaluation.* IEEE Sensors Journal, 18*(13), 5213–5218, https://doi.org/10.1109/jsen.2018.2834953.*

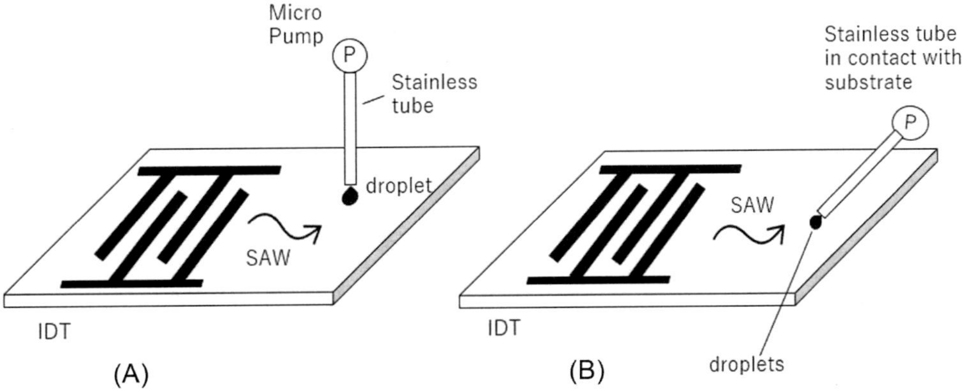

FIGURE 15.24 **Configuration of SAW device and tube**. (A) tube perpendicular to substrate and (B) tube along substrate. *SAW*, Surface acoustic wave.

15.6.3 Micro dispenser

The method to solve the problem illustrated in Fig. 15.25 is to jet a liquid droplet toward the piezoelectric substrate (Nakamoto, Ito, et al., 2018). The precise control of the distance is not necessary, and the liquid droplet volume is not influenced by the pump load. A device to jet a liquid droplet is called microdispenser. The volume of each droplet is typically nanoliter level. Its principle is shown in Fig. 15.26.

Its main part is electromagnetic coil. The plunger is driven by an electromagnetic coil with its sufficient power. The plunger rapidly moves to draw liquid in front of it for short ON time and then moves back to the original location to spout a tiny liquid droplet.

Principle of multicomponent olfactory display is illustrated in Fig. 15.27. Each microdispenser corresponds to each ingredient. The ratio of drive frequency determines the recipe. The droplet from each microdispenser is atomized by SAW device, followed by the air blow toward a user.

The olfactory display based on microdispensers is more stable than that based on micropumps. Its blending capability was confirmed by the sensory test. Chocolate, mint, and their blended flavor were used in the sensory test. Triangle test, that is, extraction of one sample different from other two among three samples, was performed. Note that the performance is better if the correct answer rate approaches the case of chance hit (5/15). The result of the sensory test is tabulated in Table 15.1. Although people distinguished between chocolate and blended flavor and did between mint and blended flavor, they could not distinguish between the flavor blended by the olfactory display and the chocolate-mint flavor manually blended in the liquid phase in advance. Those results were validated by the statistical test at 5%

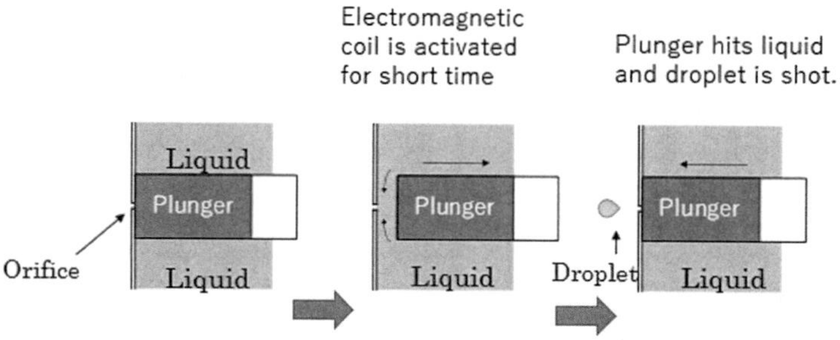

FIGURE 15.26 Principle of microdispenser based on solenoid valve. A microdispenser jets a liquid droplet.

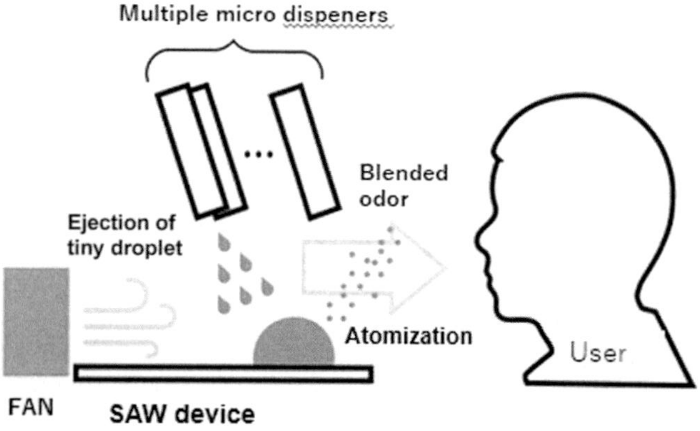

FIGURE 15.27 Principle of multicomponent olfactory display based on microdispensers. *Source: Modified from Nakamoto, T., Ito, S., Kato, S., & Qi, G. P. (2018). Multicomponent olfactory display using solenoid valves and SAW atomizer and its blending-capability evaluation.* IEEE Sensors Journal, *18(13), 5213–5218, https://doi.org/10.1109/jsen.2018.2834953.*

TABLE 15.1 Sensory test (triangle test) to check blending capability (number of correct answer/number of total subjects).

	Correct answer rate
Mint versus chocolate-mint flavor	10/15
Chocolate versus chocolate-mint flavor	11/15
Chocolate-mint flavor versus scent blended by our device	4/15

significant level. Thus it was found that blending function works well.

There was one problem to be solved. The air driven by the air pump was used to push the liquid sample to the microdispenser. However, the liquid sometimes did not enter the microdispenser because the drive power by the air pump was still weak. Thus the EO pump was used again to move the liquid to the microdispenser (Iseki et al., 2022). The drive power of EO pump is appropriate for providing the liquid sample toward the microdispenser.

One more problem occurred when EO pump was used. A droplet volume changes according to drive frequency of the microdispenser since the back pressure of the microdispenser increases in case of its low drive frequency. Thus it is necessary to adjust the EO pump voltage according to the drive frequency of the microdispenser. Although the voltage of EO pump was manually adjusted using a potentiometer at first, it took much time to adjust those of all ingredients. Thus DAC (digital-to-analog converter) was introduced to control their voltages from a computer (Hayashi et al., 2023). Final block diagram of 20-component olfactory display based on microdispensers is shown in Fig. 15.28.

PC sends the command to field programmable gate array (FPGA), which generates the command to multichannel DAC as well as trigger signal to the function generator. FPGA also generates the pulse signals to microdispensers via drive circuit. The range of DAC is between 0 and 200 (V), which is required to drive EO pump.

The voltage of each EO pump is adjusted according to drive frequency of corresponding microdispenser.

Moreover, the function generator generates RF burst signal with the duty cycle between 10% and 20%. RF burst signal is supplied to IDT in SAW atomizer via an RF power amplifier. The drive power of SAW atomizer is around 30 W. Finally, DC fan is used to convey the scent to a human's nose.

The photo of the olfactory display is shown in Fig. 15.29. A total of 20 channels were implemented into it. Each channel consists of EO pump with reservoir and microdispenser. There is an SAW device, which atomizes liquid droplets, beneath them. A DC fan can be seen in front.

15.7 Odor reproduction using olfactory display

Odor can be reproduced by blending odor components as is described in Chapter 11. Although most of the experiments in Chapter 11 were done by blending the odor components in the liquid phase, it is possible to blend odor components using the olfactory display described in Section 15.6.

A total of 20 odor components were extracted from the mass spectra of 185 essential oils. Although the mass spectrum correlation

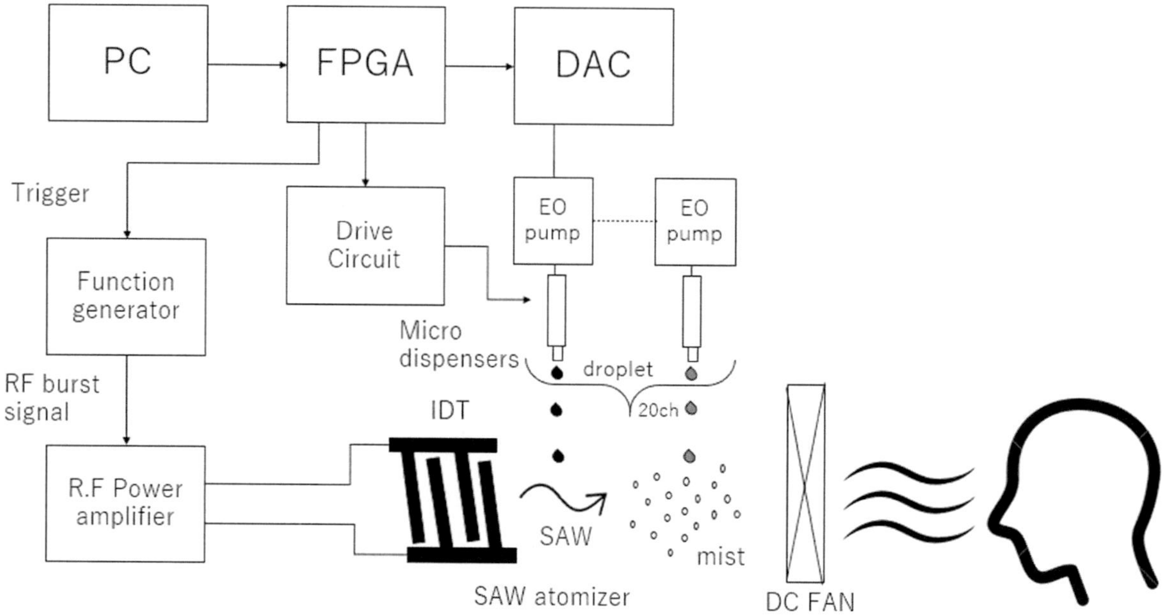

FIGURE 15.28 **Olfactory display based on microdispensers after introducing DAC**. DAC was used to control EO pump voltage. *DAC*, Digital-to-analog converter; *EO*, electroosmotic.

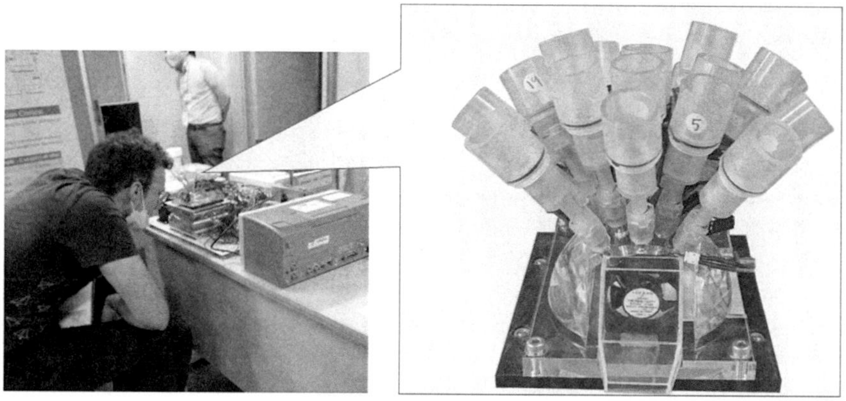

FIGURE 15.29 **A total of 20-component olfactory display using microdispensers** (left). Demo scene and (right) 20 channels composed of EO pumps and microdispensers. *EO*, Electroosmotic.

between reproduced scent and original one was very high for each essential oil, it is necessary to check the similarity at the perception level. Each odor component was set at the corresponding channel. Typical samples were taken from seven categories of essential oils, that is, citrus, exotic, spice, resin, floral, woody, and herbs.

The sensory test (triangle test) was done to check whether the blended scent was different

TABLE 15.2 Sensory test (triangle test) to discriminate between reproduced odor and original one.

	Blending in liquid phase	Blending in gas phase by olfactory display
Lemon (citrus)	3/9	8/18
Pamarose (exotic)	3/9	5/18
Carrot seed (spice)	2/9	3/18
Elemi (resin)	1/9	3/18
Lavender (floral)	2/9	5/18
Cypress (woody)	3/9	6/18
Mentha arvensis (herb)	3/9	9/18

from the original one (Iseki et al., 2022). The result of triangle test is shown in Table 15.2. The result of each case is close to the level of chance hit, that is, 1/3. The statistical test says that there was no significant difference between them (significant level: 5%) in all cases, including both cases of blending in the liquid and the gas phases. Later the sensory test for 28 essential oils, including seven samples in Table 15.2, was performed using the olfactory display. Compared with the sensory test using the liquid sample, the efficiency of the sensory test was much improved because it is not necessary to prepare for liquid samples in advance. The recipe can be changed just by sending the command to the olfactory display. Thus we can perform large-scale sensory test using the olfactory display. For 28 samples, around 10 subjects joined the sensory test of each sample. Total correct answer rate was 138/279. Although it was a little higher than 1/3, that is, chance hit, it can be said from the experiment on many samples that the odor can be reproduced with high accuracy.

Furthermore, the variation of one droplet volume among the microdispensers was investigated (Hayashi et al., 2023). There was some variation among microdispensers, and the drive frequency of each microdispenser can be compensated on the basis of the measurement of each microdispenser. The triangle test to discriminate between the raw sample and the sample with compensation was performed. The result is shown in Table 15.3. The correct answer rate was still low but slightly high for a few samples. Thus it is better to compensate for drive frequency of each channel, although significant problem was not found.

TABLE 15.3 Triangle test to discriminate between sample without compensation and one with compensation.

Essential oil	Correct answer rate
Mandarin	2/10
Lavandin	4/10
Frankincense	3/10
Nutmeg	3/10
Eucalyptus	5/10
Fennel	5/10
Niaouli	5/10

15.8 Summary

Multicomponent olfactory display with blending capability was described in this chapter. Among several types of blending methods, microdispenser with an EO pump was appropriate for realizing the specified blending ratio. Moreover, the sensory test reveals the odor reproduction capability using the olfactory display.

References

Ariyakul, Y., & Nakamoto, T. (2013). Improvement of odor blender using electroosmotic pumps and SAW atomizer for low-volatile scents. *IEEE Sensors Journal, 13*(12), 4918–4923. Available from https://doi.org/10.1109/JSEN.2013.2271912.

Grate, J. W., Ballantine, D. S., & Wohltjen, H. (1987). An automated vapor-generation and data collection instrument for the evaluation of chemical microsensors. *Sensors and Actuators, 11*(2), 173–188. Available from https://doi.org/10.1016/0250-6874(87)80015-x.

Campbell, J.J., & Jones, W.R. (1968). Proc. IEEE Ultrasonics Symposium, vol.15, 209.

Hachiyama, H., & Nakamoto, T. (2024). *Optimization of multi-component olfactory display using inkjet devices, Vol. E107-A, No.8*, 1338–1344. Available from https://doi.org/10.1587/transfun.2023EAP1051.

Hashimoto, K., & Nakamoto, T. (2016). Tiny olfactory display using surface acoustic wave device and micropumps for wearable applications. *IEEE Sensors Journal, 16*(12), 4974–4980. Available from https://doi.org/10.1109/JSEN.2016.2550486.

Hayashi, H., Prasetyawan, D., Iseki, M., & Nakamoto, T. (2023). Optimization of odor blending for 20-component olfactory display. *IEEJ Transactions on Sensors and Micromachines, 143*(11), 338–343. Available from https://doi.org/10.1541/ieejsmas.143.338.

Iseki, M., Prasetyawan, D., Yokoshiki, Y., & Nakamoto, T. (2022). A study of odor reproduction using multi-component olfactory display. *Electrical Engineering in Japan, 215* (3). Available from https://doi.org/10.1002/eej.23392.

Johnson, B. N., & Sobel, N. (2007). Methods for building an olfactometer with known concentration outcomes. *Journal of Neuroscience Methods, 160*(2), 231–245. Available from https://doi.org/10.1016/j.jneumeth.2006.09.008.

Kurosawa, M., Watanabe, T., Futami, A., & Higuchi, T. (1995). Surface acoustic wave atomizer. *Sensors and Actuators A: Physical, 50*(1–2), 69–74. Available from https://doi.org/10.1016/0924-4247(96)80086-0.

Li, H., Qi, G., Kato, S., & Nakamoto, T. (2018). Investigation and improvement of atomization efficiency based on SAW device coated with amorphous fluoropolymer film for olfactory display. *Sensors and Actuators B: Chemical, 263*, 266–273. Available from https://doi.org/10.1016/j.snb.2018.01.245.

Muñoz-Aguirre, S., Nakamoto, T., & Moriizumi, T. (2005). Study of deposition of gas sensing films on quartz crystal microbalance using an ultrasonic atomizer. *Sensors and Actuators B: Chemical, 105*(2), 144–149. Available from https://doi.org/10.1016/j.snb.2004.05.063.

Nakamoto, T., Ito, S., Kato, S., & Qi, G. P. (2018). Multicomponent olfactory display using solenoid valves and SAW atomizer and its blending-capability evaluation. *IEEE Sensors Journal, 18*(13), 5213–5218. Available from https://doi.org/10.1109/jsen.2018.2834953.

Nakamoto, T., & Minh, H. P. D. (2007). Improvement of olfactory display using solenoid valves. In *Proceedings—IEEE virtual reality*, (pp. 179–186). IEEE. Available from https://doi.org/10.1109/VR.2007.352479.

Nakamoto, T., Ollila, S., Kato, S., Li, H., & Qi, G. (2018). Enhancement method of surface acoustic wave-atomizer efficiency for olfactory display. *Journal of Visualized Experiments, 2018*(141). Available from https://doi.org/10.3791/58409.

Nakamoto, T., Takigawa, H., & Yamanaka, T. (2004). Fundamental study of odor recorder using inkJet devices for low-volatile scents. In *IEICE transactions on electronics*, (vol. E87-C, No12, pp. 2081–2086). Japan: Institute of Electronics, Information and Communication, Engineers, IEICE.

Nakamoto, T., Ustumi, S., Yamashita, N., Moriizumi, T., & Sonoda, Y. (1994). Active gas/odor sensing system using automatically controlled gas blender and numerical optimization technique. *Sensors and Actuators B: Chemical, 20*(2-3), 131–137. Available from https://doi.org/10.1016/0925-4005(93)01193-8.

Norsworthy, S. R., Schreier, R., & Temes, G. C. (1996). *Delta-sigma data converters*. IEEE Press.

Shiokawa, S., Matsui, Y., & Ueda, T. (1989). Liquid streaming and droplet formation caused by leaky Rayleigh waves. In *Ultrasonics symposium proceedings*, (pp. 643–646). IEEE. Available from https://doi.org/10.1109/ULTSYM.1989.67063.

Shoji, S., & Esashi, M. (1994). Microflow devices and systems. *Journal of Micromechanics and Microengineering, 4* (4), 157–171. Available from https://doi.org/10.1088/0960-1317/4/4/001.

Somboon, P., Wyszynski, B., & Nakamoto, T. (2007). Novel odor recorder for extending range of recordable odor. *Sensors and Actuators B: Chemical, 121*(2), 583–589. Available from https://doi.org/10.1016/j.snb.2006.04.105.

Yabe, A., Hamate, Y., Hara, M., Oguchi, H., Nagasawa, S., & Kuwano, H. (2014). A self-converging atomized mist spray device using surface acoustic wave. *Microfluidics and Nanofluidics, 17*(4), 701–710. Available from https://doi.org/10.1007/s10404-014-1358-2.

Yamanaka, T., Matsumoto, R., & Nakamoto, T. (2002). Study of odor blender using solenoid valves controlled by delta–sigma modulation method for odor recorder. *Sensors and Actuators B: Chemical, 87*(3), 457–463. Available from https://doi.org/10.1016/s0925-4005(02)00300-3.

CHAPTER

16

Wearable-olfactory displays

Takamichi Nakamoto

Institute of Integrated Research, Institute of Science Tokyo, Yokohama, Japan

16.1 Background

We have looked at various olfactory displays and multicomponent olfactory displays in Chapter 14 and 15, respectively. This chapter describes a wearable-olfactory display that is used together with HMD (head-mount display) frequently used in VR (virtual reality). This is a key device for olfactory VR. The photo of wearable-olfactory display is shown in Fig. 16.1.

Compared with desktop-olfactory display, it should be compact. A user can walk around and can sniff the scent even in the virtual environment. Since the scent generator is close to the user's nose, the delay of scent delivery is very small. Since the amount of emitted fragrance is very small, smell persistence can be minimized. On the other hand, the number of channels is typically limited due to its size constraint.

Although small olfactory displays have been introduced in Chapter 14, most of them cannot present multiple scents and do not have the function of blending scents at the specified recipe. Two examples of wearable-olfactory displays are explained in this chapter.

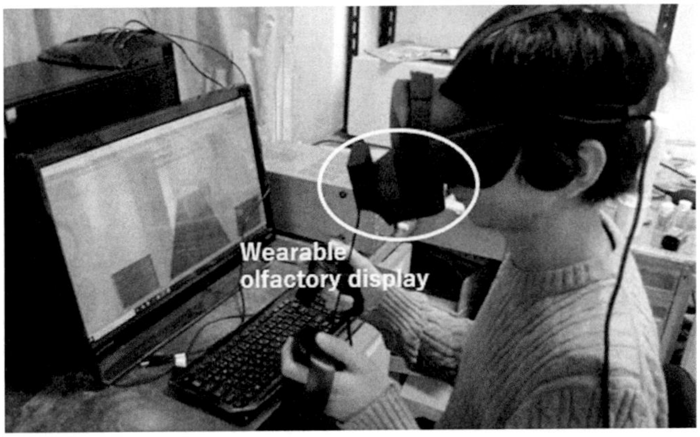

FIGURE 16.1 **Photo of wearable-olfactory display.** The wearable-olfactory display is located beneath HMD. *HMD*, Head-mount display.

Digital Technologies in Olfaction
DOI: https://doi.org/10.1016/B978-0-443-15721-9.00016-6

16.2 Wearable-olfactory display with micro pump

16.2.1 Principle

First example is a wearable-olfactory display with small pump (Hashimoto & Nakamoto, 2016). Its concept is shown in Fig. 16.2. The micro pumps deliver scented liquid from the reservoirs to the surface of SAW (surface acoustic wave) device. By controlling flow rates of the micro pumps, several scented liquids can be blended at the specified recipe. Then the SAW device atomizes the liquid instantaneously in the same manner as that in Chapter 15.6.

The olfactory display equipped with two micro pumps is shown in Fig. 16.3. The total size of the gadget is $70 \times 60 \times 33\ \text{mm}^3$, and its weight is 95 g. It is equipped with universal serial bus (USB) interface, and its power is supplied by USB bus power as well as extra 6 V direct current (DC) voltage.

16.2.2 Drive circuit of surface acoustic wave device

The driving circuit of SAW atomizer is shown in Fig. 16.4. The driving frequency

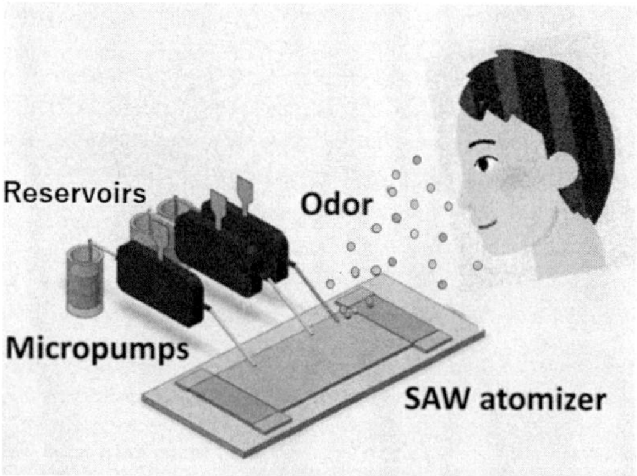

FIGURE 16.2 **Principle of wearable-olfactory display with micro pumps.** Several ingredients can be blended at any ratio by adjusting the flow rate of each micro pump.

FIGURE 16.3 **Photo of wearable-olfactory display with micro pump.** Two micro pumps are located on the left side.

should be tunable since the optimum drive frequency changes from device to device. The bulky function generator in Chapter 15 cannot be used in the wearable-olfactory display. Thus DDS (direct digital synthesizer) implemented into FPGA (field programmable gate array) was used as an RF (radio frequency) signal source (Chu, 2011). The drive frequency of SAW device was around 61 MHz, and the resolution around 10 kHz is required to adjust the drive frequency. The frequency resolution of DDS is

$$\Delta f = f_{\text{in}}/2^N, \tag{16.1}$$

where f_{in} is input frequency of DDS, and N is the bit length of DDS. f_{in} was 400 MHz, which was obtained by phase locked loop from onboard clock signal (50 MHz). When N was set to 15, Δf was 12 kHz. The output signal from DDS was supplied to RF power amplifier module via buffer circuit, that is, a transistor emitter-follower circuit, which improves the impedance matching between FPGA and the amplifier. The power of RF amplifier in this example was 30 W. Although RF power amplifier module around 61 MHz is commercially available, we cannot find an appropriate RF power amplifier module at certain frequency zone. A small-size RF power amplifier is indispensable for realizing wearable-olfactory display.

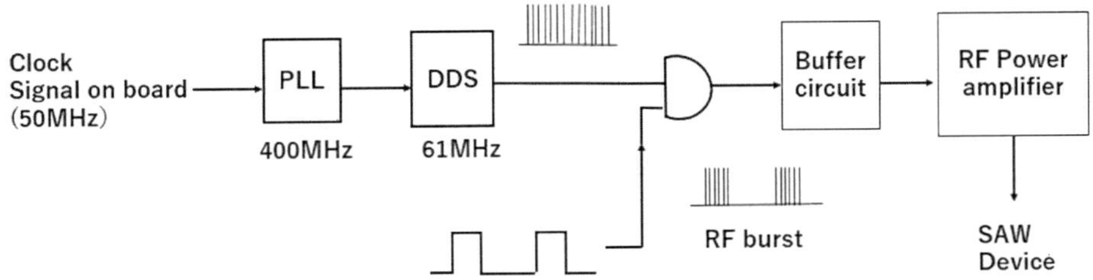

FIGURE 16.4 **Drive circuit of SAW atomizer. Drive frequency is tunable owing to its variation from device to device.** *SAW,* Surface acoustic wave.

The output signal of FPGA should not be continuous, but RF burst signal is used since the net RF power can be controlled by specifying duty cycle. If RF power is too high, the heat is generated at SAW device and the device is broken in the worst case. If RF power is too small, the atomization does not occur.

If the heat generated by SAW device is dominant, it influences resonance frequency of SAW device because $LiNbO_3$ substrate used for SAW atomizer is highly sensitive to temperature change. The temperature sensor, that is, small infrared sensor, together with feedback circuit to control drive frequency, was effective in that case (Hashimoto & Nakamoto, 2015).

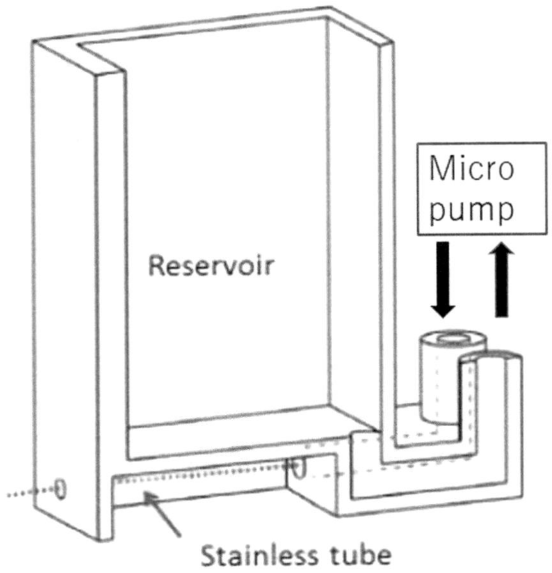

FIGURE 16.5 **Micro pump and reservoir.** Its whole size was 7 × 19 × 12.5 mm.

16.2.3 Micro pump

The micro pump and reservoir are shown in Fig. 16.5. The pump used here is a miniaturized diaphragm pump. The reservoir and the micro pump adapter were fabricated using 3D printer. Its whole size was 7 × 19 × 12.5 mm, and the reservoir volume was 400 μL. The stainless tube with inner diameter of 0.15 mm was used to supply a scented liquid to SAW device.

To drive the micro pump, alternating current (AC) high voltage is required. However, its drive circuit should be compact because of wearable device. Although the switching circuit composed of power metal oxide semiconductor field effect transistors (MOS FETs) and DC–DC converter can generate high AC voltage (Nakamoto et al., 2014), it occupies large space. Thus boost chopper circuit as is shown in Fig. 16.6 was used. The switching of the current flowing at a micro inductor L generates the high voltage at the anode of the diode, and the current flow into the capacitor C when the anode voltage is higher than that of cathode. Each pulse in a burst signal from FPGA increases the capacitor voltage step by step. Around 100 V AC voltage can be obtained even without high voltage source,

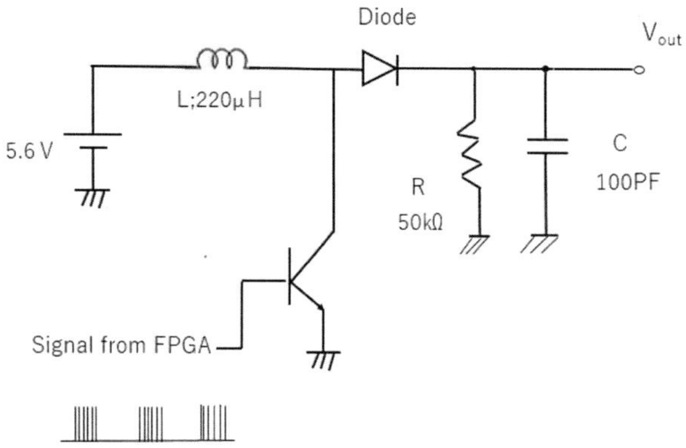

FIGURE 16.6 Boost chopper circuit. This circuit is used to generate high AC voltage for micro pump.

although only a few *V* voltage sources are necessary. During off time of the burst signal, the capacitor *C* is discharged through resistor *R*. Then its voltage increases again during on time of the burst signal. This circuit is suitable since the occupied space is much smaller than conventional MOS FET switching circuit.

Although small-size olfactory display can be realized using the micro pump, it has a problem of the instability. Although micro pump is very small, its drive power of the liquid is not enough. It often has clogging, and air bubble disturbs the liquid flow. Moreover, the amount of liquid is instable since the distance between the tip of thin stainless tube and the piezoelectric substrate cannot be controlled, which causes the variation of the liquid droplet amount (see Chapter 15.6.2). Thus another method is to use micro dispensers.

16.3 Wearable-olfactory display with micro dispenser and surface acoustic wave atomizer

If the distance between the tip of the stainless tube and the substrate is uncontrollable, how about jetting a liquid droplet to the substrate? The amount of the liquid droplet is free of that distance, and the load of the actuator is always constant. Thus the micro dispenser was used to shoot the liquid droplet.

The micro dispensers were used even in the wearable-olfactory display (Kato & Nakamoto, 2019). Its concept is illustrated in Fig. 16.7. The liquid droplet shot by the micro dispenser is atomized by SAW device. The recipe of ingredients is determined by the ratio of their drive frequencies as is described in Chapter 15.6.3. In the wearable-olfactory display, however, the scent is not spouted toward a user but it passes in front of his/her nose. The scent is collected by the adsorbent prior to the air pump. Since most of the generated scent is not inhaled by the user, that scent is collected by the wearable-olfactory display by itself to avoid spreading the scent around the user. The self-collection of the scent is possible since the wearable-olfactory display can take semiclosed structure.

The whole system of wearable-olfactory display is shown in Fig. 16.8. FPGA includes pulse width modutation (PWM) generator (PWM1) whose output signal is applied to an electro osmotic (EO) pump. EO pump moves scented liquid to micro dispenser. The duty cycle of PWM circuit corresponds to the back pressure of the micro dispenser. The function generator (function generator1) makes a pulse to drive the micro dispenser. The pulse width is determined by the device specification. Although short pulse is preferred to reduce the amount of the droplet, the plunger in the micro dispenser does not move quickly because of its response time. Typically, a few milliseconds (0.5–2 ms) of the pulse width are required. Function generator1 also determines the drive frequency. As drive frequency increases, the amount of the liquid increases.

The second function generator (function generator2) generates RF burst signal. The frequency of RF signals can be adjusted by DDS explained in the previous section. Based on the repetition cycle and duty cycle obtained from personal computer (PC), RF burst signal is

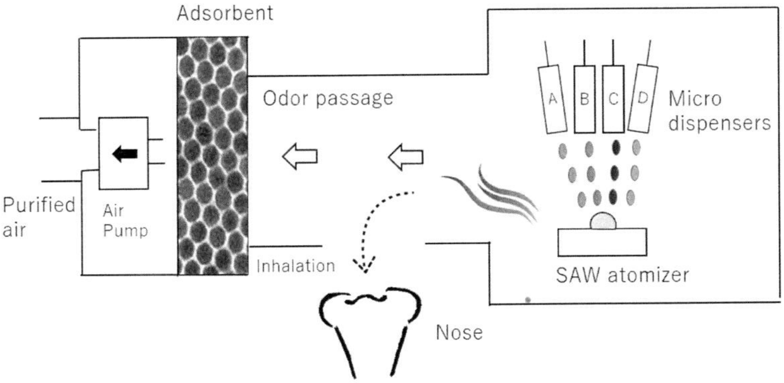

FIGURE 16.7 **Wearable-olfactory display with micro dispensers.** The scent not inhaled by a user is collected by the olfactory display by itself.

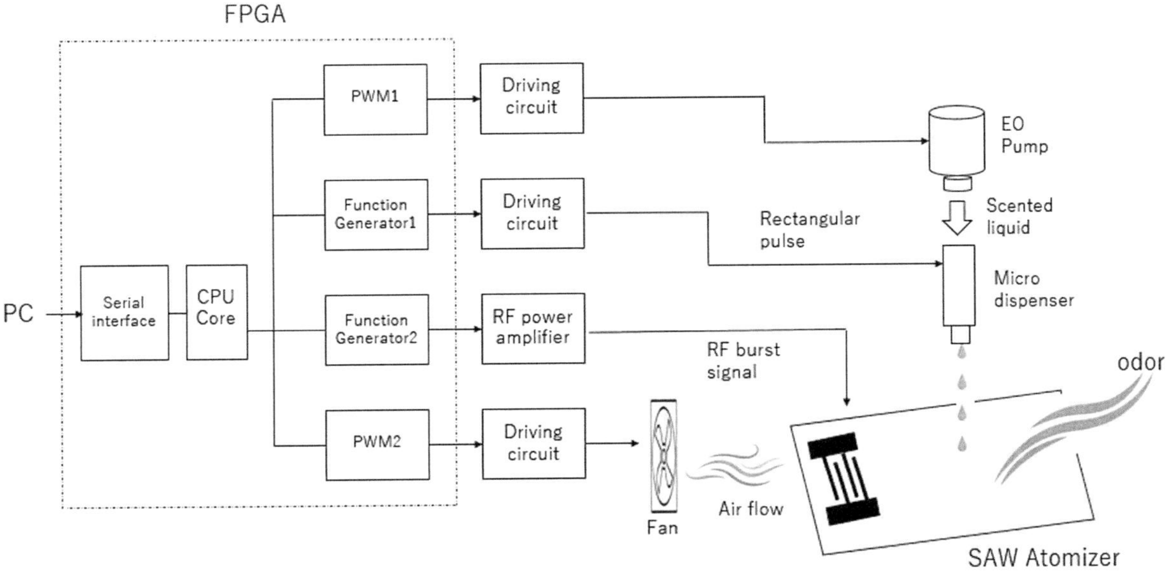

FIGURE 16.8 **Whole system of wearable-olfactory display. The circuits to control EO pumps, micro dispensers, SAW atomizer, and DC fan were implemented.** *SAW*, Surface acoustic wave.

generated here. RF burst signal is supplied to SAW atomizer through RF power amplifier. Last part is PWM circuit (PWM2) to control the intensity of fan wind. Those four parts are controlled by central processing unit (CPU) core embedded in FPGA, which is connected to PC via serial interface.

The most important part to realize a wearable-olfactory display is RF power amplifier. Although bulky commercially available RF power amplifier was used in a desktop-olfactory display (Chapter 15), it is not applicable to the wearable-olfactory display. RF power amplifier module is not available for low-frequency range, that is, 10 MHz, although it is commercially available for 61 MHz range in the previous section. 10 MHz frequency is preferable because of its strong atomization power. Thus E-class RF power amplifier was realized using discrete power MOS FET, although the development of the discrete transistor circuit is nowadays rare. Its circuit is shown in Fig. 16.9. E-class RF amplifier has high efficiency in spite of nonlinearity. Although nonlinearity is acceptable for this application, low efficiency is not allowed since the heat dissipation should be minimized in the wearable device.

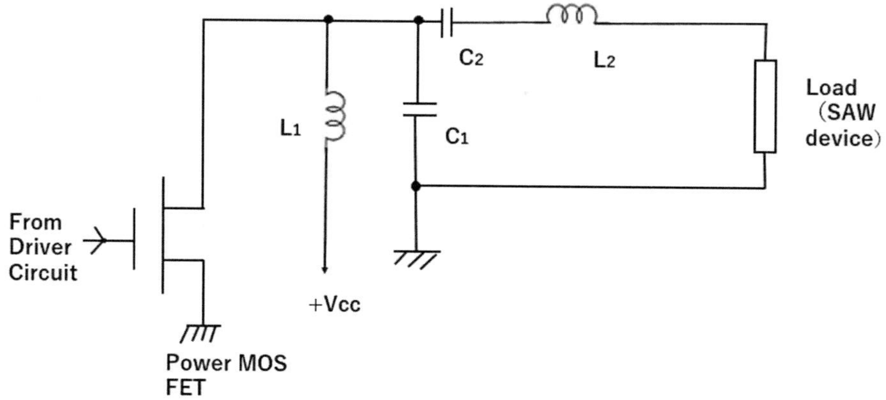

FIGURE 16.9 E-class RF power amplifier used in a wearable-olfactory display. It has high efficiency in spite of nonlinearity. *RF*, Radio frequency.

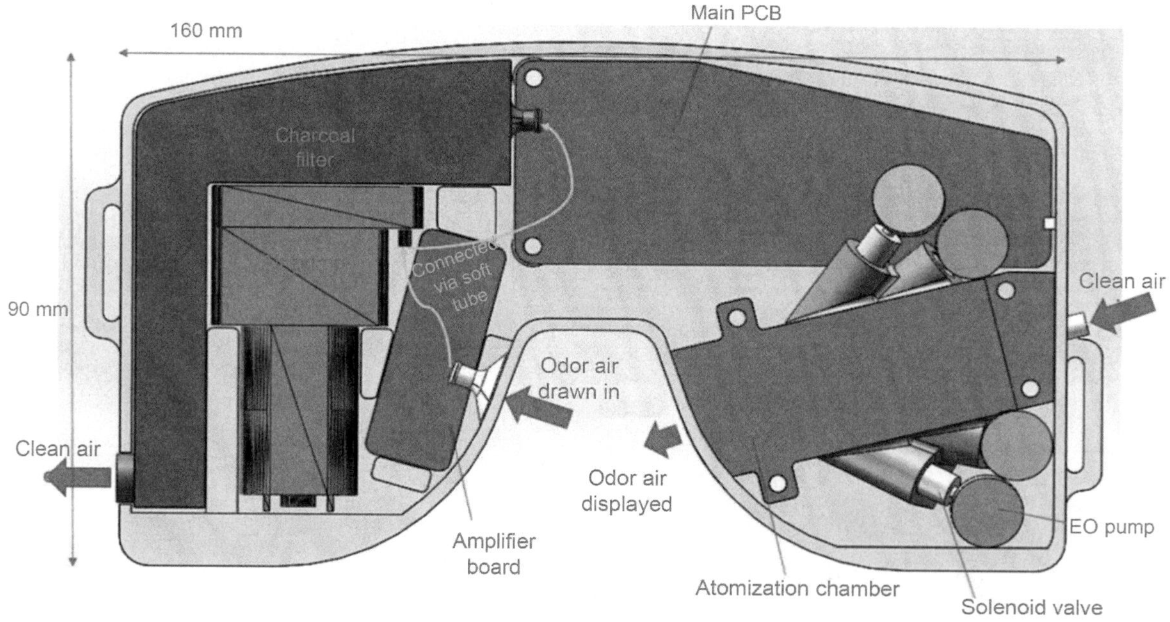

FIGURE 16.10 Layout in wearable-olfactory display. Four channels of scents were implemented. *Source: From Aiheng, L., & Nakamoto, T. (2022).* Design of wearable olfactory display using solenoid valves and electroosmotic pumps. *In IEEJ Annual Meeting (pp. 3–135).*

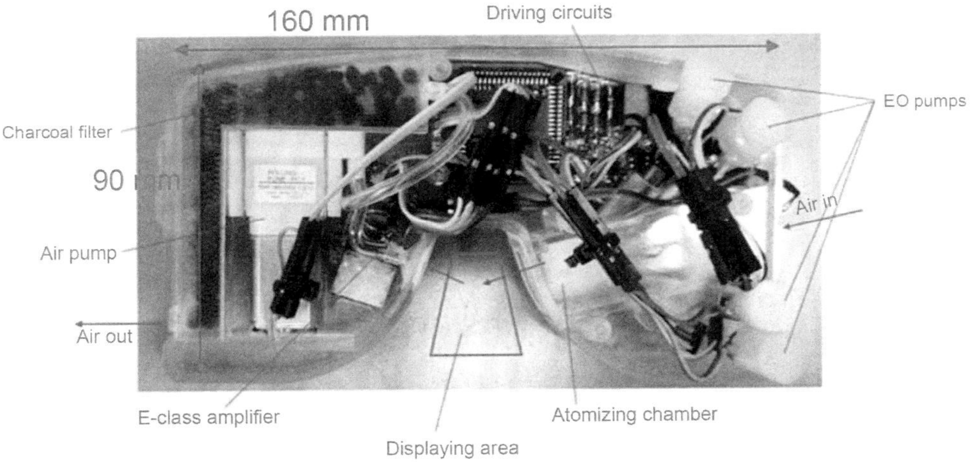

FIGURE 16.11 **Photo of internal structure of olfactory display.** Its layout corresponds to Fig. 16.10.

Prior to power MOS FET circuit, its driver circuit is necessary because of its large gate capacitance. Multiple inverter circuits connected in parallel were used to drive power MOS FET.

When we make E-class amplifier, we cannot believe the nominal value of an inductor because the inductance value has the frequency dependency due to the self-resonance. The actual inductance value deviated from the nominal value that is correct at low frequency. In this case, the coil was wound by hand and checked its value at 10 MHz using a network analyzer. Its RF power was 33 W enough to drive SAW atomizer. Its efficiency exceeded 80%, which allows us to implement other devices densely in the wearable-olfactory display.

Layout of the wearable-olfactory display is shown in Fig. 16.10 (Aiheng & Nakamoto, 2022). Four pairs of EO pumps and micro dispensers were implemented on SAW atomizer. SAW device was horizontally placed in Fig. 16.10. Jigs in the wearable-olfactory display were fabricated using 3D printer. The air pump sucked the air through charcoal filter. The scent was emitted outside the box, and flow went inside the box again and flowed toward the charcoal filter. Its photo is shown in Fig. 16.11. The wearable-olfactory display can be easily attached beneath HMD. Fig. 16.12 shows the photo of a user wearing HMD with the wearable-olfactory display. The design of wearable-olfactory display is a little different from that in Fig. 16.1. The user with the wearable-olfactory display experienced scents in the virtual environment.

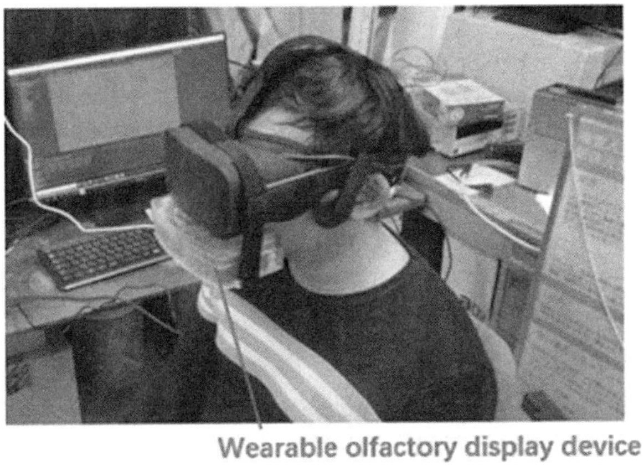

FIGURE 16.12 **Photo of user wearing HMD with wearable-olfactory display.** The user can stay in the virtual environment with scents. *HMD*, head-mount display.

16.4 Comparison of wearable-olfactory display with desktop-olfactory displays

In the olfactory contents such as movie with scents, animation with scents, and olfactory game, the speed of odor presentation and removing odor is very important. However, it is very difficult to compare the performance of olfactory display since most of them are home-made and only a developer has the device. In this section, the evaluation of several types of olfactory displays in terms of presentation speed on the same condition is described (Iseki & Nakamoto, 2021).

Three types of olfactory displays were compared. First one is based on high speed switching of solenoid valve described in Chapter 15.3.3 (abbreviation: SV). The tube is used for odor delivery. The second one is the desktop-olfactory display based on SAW atomizer (abbreviation: DS) described in Chapter 15.6.3. Third one is wearable-olfactory display with SAW atomizer described in Chapter 16.3 (abbreviation: WS). SV in this section is 13-channel olfactory display. DS has the eight-component olfactory display.

The evaluation method is the sensory test. The time when the participants started perceiving scent and the time when the odor became weaker were recorded. Three samples were selected for the experiment. 1-Propanol is a representative of highly volatile compound, isoamyl acetate is a representative of moderate-volatile compound, and alpha-pinene is a representative of low-volatile compound. Those samples were selected on the basis of the concept of perfumery, that is, top note, middle note, and last note.

Next, the odor intensities of the three samples using three olfactory displays were adjusted to become the same using the sensory evaluation. The concentration of each gas was measured and recorded by photo ionization detector.

Then the time when a subject started perceiving the smell and the time when he/she felt the odor became weaker. The smell presentation time was 15 seconds, and 10 subjects joined each test, that is, SV, DS, and WS. The measurement results of 1-propanol, isoamyl acetate, and alpha-pinene are shown in Figs. 16.13–16.15, respectively. FO time means the time required for starting perception and FW time means the time when a subject feels that smell becomes weaker. The time zero means that the computer send the command to each olfactory display. Thus FW time has dashed line at 15 seconds at which the smell generation stops.

Regarding type of olfactory display, WS was better than other olfactory displays and next one was DS. SV was the slowest. WS is fast since the position of smell generation is very close to the human's nose. Moreover, amount of consumed liquid is very small.

SV was slow in terms of both FO and FW times. Moreover, the variance was large in SV particularly isoamyl acetate and alpha-pinene. The latency and the variation are caused by odor delivery via tube. Similar to gas chromatography, it takes much time for an odorant to propagate along the tube. Moreover, the odor pulse in the tube becomes broader as well as increasing latency. The Teflon tube used in this experiment does not adsorb odorant. However, the latency occurs even if Teflon tube is used. This latency does not occur if the tube is heated to a few hundreds degree Celsius (Nakamoto, 2016). However, it is very difficult to keep the whole odor path at high temperature. Furthermore, the latency depends on the volatility of odorant. Although the latency is relatively small for highly volatile compound, it is large for low-volatile compound.

As for WS and DS, there is small dependency of sample volatility since they do not use tube for odor delivery. The variation is also small since the scented sample is forcibly and instantaneously atomized, and the concentration at the nose changes drastically. In term of

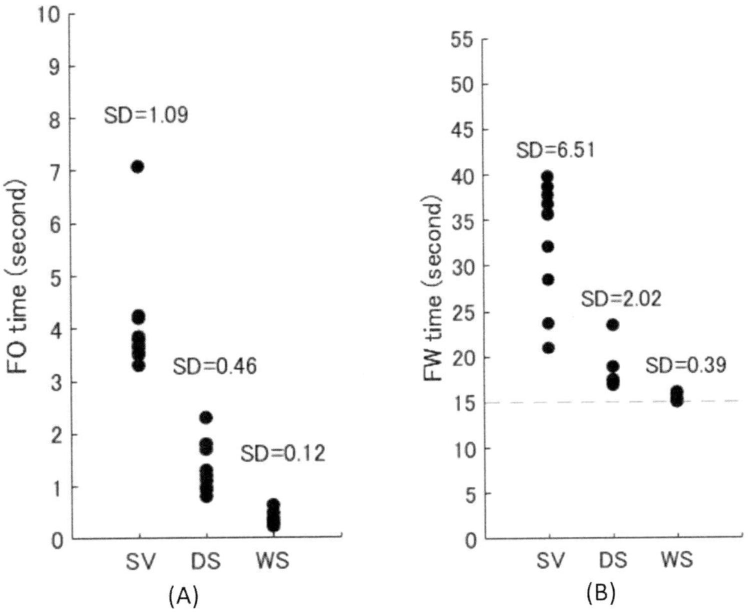

FIGURE 16.13 Temporal perception behavior for 1-propanol. (A) FO time (latency (ON)) and (B) FW time (latency (OFF)). *Source: From Iseki, M., & Nakamoto, T. (2021). Evaluation of temporal characteristics of olfactory displays with different structures.* IEICE Transactions on Fundamentals of Electronics, Communications and Computer Sciences, E104. A*(4), 744–750, https://doi.org/10.1587/transfun.2020eap1028.*

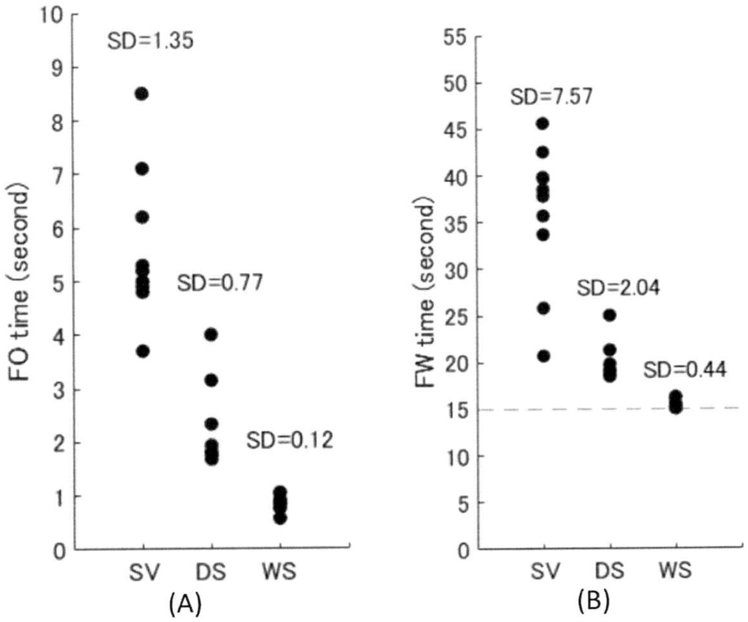

FIGURE 16.14 Temporal perception behavior for isoamyl acetate. (A) FO time (latency (ON)) and (B) FW time (latency (OFF)). *Source: From Iseki, M., & Nakamoto, T. (2021). Evaluation of temporal characteristics of olfactory displays with different structures.* IEICE Transactions on Fundamentals of Electronics, Communications and Computer Sciences, E104. A*(4), 744–750, https://doi.org/10.1587/transfun.2020eap1028.*

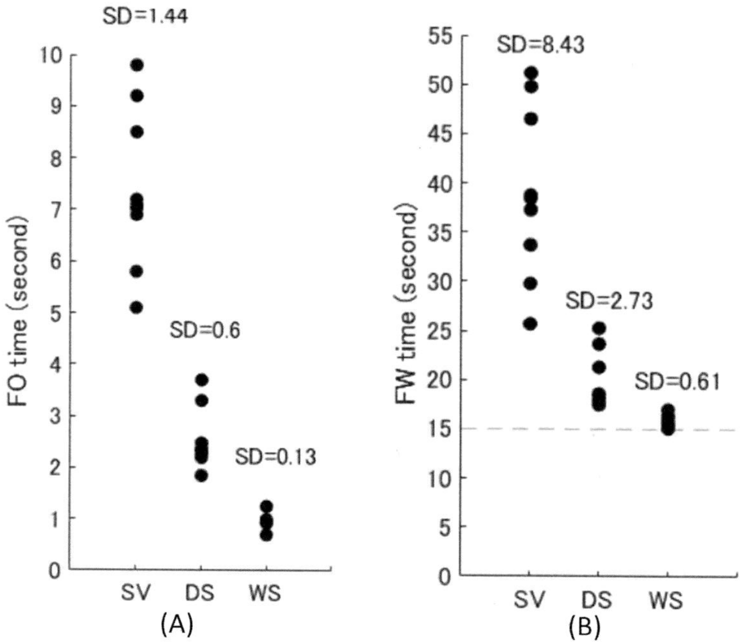

FIGURE 16.15 Temporal perception behavior for alpha-pinene. (A) FO time (latency (ON)) and (B) FW time (latency (OFF)). *Source: From Iseki, M., & Nakamoto, T. (2021). Evaluation of temporal characteristics of olfactory displays with different structures.* IEICE Transactions on Fundamentals of Electronics, Communications and Computer Sciences, E104. A*(4), 744–750, https://doi.org/10.1587/transfun.2020eap1028.*

smell persistence, FW time is important. WS seems to have advantage over other two devices.

16.5 Summary

A wearable-olfactory display is a key device to realize olfactory VR. This chapter explains its principle and the implementation method. The sophisticated implementation method is required since space available for the implementation is limited in case of wearable-olfactory display.

Although it is difficult to evaluate and compare several types of olfactory displays, three types of olfactory displays, that is, SV (high-speed switching of solenoid valve), DS (desktop olfactory display based on SAW atomizer), and WS (wearable-olfactory display based on SAW atomizer) were evaluated at the same condition. Those based on SAW atomizers were superior to that based on high-speed solenoid valve switching. Especially the wearable-olfactory display was the fastest, although many components cannot be used owing to the limited space.

References

Aiheng, L.I. U., & Nakamoto, T. (2022). *Design of wearable olfactory display using solenoid valves and electroosmotic pumps*. IEEJ Annual Meeting (pp. 3–135).

Chu, P. P. (2011). *Embedded SoPC design with Nios II processor and VHDL examples*. Wiley. Available from https://doi.org/10.1002/9781118146538.

Hashimoto, K., & Nakamoto, T. (2015). *Stabilization of SAW atomizer for a wearable olfactory display*. 2015 IEEE International Ultrasonics Symposium, IUS 2015 9781479981823 Institute of Electrical and Electronics Engineers Inc. Japan. Available from https://doi.org/10.1109/ULTSYM.2015.0355.

Hashimoto, K., & Nakamoto, T. (2016). Tiny olfactory display using surface acoustic wave device and micropumps for wearable applications. *IEEE Sensors Journal, 16*(12), 4974–4980. Available from https://doi.org/10.1109/JSEN.2016.2550486.

Iseki, M., & Nakamoto, T. (2021). Evaluation of temporal characteristics of olfactory displays with different structures. *IEICE Transactions on Fundamentals of Electronics, Communications and Computer Sciences, E104. A*(4), 744–750. Available from https://doi.org/10.1587/transfun.2020eap1028.

Kato, S., & Nakamoto, T. (2019). *Wearable olfactory display with less residual odor.* ISOEN 2019—18th International Symposium on Olfaction and Electronic Nose, Proceedings. 9781538683279 Institute of Electrical and Electronics Engineers Inc. Japan. https://doi.org/10.1109/ISOEN.2019.8823231.

Nakamoto, T. (2016). *Olfactory display and odor recorder. Essentials of machine olfaction and taste* (pp. 247–314). Wiley. Available from http://onlinelibrary.wiley.com/book/10.1002/9781118768495, https://doi.org/10.1002/9781118768495.ch7.

Nakamoto, T., Hashimoto, K., Aizawa, T., & Ariyakul, Y. (2014). *Multi-component olfactory display with a SAW atomizer and micropumps controlled by a tablet PC.* IFCS 2014—2014 IEEE International Frequency Control Symposium, Proceedings 9781479949168 IEEE Computer Society Japan. Available from https://doi.org/10.1109/FCS.2014.6859845.

CHAPTER

17

Digital contents with scents

Takamichi Nakamoto

Institute of Integrated Research, Institute of Science Tokyo, Yokohama, Japan

17.1 Background

Olfactory display is applicable to many fields. Many applications of olfactory display are oriented to entertainment. At the early stage, scents were attached to movie and animation. Then the olfactory displays were applied to game and simulator. Moreover, spatiotemporal and crossmodal effects were studied. Some people have investigated whether scents can contribute to making a narrative story. Although the approach toward simulator is described in Chapter 18, other scented contents are introduced in this chapter.

17.2 Movie/animation with scents

Typical applications of the olfactory display are scented movies and scented animations. Tomono et al. (2003) studied how many smells are required for one movie. They watched 10 cinemas and recorded the scenes they wanted to have smells. They said 3–15 types of smells were required for one cinema. People do not always want to sniff smells. It is necessary to avoid olfactory fatigue. Therefore it is important to present smell(s) effectively only in the situation when the smell is really necessary. They concluded that movie with scents can be realized since required smells are not so many.

Smells were attached to the famous cinema, Spirited Away (Nakamoto & Yoshikawa, 2006). The olfactory display using high-speed switching of solenoid valve shown in Section 15.3 was used in this content. Although it is the cinema for 2 hours, it was compressed into 4 minutes composed of 11 scenes (Nakamoto & Yoshikawa, 2006). A total of 17 participants watched the compressed movie without smell and then watched it with smells. Thereafter, every participant was asked if each scene was impressive. In scene 10, stink god with offensive smell appeared. In scene 11, stink god was soaked in medicated bath with herbal smell. The questionnaire survey reveals that the change from offensive smell in scene 10 to pleasant smell in scene 11 enhanced their impressions very much. Thus we can control that scene people have attention using smells if appropriate smell is attached to the appropriate scene.

There is possibility that drastic change of the scent provides strong impression. This is similar to the case of color where the contrast between blue and yellow often makes creative impression. If the narrative factor is added, more effect is expected.

Digital Technologies in Olfaction
DOI: https://doi.org/10.1016/B978-0-443-15721-9.00017-8

Then the smell was attached to the animation intended to make scented content. The artist group made the animations so that various objects related to smells can be included (Minh et al., 2006). Nakamoto and Nagahama made the content called "Together." It consists of nine scenes tabulated in Table 17.1, and its duration is 150 seconds. The smells used for that content are also tabulated in Table 17.1. A total of 165 participants experienced the content without smell and experienced the content with smells. The result of questionnaire survey is shown in Fig. 17.1. The questionnaire survey was performed twice at the different events. The impressions were enhanced by the smells in scenes 2 and 3. Smells of fruit and grilled meat gave participants strong impressions. On the other hand, the smell did not contribute to their impression in the scene after scene 4. Since the impression given by smells in scenes 2 and 3 was very strong, participants felt that smells became weaker after scene 3. Thus the scene people have attention can be controlled by attaching the appropriate smell to some degree. However, it is again important to attach smell(s) only when it is really necessary.

17.3 Game with scents

Although information flow is only one way in the case of movie/animation, the interaction of the player with the content can be expected when we play a scented game. Mochizuki et al. made the virtual reality (VR) game called "Fragra." Players can explore the interactive relationship between olfaction and vision (Mochizuki et al., 2004). A scent is presented to the player equipped with head-mounted display (HMD). The player tries to grasp a food and take it in front of the nose in virtual space. The player is requested to answer whether olfactory cue matches the visual one.

Nakamoto et al. (2008) made the cooking game with scents. The interactive olfactory display was developed for this purpose. The interactivity might enhance reality and immersive feeling in the virtual environment. The olfactory display using high-speed switching of solenoid valve shown in Section 15.3 was also used in this content.

In a cooking game, a player puts ingredients such as butter, garlic, onion, carrot, beef, water, red wine, curry roux, and spices in sequence in the virtual environment as is

TABLE 17.1 Scenes and scents used in the animation with scents "Together."

No.	Scene	Scent
1	A kid goes out of mushuroom house and starts walking with a dog	
2	He passes by various fruits at market	Pine and strawberry
3	He passes a meat shop	Grilled meat
4	He enters a factory	Sewing machine oil and smoke
5	He junps into sea after he goes outside the factory	Mousse de Mer (perfume)
6	A robot shoots a bullet from his breast	Smoke
7	The bullet becomes a seed, followed by flower blooming	Muguet (perfume) and rose
8	He walks on flowers	Muguet and rose
9	He goes back to home	

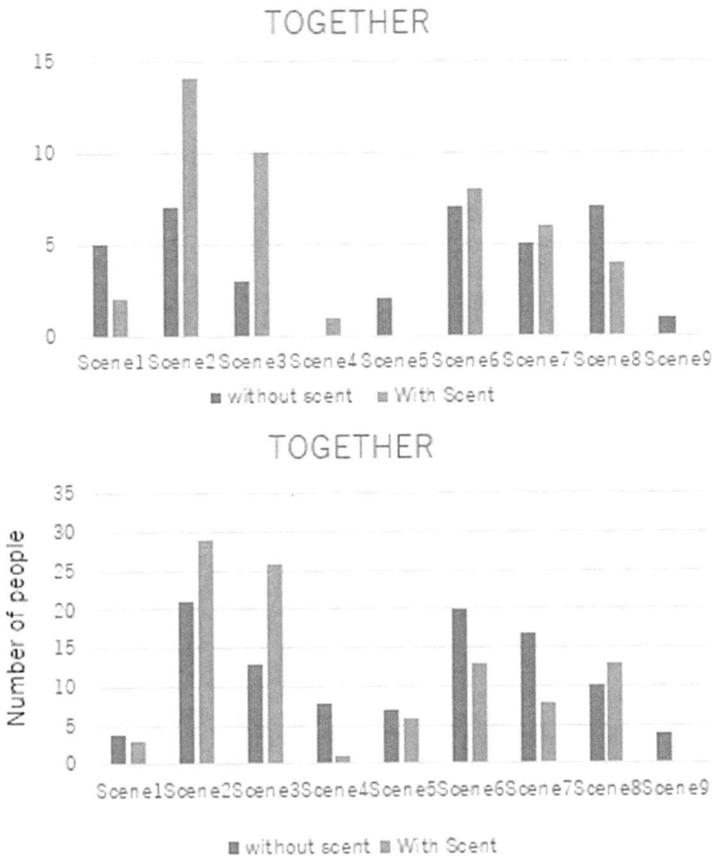

FIGURE 17.1 Result of questionnaire survey for the animation with scent "Together." Every participant was asked if each scene was impressive.

shown in Fig. 17.2. The paddle in the figure is stirring the ingredients in the pan. After he/she puts each ingredient, he/she sniffs the smell synchronously with the short movie. The smell of added ingredient is blended with those already existing in the pan. Finally, the curry with all ingredients is completed. Even a small kid can enjoy the cooking game since its operation is very easy. This game became very popular, and many people enjoyed it at several exhibitions. People wanted to go to a curry shop after they experienced the cooking game. Thus the game with scents is applicable to advertisement of food or cooking.

Iseki (2013) made virtual ice cream shop. People visit cyber virtual ice cream shop and select favorite flavors such as vanilla, milk, blueberry, mint, chocolate, coffee, rum, green tea, banana, orange, strawberry, and melon. A part of content is shown in Fig. 17.3. People can also enjoy the mixed flavor since those flavors can be mixed at the arbitrary ratio. The olfactory display using high-speed switching of solenoid valve shown in Section 15.3 was also used in this content. Its ratio can be easily changed using slide bars in the control software called EZ blender as is shown in Fig. 17.4. The blending ratio of the scent can be changed interactively according to the slide bar. Although there is no actual ice cream, people can enjoy the content as if an ice cream were there.

Other content presents the feeling when a player passes by other person. We have experience on sniffing the smell such as smell of human body or perfume when we pass by other person. The scene to pass by other person is shown in Fig. 17.5. The reality can be enhanced by sniffing the smell when we pass by other person in the virtual environment. The timing to present a smell should be carefully selected to provide a participant natural feeling.

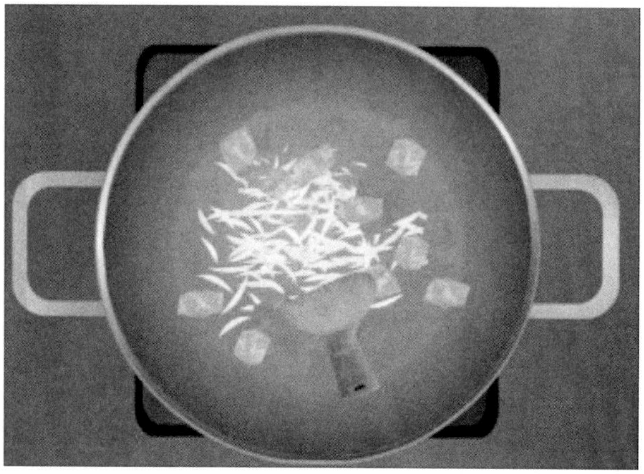

FIGURE 17.2 **Cooking game with scents**. Every participant sniffs smells synchronously with movie.

17.4 Contents using spatiotemporal effect

When we search for the object with smell in the virtual environment, it is necessary to make the odor distribution in virtual space. Hirose et al. studied the odor-source localization in the virtual environment (Yamada et al., 2006). A player equipped with small olfactory display walks around to look for virtual odor source. The odor concentration increases as a

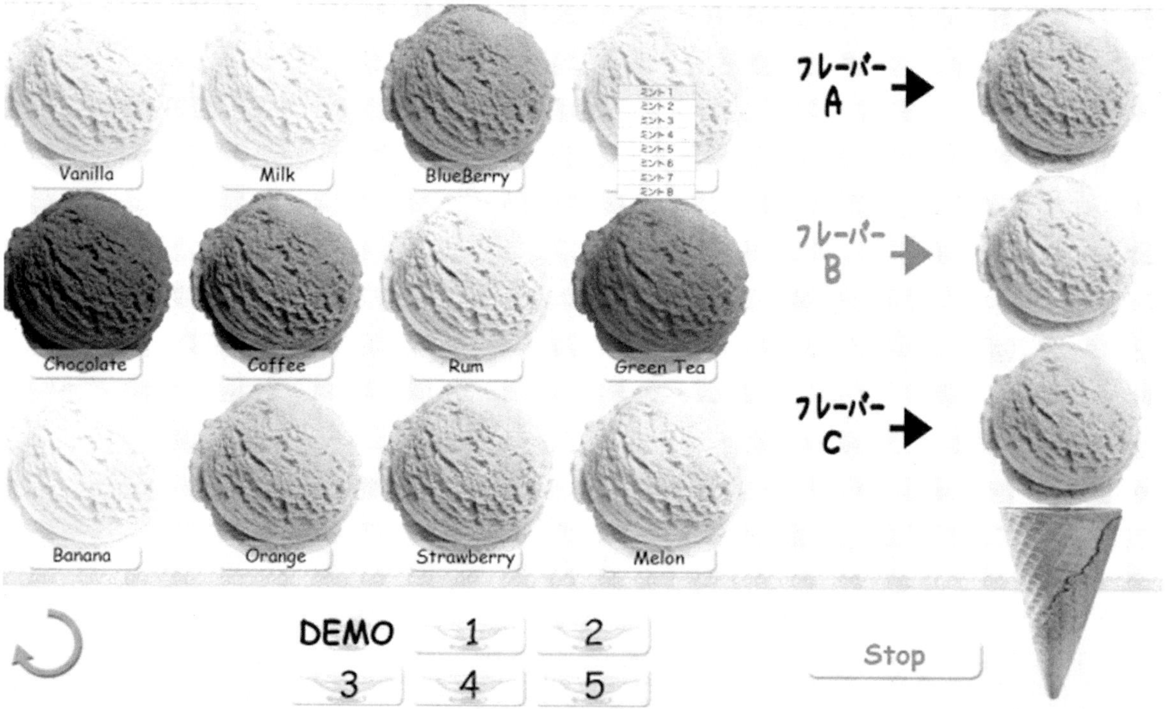

FIGURE 17.3 **Virtual ice cream shop**. A participant can enjoy the mixed flavor of ice cream.

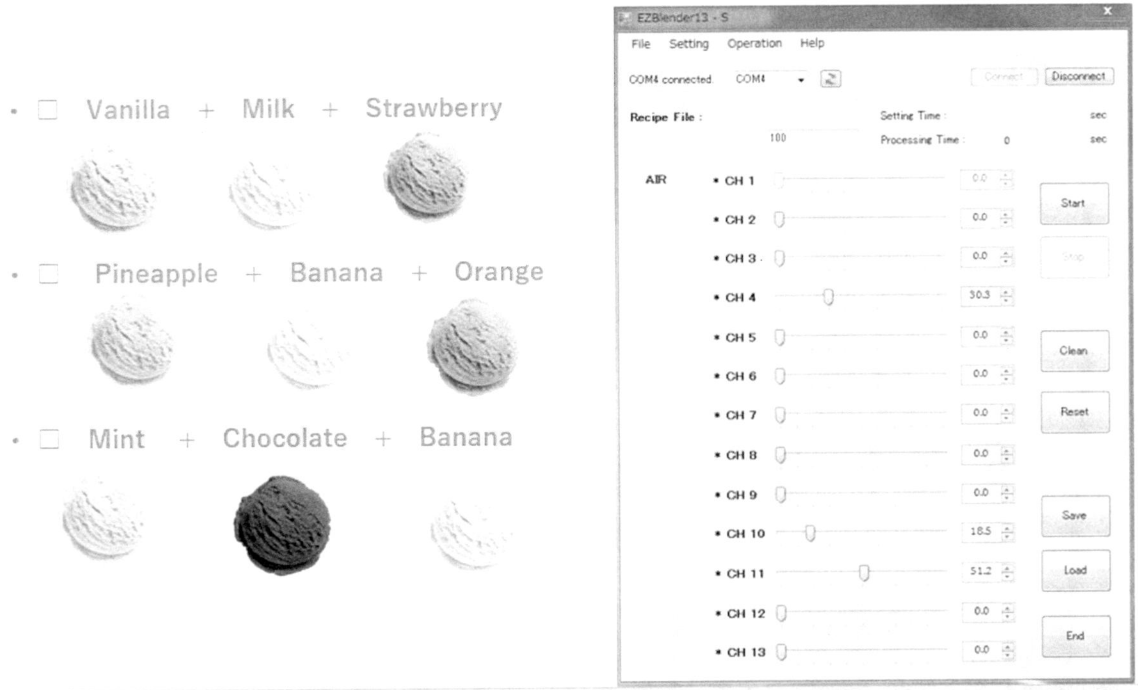

FIGURE 17.4 **Mixed flavor ice cream. Ingredient used for mixed flavor (left) and software to control blending ratio (right).** Flavors can be mixed at the arbitrary ratio.

FIGURE 17.5 **Scene to pass by other person in virtual environment.** A participant sniffs the smell when he/she passes by other person.

player approaches the odor source, while the concentration decreases as he/she moves away from the odor source. The player is requested to find the virtual odor source. The players could find the virtual odor source according to the change of the odor intensity, although the time for searching for it varied from person to person. They assumed a simple isotropic diffusion without wind and obstacles (Fig. 17.6A), although the actual situation is much complicated.

Other method to express the odor distribution is Gaussian plume model (Hinze, 1975).

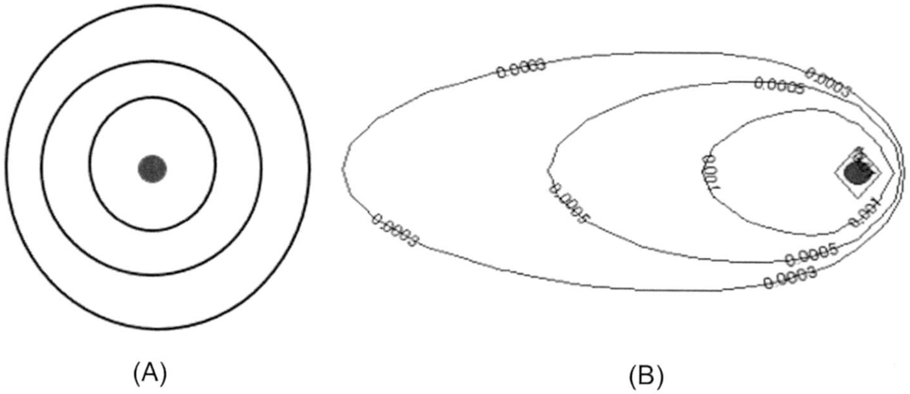

FIGURE 17.6 **Odor distribution model (A) isotropic diffusion and (B) Gaussian plume model**. Contour line means the position with the same concentration. Red point: odor source.

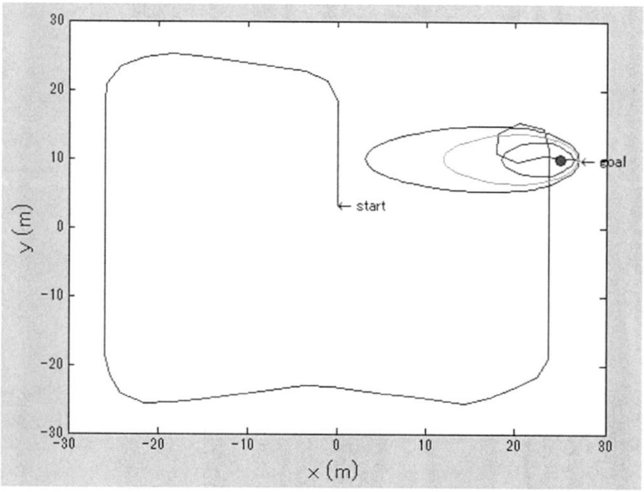

FIGURE 17.7 **Exploration of virtual odor source based on Gaussian plume model**. Trajectory to search for virtual odor source based on Gaussian plume model.

Although it was introduced into the robot algorithm to explore odor source (Ishida et al., 1998), it is also applicable to olfactory VR (Nagasaka & Nakamoto, 2018). The odor distribution is assumed as is shown in Fig. 17.6B. The uniform wind speed without obstacles is assumed in this model. One of example of virtual odor source localization using Gaussian plume model is shown in Fig. 17.7. The wind was assumed to blow from right to left. The player could reach the odor source, although he could not go straight toward the odor source.

Although it is possible to explore the odor source in the virtual environment, the models used here are too simple to realize the actual complicated situation of olfaction in the virtual environment. The isotropic diffusion or uniform wind without obstacles seldom occurs in our daily life. It is necessary to use computational fluid dynamics technique for the complicated situation. It will be described in Chapter 18.

17.5 Contents using crossmodal effect

If we use olfaction together with other senses, crossmodal interaction occurs. Most of research studies focused on interaction of olfaction with visual stimulus. Kawai and Takatsu (2002) studied crossmodal effect between olfaction and vision. They used only the mixture of several ester compounds, which has fruity scent. They presented various fruit images together with this fruit scent to the subjects and asked them their impressions. The scent impression was different from image to image, even if they used the same fruit scent. The scent impression is much influenced by visual stimulus. The experimental result obtained here suggests that a variety of scent impression can be obtained together with various images even if the number of scents is small.

Hirose et al. studied pseudo olfactory effect using vision (Nambu et al., 2010). They made olfactory map of 18 fruit flavors based on sensory test and multidimensional scaling technique and then categorized 4 groups. They picked up one as a representative in each group and presented image with that representative scent. They claimed that they can make various scents virtually from just a few scents because of the drawing effect caused by vision occurs.

Ghinea and Ademoye (2012) studied the association between olfaction and vision in multimedia. They also studied information recall task (Ademoye & Ghinea, 2013) and the synchronization of olfaction with audiovisual content (Ghinea & Ademoye, 2010). They said that olfaction ahead of audiovisual content is more tolerable than olfaction behind the audiovisual content. Murray also studied olfaction-based multisensorial media (Murray et al., 2017).

Tomono et al. studied eye catching and memory during scent presentation utilizing gaze point analysis (Tomono, 2012). The gaze moves within the entire image to obtain much information from the image. A subject, however, watched even small object with a scent when a scent and image were presented simultaneously. Moreover, memory related to the object was enhanced when the scent associated with the object was presented together with its image (Tomono et al., 2011). They also measured the biological reactions when food video with scents was presented to subjects. They claimed that the scent was effective in improving the sensation (Tomono et al., 2021).

Zellner and Whitten studied the relationship between the perceived odor intensity and the color intensity, that is, clear, light, medium, and dark. They reported that the odor intensity was influenced by color intensity independent upon the color and odor types (Zellner & Whitten, 1999; Zellner, 2013).

Kakei et al. studied artistic expression using scents together with visual information. The system called "hanahana" visualizes temporal variation of scent by projecting a flower image, which changes its shape, size, color, and degree of transparency. The flower image changes according to a metal oxide sensor output. The system called "hanahanahanahana" utilizes multiple gas sensors located in the space (Kakehi et al., 2010).

Simiscuka et al. (2023) studied olfaction-enhanced virtual reality 360-degree video. They placed four olfactory devices around the user and integrated with 360-degree video. The sensory test shows that over 70% of experiment participants noticed that the diverse scents, including apple and spices, were presented from multiple directions.

Although many crossmodal interactions between olfaction and vision have been so far introduced, there have been only a few reports related to the interaction between olfaction and auditory sense. Spense et al. studied the crossmodal effect between olfactory and auditory senses (Crisinel et al., 2013). They said that odors were preferentially matched to the musical features. For instance, the odors of candied orange and iris flower matched with significantly higher pitches than the odors of musk and roasted coffee. They also show the matching between sent and the music/sound (Spence, 2021).

Narumi et al. studied pseudotaste display using the interaction between taste and olfaction. They developed the system called "meta cookie" (Narumi et al., 2011). A user eats a plain cookie where chocolate, lomon, etc. are visually overlaid through see-through HMD. The corresponding scent is delivered to the user. He/she feels its taste as if he/she ate the cookie with corresponding taste. When we eat foods and drink beverages, we also sniff the retronasal aroma even at the oral cavity because the oral cavity is connected to the nose. When taste is synthesized by a machine using chemicals, people hesitate to eat/drink chemicals. A taste display is more easily realized with the aid of olfaction since people can naturally and passively sniff scents. Taste just consists of five basic tastes, that is, sweetness, salt, sourness, bitterness, and umami. People can enjoy a variety of flavors if those flavors are added through olfaction. A pseudogustatory display might be realistic method to perceive taste in virtual environment.

Moreover, olfaction contributes to reduction in taking sugar. A user drinks black coffee, but he/she wants to have sweetness. It is possible to feel sweet when an appropriate scent is presented. Thus people can enjoy sweet coffee without sugar if they drink black coffee with scent provided by an olfactory display. We can also enjoy sweet soda without sugar when an appropriate scent is provided (Kurihara et al., 2015). A scene of drinking sugarless beverage is shown in Fig. 17.8. The olfactory display using micro pumps and surface acoustic wave atomizer was used in this experiment. Vanilla or milk flavor was presented to the user who drank the soda without sugar. The result of questionnaire survey regarding coffee is shown in Fig. 17.9. The number of participants was 33. The questionnaire asked whether coffee with scent was delicious compared with black coffee. Its result revealed that 70% of participants answered "delicious" or "very delicious." Thus it is possible to reduce the amount of sugar in the beverage when the scent with sweet impression is presented.

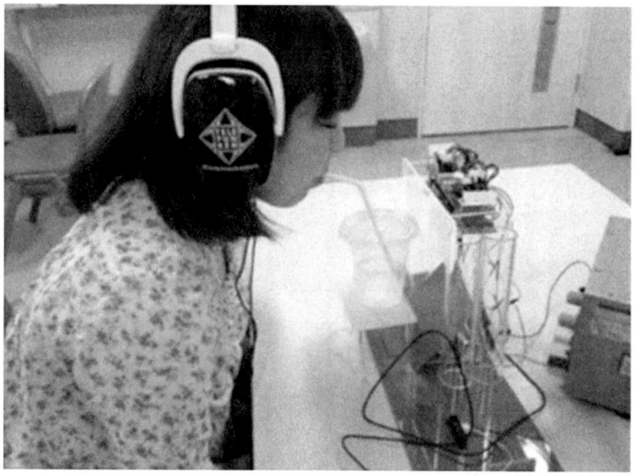

FIGURE 17.8 Scene of drinking sugarless soda together with scent presented by olfactory display. Vanilla or milk flavor was presented by olfactory display during drinking sugarless soda.

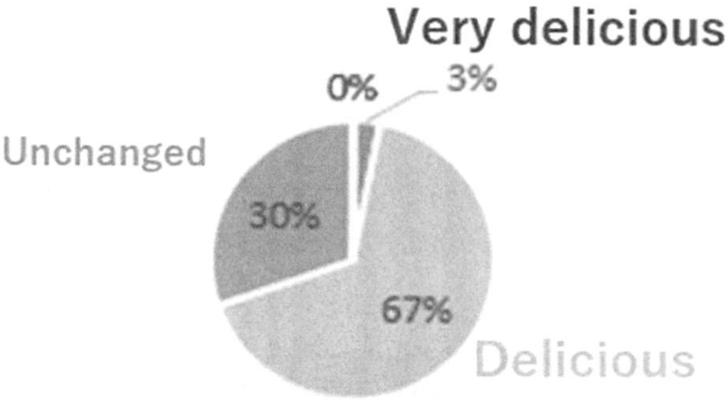

FIGURE 17.9 Result of questionnaire survey when participants drank sugarless beverage together with scent presented by olfactory display. The questionnaire asked whether beverage with scent was delicious compared with sugarless one without scent.

Next, the crossmodal effect related to haptics is described. Iwata et al. (2004) tried to express food texture in the virtual environment. They developed the haptic interface that presenting biting force. They tried to express the food texture such as cracker and cheese. Although they claimed that they could express the food hardness, it does not seem to be applicable to any food.

Ishibashi et al. made Ikebana, that is, flower arrangement, using olfactory and haptic interfaces in the virtual environment (Sithu et al., 2015). When a user approaches the flower, he/she can sniff the scent of the flower. Although haptic and olfactory devices were used simultaneously, the interaction between haptics and olfaction seems to be the future problem.

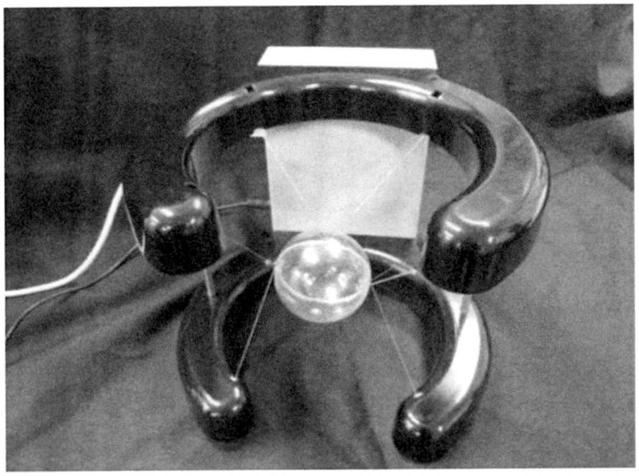

FIGURE 17.10 **Photo of haptic display SPIDAR**. Interactive haptic display with force feedback.

Akahane et al. (2019) studied multimodal VR environment using olfactory and haptic displays. They used the haptic display called "SPIDAR" (Jeong et al., 2004) and the olfactory display using high speed switching of solenoid valves (Nakamoto & Minh, 2007). The photo of the haptic display SPIDAR is shown in Fig. 17.10. It is the interactive wire-driven haptic display with force feedback. The olfactory display used here was described in Section 15.3.

They made several contents. First one is to bake a cake using an oven. The second one is to have a glass with pineapple juice, followed by moving it closer to the nose or away from the nose. However, only one sense was dominant over other sense when we made the sensory test. Then we made the content to grate an apple as is shown in Fig. 17.11. White circle is a grater, and red one on its right side is the apple. When we grate the apple, we can sniff the apple scent from the olfactory display as well as feel the rugged surface of the grater

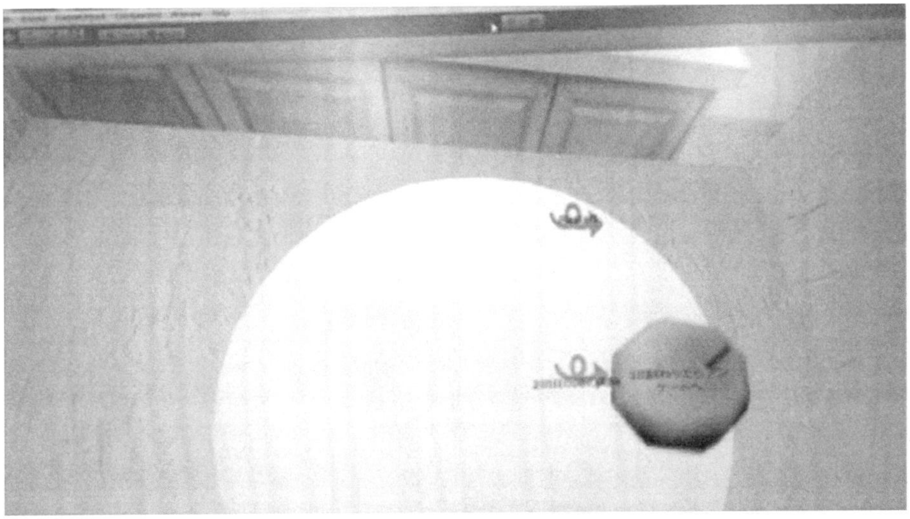

FIGURE 17.11 **Screen shot of grating an apple**. We can sniff apple scent when we grate the apple.

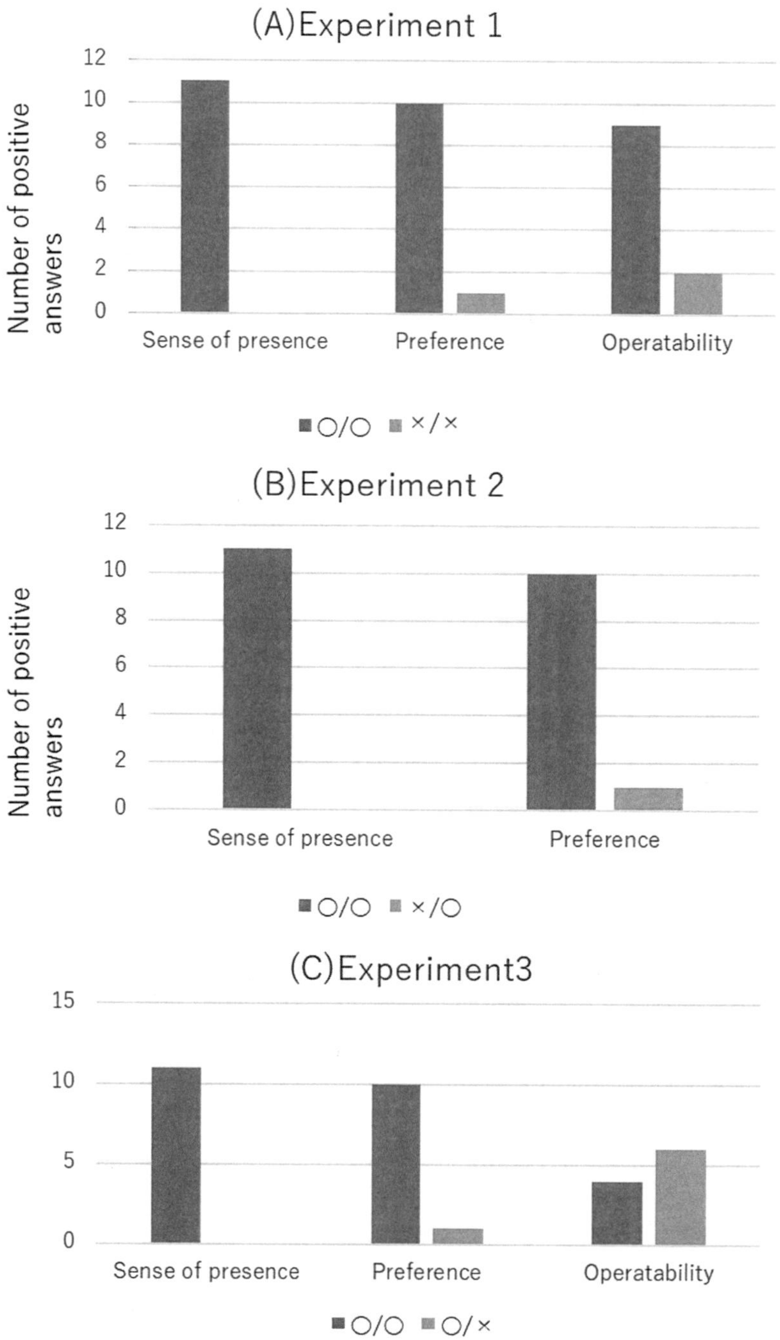

FIGURE 17.12 Experiment on grating an apple. (A) Both senses are inactive, (B) only haptic is active, and (C) only olfaction is active.

using the haptic device. We did the sensory evaluation for the following three cases:

Experiment 1: Comparison of O/O with X/X.
Experiment 2: Comparison of O/O with X/O.
Experiment 3: Comparison of O/O with O/X.

The former symbol is for olfaction and the latter for haptics, and O means the corresponding sense is active and X means that it is inactive. For example, "O/X" means olfaction is active, but haptics is inactive. Eleven people joined the experiment. Fig. 17.12A–C shows the results of three experiments. The vertical axis means the number of positive answers in terms of sense of presence, preference, and operability. These experiments reveal that the multimodal sensation is dominant over either of two senses and is dominant over none of them. In this content, the contributions of both olfactory and haptic senses are important.

17.6 Narrative contents with scents

We can make a narrative content with story using vision and/or auditory sense. Is it possible to make it using olfaction? As a first step to make a story, Iseki et al. investigated the association between scent and landscape. Question here is whether people can associate a landscape scene with given scent under the situation without vision and sound (Iseki & Nakamoto, 2013). We can say that this is a modernized Kodo, that is, traditional incense-smelling game in Japan.

People only sniff a set of certain scents and select one landscape scene that they think most suitable on personal digital assistance panel. Then they experience the scent of the scene with short movie and sound. Olfactory display using high speed switching of solenoid valves was used in this study (see Section 15.3). Six landscape scenes, such as tea ceremony

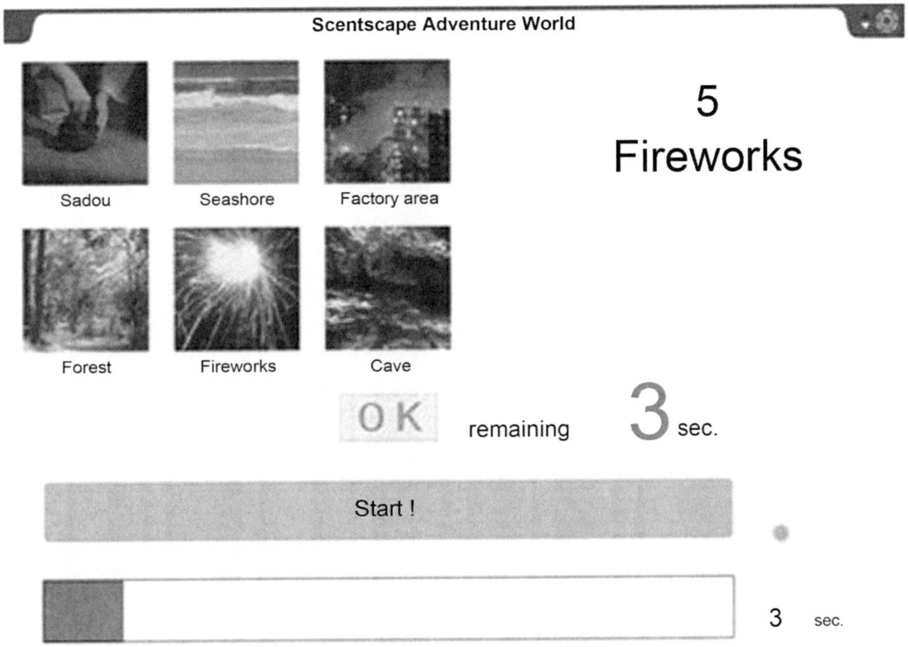

FIGURE 17.13 **Screen of scent-landscape association game**. People were asked to associate a landscape with given scent without vision/sound.

TABLE 17.2 Confusion matrix of scent-landscape association.

Answer/ presentation	Tea ceremony	Forest	Seashore	Factory	Fireworks	Cave	Time runs out	Correct answer rate	Total number
Tea ceremony	10	7	0	3	0	1	0	0.476	21
Forest	3	24	0	5	2	1	1	0.667	36
Seashore	0	10	3	3	3	5	1	0.120	25
Factory	0	6	5	6	2	1	3	0.261	23
Fireworks	1	8	7	0	2	3	1	0.091	22
Cave	0	14	0	0	2	2	5	0.087	23

(Sadou), seashore, factory area, forest, fireworks, and cave, were prepared as is shown in Fig. 17.13. Confusion matrix to show how many people succeeded in correct landscape is shown in Table 17.2. It was easy to associate the landscape with scent if they have much experience, for example, forest. However, it was difficult to do the association if they seldom experience the landscape, for example, factory area and cave. Moreover, that association was difficult if people could not grasp the scent feature of corresponding landscape. Since the olfactory display can change several smells in sequence, we can do that association test easily. More detailed investigation is required in the future.

Ranasinghe et al. (2019) developed the game called "Tainted" that enables players to experience traditional horror folklore using visual–auditory–olfactory interactions. The information provided by olfactory stimuli delivers an immersive game experience. Moreover, smell stimuli induce emotional responses such as fear and curiosity. The demo scene with scent emitted module is shown in Fig. 17.14. They used four smells, that is, jasmine, pineapple, mango, and banana. Attributing to the original folklore, when the banana scent is diffused, it indicates that a ghost is nearby. It reoccurs every time the ghost is in close proximity. The other scent triggers are static and activate only once when a player is within proximity to designated areas. Mementos may be collected in any order. These collectable items deliver the game narrative and are critical to "survive" (or complete) the game. They measured user enjoyment and experience with and without the added smell feedback using an adapted questionnaire from the GameFlow model (Sweetser & Wyeth, 2005). The analysis of the questionnaire survey reveals that scents can be selectively used for weaving narratives into a system for an interactive multimodal experience.

Azuma et al. (2021) studied storytelling using olfactory information. A user becomes a main character, and he/she is a detective. Other three people using different perfumes appear in the story. Criminal(s) are included in those three people. When a character comes close to the user or latest sign of his/her staying at that place exists, the user sniffs the corresponding scent. The user, that is, the main character, should identify the criminal(s). In their scenario, the user cannot identify the criminal(s) without olfactory information. They used the olfactory display called Aroma Shooter (n.d.) together with visual olfactory information. They said that most of participants enjoyed that content but some of them felt difficulty in memorizing the scent.

Although the narrative contents with scent are not so many at the current stage, more contents with various directions are expected to appear in the near future.

FIGURE 17.14 Participants experience game "Tainted" during a demo with the smell-emitting module. Game with visual–auditory–olfactory interactions. *Source: From Ranasinghe, N., Koh, K. C. R., Tram, N. T. N., Liangkun, Y., Shamaiah, K., Choo, S. G., Tolley, D., Karwita, S., Chew, B., Chua, D., & Yi-Luen Do, E. (2019). Tainted: An olfaction-enhanced game narrative for smelling virtual ghosts.* International Journal of Human-Computer Studie, 125, *7–18, https://doi.org/10.1016/j.ijhcs.2018.11.011.*

17.7 Summary

Initially, movies/animations with scents were studied, and then games with scents have been so far studied. Relatively many researchers have studied the crossmodal interaction between olfaction and other senses. Narrative contents with scents are relatively new. The digital contents with scents mean the application of digital olfaction technology, which includes just small number of examples at the current stage. More and more people can join this area to expand digital olfaction.

References

Ademoye, O. A., & Ghinea, G. (2013). Information recall task impact in olfaction-enhanced multimedia. *ACM Transactions on Multimedia Computing, Communications, and Applications*, *9*(3), 1–16. Available from https://doi.org/10.1145/2487268.2487270.

Akahane, K., Takahashi, M., Sato, M., & Nakamoto, T. (2019). Development of multi-modal VR environment using olfactory and haptic displays. *TVSRJ*, *24*, 69–76.

Azuma, M., Sawahata, Y., & Komine, K. (2021). *A study on storytelling using olfactory information for the media with five-sense information*. In Annual Meeting of VRSJ.

Crisinel, A.-S., Jacquier, C., Deroy, O., & Spence, C. (2013). Composing with cross-modal correspondences: Music and odors in concert. *Chemosensory Perception*, *6*(1), 45–52. Available from https://doi.org/10.1007/s12078-012-9138-4.

Ghinea, G., & Ademoye, O. (2012). User perception of media content association in olfaction-enhanced multimedia. *ACM Transactions on Multimedia Computing, Communications, and Applications*, *8*(4), 1–19. Available from https://doi.org/10.1145/2379790.2379794.

Ghinea, G., & Ademoye, O. A. (2010). Perceived synchronization of olfactory multimedia. *IEEE Transactions on Systems, Man, and Cybernetics—Part A: Systems and Humans*, *40*(4), 657–663. Available from https://doi.org/10.1109/tsmca.2010.2041224.

Hinze, J. O. (1975). *Turbulence*. McGraw–Hill.

Iseki, M. (2013). *Virtual ice cream shop*. In Abstract of Digital Olfaction Congress (p. 29).

Iseki, M., & Nakamoto, T. (2013) *Cyber incense smelling with olfactory display*. In Proceedings of the International Conference on Artificial Reality and Telexistence.

Ishida, H., Nakamoto, T., & Moriizumi, T. (1998). Remote sensing of gas/odor source location and concentration distribution using mobile system. *Sensors and Actuators B: Chemical*, *49*(1–2), 52–57. Available from https://doi.org/10.1016/s0925-4005(98)00036-7.

Iwata H., Yano H., Uemura T., & Moriya T. (2004). *Food simulator: A haptic interface for biting*. In Proceedings—Virtual Reality Annual International Symposium (pp. 51–57), https://doi.org/10.1109/VR.2004.1310055

Jeong, S., Hashimoto, N., & Makoto, S. (2004). A novel interaction system with force feedback between real and virtual human. *ACM International Conference Proceeding Series*, *74*, 61–66. Available from https://doi.org/10.1145/1067343.1067350.

Kakehi, Y., Chikamori, M., & Kunoh, K. (2010). hanahanahana. *Leonardo*, *43*(4), 396–397. Available from https://doi.org/10.1162/leon_a_00019.

Kawai, T., & Takatsu, T. (2002). Development and evaluation of fragrance presentation system using visual information. *Aroma Research*, *3*(4), 26–30.

Kurihara, R., Iseki, M., & Nakamoto, T. (2015). *Development of olfactory display using surface acoustic eave device and its application to experience on drinking beverage*. In Annual Meeting of VRSJ.

Minh, H. P. D., Nakamoto, T., Nagahama, M., Hanzaki, S., & Ohnishi, K. (2006). *Evaluation of animation with smell using virtual olfaction display*. In IEICE General Conference.

Mochizuki, A., Amada, T., Sawa, S., Takeda, T., Motoyashiki, S., Kohyama, K., Imura, M., & Chihara, K. (2004). *Fragra: A visual-olfactory VR game*. In ACM SIGGRAPH 2004 Sketches, SIGGRAPH'04. Available from https://doi.org/10.1145/1186223.1186377.

Murray, N., Ademoye, O. A., Ghinea, G., & Muntean, G. M. (2017). A tutorial for olfaction-based multisensorial media application design and evaluation. *ACM Computing Surveys*, *50*(5). Available from https://doi.org/10.1145/3108243.

Nagasaka, R., & Nakamoto, T. (2018). *Virtual odor-source localisation using surface acoustic wave atomizer*. In Annual Meeting of IEEJ (pp. 3–187).

Nakamoto T., & Minh H. P. D. (2007). *Improvement of olfactory display using solenoid valves*. In *Proceedings—IEEE Virtual Reality* (pp. 179–186), Available from https://doi.org/10.1109/VR.2007.352479

Nakamoto, T., Otaguro, S., Kinoshita, M., Nagahama, M., Ohinishi, K., & Ishida, T. (2008). Cooking up an interactive olfactory game display. *IEEE Computer Graphics and Applications*, *28*(1), 75–78. Available from https://doi.org/10.1109/MCG.2008.3.

Nakamoto, T., & Yoshikawa, K. (2006). Movie with scents generated by olfactory display using solenoid valves. *IEICE Transactions on Fundamentals of Electronics, Communications and Computer Sciences*, *E89-A*(11), 3327–3332. Available from https://doi.org/10.1093/ietfec/e89-a.11.3327.

Nambu A., Narumi T., Nishimura K., Tanikawa T., Hirose M. (2010). *Visual-olfactory display using olfactory sensory map*. In Proceedings—IEEE Virtual Reality (pp. 39–42), Available from https://doi.org/10.1109/VR.2010.5444817

Narumi T., Kajinami T., Nishizaka S., Tanikawa T., Hirose M. (2011). *Pseudo-gustatory display system based on cross-modal integration of vision, olfaction and gustation*. In Proceedings—IEEE Virtual Reality (pp. 127–130), Available from https://doi.org/10.1109/VR.2011.5759450

Ranasinghe, N., Koh, K. C. R., Tram, N. T. N., Liangkun, Y., Shamaiah, K., Choo, S. G., Tolley, D., Karwita, S., Chew, B., Chua, D., & Yi-Luen Do, E. (2019). Tainted: An olfaction-enhanced game narrative for smelling virtual ghosts. *International Journal of Human-Computer Studie*, *125*, 7–18. Available from https://doi.org/10.1016/j.ijhcs.2018.11.011.

Simiscuka, A. A., Ghadge, D. A., & Muntean, G.-M. (2023). OmniScent: An omnidirectional olfaction-enhanced virtual reality 360° video delivery solution for increasing viewer quality of experience. *IEEE Transactions on Broadcasting*, *69*(4), 941–950. Available from https://doi.org/10.1109/tbc.2023.3277215.

Sithu, M., Ishibashi, Y., Huang, P., Fukushima, N. Ikebana competition in networked virtual environment with haptic and olfactory senses, 2014 13th Annual Workshop on Network and Systems Support for Games, Nagoya, Japan, 2014, pp. 1–3, https://doi.org/10.1109/NetGames.2014.7008956.

Spence, C. (2021). Musical scents: On the surprising absence of scented musical/auditory events, entertainments, and experiences. *i-Perception*, *12*(5). Available from https://doi.org/10.1177/20416695211038747.

Sweetser, P., & Wyeth, P. (2005). GameFlow: a model for evaluating player enjoyment in games. *Computers in Entertainment*, *3*(3), 3. Available from https://doi.org/10.1145/1077246.1077253.

Tomono, A. (2012). Display technology of images with scents and its psychological evaluation. *Human Olfactory Displays and Interfaces: Odor Sensing and Presentation*, 429–445. Available from https://doi.org/10.4018/978-1-4666-2521-1.ch022.

Tomono, A., Kanda, K., & Otake, S. (2011). Effect of smell presentation on individuals with regard to eye catching and memory. *Electronics and Communications in Japan*, *94*(3), 9–19. Available from https://doi.org/10.1002/ecj.10319.

Tomono, A., Yanagida, Y., & Testutani, N. (2003). A scent-of-smell stimulus presentation method for making visual media with scent. *SICE*, 373–374.

Tomono, K., Ishibashi, Y., & Tomono, A. (2021). Evaluation of realistic sensation using biological reaction measurements for food videos presented by KMMD. *Sensors*, *21*(22). Available from https://doi.org/10.3390/s21227670.

Yamada T., Yokoyama S., Tanikawa T., Hirota K., Hirose M. (2006). *Wearable olfactory display: Using odor in outdoor environment*. In Proceedings—IEEE Virtual Reality (pp. 199–206), Available from https://doi.org/10.1109/VR.2006.147

Zellner, D. A. (2013). Color–odor interactions: A review and model. *Chemosensory Perception*, *6*(4), 155–169. Available from https://doi.org/10.1007/s12078-013-9154-z.

Zellner, D. A., & Whitten, L. A. (1999). The effect of color intensity and appropriateness on color-induced odor enhancement. *The American Journal of Psychology*, *112*(4). Available from https://doi.org/10.2307/1423652.

CHAPTER

18

Olfactory VR environment with computational fluid dynamics

Takamichi Nakamoto

Institute of Integrated Research, Institute of Science Tokyo, Yokohama, Japan

18.1 Introduction

In Chapter 17, contents using spatiotemporal effect in olfaction were described. The spatiotemporal effect as well as odor reproduction is important to realize virtual olfactory environment. Since actual behavior of the plume, that is, aerial trail of odor molecules, is very complicated, the simulator to calculate the odor concentration at the specified point is indispensable for realizing the olfactory environment in virtual reality (VR). The preliminary studies to use isotropic diffusion and Gaussian plume model were described in Chapter 17.4. However, the isotropic diffusion assumes no wind, and the Gaussian plume model assumes the uniform wind. Obstacles in the wind field are not allowed in both cases. However, the situation in the real world is much more complicated. Therefore CFD (computational fluid dynamics) simulation is useful to construct the virtual environment with spatiotemporal effect. The multiple obstacles can be placed in the virtual space, and the complicated wind condition can be treated as is shown in Fig. 18.1. Moreover, the influence of the temperature can be considered since the heat also influences the gas distribution owing to the convection.

On the other hand, the computational load of CFD is very heavy since the huge number of lattices are used in CFD. In CFD calculation, even a supercomputer is often used. There is the trade-off between calculation time and its accuracy. Moreover, the area used in the simulation is limited since the number of lattices used in CFD increases.

Although we have several issues related to CFD in olfactory VR to be solved, it is attractive to realize spatiotemporal effect in olfactory VR. Several attempts to use CFD in olfactory VR are introduced in this chapter.

18.2 Basic equations of fluid dynamics

The air flow dynamics are governed by Navier–Stokes (NS) equation. NS equation for weakly compressible fluids is given in the following equation:

Digital Technologies in Olfaction
DOI: https://doi.org/10.1016/B978-0-443-15721-9.00018-X

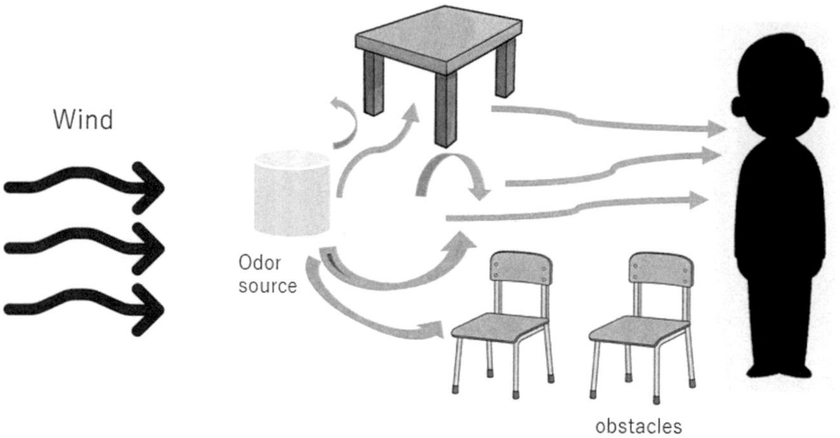

FIGURE 18.1 Complicated air flow caused by obstacles. Temperature also influences gas distribution.

$$\frac{\partial \boldsymbol{u}}{\partial t} + \nabla \cdot (\boldsymbol{uu}) = -\nabla p + v\nabla^2 u + f \quad (18.1)$$

In the NS equation, the flow velocity is $\boldsymbol{u}$, the pressure is p, the kinematic viscosity is v, and the external force per unit volume is $\boldsymbol{f}$.

The pressure required to solve the NS equation is obtained by the pressure development equation shown in the following equation:

$$\frac{\partial p}{\partial t} = -c_s^2 \nabla \cdot u + v\nabla^2 p \quad (18.2)$$

Note that the speed of sound is set as c_s, and in the case of weakly compressible fluid calculations, the speed of sound is set to about 1/10 of the real speed to increase the time step.

Since the second term in Eq. (18.2) takes a smaller value than the first term, it can be ignored.

While the NS equation is a conservation law of fluid momentum, the energy conservation law for incompressible fluids yields the following transport equation for temperature in the following equation:

$$\frac{\partial T}{\partial t} + \nabla \cdot (\boldsymbol{u}T) = \alpha\nabla^2 T \quad (18.3)$$

Note that the temperature is T, and the thermal diffusion coefficient is α.

The transport equation of concentration for determining the odor concentration is given in the following equation:

$$\frac{\partial Y}{\partial t} + \nabla \cdot (\boldsymbol{u}Y) = D\nabla^2 Y \quad (18.4)$$

Note that the concentration (mass fraction of the odorant) is Y, and the diffusion coefficient in material diffusion is D.

The concentration distribution at a certain time can be obtained when Eqs. (18.1–18.4) are numerically calculated. Although the numerical calculation method is omitted here, NS equation together with pressure development equation, transport equation for temperature, and transport equation for concentration are basic equations to govern spatiotemporal effect. Readers can refer to the reference (Clausen, 2013).

18.3 Analysis of odor distribution and its application to VR

In odor sensing technology, one of research topics is an odor source localization (Ishida et al., 1994; Lilienthal et al., 2001; Terutsuki et al., 2021). Although the experimental approaches were taken in most of cases, the algorithm to determine the direction to be searched is very important. It is efficient to study the algorithm of the odor source localization using the simulation based on CFD (Kim et al., 2019; Muhtadi & Nakamoto, 2018). But CFD is also applicable to olfactory VR if the viewpoint to be focused on is changed.

Let us imagine that the smoke flow in the ambient air. The molecules of chemical substances included in the odor vapor exhibit random Brownian motion and, therefore, are gradually transported from a region of high concentration to one of a lower concentration if there is no wind. Unless there is an obstacle or temperature gradation, the odor vapor diffuses isotropically in the environment. However, this happens only in the environment with no significant airflow. The diffusion of odor molecules into the air is an extremely slow process in this case. On the other hand, the slight airflow is almost always found even in closed indoor environments. Even though air conditioners and ventilators are turned off, slight temperature variations in the room induce airflow. The odor molecules are carried by the air flow. The transport of the odor molecules by the resultant convective airflow field dominates the slow molecular diffusion in most of cases.

Moreover, there is the turbulence of airflow we encounter in our daily lives in most of cases. Since it contains a lot of eddies with a variety of sizes, the aerial trail of the odor vapor extending from its source comes to have a patchy structure. The trail meanders randomly due to the large eddies contained in the turbulent flow. The intensities of the odors we perceive in actual situations are, therefore, always fluctuating.

Since the odor concentration distribution is mainly determined by the airflow and its turbulent intensities, it is governed by the theories of fluid dynamics. Therefore the fluid dynamic simulation is effective in predicting the odor distribution in an unknown environment.

An example of CFD calculation is explained here. CFD calculation was performed using the commercially available software, that is, ANSYS Fluent 17.2. The environment used here is depicted in Fig. 18.2 (Muhtadi & Nakamoto, 2018). It is the closed room with three screens extended from the one-screen model (Ishida et al., 2008). Three screens (180 cm high and 50 cm wide) are placed as obstacles. The odor source is an outlet with surface of 25 cm^2 emitting ethanol continuously at 50 mL/minute. The temperature at the window is 10°C, and it is 15°C at other places. That temperature difference induces the convection in this closed room even if there is no wind source in Fig. 18.2. The flow model used here is standard k-ε model often used in CFD calculation . Mesh used in CFD analysis is shown in Fig. 18.3.

The simulation was performed in two phases: First, the simulation was conducted without odor release for 600 seconds. During this time, natural convective airflow was gradually developed due to the temperature difference between the window (10°C) and other parts of the room (15°C). Next, it was continued with odor release from the gas outlet for 600 seconds. These two phases generated the

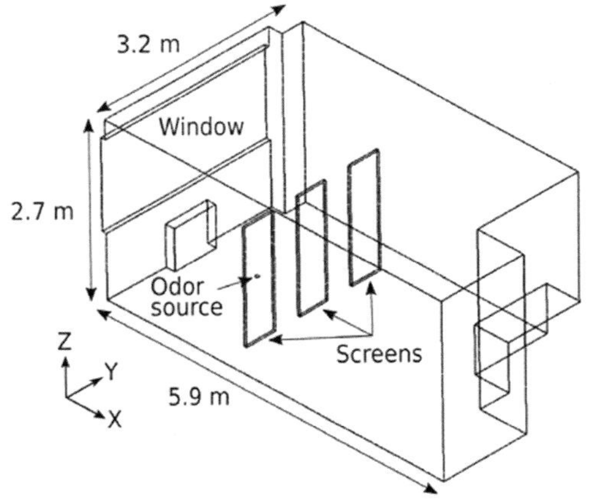

FIGURE 18.2 **Closed room model used in CFD.** Three screens were placed in a room. *CFD*, Computational fluid dynamics. *Source: From Muhtadi, M., & Nakamoto, T. (2018). Plume tracking strategy in turbulent environment using odor sensor with time constant. Sensors and Materials, 30(9). https://doi.org/10.18494/sam.2018.1950.*

time series of 3D data for 1200 seconds. That data includes odor concentration (ethanol mass fraction) and flow velocity irregularly distributed in space.

Fig. 18.4A shows the air flow profile and odor distribution in that room at 700 seconds. The obstacles (three screens) disturb the air flow and create the turbulence at their downwind side. Fig. 18.4B shows the temporal change of the gas concentration at point A in Fig. 18.4A. The gas concentration changes rapidly and irregularly due to the turbulence.

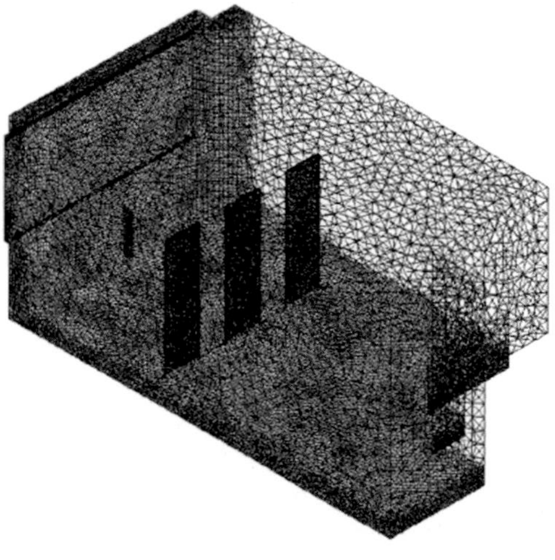

FIGURE 18.3 Mesh of closed room with three screens. Mesh was used in CFD analysis. *CFD*, Computational fluid dynamics.

This CFD calculation was applied to the virtual reality. Ishida et al. studied the model of the closed room with one screen as an obstacle. Teapot was hidden behind the screen, but smell reached the used even if he/she could not see the teapot. Since it takes much time to perform CFD analysis, its odor distribution was calculated in advance prior to the VR experience (Ishida et al., 2011; Matsukura et al., 2009). The software package used for CFD calculation was CFD 2000 (Adaptive Research).

Its CFD calculation result is shown in Fig. 18.5 (Nakamoto et al., 2011). The user took the route from point A to point D close to the odor source. The scene at each point in the virtual environment is shown in Fig. 18.6. Initially, the user at point A just sees the screen, but later he looks at the teapot. The concentration change obtained from CFD simulation is shown in Fig. 18.7. The olfactory display with high speed switching of solenoid valve described in Section 15.3 was used here.

Five students participated in this virtual experience. The movie clip was shown on the

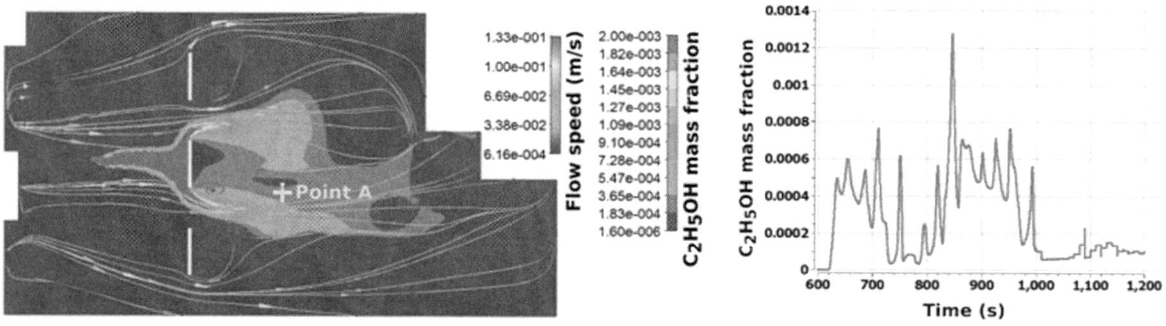

(A) (B)

FIGURE 18.4 CFD calculation result of closed room with three screens. (A) Airflow profile and gas distribution and (B) temporal change of gas concentration. *CFD*, Computational fluid dynamics. *Source: From Muhtadi, M., & Nakamoto, T. (2018). Plume tracking strategy in turbulent environment using odor sensor with time constant. Sensors and Materials, 30(9). https://doi.org/10.18494/sam.2018.1950.*

computer screen, and the smell of peach with appropriate concentration was presented to the user from the olfactory display. After experiencing the odor presentation, the student was asked if he/she felt that the peach smell was emanating from the teapot. Four out of the five students said that they felt stronger smell when the viewpoint of the movie clip approached the teapot. Most of the users successfully perceived the intended change in the odor intensity.

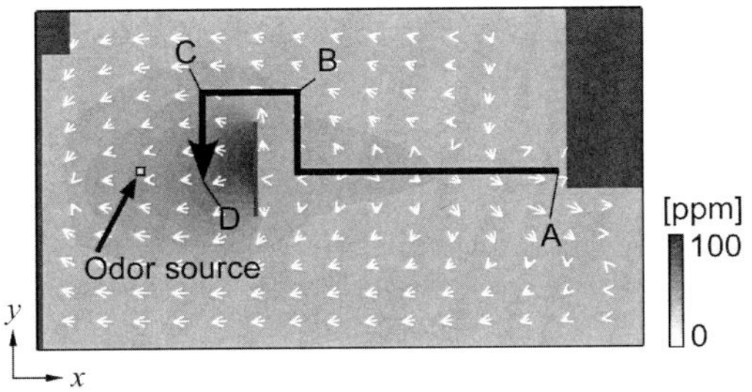

FIGURE 18.5 Result of CFD calculation of one-screen closed room model (top view). Airflow field and gas distribution in the horizontal plane at 80 cm from the floor. The maximum airflow velocity on this plane was 36 cm/s. *CFD*, Computational fluid dynamics. *Source: From Nakamoto, T., Ishida, H. & Matsukura, H. (2011). Olfactory display using solenoid valves and fluid dynamics simulation. In Multiple sensorial media advances and applications: New developments in MulSeMedia (pp. 140–163). IGI Global. https://doi.org/10.4018/978-1-60960-821-7.ch007.*

18.4 Computational fluid dynamics application to disaster drill

Although the example of CFD to VR application was at the preliminary stage in the previous section, it was extended to a disaster drill (Nakamoto et al., 2020). When we encounter fire at the early stage, smell is the most important information available for its localization. If we feel the temperature increase, it is too late and we should escape from the fire as soon as possible. If an optical sensor detects the light intensity change caused by the smoke from the fire, which is typically equipped with a fire alarm, it is also too late. We first sniff the smell of smoke at the early stage of fire even if temperature change or light intensity change cannot be detected.

The application of olfactory VR to the disaster drill is very important since it is difficult to make an actual disaster situation. Therefore the simulator is useful for the disaster drill, which

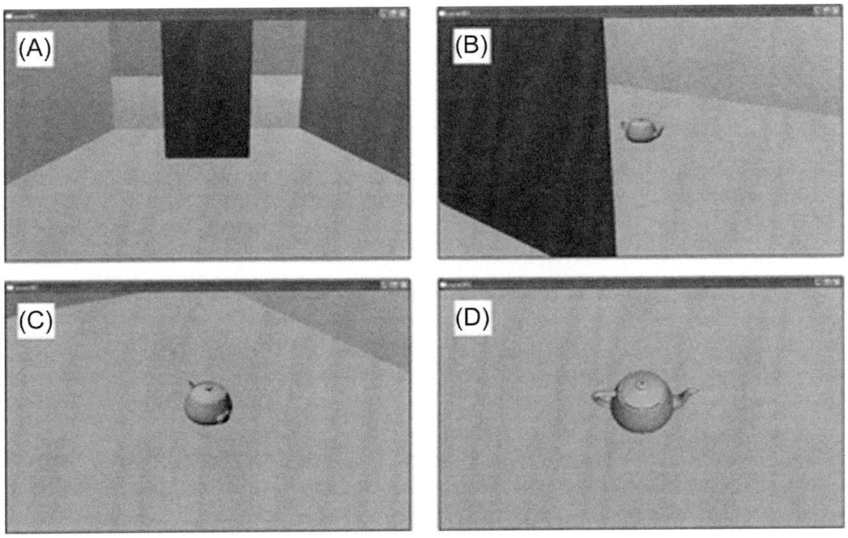

FIGURE 18.6 Scene in virtual environment. Scenes A–D correspond to points A–D in Fig. 18.5, respectively. *Source: From Nakamoto, T., Ishida, H. & Matsukura, H. (2011). Olfactory display using solenoid valves and fluid dynamics simulation. In Multiple sensorial media advances and applications: New developments in MulSeMedia (pp. 140–163). IGI Global. https://doi.org/10.4018/978-1-60960-821-7.ch007.*

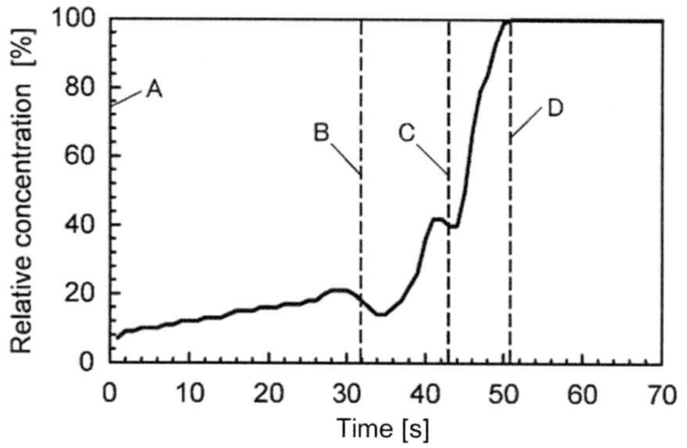

FIGURE 18.7 Variation in relative odor concentration observed on the path shown in Fig. 18.5. Points A–D correspond to A–D in Fig. 18.5, respectively. *Source: From Nakamoto, T., Ishida, H. & Matsukura, H. (2011). Olfactory display using solenoid valves and fluid dynamics simulation. In Multiple sensorial media advances and applications: New developments in MulSeMedia (pp. 140–163). IGI Global. https://doi.org/10.4018/978–1-60960–821-7.ch007.*

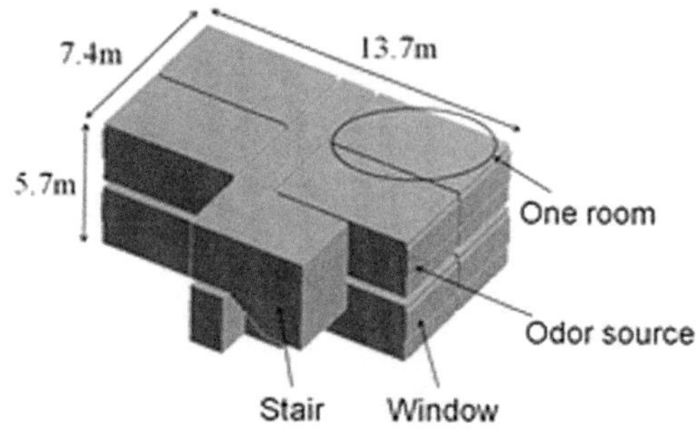

FIGURE 18.8 Model of two-story building used in a disaster drill. Each floor has four rooms. Totally eight rooms were used for CFD simulation. *CFD*, Computational fluid dynamics. *Source: From Nakamoto, T., Hirasawa, T., & Hanyu, Y. (2020). Virtual environment with smell using wearable olfactory display and computational fluid dynamics simulation. In Proceedings—2020 IEEE Conference on Virtual Reality and 3D User Interfaces, VR 2020 (pp. 713–720). Institute of Electrical and Electronics Engineers. https://doi.org/10.1109/VR46266.2020.1581303036097.*

is different from entertainment we frequently find in olfactory-VR application. Although several researchers proposed the fire training simulators (Backlund et al., 2007; Cha et al., 2012; Nahavandi et al., 2019), there have been so far no fire training simulators, including olfactory experience.

In this scenario, a user is asked to search for the origin of the fire according to the odor intensity of burning smell in a house even if he/she cannot localize the source just using visual and audio information.

The model used in this study is shown in Fig. 18.8. It has a two-story building composed of eight rooms, that is, four rooms at each floor. The two rooms are located on both sides of corridor, and the door of each room at the corridor is open. Each room has a window. The corridor of the second floor is connected to the first floor through stairs. The 5 cm × 5 cm odor source emitting the vapor continuously is located at one of four rooms at the second floor.

CFD calculation was performed using the CFD software ANSYS18.1 working on the supercomputer TSUBAME 3.0 at Institute of Science Tokyo. Fig. 18.9 shows the mesh setting for the simulation. The total number of meshes was 1036,530.

The turbulence model used here was $k-\varepsilon$ model, which is the same as that in the previous section. Although the actual odor was the smoke smell, ethanol was used in the simulation. Since the odor is carried by the wind, the difference among vapors is subtle, whereas the vapor type influences the distribution under the environment governed by diffusion. The vapor was spouted in the room at the rate of 8.325×10^{-7} m^3/second.

The temperature at the window was set at 5°C, whereas the temperature at other place in the building was set at 15°C. This temperature difference causes the air convection, and thus the generated wind carries the smell from the odor source. The building itself is closed, and there is no outlet of the house.

There was no odor for first 600 seconds to develop the air flow, and then the odor began to be emitted at 600 seconds, which continued for next 600 seconds. The odor distribution was obtained at the interval of 1 second. It took

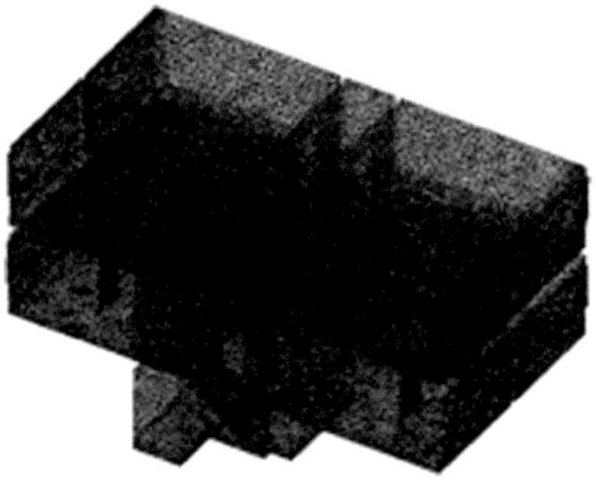

FIGURE 18.9 **Mesh setting of two-story building.** It was used for CFD simulation in disaster drill. *CFD*, Computational fluid dynamics. *Source: From Nakamoto, T., Hirasawa, T., & Hanyu, Y. (2020). Virtual environment with smell using wearable olfactory display and computational fluid dynamics simulation. In Proceedings—2020 IEEE Conference on Virtual Reality and 3D User Interfaces, VR 2020 (pp. 713–720). Institute of Electrical and Electronics Engineers. https://doi.org/10.1109/VR46266.2020.1581303036097.*

7 hours and 35 minutes for CFD simulation even using the supercomputer TSUBAME before exploration of the odor source.

The odor distribution is fixed after CFD simulation. The layout and geometry of each room cannot be changed when the user explores the odor source. The calculated odor distribution 600 seconds after the start of the odor emission is shown in Fig. 18.10A and B. The odor source was located at one of the rooms at the second floor. It spread and went to the corridor but went down to the first floor instead of entering the room in front of the room with the odor source.

Based on CFD calculation, small fire at early stage was simulated. A user equipped with head-mounted display (HMD) and wearable olfactory display beneath it can freely walk around in this building. HMD used here was Oculus Rift, and controller to move in the virtual space was Oculus Touch.

The system configuration implemented using Unity is shown in Fig. 18.11. The result of CFD simulation was in advance stored into the database, and the platform to realize the virtual environment is Unity. Once the user location in the virtual space was obtained using the function of HMD, the concentration at the user's location was obtained from the simulation database. Then the concentration was sent to the wearable olfactory display via serial communication interface. The wearable olfactory display was explained in Chapter 16.3, where the odor concentration is controlled by the frequency of the micro dispenser with liquid droplet ejected toward SAW atomizer.

When the user sniffed the smell, he/she moved to the next location in the virtual space using the controller. The user typically determined the direction to be moved according to the concentration change at the nose. This process was repeatedly performed until the user declared the place of the odor source.

Fig. 18.12 shows the photo of the user equipped with a wearable olfactory display. The screen shot of the virtual space is shown in Fig. 18.13. The user walks in the corridor with the entrances of the rooms on both sides to search for the odor source. The visibility was poor because of the smoke. However, the user cannot identify the source using the smoke density since the visual smoke distribution is independent of the odor source. The small box was placed at the floor of each room. One of them was the odor source to be searched for although the user could not visually identify it.

Then the experiment on searching for the odor source in the virtual environment where a user encountered a small fire was performed.

The example of the trajectory of searching for the odor source in virtual environment is shown in Fig. 18.14A. The temporal change of the odor concentration in the simulation in Fig. 18.14A is also shown in Fig. 18.14B. The user started from the point close to the stairs at the second floor. Although a box was located in each room, only the box in Room1 contained the odor source. The user was requested to find which box had the odor source. The user entered Room1, went to Room2, went back to

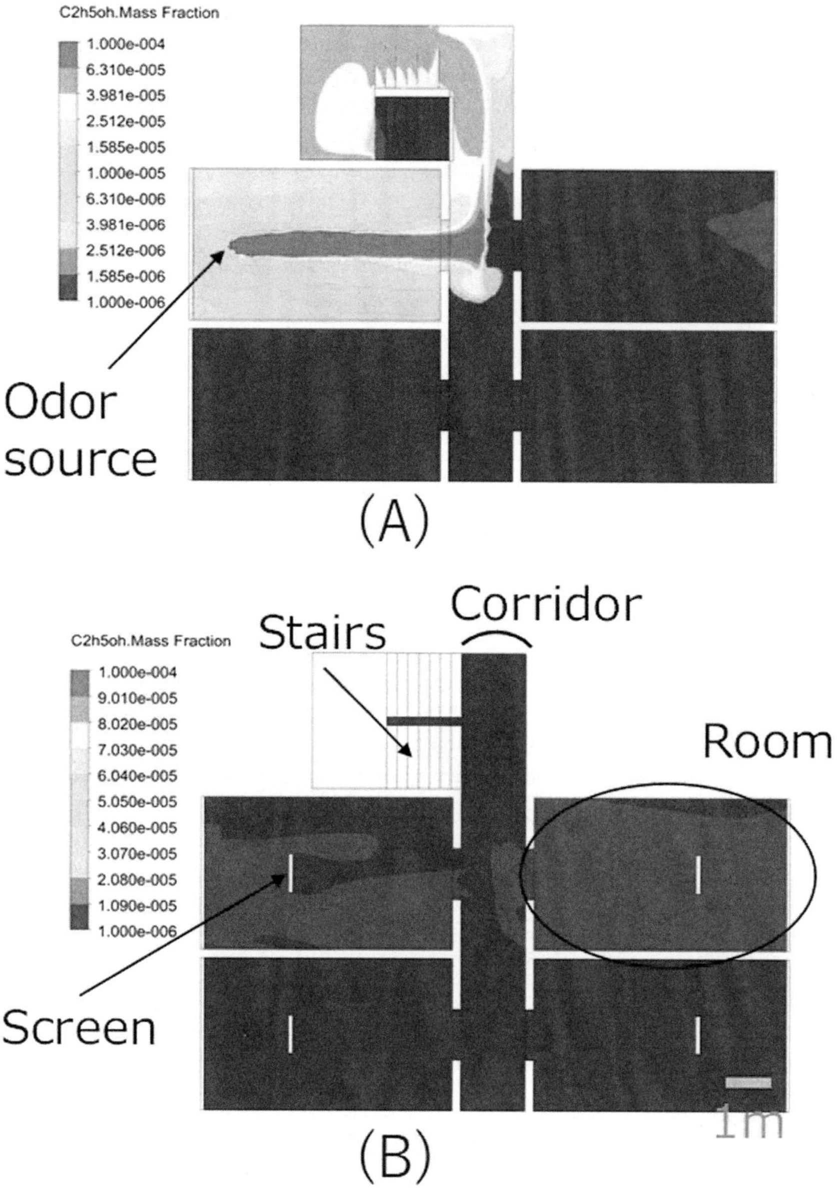

FIGURE 18.10 CFD simulation result in a disaster drill: Odor distribution 600 s after the odor emission started. (A) Second floor and (B) first floor. *CFD*, Computational fluid dynamics. *Source: From Nakamoto, T., Hirasawa, T., & Hanyu, Y. (2020). Virtual environment with smell using wearable olfactory display and computational fluid dynamics simulation. In Proceedings—2020 IEEE Conference on Virtual Reality and 3D User Interfaces, VR 2020 (pp. 713–720). Institute of Electrical and Electronics Engineers. https://doi.org/10.1109/VR46266.2020.1581303036097.*

Room1 again and declared that the odor source existed there. Since the user quickly found the odor source in this example, it took only 42 s to localize it. When a user visited more rooms, the time to search for the odor source became longer.

When 10 participants joined the experiment using the wearable olfactory display, the correct answer rate was 50%. The average time to search for the odor source was 117.1 seconds, and its standard deviation 21.3 seconds. Thus it can be used as a fire training simulator at the early stage.

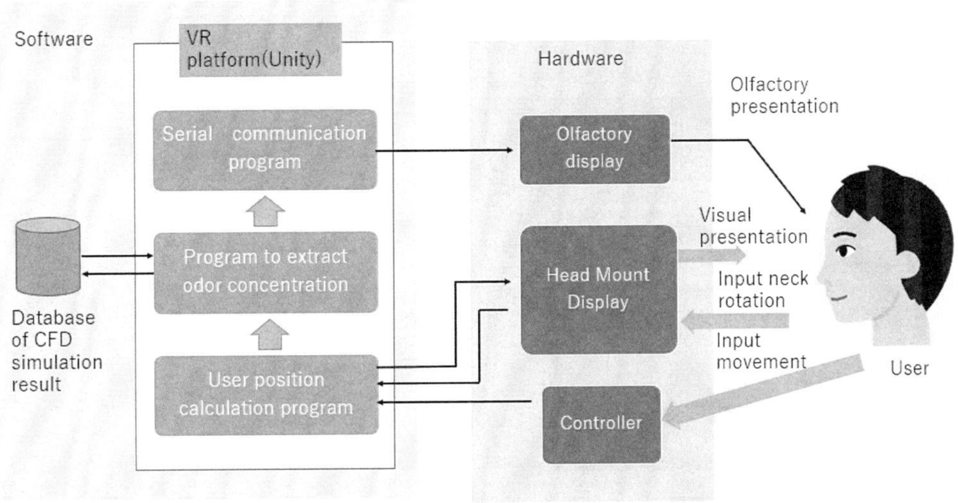

FIGURE 18.11 **System configuration diagram to realize disaster drill**. Virtual environment platform is Unity.

FIGURE 18.12 **Photo of a user with wearable olfactory display**. Wearable olfactory display was explained in Chapter 16.3.

FIGURE 18.13 **Screen shot of the virtual space in disaster drill**. Corridor in the virtual space with smoke.

18.5 Summary

Although there has been so far no fire training simulator using an olfactory stimulus, the olfaction offers the important cue to find the fire origin at the early stage. Thus olfactory VR is important for disaster drill. Moreover, CFD is indispensable for the simulation since the direction of the odor plume is complicated and unpredictable in most of cases.

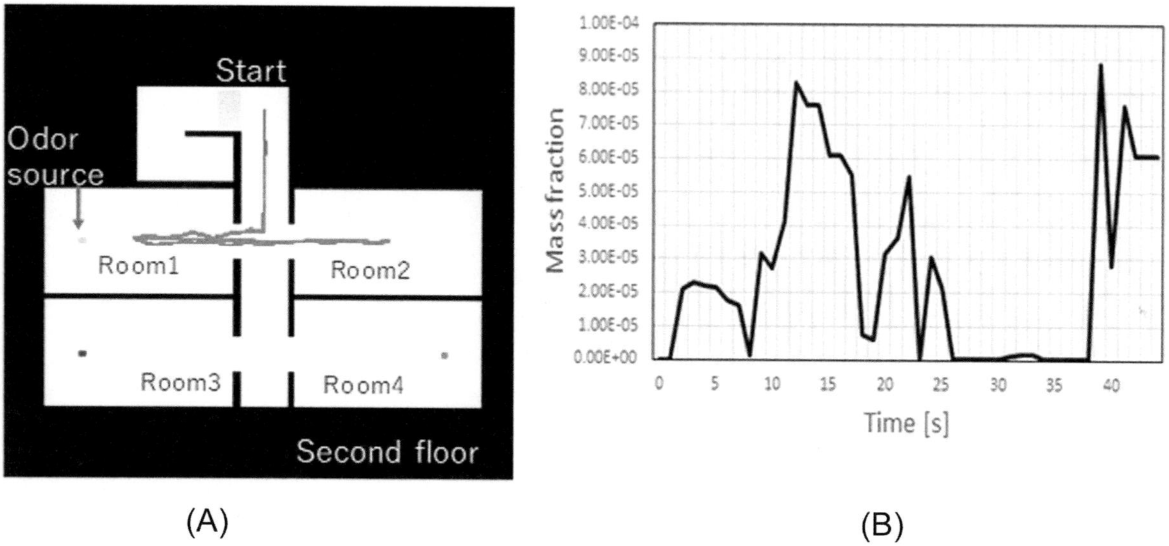

FIGURE 18.14 Example of a user's trajectory searching for odor source in virtual environment. (A) Route that user took (top view) and (B) Temporal change of concentration along his route. *Source: From Nakamoto, T., Hirasawa, T., & Hanyu, Y. (2020). Virtual environment with smell using wearable olfactory display and computational fluid dynamics simulation. In Proceedings—2020 IEEE Conference on Virtual Reality and 3D User Interfaces, VR 2020 (pp. 713–720). Institute of Electrical and Electronics Engineers. https://doi.org/10.1109/VR46266.2020.1581303036097.*

A issue to be solved is to realize real-time CFD in olfactory VR. Although a user him/herself is an obstacle against the air flow, that obstacle must be ignored in this simulation. The user acts as if he/she were a transparent human, although it actually influences to the odor concentration at human's nose. This is the limitation of the batch style CFD simulation. Since the user movement cannot be predicted in advance, real-time CFD will enable us to provide more reality in the near future.

References

Backlund, P., Engström, H., Hammar, C., Johannesson, M., & Lebram, M. (2007). *Sidh—A game based firefighter training simulation.* In Proceedings of the International Conference on Information Visualisation (pp. 899–907). Available from https://doi.org/10.1109/IV.2007.100.

Cha, M., Han, S., Lee, J., & Choi, B. (2012). A virtual reality based fire training simulator integrated with fire dynamics data. *Fire Safety Journal, 50*, 12–24. Available from https://doi.org/10.1016/j.firesaf.2012.01.004.

Clausen, J. R. (2013). Entropically damped form of artificial compressibility for explicit simulation of incompressible flow. *Physical Review E, 87*(1). Available from https://doi.org/10.1103/physreve.87.013309.

Ishida, H., Matsukura, H., Yoshida, H., & Nakamoto, T. (2008). *Application of computational fluid dynamics simulation to olfactory display.* In Proc. ICAT (pp. 285–288).

Ishida, H., Suetsugu, K., Nakamoto, T., & Moriizumi, T. (1994). Study of autonomous mobile sensing system for localization of odor source using gas sensors and anemometric sensors. *Sensors and Actuators A: Physical, 45*(2), 153–157. Available from https://doi.org/10.1016/0924-4247(94)00829-9.

Ishida, H., Yoshida, H., & Nakamoto, T. (2011). Introducing computational fluid dynamics simulation into olfactory display. *Electrical Engineering in Japan, 177*(1), 65–72. Available from https://doi.org/10.1002/eej.21087.

Kim, H., Park, M., Kim, C. W., & Shin, D. (2019). Source localization for hazardous material release in an outdoor chemical plant via a combination of LSTM-RNN and CFD simulation. *Computers & Chemical Engineering, 125*, 476–489. Available from https://doi.org/10.1016/j.compchemeng.2019.03.012.

Lilienthal, A. J., Zell, A., Wandel, M. R., & Weimar, U. (2001). Experiences using gas sensors on an autonomous mobile robot. In EUROBOT2001, 4th European

Workshop on Advanced Mobile Robots (pp. 1–8). https://www.researchgate.net/publication/2376313_Experiences_Using_Gas_Sensors_On_An_Autonomous_Mobile_Robot#full-text

Matsukura, H., Yoshida, H., Ishida, H., Saitoh, A., & Nakamoto T. (2009). Odor presentation with a vivid sense of reality: Incorporating fluid dynamics simulation into olfactory display. In Proceedings—IEEE Virtual Reality (pp. 295–296). Available from https://doi.org/10.1109/VR.2009.4811062

Muhtadi, M., & Nakamoto, T. (2018). Plume tracking strategy in turbulent environment using odor sensor with time constant. *Sensors and Materials*, *30*(9). Available from https://doi.org/10.18494/sam.2018.1950.

Nahavandi, S., Wei, L., Mullins, J., Fielding, M., Deshpande, S., Watson, M., Korany, S., Nahavandi, D., Hettiarachchi, I., Najdovski, Z., Jones, R., Mullins, A., & Carter, A. (2019). *Haptically-enabled VR-based immersive fire fighting training simulator* (997, pp. 11–21). Springer Science and Business Media LLC. Available from https://doi.org/10.1007/978-3-030-22871-2_2.

Nakamoto, T., Hirasawa, T., & Hanyu, Y. (2020). *Virtual environment with smell using wearable olfactory display and computational fluid dynamics simulation*. In Proceedings—2020 IEEE Conference on Virtual Reality and 3D User Interfaces, VR 2020 (pp. 713–720). Institute of Electrical and Electronics Engineers Inc. Available from https://doi.org/10.1109/VR46266.2020.1581303036097,

Nakamoto, T., Ishida, H., Matsukura, H. (2011). Olfactory display using solenoid valves and fluid dynamics simulation. In Multiple sensorial media advances and applications: New developments in MulSeMedia (pp. 140–163). IGI Global. http://www.igi-global.com/book/multiple-sensorial-media-advances-applications/50515. Available from https://doi.org/10.4018/978-1-60960-821-7.ch007.

Terutsuki, D., Uchida, T., Fukui, C., Sukekawa, Y., Okamoto, Y., & Kanzaki, R. (2021). Real-time odor concentration and direction recognition for efficient odor source localization using a small bio-hybrid drone. *Sensors and Actuators B: Chemical*, *339*. Available from https://doi.org/10.1016/j.snb.2021.129770.

CHAPTER

19

Olfactory displays for teleolfaction

Takamichi Nakamoto

Institute of Integrated Research, Institute of Science Tokyo, Yokohama, Japan

The fundamental method and experimental result of smell transmission to the remote site are explained.

19.1 Introduction

Nowadays, visual and auditory information can be transmitted to remote site quite easily. Teleolfaction means the scent transmission in the same manner as those of television and telephone. Is it possible to realize teleolfaction? Although a complete teleolfaction system has not so far appeared yet, its early version models have been proposed (Fabrizio & Lundström, 2001). If the teleolfaction is feasible, it can be used for on-line shopping, direct sales remotely from the farm, etc. When we watch advertisements for soap, detergent, air refreshner, food, beverage, etc. on television, visual alternative, that is, drawing of the patchy plume, is sometimes used since we cannot transmit smells. However, we have remarkable reality if the smell is presented on the television.

The concept of teleolfaction is illustrated in Fig. 19.1A and B. There are two types of teleolfaction. The first type does not have sensing system. If the command comes from the remote site, the corresponding smell is generated from an olfactory display as is shown in Fig. 19.1A. It should be synchronized with visual information if the movie is used. The second type of teleolfaction includes the odor-sensing system. It detects the smell, identifies it, and then sends that information to the olfactory display at the remote site. In some cases, mixture composition is quantified at the sensing system, and it is transferred to the remotely located olfactory display. The smell information should also be synchronized with visual one if the movie file is used. Its variant is to send the sensor data to the remote site, and the olfactory display generates the smell with the same sensing data. The mixture quantification is useful if we transmit the unknown smell to the remote site. Since we encounter unknown smells frequently in our daily life, the mixture quantification method is useful. In the next section, several examples are explained.

19.2 Approach toward odor transmission

Most of previous examples belong to the type in Fig. 19.1A. Therefore they did not have the sensing part.

Digital Technologies in Olfaction
DOI: https://doi.org/10.1016/B978-0-443-15721-9.00019-1

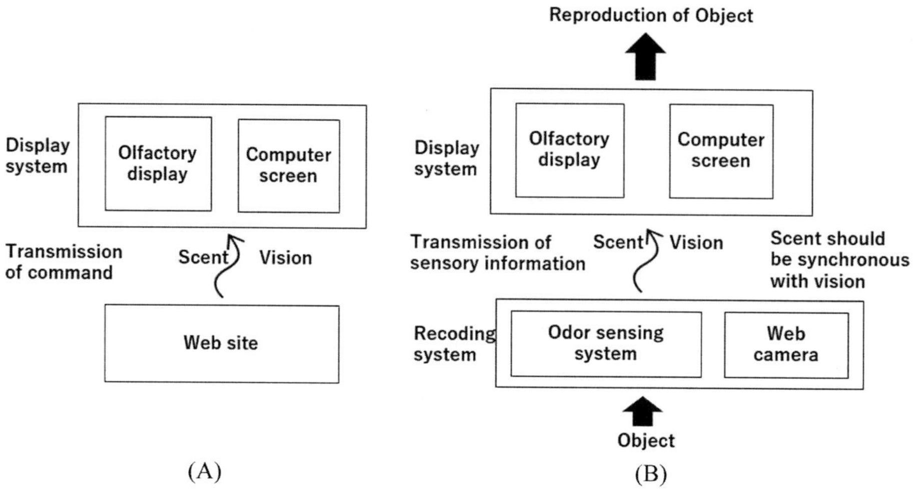

FIGURE 19.1 **Concept of teleolfaction.** (A) Teleolfaction with command, (B) teleolfaction with sensing system.

iSmell was a small personal smell diffuser developed by DigiScents Inc. Although it became famous in early 2000, the company did not last for a long time (Williams, 2001).

France Telecom developed the scent diffuser (Manabe et al., 2007). A user could select one of several smells. They made the content to get around the wineries to enjoy the scents of wines. Tsuji used that diffuser and made the web-based gourmet content to enjoy the scents from various dishes (Tsuji, 2008).

Other company developed the content to present the smells. They used that system for room fragrance for hotel and cinema (Hamada, 2007). Moreover, they did the digital signage experiment on collecting customers by diffusing the smells at the entrance of pub. Although those activities occurred around 2007, they disappeared soon. Although the user could select one of several smells prepared in advance, smell types are almost infinite. The system should be more flexible to present a variety of smells. Moreover, the sensing part is not included. Thus the system, including the odor-sensing system (Nakamoto, Nimsuk, et al., 2008), was developed and is explained in the next section.

19.3 Experimental setup for teleolfaction

The experimental setup (Nakamoto, Cho, et al., 2008) based on Fig. 19.1B is illustrated in Fig. 19.2. Although the odor sensing together with image capturing was studied (Yamanaka et al., 2007), the odor data transmission was not considered. It has a sensing system made up of QCM (quartz crystal microbalance) sensor array and a neural network circuit implemented into FPGA (field programmable gate array) (Dubey, 2009; Yoshino & Nakamoto, 2014). The classification result instead of raw sensor data was sent to the user's site.

The odor sources are placed on the turn table. The tablet personal computer (PC)1 is also used to capture the image around the turn table. That PC1 at the sensing site is connected to other tablet PC2 located at the user's site via Internet. That tablet PC2 controls the olfactory display to generate the smells based on the classification results submitted from the odor-sensing system. The user can see the object on the turn table since the image from the tablet PC1 with the odor-sensing system is transferred to tablet PC2 at the user's site. The user can also remotely control the location of the sensing

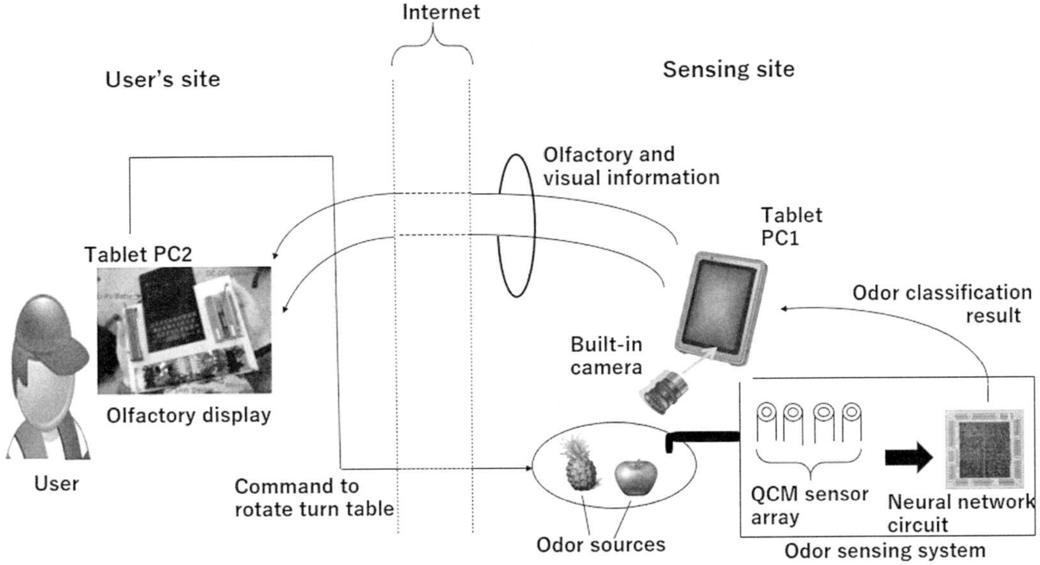

FIGURE 19.2 **Experimental setup of teleolfaction system.** Odor-sensing system at sensing site and olfactory display at remote site.

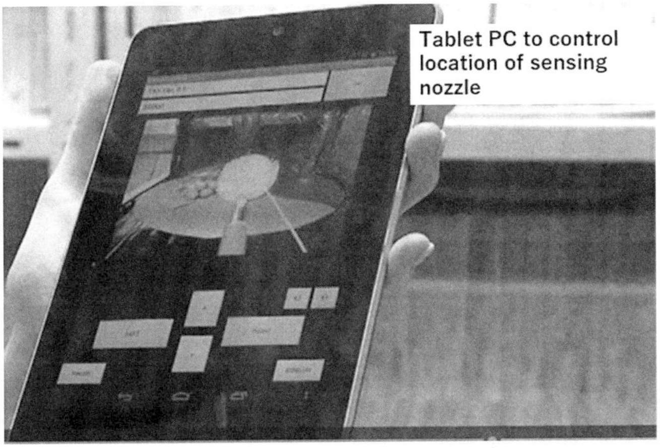

FIGURE 19.3 **Photo of tablet PC2 controlled by a user in teleolfaction.** A user can control the location of sensing nozzle using tablet PC2.

nozzle at the turn table and search for the target smell. The photo of Tablet PC2 is shown in Fig. 19.3. The olfactory display was placed at the place close to a user. The image on the tablet PC2 was the remotely located turn table.

The odor information is combined with the visual information taken by the build-in camera of tablet PC1 and is transmitted to the user's site via the Internet. At the user's site, the odor information is separated from the visual one and is presented using the olfactory display, whereas the visual information is presented at the screen of tablet PC2. It is necessary to send both visual and olfactory information together because the temporal changes of odor type and concentration should be synchronized with image.

The odor-sensing system is explained in detail. The four QCM sensors with different coatings were set at the sensor cell, and the air pump sucked the air. Data flow of odor-sensing system is shown in Fig. 19.4. Each QCM measurement circuit reads out the shift of the oscillation frequency. It consists of a reciprocal counter to achieve high sampling rate without the deterioration of the frequency resolution (Segawa et al., 2002). The high sampling rate of the frequency counter is required to accomplish real-time frequency analysis. The sampling rate of the reciprocal counter was eight samplings per second. Since the odor concentration is highly fluctuated in the ambient air due to the turbulence, its feature can be extracted using the frequency analysis, that is, short-time Fourier transform (STFT) (Nimsuk & Nakamoto, 2007; Nimsuk & Nakamoto, 2008). STFT uses moving window to achieve real-time frequency analysis. Its window function is Hann window. The classification accuracy decreased very much when the raw data were input to the neural network circuit. Fourier analysis can extract the information related to response/recovery time even if the temporal change of the odor concentration is highly fluctuated. Thus the dedicated digital circuit to accomplish STFT in real time was developed.

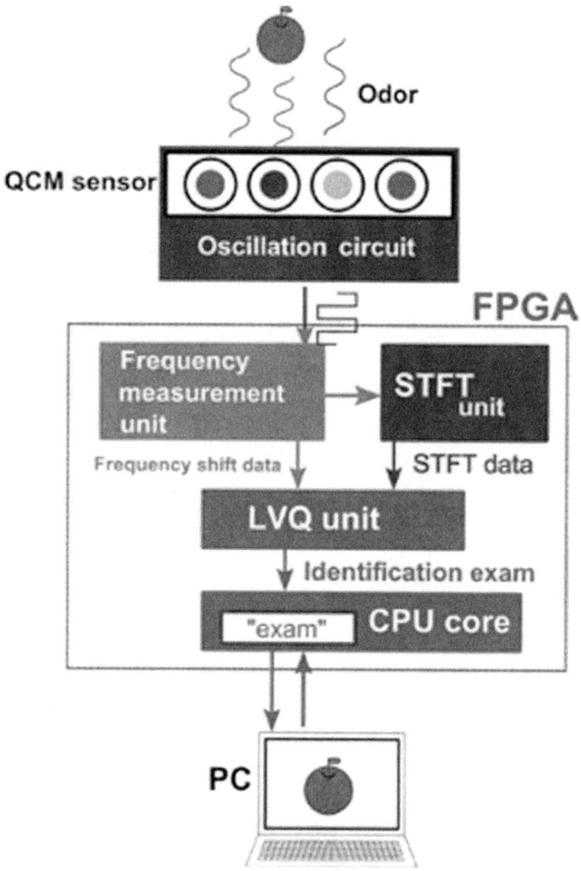

FIGURE 19.4 **Data flow of odor-sensing system used for teleolfaction experiment** (Yoshino & Nakamoto, 2014). Odor recognition chip implemented into FPGA classifies odors. *FPGA*, Field programmable gate array. Source: *From Yoshino, F., & Nakamoto, T. (2014). Odor recognition system using embedded leaning vector quantization circuit.* Sensors and Materials, *26(3), 137–147.*

Then frequency components together with signal amplitudes are input to LVQ (learning vector quantization) circuit. LVQ is one of the neural networks (Kohonen, 2012). Since its classification performance is good in spite of simple algorithm, it is suitable for hardware implementation. LVQ works in the form of hardware, whereas most of the cases are software implementations. In this system, the hardware implementation was adopted since the sensing system becomes smaller, keeping the high-speed calculation for the neural network training, although the software implementation is used in most of the cases. Central processing unit (CPU) core was implemented into FPGA. However, CPU core itself does not do the complicated calculation but manipulates the data so that the data can be used for training the neural network.

Eight-dimensional data was input to LVQ circuit: Four-dimensional data come from the frequency measurement unit, and the remaining four-dimensional data are from STFT unit. STFT unit discrete Fourier transform using the latest 32 points for 4 seconds. The component of 0.5 Hz was picked up since the previous experiment reveals that 0.5 Hz component contributed to odor classification (Nimsuk & Nakamoto, 2007).

19.4 Experimental result of teleolfaction

The sensing films for QCM sensors (AT-CUT, 20 MHz) were tricresyl phosphate, silicone OV-17, Apiezon L, and polyethylene glycol 1000 that were the same as those in the reference (Nimsuk & Nakamoto, 2007). Those are the materials used as Gas Chromatography stationary phases. The samples used here were the flavors of apple, mint, melon, and citrus. Citrus is blended essential oil named "Refresh" purchased from Absolute Aromas. Each sample was put into the petri dish and was released in the ambient air. The air was sucked and flowed to the sensor cell by the air pump. The distance between the inlet (tip of Teflon tube) of the odor-sensing system and the petri dish was about 3 cm.

LVQ circuit together with the reciprocal frequency counters and STFT circuits was implemented into FPGA. Sixteen reference vectors were used for each category. The olfactory display used here is based on high-speed switching of the solenoid valve as is explained in Section 15.3.

The teleolfaction system was demonstrated at an International Conference on Artificial Reality and Telexistence (ICAT 2013; Nozaki et al., 2013). The sensing site was located at Midori-ku, Yokohama, Japan, and the user stayed at Odaiba, Koto-ku, Tokyo. The distance between them is about 30 km.

The experimental results are shown in Tables 19.1–19.3. Table 19.1 shows that most of people could identify the scents of four samples since the overall classification accuracy exceeded 90%. In Table 19.2, each sample was measured 20 times using the odor-sensing system. Its overall classification accuracy was 85%. The classification accuracy was smaller than that of the closed system owing to the sensor data fluctuation due to the turbulence even if the real-time frequency analysis method was introduced. However, the teleolfaction system enhances the reality even if the classification accuracy at the sensing system slightly deceases.

Finally, the classification result of human perception at remote site is shown in Table 19.3. The classification accuracy was similar to that of the odor-sensing system itself. Although there is

TABLE 19.1 Classification accuracy of human perception using olfactory display.

Identified sample/sample	Apple	Mint	Melon	Citrus (blended essential oil)
Apple	100%	0	0	0
Mint	0	100%	0	0
Melon	0	10%	90%	0
Citrus (blended essential oil)	0	0	0	100%

TABLE 19.2 Confusion matrix of odor-sensing system using teleolfaction system.

Identified sample/sample	Apple (%)	Mint (%)	Melon (%)	Citrus (blended essential oil) (%)
Apple	90	5	5	0
Mint	0	90	5	5
Melon	5	10	80	5
Citrus (blended essential oil)	5	15	0	80

TABLE 19.3 Confusion matrix of human perception at remote site when teleolfaction was used.

	Apple	Mint	Melon	Citrus (blended essential oil)
Apple	6	0	1	0
Mint	0	8	0	0
Melon	0	2	5	0
Citrus (blended essential oil)	0	1	1	4

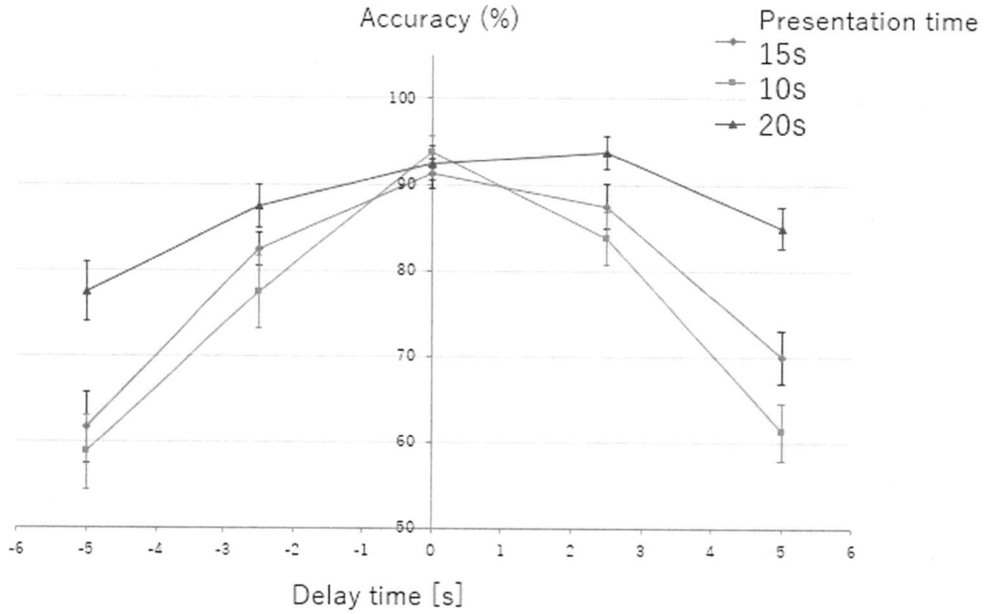

FIGURE 19.5 **Relationship between the delay difference and the classification accuracy**. Time difference between image and smell influences classification accuracy.

still room for the improvement, the feasibility of the teleolfaction was confirmed.

Moreover, the synchronization of the movie with scent was studied. Since Fig. 19.5 shows the relationship between the time difference and the classification accuracy of the human, the positive delay means the image is advanced than the smell, whereas the negative delay means that the smell is advanced than the image. If there was no delay, the classification accuracy was more than 90%. Its accuracy deteriorated as the delay increased. Although the classification accuracy did not change so much regardless of the delay in the case of the long presentation time, it was influenced by the delay when the presentation time of the scent was short.

19.5 Teleolfaction without sensing system

Teleolfaction without sensing system is also possible when the remote meeting system, such as ZOOM and Teams, is used. The command can be remotely given to the computer. Fig. 19.6 shows the scene to control the olfactory display. The scent expert stays at the remote site and is away from the olfactory display. Since he is very busy, he cannot bring the smell to the customer. However, he can remotely operate olfactory display to present smells according to the customer's request.

In Fig. 19.6, he remotely selected the smell presented the smell using the olfactory display with 20 odor components. The olfactory display used here is described in Chapter 15. The web site called 100 essential oils selection based on the essential oil library was made as is shown in Fig. 19.7. The top menu shows the seven main categories of essential oils, that is, citrus, exotic, spice, resin, flower, woody, and herb. Then click the citrus, and finally orange smell is presented in this figure. The scent expert can select the smell and present it remotely according to the customer's preference. Therefore the interactive communication using the smell is possible even if there is no odor-sensing system.

Moreover, we can easily use the object recognition software such as YOLO recently (Jiang et al., 2022). It is possible to recognize objects, that is, orange, lemon, melon, etc., embedded in the movie file. Thus that technique can be used to generate the smell at the remote site even if there is no odor-sensing system. Since that technique does not include the odor-sensing system, the situation to use the object recognition is limited. However, we can avoid the sensing system development, whose reliability is often not so high. Those techniques introduced in this section are the intermediate system of teleolfaction, but those might be useful in certain situations.

19.6 Summary

Fundamental experiment on teleolfaction was explained in this chapter. The teleolfaction system used in Sections 19.3 and 19.4 is based on the classification system of odors. If the smell to be transmitted is not included in the

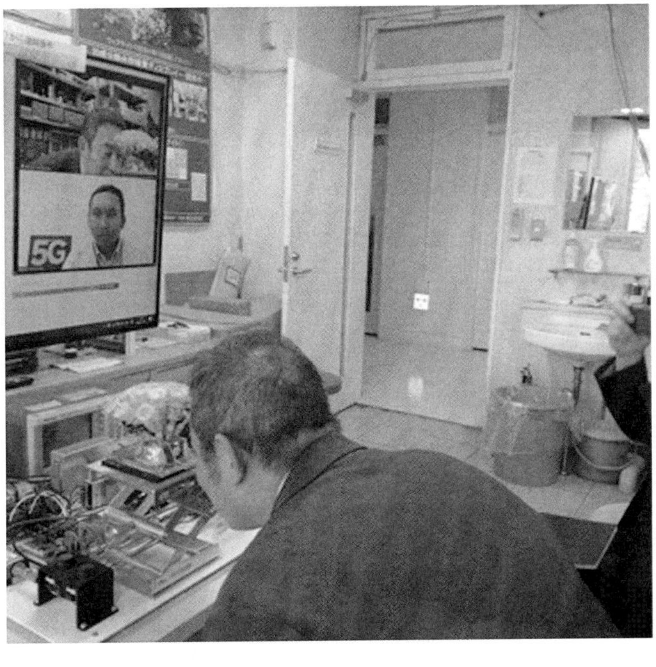

FIGURE 19.6 **Scene of operating olfactory display at remote site**. Olfactory display was remotely controled.

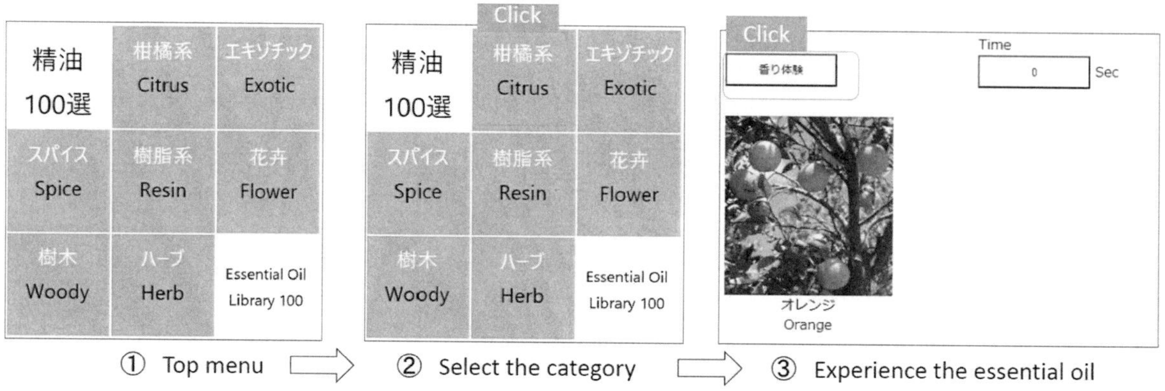

FIGURE 19.7 Web site called 100 selections of essential oils. (Left) Essential oil category and (middle) citrus category were selected, and (right) orange was finally selected.

training dataset of the sensing data, it is not possible to transmit the smell.

The odor typically consists of multiple ingredients. The teleolfaction system becomes more flexible if the mixture composition is quantified at one site and transmit the mixture composition to the remote site, followed by reproduction of the odor using olfactory display at the remote site. This type of system is called odor recorder (Nakamoto et al., 2001; Nakamoto, 2005; Yamanaka & Nakamoto, 2003; Yamanaka & Matsumoto, & Nakamoto, 2003). Since the mixture quantification is more sensitive to the disturbance than classification, it is difficult to quantify the mixture accurately in the ambient air. However, the range to cover the smell increases if the mixture quantification is transmitted.

Furthermore, it is also possible to use a set of odor components, which appears in Chapter 11, to transmit the smell. The mixture composition of odor components can be transmitted from one site to the remote site. It is more generalized method to use odor components.

Although the capability of the odor-sensing system is limited at the current stage, the improvement of the sensing system capability is indispensable for realizing the teleolfaction for more general purpose.

References

Dubey, R. (2009). *Introduction to embedded system design using field programmable gate arrays* (pp. 1–154). Springer London. Available from https://doi.org/10.1007/978-1-84882-016-6.

Fabrizio, H. M., & Lundström, I. (2001). 12 Virtual olfactory interfaces: Electronic noses and olfactory displays. In *Communications through virtual technology: Identity community and technology in the internet age*, IOS Press.

Hamada, S. (2007). Scent information communication. In M. Tonoike (Ed.), *Information and communication technology of olfaction* (pp. 240–254). Tokyo: Fragrance Journal.

Jiang, P., Ergu, D., Liu, F., Cai, Y., & Ma, B. (2022). A review of Yolo algorithm developments. *Procedia Computer Science, 199*, 1066–1073. Available from https://doi.org/10.1016/j.procs.2022.01.135.

Kohonen, T. (2012). *Self-organization and associative memory* (8). Springer Science & Business Media.

Manabe, H., Ueda, K., & Kumagai, S. (2007). On-line distibution of scent. In M. Tonoike (Ed.), *Information and communication technology of olfaction* (pp. 226–239). Tokyo: Fragrance.

Nakamoto, T. (2005). Record of dynamic changes of odors using an odor recorder. *Sensors and Materials, 17*(7), 365–383. Available from https://sensors.myu-group.co.jp/sm_pdf/SM614.pdf.

Nakamoto, T., Nakahira, Y., Hiramatsu, H., & Moriizumi, T. (2001). Odor recorder using active odor sensing system. *Sensors and Actuators B: Chemical, 76*(1-3), 465–469. Available from https://doi.org/10.1016/s0925-4005(01)00587-1.

Nimsuk, N., & Nakamoto, T. (2007). Improvement of capability for classifying odors in dynamically changing concentration using QCM sensor array and short-time Fourier transform. *Sensors and Actuators B: Chemical, 127* (2), 491–496. Available from https://doi.org/10.1016/j.snb.2007.05.009.

Nimsuk, N., & Nakamoto, T. (2008). Study on the odor classification in dynamical concentration robust against humidity and temperature changes. *Sensors and Actuators, B: Chemical, 134*(1), 252–257. Available from https://doi.org/10.1016/j.snb.2008.04.047.

Nakamoto, T., Nimsuk, N., Wyszynski, B., Takushima, H., Kinoshita, M., & Cho, N. (2008). *Reproduction of scent and video at remote site using odor sensing system and olfactory display together with camera*. In Proceedings of IEEE Sensors (pp. 799–802), Available from https://doi.org/10.1109/ICSENS.2008.4716562.

Nakamoto, T., Cho, N., Nitikarn, N., Wyszynski, B., Takushima, H., & Kinoshita, M. (2008). *Experiment on teleolfaction using odor sensing system and olfactory display synchronous with visual information*. Proc. ICAT (Vol. 8, pp. 85–92).

Nozaki, Y., Yoshino, F., Wyszynski, B., & Nakamoto, T. (2013). *Demonstration of improved teleolfaction system*. In Proceedings of 23rd International Conference on Artificial Reality and Telexistence (pp. 179–180).

Segawa, N., Tokuhiro, T., Nakamoto, T., & Moriizumi, T. (2002). Multi-channel frequency shift measurement circuit with high sampling rate for QCM odor sensors. *IEEJ Transactions on Sensors and Micromachines, 122*(1), 16–22. Available from https://doi.org/10.1541/ieejsmas.122.16.

Tsuji, M. (2008). Toward industrialization of scent web. In T. Nakamoto (Ed.), *Olfactory display* (pp. 235–245). Fragrance Journal.

Williams, M. (2001). Company seeks to bring scents to computer sceneAvailable from https://dailybruin.com/2001/03/01/company-seeks-to-bring-scents. (2001).

Yamanaka, T., Matsumoto, R., & Nakamoto, T. (2003). Fundamental study of odor recorder for multicomponent odor using recipe exploration method based on singular value decomposition. *IEEE Sensors Journal, 3*(4), 468–474. Available from https://doi.org/10.1109/JSEN.2003.815778.

Yamanaka, T., & Nakamoto, T. (2003). Real-time reference method in odor recorder under environmental change. *Sensors and Actuators B: Chemical, 93*(1-3), 51–56. Available from https://doi.org/10.1016/s0925-4005(03)00202-8.

Yamanaka, T., Nimsuk, N., & Nakamoto, T. (2007). Concurrent recording and regeneration of visual and olfactory information using odor sensor. *Presence: Teleoperators and Virtual Environments, 16*(3), 307–317. Available from https://doi.org/10.1162/pres.16.3.307.

Yoshino, F., & Nakamoto, T. (2014). Odor recognition system using embedded leaning vector quantization circuit. *Sensors and Materials, 26*(3), 137–147. Available from https://doi.org/10.18494/sam.2014.949.

CHAPTER

20

Olfactory Art

Nathan Cohen

Art and Science, University of the Arts London, London, United Kingdom

20.1 Olfactory art, science, and technology—a transdisciplinary exploration

20.1.1 Introduction

In this chapter, olfactory projects and technological applications developed by researchers from the fields of art, engineering, science, psychology, and education, collaborating transdisciplinarily, are discussed. This includes nondigital, hybrid, and digital forms of olfactory display illustrating the development of this research, including exploration of narrative, nostalgia, and memory in relation to smell, to create artworks and artifacts designed to enhance user well-being. Historical and contemporary context for the use of smell is also considered, including different ways in which it has inspired human culture and creativity.

Smell surrounds us, it is found in nature, and we have the ability to detect it (Buck & Axel, 1991). Yet, for the most part, it is invisible. Plants, animals, and organic materials provide natural olfactory sources that we can detect (Bentley et al., 2023), together with artificial and chemical sources generating a wide spectrum of smells, among a rich array of airborne particles, we are continuously immersed in. Over centuries, we have devised ever more sophisticated ways of extracting and preserving their essential oils (Dhifi et al., 2016) and volatile molecules for later use and for chemically synthesizing new smells (Surburg & Panten, 2006). That we have found this to be both desirable and worthy of the effort suggests that our sense of smell, and how our experience of it may be enhanced through the cultivation and preserving of a wide range of different types, performs a role of psychological significance (Shiner, 2020a). In our contemporary world, many different smells are now applied to a wide range of daily applications, including perfume, hygiene, well-being, and the manufacture of food and cleaning products.

Pleasure in relation to the stimulus of our olfactory sense can be a significant incentive to engage with it. Eating food and drinking liquors of all sorts would be rendered far less appetizing were it not for our sense of smell, and as their consumption enhances social interaction, it may be seen to connect with many of the developments that have led us to where we are socially today (Mintz & Du Bois, 2002). Even our preferences for who we may wish to interact with more intimately are, in part, determined by our

Digital Technologies in Olfaction
DOI: https://doi.org/10.1016/B978-0-443-15721-9.00020-8

sense of smell and how we respond to that of others (Rikowski & Grammer, 1999).

Smell is a vital sense that enables us to navigate and interpret our environment, and it is expressed in different cultures and forms globally. More intense and pleasurable smells have come to be appreciated and developed for consumption. The preciousness of aromatic compounds often meant they were only accessible to those wealthy enough to possess them. Rarer plants and trees that yield myrrh, frankincense, sandalwood, and saffron were much sought after with the Romans importing some of these by the ton to meet demand. Nutmeg was initially harvested and shipped to the West from the Banda Islands of Indonesia and, in the 17th century, was in high demand by those who could afford it for its perceived medicinal properties as, along with its potent aroma, it also gained a reputation, albeit false, as a protection from the Plague (Milton, 1999). Today, the aromas and taste of tea and coffee are now appreciated daily by millions worldwide.

The degree to which we use language to describe smells is indicative of how we comprehend and compare them (Majid, 2020). Whether it is the individual trying to explain their responses to smells, or the perfumer composing with different aromatic solutions and compounds to arrive at reproducible formulae for creating particular smell combinations, lexicons of olfactory terminology continue to evolve relating to the types of smell and their blending (Nikitina & Furuoka, 2023), while the advertising of perfumed products as aspirational drives the economics of manufacturers and their brands internationally.

The reasons for releasing smell into the surrounding environment are many and varied. Those related to ritual find a place in religious ceremonies. The dynastic Egyptians used scented resins and oils to embalm and preserve human tissue as well as covered the walls of their tombs with images of gardens, to enable the spirit of the dead to live a fulfilled afterlife (Sullivan, 2023). Incense had both a spiritual and practical use in Christian churches, in part evoking a more heavenly presence on Earth while to some extent masking the smells of human decay emanating from the crypts below, while in China, Japan, India, and the Middle East, there is a long history of creating vessels for the burning of aromatic compounds in both ritual and domestic settings, including as a means to measure time. Scent has also been used in European theater (Shiner, 2020b), while other, more contemporary examples include its use in cinemas (Spence, 2020) and public transport (Spence, 2021). Other uses of scent include those that are for more personal application in the form of perfume, cosmetics, and medicinal purposes, serving functions in enhancing health and well-being, social prestige, formal, and more intimate contact.

Aromatic compounds are, by their very nature, volatile to greater or lesser degrees. All types need to be contained in vessels that will both preserve them against the effects of oxidation and light that will degrade the aroma over time and enable them to be accessed when required. This has resulted in the creation over millennia of beautiful jars, bottles, vials, pomanders, and other containers for this purpose that are, in themselves, works of craftsmanship and art (North, 2006). Smell, as a sense, is also distinctive in relation to vision and touch, and its invisibility challenging understanding of the way we detect it (Billesbølle et al., 2023) and how it impacts on our perceptions. It is, perhaps, when it is most absent that we truly begin to comprehend its significance. An inability to smell became more widely prevalent, albeit for short periods, for many who caught the coronavirus, although for some people impairment of the ability to detect smell is more persistent and requires treatment.

20.2 Science, art, and olfaction—research projects

Utilizing creative thought processes from both art and science, while applying observation in the generation of "systematic and formulated knowledge" (Allen, 1986), has the potential to yield novel and innovative outcomes. Problem solving can be achieved through collaborative experimentation (Paterson et al., 2020). Sharing ideas and exploring technology across disciplines allows scope for invention and interpretation, as artists and scientists are both intellectually and technically curious (Schnugg, 2022). As our knowledge and development of new materials have developed, so have their use in the creation of art, where their transformation into images, objects, and interactive environments contribute to artistic endeavor (Cohen et al., 2021). This also relates to transdisciplinary research and education (Carter et al., 2021), for using new materials to create artworks and artifacts may also be enhanced by collaborating with scientists, technologists, and engineers. A completed work of art carries within its form indications of its creator's purpose and intention. Even if we are unfamiliar with its origins and the context within which it was made, we may still be able to interpret and comprehend how it can relate to our own experiences. The form that it takes affects the way we may interact with it as we invent with various media in ways that affect our perceptions. Digital media has become increasingly prevalent as computing technology and programming has advanced and become more accessible, providing new tools and another medium that can be put to use by creatives, although artists create artworks that encourage interaction and challenge perceptions that utilize a range of digital, hybrid, and nondigital media to achieve this, as indicated in this chapter.

Among the many factors considered when designing olfactory technology includes what its purpose is for, what functions it should be capable of performing, and how it might impact our sensory perception. The human body is a mobile mass of receptors evolved to relay information to the central nervous system through touch, heat, vision, sound, smell, and taste. These senses interact to enable us to navigate our environment at any given moment. As we interpret what we encounter through our senses and compare experiences over time, we can develop a deeper understanding of the world around us and our interactions with it (Bossomaier, 2012).

20.2.1 Olfactory project 1. Olfactory art and science, memory, and well-being

In 2017 an international olfactory research project commenced to explore how art practice and scientific knowledge could elide to create artworks and artifacts that could address a clinical need in gerontology, to enhance well-being for elderly residents in care homes by improving the quality of smell in their environment (Iwasaki & Cohen, 2019). This includes drawing on psychological research that examines how particular smells trigger positive associations with autobiographical memories that can be beneficial in achieving a sense of well-being (Sullivan et al., 2015).

The artworks the Japanese artist Reid et al. (2014) have made together as our contribution toward this research are designed through an iterative process and take into consideration psychologist's research into which scents stimulate *natsukashii* (nostalgic) feelings. Smell can induce recollection of past experiences (van Campen, 2014) and can be used in conjunction with related visual forms to enhance memory associations. In our artwork, the narrative content is explored through interactions between visual and haptic stimuli that relate to, and are augmented by, the olfactory component embedded within each piece. As the project developed, we also realized that it would be desirable to design the artworks in such a way that they could be personalized with narrative content particular to the individuals interacting with them. Consequently, we have been creating easy-to-handle olfactory displays that include narratives that encourage interpersonal association and the sharing of memories, such as places, the memories of which people are fond of (Fig. 20.1).

Comparisons can be made between this and the constructions of memorabilia found in the artworks of Joseph Cornell and earlier cabinets of curiosity (Lea & Hartigan, 2015). We were also intrigued by the pop-up books of Vojtěch Kubašta (Findlay et al., 2005), whose constructions enable a two-dimensional (2D) image to become three-dimensional (3D) as the pages are turned. There is a playfulness to this that is engaging, and we adapted this for our first boxed artwork *Memory Box* (Cohen & Kubota, 2017a), where elements of a landscape expand as the box lid is opened. As an olfactory display, we considered this to be expressive of the expansiveness of smell within the atmosphere, and the opening of the box lid releasing a draft of the nostalgic scent contained within.

In making these pieces for an international research project, we are also interested in exploring the relationships among visual, olfactory, and haptic stimuli at both the individual and cultural levels and their implications for well-being. So, while the esthetic of these olfactory artworks is situated partly within a Western art and science tradition, the concept of *natsukashii* (nostalgic) that informs them also relates to Japanese art, literary, and philosophical traditions. Psychologists characterize involuntary memory as Proustian (a reference to the French writer Marcel Proust) (Chu & Downes, 2000; Green et al., 2023), while in Japanese culture, animism (Rambelli, 2019) and a melding of the contemporary (Yoneyama, 2021) with the traditional are found in various art forms and practices that embrace smell, contemplation, memory, narrative, interaction, and location. These include Kōdō (Way of Fragrance) (Horiguchi & Jung, 2013), Sadō (Way of Tea) (Kakuzo & Foreword Grilli, 1989), and Nōh theater (Pinnington, 2019; Salz, 2016). These art forms distill the essence of human experience through ritualized interaction, which has informed the thinking behind the artworks we have created for this project.

A gesture in Nōh theater becomes symbolic in the telling of a story, while the intimacy of the tea ceremony, set within a specially constructed architectural space, is of the moment and evocative, its performance an abstracted form of reenacted narrative codifying human interactions. Our artworks *Viola Odorata* (hybrid) (Cohen & Kubota, 2017c) and *Scentscape* (hybrid) (Cohen & Kubota, 2019) explore this, combining the temporal and the contemporary, while *Memory Box* (nondigital)

FIGURE 20.1 Memory Box, shown both closed and open with contents. https://www.olfactoryresearch.net/memory-box. (Cohen, and Kubota, 2017a).

(Cohen & Kubota, 2017a) explores the relationship between memories of a landscape and the immediacy of the scent that expresses it. Kōdō relates both to familiarity and discovery through the act of determining what a particular scent might be, a process reflected in *Pomander* (nondigital) (Cohen & Kubota, 2017b) and *Game Box* (nondigital) (Fig. 20.2). As the project and artifacts have evolved so we have had the opportunity to observe how people interact with them within a gallery context (Cohen & Kubota, 2017b).

20.2.2 Olfactory project 2

The next stage of the project was to create artifacts that meld art with olfaction in a way that elderly people's responses to them in a nongallery setting could be assessed. To this end, we have developed a set of six artifacts, with advice from psychologists Prof. Takefumi Kobayashi and Prof. Kohsuke Yamamoto, who recommended that we design them in a way that reduces unnecessary complexity so that individual's responses can be measured and analyzed accurately to determine their well-being effect both with and without smell. For both practical and esthetic reasons, intimacy forms a key part of what we are trying to achieve in their design and presentation of smell within the artifacts. Having observed how people interacted with an installation (*Viola Odorata*) and the boxed artworks, which might not be so practical for the elderly to interact with in care home settings, we were determined to incorporate those aspects of the design that would offer forms *suggestive* of things seen and experienced in association with a smell source, but in a way that would also be more compact and usable (Fig. 20.3).

FIGURE 20.2 *Viola Odorata*, 2017 (left) and Scentscape, 2019 (right), both examples of hybrid olfactory displays, including digital and nondigital elements.

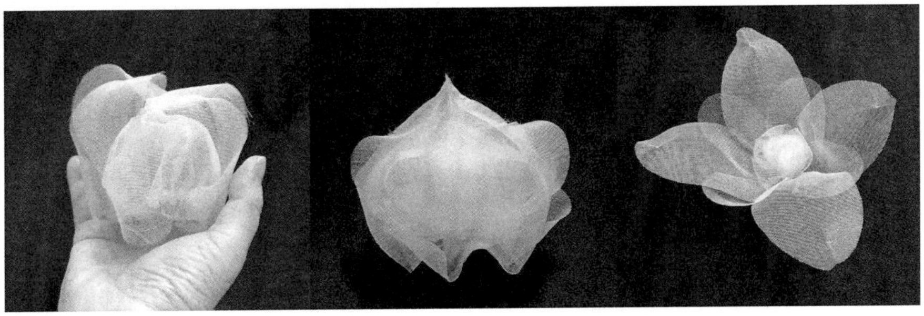

FIGURE 20.3 Aero Odorata, showing three of the six artifacts with different scents embedded in each.

Following experimentation in our studio laboratory with various light-weight media and designs, a set of artifacts were subsequently made by Reiko Kubota, titled *Aero Odorata* (Kubota & Cohen, 2021). These include six differently shaped but related artifacts made from organza fabric, cut, folded, and stitched to create the various shapes, with an accessible pocket for a different olfactory sample embedded and concealed from direct view within each of them. They are small enough to be held in the palm of one hand and moved easily by the observer or their caregiver interacting directly with them, the air passing through the porous material releasing the smell in each piece. Two sets of artifacts have been created, one with the smell sources embedded and the other without. These have now been tested by the psychologists who explain the method used and the results obtained.

'The study employed artifacts to investigate whether pairing an olfactory stimulus with each influences psychological assessments, including emotional arousal. Initially, a within-subject design experiment was conducted with Japanese university students. They rated, on a 10-point scale, the change in emotional arousal while viewing the artifact with and without the odor of mandarin, rose, soap, cinnamon, and kyara. Results indicated higher emotional arousal, nostalgia, a sense of being transported back to the past, and vividness of memories in the with-odor condition compared to the without-odor condition (Shirai et al., 2022). Additionally, a recent experiment involving elderly participants, utilizing a more strictly controlled, between-subject design, demonstrated that the odor-paired group correlated with higher arousal, greater vividness of events, and a heightened sense of being pulled back to the past compared to the nonodor condition (Yamamoto et al., 2024). Consequently, empirical evidence from the psychological experiments supports the efficacy of pairing olfactory stimulus in the evaluation of the artifacts.'

20.3 Olfactory game—digital developments

In parallel with the research described above, the author has been exploring digital olfactory display technology toward enhancing user experience, combining visual and olfactory stimuli within virtual reality (VR) environments (Tokyo Tech World Research Hub Initiative WRHI, 2023). This has been progressed by working with Nakamoto (2012) whose research focuses on olfactory sensors and digital displays. The research has involved developing digital olfactory games, initially with the assistance of Saya Onai in the Nakamoto Laboratory, that can also have well-being applications.

To achieve this, we selected a digital olfactory display device, manufactured by Ono-Denki Digital Olfactory Display device (2024), that allows up to 12 different olfactory samples to be used simultaneously. This device can be programmed to release each smell sample in a timed way in specific quantities through a solenoid valve assembly driven by an air pump. Smells can be delivered individually and in combination. Due to the nature of the digital device construction, the impact of smell cross-contamination is controlled by each smell having its own container, distributed through a nylon tube attached to a head-mounted disseminator or display that sits just below the player's nose. We chose this device for its reliability and accuracy as the release of the smell is controlled through a computer in real time in response to player interactions with the game environment.

20.3.1 Aromatic Garden—two-dimensional versions

Initially, we set out to design a 2D olfactory game environment that a player navigates to find smells using an avatar with movement limited to up, down, and sideways as the player seeks out each smell source embedded within the landscape. The game space is pictorial, an image in 2D depicting a 3D space within which the player can search, with a foreground, middle distance, and background to enhance the illusion of depth. This enabled testing player interaction without the complexities of 3D orientation. An early source of inspiration for the game imagery was the triptych painting *The Garden of Earthly Delights* (c.1495) by Hieronymus Bosch (Prado Museum, 2024). The central image depicts a surreal imaginary garden full of potential smell sources, including fruits and plants. Initially, we adapted this for our game, and to encourage the player to concentrate on seeking out each smell source by concentrating on the sought for smell's intensity, which increases relative to the proximity of the player's avatar, the whole game landscape was covered with a grid of neutrally colored tiles. As the player moves their avatar in search of each smell source, the tiles disappear for a few seconds revealing the landscape being traversed. When the player arrives at the smell source, guided by its intensity, they can collect it. Once all 12 smell sources, generated to be searchable one at a time in a preprogrammed sequence, have been collected, the game is complete (Fig. 20.4).

20.3.2 Selecting smells

In determining which smells are to be used for the game, there are many to choose from, but some are more recognizable than others, and many can share similar aromatic characteristics, so may be more difficult distinguish from one to another. Initially, smell selection related closely to the imagery depicted in the virtual landscape; a strawberry, or an orange, flower smells, etc. But it was necessary to go beyond the options available in the Bosch painting to offer a wider range relating to more contemporary experience, including a spectrum of pleasant to less pleasant smells. To do this, we created our own landscapes that players could navigate between to collect the 12 smell samples. Two of these are landscapes that contain more pleasant smells, and the other two include smells more associated with volcanic and polluted landscapes. We called our new virtual game environment *The Aromatic Garden*, partly in recognition of the original source of inspiration for the game imagery.

While the pleasant landscape odors focus on more familiar and distinctive smells of

FIGURE 20.4 Aromatic Garden (2021) 2-dimensional version showing four landscape types: Pleasant (Natural), Pleasant (Artificial), Unpleasant (Natural), and Unpleasant (Artificial). .

particular herbs, fruit, and flowers, the polluted and volcanic landscape smell selections were more challenging, as many of the smells that are associated with these environments are not so individually recognizable and can be toxic. While we wish to offer a dynamic smell contrast across the four landscapes, we also want players to enjoy their experience of playing the game and searching for the different smells. If they are too noxious, this could have the undesirable effect of inducing nausea or negative responses to continuing until all of the smells have been located. Noxious smells are also more problematic for the olfactory device to deal with, lingering longer in the nylon tube delivering the smell to the game player, and potentially contaminating other smells in the process. Consequently, selecting the smells to be used in the game required a combination of prior olfactory knowledge, intuition, experimentation, and testing these to determine user response.

Given that smells have different characteristics, and individual perception can also vary, we devised a triangular chart with the sides labeled Sweet (Pleasant), Bad (Unpleasant), and Natural. The test participants were asked to place a dot where they felt each smell is located relative to these 3 characteristics, following which all of the results could be collected and visual comparisons made for each of the smells, and between the different smells (Fig. 20.5).

The results revealed that many of the test participants shared common perceptions but to varying degrees, and where the results most clearly indicated a shared perception, this was useful for selecting the smells for each of the four game environments: Pleasant (Natural) landscape—flowers and fruit smells; Pleasant (Artificial) landscape—food and more pleasant synthetic smells; Unpleasant (Natural) landscape—sulfurous, decaying smells; and Unpleasant (Artificial) landscape—pollution and bad artificial smells.

Another aspect we wished to explore was the combination of paired smells over time. Having created four different landscapes with different smell properties, we needed to enable the player to move from one to the other to sample and collect all of the 12 smells embedded collectively within them. To do this, we created scent "tunnels," whereby a player arriving at a particular smell in one landscape would then be transported automatically to a different smell paired with it in one of the contrasting landscapes (Fig. 20.6). As the player enters the scent tunnel, they smell 100% of the first smell, with the ratio changing evenly over a 20 + second duration of the journey to 100% of the second smell on arrival at the new landscape. This results in the player experiencing a temporal shift in the smell combination with a third smell sensation with 50%–50% mix of the smells occurring halfway through the tunnel journey.

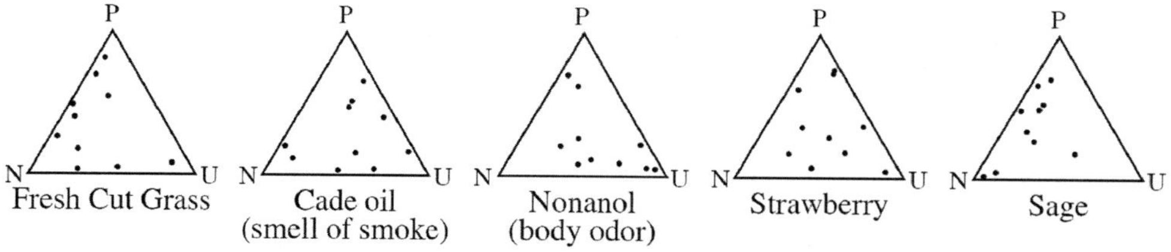

FIGURE 20.5 Chart illustrating triangular olfactory test results where P = Pleasant, U = Unpleasant, and N = Natural. *Source:From Cohen, N., Onai, S., & Nakamoto, T. (2021). The Aromatic Garden: Exploring new ways to interactively interpret narratives combining olfaction and vision including temporal change of scents using olfactory display.* SA '21: SIGGRAPH Asia 2021 Emerging Technologies, 17, 1–3. *https://doi.org/10.1145/3476122.3484840.*

FIGURE 20.6 Locations of 12 odor sources to be collected across the 4 landscapes of the 2-dimensional olfactory game version with arrows indicating "scent tunnel" connections between landscapes. *Source: From Cohen, N., Onai, S., & Nakamoto, T. (2021). The Aromatic Garden: Exploring new ways to interactively interpret narratives combining olfaction and vision including temporal change of scents using olfactory display.* SA '21: SIGGRAPH Asia 2021 Emerging Technologies, 17, *1–3. https://doi.org/10.1145/3476122.3484840.*

One of the challenges for digital olfactory devices is the limitation of the number of smells they can carry. Currently, body-worn devices are even more limited in capacity compared with desk-top versions due to the constraints of space and weight limiting the number of scent containers and the dispensing mechanics and electronics required for diffusion, although the technology is advancing to increase capacity at smaller scales (Nakamoto et al., 2020) that will be of particular benefit for incorporating into VR headsets. Creating distinctive odors through the pairing of different odors increases the range available and potentially generates novel smell experiences. As part of this research, we have been investigating pairs of odors that display nonlinear effects, where each mixture provides an impression different from both of the base odors used to create it (Ma et al., 2020). The "scent tunnel" we have included in our game design, enabling a link between the virtual landscape environments and the pairing of smells with contrasting characteristics that generates a distinctive third smell over time, increases the smell sensations experienced while playing the game, although further research is required to produce combinations that are more predictable in how they are perceived.

Having demonstrated the first version of the Aromatic Garden at Siggraph Asia 2021 (Onai, 2021; Cohen et al., 2021), we realized a number of improvements could be made to both the design and navigation of the game. For those players who volunteered to complete a short feedback form, the general impression was that it took too long to locate all of the smells across the four landscapes to be able to complete the game in a reasonable time (this also leads to the question of what a player is playing the game for, an issue addressed in part by the latest iteration *Interactive Smellscape*). For others, the scent tunnels took too long to travel through, resulting in some confusion as one smell changed into the other. However, the overall impression was one of enjoyment at being able to play a game that included olfactory stimuli that respond to real-time intervention.

While the duration of the game was frustrating for some, one aspect that would be interesting to explore is the construction of a game that intentionally takes longer to play. If the game environment encompasses a landscape that includes a meadow, lake, hills, and mountains, within which the olfactory sources are embedded, then the speed at which a player navigates can be varied depending on how they are enabled virtually to travel. At walking pace, it might take hours to wander through the landscape, just as one would do on a real trek through the wilderness. Enabled to glide or fly at greater speed, the journey could be reduced to minutes. From a well-being perspective, there may be some players who would enjoy a longer virtual ramble and gain from the experience, particularly those whose mobility may be impaired. For others, wishing to complete the task of collecting all of the embedded smells, reward might be measured by speed of achievement. The game environment allows for both approaches depending on how it is programmed.

20.3.3 Aromatic Garden (three-dimensional version)

In response to the observations made by the players, and in pursuit of enhancing the experience of detecting smells within different virtual landscape settings, we moved from a 2D to a 3D virtual game environment, allowing us to play with the scale of field and providing depth to move around objects. Smell sources could be hidden in more locations, not just those always visible from any angle, encouraging players to navigate the virtual landscapes guided by the strength of the smell source they are searching for, with intensity of the smell relating to proximity, becoming more detectable the closer a player moves toward it. We could also dispense with avatars, placing the player directly within the landscape as they navigate toward each smell. In response to the need to simplify, we reduced the landscapes to two from four, one with the more pleasant smells (meadow) contrasting with the other having the less pleasant smells (volcanic). We retained the scent tunnels linking the two landscapes but shortened the journey times and began the process of testing pairs of smells that would be more recognizably distinctive in comparison with each other and yet yield an interesting third smell in combination Fig. 20.7.

As the design for the game progresses, we have learned that several factors need to be considered to create the overall game experience and make it playable. These include designing virtual environments that are stimulating to engage with, enabling the player quickly to familiarize themselves with how the

FIGURE 20.7 Screenshot from the 3-dimensional version of the Aromatic Garden showing the orange scent smell source embedded in the virtual landscape.

game works when they enter the game space, and ensuring that once the game commences, the process of searching for and locating each smell can be achieved relatively quickly while also allowing sufficient input from the player to make it enjoyably challenging. It is also necessary to present smells that cover a range of olfactory types, are contrasting, and ideally also quite familiar. To varying degree, depending on smell type, it is usually quite difficult to identify particular smells without some additional information being provided. It is therefore important that the imagery complement and enhance the process of searching for each smell. The visual design needs to achieve a fine balance between illustrating what is available and encouraging the player to explore, with time a factor in determining comparative success in locating the smell sources: too long and the player starts to forget or become less sensitive to the smell being sought, with a subsequent loss of concentration and focus to complete the game tasks.

20.4 Interactive Smellscape

At CEATEC 2023, we had the opportunity to present the most recent version of the game, *Interactive Smellscape* (Fig. 20.8). The emphasis has shifted to creating a version that can assist those with anosmia and other smell impairment conditions (Dudine et al., 2021) to relearn what particular smells are like, while also remaining of interest for game players generally. A test version was displayed, and many CEATEC visitors were able to interact with it and provide their feedback. Laboratory testing of the smells to be used and of the game itself prior to its public presentation is essential in developing the many aspects of the game's contents and design required for it to function, but when it is open to scrutiny by a wider audience, their interactions enable us to find out more about what needs to be improved to meet players expectations.

A key decision in developing the current iteration of the game is that it should have a clearly defined purpose toward enhancing well-being, with measurable outcomes. This involves creating a game that will enhance players' ability to recognize and remember particular smells (Pieniak et al., 2022). The earlier digital versions addressed this more broadly, creating a range of environments and forms of interaction that channeled olfactory stimuli as reward for completing the game. With this most recent version, we have made a number of changes to the game structure and layout. Players still need to collect smells to complete the game. However, the emphasis with the *Interactive Smellscape* design is to encourage players to memorize and differentiate each of the key smells (Olofsson et al., 2021).

A player begins the game by reading the rules of engagement, and then they enter the

FIGURE 20.8 Interactive Smellscape landscape with tōrō Japanese stone lantern (key smell source) and tatefuda (Japanese wooden notice boards) presenting three smell options. *Image from olfactory game "Interactive Smellscape"*

virtual landscape where they are required to locate a tōrō (Japanese stone lantern) from which a particular key smell is emanating. On touching the tōrō, three smell options appear close by represented by traditional Japanese wooden notice boards (tatefuda), which are in keeping with the Japanese themed styling for this version of the game (created with the assistance of Ryota Sunami, Nakamoto Laboratory). The player can sample each of the different smells at each of the three notice boards and then select the one they think is the key smell they are searching for. If they identify this correctly, an image representing the smell is projected onto the corresponding notice board, and they return to the tōrō to collect the smell before moving on to find the next smell. If their first or second attempts are incorrect they continue to receive the smell options until they select the correct one. While we use 12 smells in total to play this game, only 4 are the key smells being sought, with the other smells acting as contrasting smells they compare the key smells with. At each attempt to identify the key smell, 2 of the other smells are presented to select from, with different pairings for each of the 4 key smells, utilizing all 12 smells to complete this level of the game. By introducing the key smell to start each sequence, the player needs to try and memorize that smell, so when they have the other two smells to compare them with, they can select the correct smell. For players with poor smell sense, it is anticipated that this will require playing the game several times until they begin to recognize the key smells they are seeking. Consequently, the location of the smells is changed for each game session preventing the player predicting that is the correct smell source.

To create this, the graphics and the way players orient and move within the landscape have changed from previous versions of the game. The player is also guided closer to the smell sources before they commence their search reducing the time required to remember each of the key smells. It has also been necessary to select a new set of 12 smells, the 4 key smells being searched for and 4 pairs of contrasting smells, one pair for each key smell. Each of the key smells should be quite familiar and consequently memorable for the players, while the pairs of smells need to contrast with these so that they are quite easily distinguishable. This has required experimentation in the laboratory to find optimal pairings and to determine the concentrations at which each of the smells is most clearly discernible. Too weak, and the smells are not so easily detectable leading to confusion; too strong, and they dominate the contrasting smells and impair comparison and recollection. The next stage of development for the game will be based on ongoing testing of players' responses in a controlled playing environment. The psychologists Prof. T Kobayashi and Prof. K Yamamoto are conducting experiments with volunteer players to determine how they respond to the game and to establish if it does achieve its purpose in enabling and enhancing recollection of the key smells. They observe the following:

> As senior citizens often experience a decline in various cognitive functions including olfactory function, utilizing such games are expected to implement cognitive training to prevent dementia. Accordingly, it is necessary to conduct experimental investigations to test whether this olfactory game contributes to enhancing cognitive function in the elderly. Specifically, cognitive function tests such as the Mini-Mental State Examination (MMSE) are being administered before and after playing this game. If the olfactory game enhances cognitive functions, it is expected that cognitive components like memory function will show improvement from pre- to post-game. Additionally, a precise assessment of the olfactory game's impact on cognitive functions should involve a comparison with a control group where the game excludes the "odor" component.

The initial test sessions included healthy elderly individuals aged between 63 and 75 years who each participated in two sessions of the game. Cognitive tests, derived from multiple cognitive function assessments, were administered before and after the sessions to compare changes in cognitive performance. The aim is to identify that cognitive abilities could be positively influenced by the game, as well as to aid in future experimental design considerations. Interestingly, although many of the participants reported weak olfactory intensity during the game, their actual performance on olfactory identification was relatively high (Fig. 20.9).

20.5 Transdisciplinary olfactory display research

"The framework of a work of art is also it's most secret and its deepest poetry." Villon and Bouleau (1963).

As artists, the rationale for working with psychologists and scientists is that it enables us to consider what we are seeking to discover and trying to create from different perspectives. Each brings expertise and insights gained from problem solving within their chosen disciplines, offering the potential to synthesize this knowledge to create new olfactory artworks and artifacts that might only be created in this way. Sharing ideas across disciplines also broadens the scope for what they might be made for and how they function within the public domain. The esthetic choices in relation to the imagery and objects we create as artists working transdisciplinarily are also aligned with the technology and utility intended for those who will interact with them. In the examples given in this chapter, this includes creating art forms that address narrative, memory, smell, and well-being, for a range of different audiences.

All of this research has been conducted over a period of years. It takes time to undertake the work and to forge the relationships necessary across the different disciplines required. This has required consideration of a number of factors that have proven to be necessary to undertake these olfactory projects, explore the technology that is being invented to enable this and to disseminate our findings (Nakamoto & Cohen, 2024). For instance, it is necessary to remain openminded about what the shared project aims might lead to at the outset and to define these more clearly as the project proceeds. This allows time for exploration and imagination to form part of the project's evolutionary process. A willingness to contribute to the process of sharing ideas and to collaborate toward advancing research in new ways also

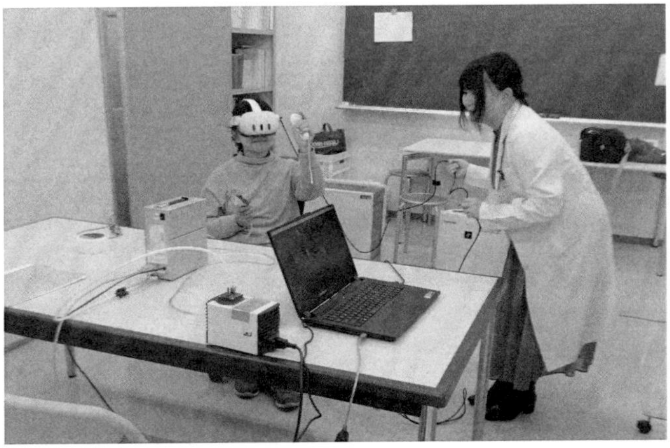

FIGURE 20.9 Testing the Interactive Smellscape olfactory game April 2024. An image taken during testing sessions of the Interactive Smellscape olfactory game.

requires effective communication and sensitivity to different disciplinary and cultural approaches, including those of educational and research institutions across countries and continents, whose regulations and ways of disseminating knowledge can impact how researchers behave when engaging with each other. Different views also need to be embraced constructively to allow for discourse and debate enabling the project outcomes to reflect diversity in thought and approach toward achieving coherent outcomes. The transdisciplinary collaboration presented in this chapter is an ongoing endeavor. It is exciting to anticipate what is yet to come by working together to create new ways of inventing with and comprehending what olfactory displays enable people to do to enhance their awareness of smell as a vital sense, with the intended positive consequences for their enjoyment and well-being.

References

Allen, R. E. (1986). In R. E. Allen (Ed.), *The Oxford Dictionary of Current English*. Oxford University Press.

Bentley, P. R., Fisher, J. C., Dallimer, M., Fish, R. D., Austen, G. E., Irvine, K. N., & Davies, Z. G. (2023). Nature, smells, and human wellbeing. *Ambio, 52*, 1–14. Available from https://doi.org/10.1007/s13280-022-01760-w, https://link.springer.com/article/10.1007/s13280-022-01760-w.

Billesbølle, C. B., de March, C. A., van der Velden, W. J. C., Ma, N., Tewari, J., del Torrent, C. L., Li, L., Faust, B., Vaidehi, N., Matsunami, H., & Manglik, A. (2023). Structural basis of odorant recognition by a human odorant receptor. *Nature, 615*(23 March), 742–749. Available from https://doi.org/10.1038/s41586-023-05798-y.

Bossomaier, T. J. (2012). *Introduction to the senses from biology to computer science*. Cambridge University Press. Available from https://doi.org/10.1017/CBO9781139016001.

Buck, L., & Axel, R. (1991). A novel multigene family may encode odorant receptors: A molecular basis for odor recognition. *Cell, 65*(5th April), 175–187.

Campen, C. van (2014). *The Proust effect: The senses as doorways to lost memories*. Oxford University Press. Available from https://doi.org/10.1093/acprof:oso/9780199685875.001.0001.

Carter, C., Cohen, N., Durall, E., Knack, F., Lordick, D., et al. (2021). Defining STEAM approaches for higher education. *European Journal of STEM Education, 6*(1), 16. Available from https://doi.org/10.20897/ejsteme/11354, https://www.researchgate.net/publication/356361273_Defining_STEAM_Approaches_for_Higher_Education.

Chu, S., & Downes, J. J. (2000). Odour-evoked autobiographical memories: Psychological investigations of Proustian phenomena. *Chemical Senses, 25*(1), 111–116. Available from https://doi.org/10.1093/chemse/25.1.111, https://academic.oup.com/chemse/article/25/1/111/345723.

Cohen, N., & Kubota, R. (2017a). *Olfactory research memory box*. https://www.olfactoryresearch.net/memory-box.

Cohen, N., & Kubota, R. (2017b). *Pomander*. https://www.olfactoryresearch.net/pomander.

Cohen, N., & Kubota, R. (2017c). *Viola odorata*. https://www.olfactoryresearch.net/viola-odorata.

Cohen, N., & Kubota, R. (2019). *Game box and scentscape*. https://www.olfactoryresearch.net/game-box-and-scentscape.

Cohen, N., Onai, S., Nakamoto, T. (2021). The Aromatic Garden: Exploring new ways to interactively interpret narratives combining olfaction and vision including temporal change of scents using olfactory display. In *SA '21: SIGGRAPH Asia 2021 Emerging Technologies* (Vol. 17, pp. 1–3). https://doi.org/10.1145/3476122.3484840.

Cohen, N., Holme, A., Bright, R., Calvert, S., & Barnett, H. (2021). *The Art of Science*. Welbeck.

Dhifi, W., Bellili, S., Jazi, S., Bahloul, N., & Mnif, W. (2016). Essential oils' chemical characterization and investigation of some biological activities: A critical review. *Medicines (Basel), 3*(4), 25. Available from https://doi.org/10.3390/medicines3040025, https://www.ncbi.nlm.nih.gov/pmc/articles/PMC5456241/.

Dudine, L., Canaletti, C., Giudici, F., Lunardelli, A., Abram, G., Santini, I., Baroni, V., Paris, M., Pesavento, V., Manganotti, P., Ronchese, F., Gregoretti, B., & Negro, C. (2021). Investigation on the loss of taste and smell and consequent psychological effects: A cross-sectional study on healthcare workers who contracted the COVID-19 infection. *Frontiers in Public Health, 9*. Available from https://doi.org/10.3389/fpubh.2021.666442, https://www.frontiersin.org/articles/10.3389/fpubh.2021.666442/full.

Findlay, J., Rubin, E., & Kubašta, V. (2005). *Pop-ups, illustrated books, and graphic designs of Czech artist and paper engineer, Vojtěch*. In *Kubašta*, (pp. 1914–1992). Bienes Center for the Literary Arts.

Green, J. D., Reid, C. A., Kneuer, M. A., & Hedgebeth, M. V. (2023). The Proust effect: Scents, food, and nostalgia. *Current Opinion in Psychology, 50*(101562). Available from https://doi.org/10.1016/j.copsyc.2023.101562.

Horiguchi, S., & Jung, D. (2013). Kōdō—Its spiritual and game elements and its interrelations with the Japanese literary arts. *Journal of the Royal Asiatic Society*, *23*(1), 69–84. Available from https://doi.org/10.1017/S1356186313000011.

Iwasakia, Y, & Cohen, N (2019). Olfactory art and science, memory and wellbeingAvailable from https://www.olfactoryresearch.net/research. (2019).

Kakuzo, O., & (Foreword) Grilli, E. (1989). *The book of tea*. Tuttle Publishing.

Kubota, R, & Cohen, N (2021). *Aero odorata: Creating new olfacto-narrative forms. Olfactory Research*. Available from https://www.olfactoryresearch.net/aero-odorata.

Lea, S., & Hartigan, L. (2015). *Joseph Cornell: Wanderlust*. Royal Academy of Arts London.

Majid, A. (2020). Human olfaction at the intersection of language, culture, and biology. *Trends in Cognitive Sciences*, *25*(2), 111–123. Available from https://doi.org/10.1016/j.tics.2020.11.005.

Ma, Y., Tang, K., Thomas-Danguin, T., & Xu, Y. (2020). Pleasantness of binary odor mixtures: Rules and prediction. *Chemical Senses*, *45*(4), 303–311. Available from https://doi.org/10.1093/chemse/bjaa020, https://academic.oup.com/chemse/article/45/4/303/5809744.

Milton, G. (1999). *Nathaniel's nutmeg: How one man's courage changed the course of history*. Hodder & Stoughton. Available from urn:lcp:howonemanscourag0000nath:lcpdf:4be45c1f-22fd-40a7-9dd9-626d1d2c5f8b.

Mintz, S., & Du Bois, C. (2002). The anthropology of food and eating. *Annual Review of Anthropology*, *31*, 99–119. Available from https://doi.org/10.1146/annurev.anthro.32.032702.131011.

Nakamoto, T. (2012). *Human olfactory displays and interfaces: Odor sensing and presentation* (1). Information Science Reference. Available from https://doi.org/10.4018/978-1-4666-2521-1.

Nakamoto, T., & Cohen, N. (2024). *International olfactory display workshop*. Available from http://silvia.mn.ee.titech.ac.jp/html_en/Location8.pdf.

Nakamoto, T., Hirasawa, T., & Hanyu, Y. (2020). *Virtual environment with smell using wearable olfactory display and computational fluid dynamics simulation*. In IEEE—Conference on Virtual Reality and 3D User Interfaces (VR). Available from https://ieeexplore.ieee.org/document/9089617.

Nikitina, L., & Furuoka, F. (2023). Words of scents: A linguistic analysis of online perfume reviews. *Linguistics Vanguard*. Available from https://doi.org/10.1515/lingvan-2022-0135.

North, J. (2006). *Perfume, cologne, and scent bottles* (4). Schiffer Publications Ltd.

Olofsson, J. K., Ekström, I., Lindström, J., Syrjänen, E., Stigsdotter-Neely, A., Nyberg, L., Jonsson, S., & Larsson, M. (2021). Smell-based memory training: Evidence of olfactory learning and transfer to the visual domain. *Chemical Senses*, *45*(7), 593–600. Available from https://doi.org/10.1093/chemse/bjab015, https://www.ncbi.nlm.nih.gov/pmc/articles/PMC8345828/.

Ono-Denki Digital Olfactory Display device. (2024). https://ono-denki.com/development/.

Onai, S. (2021). *Olfactory research demonstration at the SIGGRAPH ASIA*. Available from http://silvia.mn.ee.titech.ac.jp/html/display10.html.

Paterson, S. K., Tissier, M. L., Whyte, H., Robinson, L. B., Thielking, K., Ingram, M., & McCord, J. (2020). Examining the potential of art-science collaborations in the Anthropocene: A case study of catching a wave. *Frontiers in Marine Science*, *7*. Available from https://doi.org/10.3389/fmars.2020.00340, https://www.frontiersin.org/articles/10.3389/fmars.2020.00340/full.

Pieniak, M., Oleszkiewicz, A., Avaro, V., Calegari, F., & Hummel, T. (2022). Olfactory training—Thirteen years of research reviewed. *Neuroscience & Biobehavioral Reviews*, *0149–7634*, 141. Available from https://doi.org/10.1016/j.neubiorev.2022.104853, https://www.sciencedirect.com/science/article/pii/S0149763422003426.

Pinnington, N. J. (2019). *A new history of medieval Japanese theatre Palgrave studies in theatre and performance history* (1). Palgrave Macmillan. Available from https://doi.org/10.1007/978-3-030-06140-1.

Prado Museum. (2024). *d*, 1490–1500. Available from https://www.museodelprado.es/en/the-collection/art-work/the-garden-of-earthly-delights-triptych/02388242-6d6a-4e9e-a992-e1311eab3609?search-id = e0b0ff98-a0f1-134d-8d97-f6c4d3ffe4a8.

(Preface) Villon, J., & Bouleau, C. (1963). *The painter's secret geometry—A study of composition in art*. Thames and Hudson - Harcourt, Brace & World, Inc.

Rambelli, F. (2019). In F. Rambelli (Ed.), *Spirits and animism in contemporary Japan: The invisible Empire*. Bloomsbury Academic. Available from https://doi.org/10.1525/nr.2020.24.2.107.

Reid, C. A., Green, J. D., Wildschut, T., & Sedikides, C. (2014). Scent-evoked nostalgia. *Memory (Hove, England)*, 1–10. Available from https://doi.org/10.1080/09658211.2013.876048.

Rikowski, A., & Grammer, K. (1999). Human body odour, symmetry and attractiveness. *Proceedings of the Royal Society B*, *266*(1422). Available from https://doi.org/10.1098/rspb.1999.0717.

Salz, J. (2016). In J. Salz (Ed.), *A history of Japanese theatre Noh and Muromachi culture*. University Press. Available from https://doi.org/10.1017/CBO9781139525336.007.

Schnugg, C. (2022). *Art + water artscience collaboration: Serendipity and beyond innovation*. V2_Publishing. https://www.researchgate.net/publication/366065426_ArtScience_Collaboration_Serendipity_and_Beyond_Innovation.

Shiner, L. (2020a). The neuroscience and psychology of smell I: What the nose can do. *Art Scents: Exploring the Aesthetics of Smell and the Olfactory Arts*, 37. Available from https://doi.org/10.1093/oso/9780190089818.003.0006.

Shiner, L. (2020b). Toward a total work of art: Smell in theater, film, and music. *Art scents: exploring the aesthetics of smell and the olfactory arts*. Oxford University Press. Available from https://doi.org/10.1093/oso/9780190089818.003.0020.

Spence, C. (2021). Scent in motion: On the multiple uses of ambient scent in the context of passenger transport. *Frontiers in Psychology*, *12*, 1–19. Available from https://doi.org/10.3389/fpsyg.2021.702517, https://www.researchgate.net/publication/353196597_Scent_in_Motion_On_the_Multiple_Uses_of_Ambient_Scent_in_the_Context_of_Passenger_Transport.

Shirai, M, Kobayashi, T, Kunieda, S, Yamamoto, K, Cohen, N, & Kubota, R (2022). Investigation on psychological effects of artworks with nostalgic odor. *The 86th Annual Convention of the Japanese Psychological Association*4AM-21-PH.

Spence, C. (2020). Scent and the cinema. *i-Perception*, *11*(6), 1–22. Available from https://doi.org/10.1177/2041669520969710, https://journals.sagepub.com/doi/10.1177/2041669520969710.

Sullivan, E. A. (2023). The senses & the sacred: A multisensory and digital approach to examining an ancient Egyptian funerary landscape. In G. Landeschi, & E. Betts (Eds.), *Capturing the Senses. Quantitative Methods in the Humanities and Social Sciences*. Springer. Available from https://doi.org/10.1007/978-3-031-23133-9_3.

Sullivan, R. M., Wilson, D. A., Ravel, N., & Mouly, A.-M. (2015). Olfactory memory networks: From emotional learning to social behaviors. *Frontiers in Behavioral Neuroscience*, *9*, 1662–5153. Available from https://doi.org/10.3389/fnbeh.2015.00036, https://www.frontiersin.org/articles/10.3389/fnbeh.2015.00036/full.

Surburg, H., & Panten, J. (2006). *Common fragrance and flavor materials: Preparation, properties and uses*. Wiley-VCH Verlag GmbH & Co. Available from https://doi.org/10.1002/3527608214.

Tokyo Tech World Research Hub Initiative (WRHI). (2023). https://www.irfi.titech.ac.jp/wrhi-archive/en/people/nathan-cohen/index.html.

Yamamoto, K, Kobayashi, T, Kunieda, S, Cohen, N, & Kubota, R (2024). A study on the psychological effects of artifacts with nostalgic odors (2). *The 35th Annual Meeting for the Japan Society of Developmental Psychology*8PM1-P-PS11.

Yoneyama, S. (2021). Miyazaki Hayao's animism and the Anthropocene. *Theory, Culture & Society*, *38*(7/8), 251–266. Available from https://doi.org/10.1177/02632764211030550.

SECTION 5

Conclusions

CHAPTER

21

Conclusions

Takamichi Nakamoto
Institute of Integrated Research, Institute of Science Tokyo, Yokohama, Japan

21.1 Conclusions

We have looked at digital technologies in olfaction at the current stage in 2024. Regarding odor-sensing system, we have seen many sensor devices with different principles. Quarz crystal microbalance (QCM) sensors are very simple and easy-to-use. Odor biosensors have potential to have high capability. However, a breakthrough is indispensable for practical use in the actual environment even if those sensors can be used in a laboratory.

Cheminformatics approach goes to a new stage. Although that technique was previously used to classify the pattern from a sensor array in the odor-sensing system, odor reproduction technique using odor components is now evolving. It is completely based on numerical analysis, which is different from previous hypothesis of the primary smells (Amoore et al., 1964). The original algorithm became a little sophisticated in terms of a measure of pattern difference and removal of interference. Moreover, odor perception, which is more abstract than previous odor classification and mixture quantification, has been studied in recent years. Artificial intelligence and deep learning will contribute to the progress in cheminformatics.

Olfactory displays have history shorter than odor-sensing system. Although its community is still very small, the number of papers is gradually increasing. Most of olfactory-display studies have focused on hardware development. Simple structures of olfactory displays have been so far developed so that many non-expert people can make and use them. However, the number of odors to be used is limited. Thus the multicomponent olfactory display in combination with odor components for odor reproduction is important to present a variety of smells.

Recently, digital contents using an olfactory display appeared. Appropriate contents enhance the application of olfactory display. Initially, movie with scents and game with scents were reported. But simulators with scents and narrative contents with scents have been studied in recent years. The olfactory contents can be extended even to artistic expression. Finally, teleolfaction at the preliminary level was studied. Although the technique at the current stage is to select one of odors prepared in advance, odor reproduction technology enables us to use a variety of odors. Thus scent transmission is becoming feasible.

Digital Technologies in Olfaction
DOI: https://doi.org/10.1016/B978-0-443-15721-9.00021-X

Previous machine olfaction technologies are currently changing toward more sophisticated level. Olfaction is appearing in cyber space in the same manners as vision and audition in 2024.

References

Amoore, J. E., Johnston, J. W., & Rubin, M. (1964). The stereochemical theory of odor. *Scientific American, 210*(2), 42–49. Available from https://doi.org/10.1038/scientificamerican0264-42.

5. Conclusions

CHAPTER

22

Future of machine olfaction

Takamichi Nakamoto

Institute of Integrated Research, Institute of Science Tokyo, Yokohama, Japan

22.1 Future of machine olfaction

Digital technologies in olfaction are now changing study of machine olfaction. However, it is significantly important to have good sensors for odor-sensing system. Even if we often use the word "electronic nose," we have not obtained true electronic nose yet. In terms of sensitivity, selectivity, and stability, excellent sensors are really necessary. A sensor device typically consists of sensing film and a transducer. Although we can have a highly sensitive sensor owing to a transducer characteristic in some cases, most of sensors should be more selective. The selectivity depends on a sensing film. It does not mean a sensor specific to a certain compound in the same way as enzyme or antibody. Many sensors are required if sensors' characteristics are too specific. However, most of current sensors do not reach even appropriate level of selectivity. Although pure biological material has high selectivity, issues to be solved are life time and stability. Therefore sensing film with high stability and long life time together with appropriate selectivity is desired.

Regarding cheminformatics, more abstract information should be handled. It should predict sensory information accurately even if the number of the data used for the training is small. That problem is the same as that of image classification to detect disease since researchers cannot obtain many image data judged by a medical doctor. In the case of sensory test of olfaction, not so many data are usually obtained. In addition to the progress in cheminformatics, however, large-scale reliable sensory test should be conducted. Many researchers used the data obtained from Dravnieks (1985) who performed the sensory test in 1985! Moreover, sophisticated model dedicated to cheminformatics can appear since most of the algorithms used in cheminformatics come from other areas.

Regarding odor components, more data are required to expand the range of smells to be reproduced. Although the current technology of odor reproduction is based on mass spectrum data, odors can be reproduced on the basis of human perceptual data in spite of the high nonlinearity. Moreover, odor reproduction based on sensor array data might be possible if the good sensors are realized. If it is possible to reproduce the smells based on the sensor array data, the odor reproduction in real time using cheap sensors will be feasible. Teleolfaction in Chapter 19 becomes more realistic in this case.

Digital Technologies in Olfaction
DOI: https://doi.org/10.1016/B978-0-443-15721-9.00022-1

Regarding olfactory display, multicomponent olfactory display should be smaller and cheaper. Then more applications will appear. The devices used in the olfactory display might be newly developed, although the devices commercially available are typically used there.

The contents of the olfactory displays are not so many at the current stage. However, many contents will appear if the range of the smell to be reproduced is expanded. A content developer or even an IT (Information Technology) engineer without the knowledge of smell can easily develop their own content with smell just by selecting odor type from the library. The smell can be embedded into the narrative story and contribute to the activation of the imagination. The smell can be also used as artistic expression. The olfactory art in the form of digital content might appear in the future. Moreover, the application toward healthcare will become more important because of aged society. Olfaction can be used to detect the decrease in olfactory capability or to increase the memory capability through olfactory training.

Digital technologies in olfaction can be extended to scent creation (Hasebe et al., 2022), although this point was not described in this chapter. Intended scent can be created if we combine prediction technique of odor impression with odor reproduction and olfactory display. AI (Artificial Intelligence) perfumer is one of the directions in the future.

Digital technologies in olfaction will contribute to a variety of the aspects in our daily life. Current digital technologies will change in the future when olfaction is added to digital contents together with vision and audition.

References

Hasebe, D., Alexandre, M., Nakamoto, T., & Mohammadzadeh, A. (2022). Exploration of sensing data to realize intended odor impression using mass spectrum of odor mixture. *PLoS ONE*, *17*(8). Available from https://doi.org/10.1371/journal.pone.0273011.

Dravnieks, A. (1985). *Atlas of odor character profiles*. DS61 ASTM.

Index

Note: Page numbers followed by "*f*" and "*t*" refer to figures and tables, respectively.

E

F

Q

R